SPRINGER STUDY EDITION

Lars Hörmander

The Analysis of Linear Partial Differential Operators I

Distribution Theory and Fourier Analysis

Second Edition

Springer-Verlag
Berlin Heidelberg New York
London Paris Tokyo Hong Kong

Lars Hörmander
Lunds Universitet
Matematiska Institutionen
Box 118
S-22100 Lund
Sweden

The hard cover edition of this book has been published as:
Grundlehren der mathematischen Wissenschaften, Band 256,
with the
ISBN 3-540-52345-6 Springer-Verlag Berlin Heidelberg New York
ISBN 0-387-52345-6 Springer-Verlag New York Berlin Heidelberg

With 5 Figures

Mathematics Subject Classification (1980):
46F, 46E, 26A, 42A, 35A, 35E, 35J, 35L

ISBN 3-540-52343-X Springer-Verlag Berlin Heidelberg New York
ISBN 0-387-52343-X Springer-Verlag New York Berlin Heidelberg

Library of Congress Catalog Card Number 89-26134

Typesetting: Universitätsdruckerei H. Stürtz AG, Würzburg
2141/3111 - 54321 – Printed on acid-free paper

Preface to the Second Edition

The main change in this edition is the inclusion of exercises with answers and hints. This is meant to emphasize that this volume has been written as a general course in modern analysis on a graduate student level and not only as the beginning of a specialized course in partial differential equations. In particular, it could also serve as an introduction to harmonic analysis. Exercises are given primarily to the sections of general interest; there are none to the last two chapters.

Most of the exercises are just routine problems meant to give some familiarity with standard use of the tools introduced in the text. Others are extensions of the theory presented there. As a rule rather complete though brief solutions are then given in the answers and hints.

To a large extent the exercises have been taken over from courses or examinations given by Anders Melin or myself at the University of Lund. I am grateful to Anders Melin for letting me use the problems originating from him and for numerous valuable comments on this collection.

As in the revised printing of Volume II, a number of minor flaws have also been corrected in this edition. Many of these have been called to my attention by the Russian translators of the first edition, and I wish to thank them for our excellent collaboration.

Lund, October 1989 Lars Hörmander

Preface

In 1963 my book entitled "Linear partial differential operators" was published in the Grundlehren series. Some parts of it have aged well but others have been made obsolete for quite some time by techniques using pseudo-differential and Fourier integral operators. The rapid development has made it difficult to bring the book up to date. However, the new methods seem to have matured enough now to make an attempt worth while.

The progress in the theory of linear partial differential equations during the past 30 years owes much to the theory of distributions created by Laurent Schwartz at the end of the 1940's. It summed up a great deal of the experience accumulated in the study of partial differential equations up to that time, and it has provided an ideal framework for later developments. "Linear partial differential operators" began with a brief summary of distribution theory for this was still unfamiliar to many analysts 20 years ago. The presentation then proceeded directly to the most general results available on partial differential operators. Thus the reader was expected to have some prior familiarity with the classical theory although it was not appealed to explicitly. Today it may no longer be necessary to include basic distribution theory but it does not seem reasonable to assume a classical background in the theory of partial differential equations since modern treatments are rare. Now the techniques developed in the study of singularities of solutions of differential equations make it possible to regard a fair amount of this material as consequences of extensions of distribution theory. Rather than omitting distribution theory I have therefore decided to make the first volume of this book a greatly expanded treatment of it. The title has been modified so that it indicates the general analytical contents of this volume. Special emphasis is put on Fourier analysis, particularly results related to the stationary phase method and Fourier analysis of singularities. The theory is illustrated throughout with examples taken from the theory of partial differential equations. These scattered examples should give a sufficient knowledge of the classical theory to serve as an introduction to the systematic study in the later volumes. Volume I should also be a useful introduction to harmonic analysis. A chapter on hyperfunctions at the end is intended to give an introduction in the spirit of Schwartz distributions to this subject and to the analytic theory of partial differential equations. The great progress in this area due primarily to the school of Sato is beyond the scope of this book, however.

The second and the third volumes will be devoted to the theory of differential equations with constant and with variable coefficients respectively. Their prefaces will describe their contents in greater detail. Volume II will appear almost simultaneously with Volume I, and Volume III will hopefully be published not much more than two years later.

In a work of this kind it is not easy to provide adequate references. Many ideas and methods have evolved slowly for centuries, and it is a task for a historian of mathematics to uncover the development completely. Also the more recent history provides of course considerable difficulties in establishing priorities correctly, and these problems tend to be emotionally charged. All this makes it tempting to omit references altogether. However, rather than doing so I have chosen to give at the end of each chapter a number of references indicating recent sources for the material presented or closely related topics. Some references to the earlier literature are also given. I hope this will be helpful to the reader interested in examining the background of the results presented, and I also hope to be informed when my references are found quite inadequate so that they can be improved in a later edition.

Many colleagues and students have helped to improve this book, and I should like to thank them all. The discussion of the analytic wave front sets owes much to remarks by Louis Boutet de Monvel, Pierre Schapira and Johannes Sjöstrand. A large part of the manuscript was read and commented on by Anders Melin and Ragnar Sigurdsson in Lund, and Professor Wang Rou-hwai of Jilin University has read a large part of the proofs. The detailed and constructive criticism given by the participants in a seminar on the book conducted by Gerd Grubb at the University of Copenhagen has been a very great help. Niels Jørgen Kokholm took very active part in the seminar and has also read all the proofs. In doing so he has found a number of mistakes and suggested many improvements. His help has been invaluable to me.

Finally, I wish to express my gratitude to the Springer Verlag for encouraging me over a period of years to undertake this project and for first rate and patient technical help in its execution.

Lund, January 1983 Lars Hörmander

Contents

Introduction

In differential calculus one encounters immediately the unpleasant fact that not every function is differentiable. The purpose of distribution theory is to remedy this flaw; indeed, the space of distributions is essentially the smallest extension of the space of continuous functions where differentiation is always well defined. Perhaps it is therefore self evident that it is desirable to make such an extension, but let us anyway discuss some examples of how awkward it is not to be allowed to differentiate.

Our first example is the Fourier transformation which will be studied in Chapter VII. If v is an integrable function on the real line then the Fourier transform Fv is the continuous function defined by

$$(Fv)(\xi)=\int_{-\infty}^{\infty} e^{-ix\xi} v(x)\,dx, \qquad \xi\in\mathbb{R}.$$

It has the important property that

$$F(Dv)=MFv, \qquad F(Mv)=-DFv \tag{1}$$

whenever both sides are defined; here $Dv(x)=-i\,dv/dx$ and $Mv(x)=xv(x)$. In the first formula the multiplication operator M is always well defined so the same ought to be true for D. Incidentally the second formula (1) then suggests that one should also define F for functions of polynomial increase.

Next we shall examine some examples from the theory of partial differential equations which also show the need for a more general definition of derivatives. Classical solutions of the Laplace equation

$$\partial^2 u/\partial x^2+\partial^2 u/\partial y^2=0 \tag{2}$$

or the wave equation (in two variables)

$$\partial^2 v/\partial x^2-\partial^2 v/\partial y^2=0 \tag{3}$$

are twice continuously differentiable functions satisfying the equations everywhere. It is easily shown that uniform limits of classical solutions of the Laplace equation are classical solutions. On the other hand, the classical solutions of the wave equation are all functions of the form

$$v(x, y)=f(x+y)+g(x-y) \tag{4}$$

with twice continuously differentiable f and g, and they have as uniform limits all functions of the form (4) with f and g continuous. All such functions ought therefore to be recognized as solutions of (3) so the definition of a classical solution is too restrictive.

Let us now consider the corresponding inhomogeneous equations

$$\partial^2 u/\partial x^2+\partial^2 u/\partial y^2=F, \tag{5}$$

$$\partial^2 v/\partial x^2-\partial^2 v/\partial y^2=F \tag{6}$$

where F is a continuous function vanishing outside a bounded set. If F is continuously differentiable a solution of (6) is given by

$$v(x, y)=\iint_{\eta-y+|x-\xi|<0} -F(\xi, \eta)\, d\xi\, d\eta/2. \tag{7}$$

However, (7) defines a continuously differentiable function v even if F is just continuous. Clearly we must accept v as a solution of (6) even if second order derivatives do not exist. Similarly (5) has the classical solution

$$u(x, y)=(4\pi)^{-1}\iint F(\xi, \eta)\log((x-\xi)^2+(y-\eta)^2)\, d\xi\, d\eta \tag{8}$$

provided that F is continuously differentiable. Again (8) defines a continuously differentiable function u even if F is just continuous, and we should be able to accept u as a solution of (5) also in that case.

The difficulties which are illustrated in their simplest form by the preceding examples were eliminated already by the concept of *weak solution* which preceded distribution theory. The idea is to rewrite the equation considered in a form where the unknown function u is no longer differentiated. Consider as an example the equation (6). If u is a classical solution it follows that

$$\iint(\partial^2 u/\partial x^2-\partial^2 u/\partial y^2)\,\phi\, dx\, dy=\iint F\phi\, dx\, dy \tag{6$'$}$$

for every continuous function ϕ vanishing outside a compact, that is, closed and bounded, set. Conversely, if (6)$'$ is fulfilled for all such ϕ which are say twice continuously differentiable then (6) is fulfilled. In fact, if (6) were not satisfied at a point (x_0, y_0) we could take ϕ non-negative and 0 outside a small neighborhood of (x_0, y_0) and conclude that (6)$'$ is not fulfilled either. For such "test functions" ϕ we can integrate by parts twice in the left-hand side of (6)$'$ which gives the equivalent formula

(6)″ $$\iint u(\partial^2\phi/\partial x^2 - \partial^2\phi/\partial y^2)\,dx\,dy = \iint F\phi\,dx\,dy.$$

Summing up, if u is twice continuously differentiable then (6) is equivalent to the validity of (6)″ for all test functions ϕ, that is, twice continuously differentiable functions ϕ vanishing outside a compact set. However, (6)″ has a meaning if u is just continuous, and one calls u a weak solution of (6) when (6)″ is valid.[1] It is easily verified that the flaws of the classical solutions pointed out above disappear if one accepts weak continuous solutions.

The function F is uniquely determined by u when (6)″ is fulfilled. However, for an arbitrary continuous function u there may be no continuous function F such that

(9) $$L(\phi) = \iint u(\partial^2\phi/\partial x^2 - \partial^2\phi/\partial y^2)\,dx\,dy$$

can be written in the form

(10) $$L(\phi) = \iint F\phi\,dx\,dy.$$

Distribution theory goes beyond the definition of weak solutions by accepting for study expressions L of the form (9) even when they are not of the form (10). A distribution is any such expression which depends linearly on a test function ϕ (and its derivatives). When it can be written in the form (10) it is identified with the function F. It turns out that one can extend the basic operations of calculus to distributions; in particular differentiation is always defined.

Let us also consider some examples of similar expressions occurring in physics. First consider a point mass with weight 1 at a point a on the real axis. This can be considered as a limiting case of a mass distribution with uniform density $1/2\varepsilon$ in the interval $(a-\varepsilon, a+\varepsilon)$ as $\varepsilon \to 0$. The corresponding functional is

$$L_\varepsilon(\phi) = \int_{a-\varepsilon}^{a+\varepsilon} \phi(x)\,dx/2\varepsilon.$$

When $\varepsilon \to 0$ we have $L_\varepsilon(\phi) \to \phi(a)$, so $L(\phi) = \phi(a)$ should represent the unit mass at a. This interpretation is of course standard in measure theory.

Next we consider a dipole at 0 with moment 1. This may be defined as the limit of the pointmass $1/\delta$ at δ and $-1/\delta$ at 0 as $\delta \to 0$. Thus we must consider the limit of the functional

$$M_\delta(\phi) = \delta^{-1}\phi(\delta) - \delta^{-1}\phi(0)$$

---

[1] Note that differential equations appear naturally in a weak form in the calculus of variations.

when $\delta \to 0$, which is $M(\phi)=\phi'(0)$. This functional is therefore the appropriate description of the dipole.

It is possible to pursue this development and define distributions as limits of functionals of the form (10). However, we shall not do so but rather follow the path suggested by the definition of weak derivatives. This is the original definition of Schwartz and it has the advantage of avoiding the question which limits define the same distribution.

The formal definition of distributions is given in Chapter II after properties of test functions have been discussed at some length in Chapter I. Differentiation of distributions is then studied in Chapter III; it is shown in Section 4.4 that we have indeed obtained a minimal extension of the space of continuous functions where differentiation is always possible. In Chapters IV, V, VI we extend convolution, direct product and composition from functions to distributions. Chapter VII is devoted to Fourier analysis of functions and distributions. The choice of material differs a great deal from standard texts since it is dictated by what is required in the later parts. The method of stationary phase is given a particularly thorough treatment. In Chapter VIII we discuss the Fourier analysis of singularities of distributions. This turns out to be a local problem so it can be discussed also for distributions on manifolds. The phrase "singularity" above is deliberately vague; in fact we shall consider singularities both from a C^∞ and from an analytic point of view. The results lead to important extensions of the distribution theory in Chapters III–VI. For instance, one can define the restriction of a distribution u to a submanifold Y if u has no singularity at a normal to Y. Many applications to regularity and uniqueness of solutions of differential equations are also given. The analytic theory is continued in Chapter IX which is devoted to hyperfunctions. These are defined just as distributions but with real analytic test functions. The main new difficulty stems from the fact that there are no such test functions vanishing outside a compact set.

Chapter I. Test Functions

Summary

As indicated in the introduction one must work consistently with smooth "test functions" in the theory of distributions. In this chapter we have collected the basic facts that one needs to know about such functions. As an introduction a brief summary of differential calculus is given in Section 1.1. It is written with a reader in mind who has seen the material before but perhaps with different emphasis and different notation. The reader who finds the presentation hard to follow is recommended to study first a more thorough modern treatment of differential calculus in several variables, and experienced readers should proceed directly to Section 1.2. In addition to the basic indispensible facts we have included in Sections 1.3 and 1.4 some more refined constructions which will be useful some time in this book but are not important for the main theme. The reader in a hurry may therefore wish to omit Section 1.3 from Theorem 1.3.5 on and also Theorem 1.4.2, Lemma 1.4.3 and the rest of Section 1.4 from Theorem 1.4.6 on.

1.1. A Review of Differential Calculus

At first we shall consider functions of a single real variable but permit values in a Banach space. Thus let I be an open interval on the real line $\mathbb{R}$ and let V be a Banach space with norm denoted $\| \ \|$. A map f: $I \to V$ is called differentiable at $x \in I$, with derivative $f'(x) \in V$, if

$$\|(f(x+h)-f(x))/h - f'(x)\| \to 0 \quad \text{when } h \to 0. \tag{1.1.1}$$

We can write (1.1.1) in the equivalent form

$$\|f(x+h)-f(x)-f'(x)\,h\| = o(|h|) \quad \text{when } h \to 0. \tag{1.1.1$'$}$$

If $V = \mathbb{R}^n$ and we write $f = (f_1, \dots, f_n)$ then differentiability of f is of course equivalent to differentiability of each component f_j. For vector valued functions the mean value theorem must be replaced by the following

Theorem 1.1.1. *If $f\colon I \to V$ is differentiable at every point in I, then*

$$\|f(y)-f(x)\| \leqq |y-x| \sup\{\|f'(x+t(y-x))\|,\ 0 \leqq t \leqq 1\}; \quad x, y \in I. \tag{1.1.2}$$

Proof. Let $M > \sup\{\|f'(x+t(y-x))\|, 0 \leqq t \leqq 1\}$ and set

$$E = \{t;\ 0 \leqq t \leqq 1,\ \|f(x+t(y-x))-f(x)\| \leqq Mt|x-y|\}.$$

E is closed since f is continuous, and $0 \in E$, so E has a largest element s. If $t > s$ and $t-s$ is sufficiently small we have

$$\begin{aligned} &\|f(x+t(y-x))-f(x)\| \\ &\quad \leqq \|f(x+t(y-x))-f(x+s(y-x))\| + \|f(x+s(y-x))-f(x)\| \\ &\quad \leqq M|(t-s)(y-x)| + Ms|y-x| = Mt|y-x|. \end{aligned}$$

Hence $s=1$ which proves the theorem.

Remark. If f is just continuous in $[x, y]$ and differentiable in the interior we obtain (1.1.2) with supremum for $0<t<1$ as a limit of (1.1.2) applied to smaller intervals. If $v \in V$ an application of (1.1.2) to $x \to f(x) - xv$ gives

$$\|f(y)-f(x)-v(y-x)\| \leqq |y-x| \sup_{0<t<1} \|f'(x+t(y-x))-v\| \tag{1.1.2$'$}$$

which is often more useful, particularly with $v = f'(x)$.

Corollary 1.1.2. *Let f be continuous in I and differentiable outside a closed subset F where $f=0$. If $x \in F$ and $f'(y) \to 0$ when $I \setminus F \ni y \to x$, then $f'(x)$ exists and is equal to 0.*

Proof. If $y \in F$ then $f(y)-f(x)=0$. Otherwise let z be the point in $F \cap [x, y]$ closest to y. Then (1.1.2)′ gives

$$\|f(y)-f(x)\| = \|f(y)-f(z)\| \leqq |y-z| \sup_{0<t<1} \|f'(z+t(y-z))\|$$

which is $o(|y-x|)$ as $y \to x$.

Example 1.1.3. If P is a polynomial and $f(x) = P(1/x)\, e^{-1/x}$, $x>0$, $f(x)=0$, $x \leqq 0$, then f is continuous. The derivative for $x \neq 0$ is of the same form with $P(1/x)$ replaced by $(P(1/x)-P'(1/x))/x^2$ so $f'(0)$ exists and is equal to 0.

Let U be another Banach space, X an open subset of U. If f is a function from X to V we define differentiability by an analogue of (1.1.1)′:

Definition 1.1.4. f is called differentiable at $x\in X$ if there is an element $f'(x)\in L(U,V)$ such that

$$\|f(x+h)-f(x)-f'(x)\,h\|=o(\|h\|),\qquad h\to 0. \tag{1.1.1$''$}$$

Here $L(U,V)$ is the space of continuous linear transformations from U to V, which is a Banach space with the norm

$$\|T\|=\sup_{\|x\|<1}\|Tx\|,\qquad T\in L(U,V).$$

By $C^1(X,V)$ we denote the set of continuously differentiable functions from X to V, that is, the set of functions f which are differentiable at every point and for which $X\ni x\to f'(x)\in L(U,V)$ is continuous.

If f is just differentiable at every point on the line segment $[x,y]=\{x+t(y-x);\ 0\leqq t\leqq 1\}$ then (1.1.2)′ gives for every $T\in L(U,V)$

$$\|f(y)-f(x)-T(y-x)\|\leqq\|y-x\|\sup_{0<t<1}\|f'(x+t(y-x))-T\| \tag{1.1.2$''$}$$

for $f(x+t(y-x))-Tt(y-x)$ is differentiable in t on $[0,1]$ with derivative

$$f'(x+t(y-x))(y-x)-T(y-x).$$

Theorem 1.1.5. *If $f_j\in C^1(X,V)$ and $f_j\to f$, $f_j'\to g$ locally uniformly in X, then $f\in C^1(X,V)$ and $f'=g$.*

Proof. If we apply (1.1.2)″ to f_j with $T=f_j'(x)$ we obtain when $j\to\infty$

$$\|f(y)-f(x)-g(x)(y-x)\|\leqq\|y-x\|\sup_{0<t<1}\|g(x+t(y-x))-g(x)\|$$

which proves that f is differentiable at x with $f'(x)=g(x)$. Since g is continuous this proves the theorem.

Theorem 1.1.6. *If f and g are continuous functions in X with values in V and $L(U,V)$ respectively, and $t\to f(x+ty)$ is for all x, $y\in U$ differentiable with respect to t with derivative $g(x+ty)\,y$ when $x+ty\in X$, then $f\in C^1$ and $f'=g$. It suffices to make the hypothesis for all y in a set $Y\subset U$ with closed linear hull equal to U.*

Proof. (1.1.2)′ gives for small $\|y\|$

$$\|f(x+y)-f(x)-g(x)\,y\|\leqq\|y\|\sup_{0<t<1}\|g(x+ty)-g(x)\|$$

which proves the first statement. To prove the second one it suffices to show that the set of all y for which the hypothesis holds is linear and closed. This follows from (1.1.2)′; the details are left for the reader.

If f is a linear map $U \to V$ then f is of course differentiable and $f'(x)=f$ for every x. More generally, let $U_1, \ldots, U_k$ be Banach spaces and $L(U_1, \ldots, U_k; V)$ the space of multilinear maps

$$U_1 \times \ldots \times U_k \ni (x_1, \ldots, x_k) \to f(x_1, \ldots, x_k) \in V$$

which are continuous, that is,

$$\|f\| = \sup_{\|x_j\|<1} \|f(x_1, \ldots, x_k)\| < \infty.$$

With this norm $L(U_1, \ldots, U_k; V)$ is a Banach space. The map

$$U_1 \oplus \ldots \oplus U_k \ni (x_1, \ldots, x_k) \to f(x_1, \ldots, x_k) \in V$$

is differentiable for every $(x_1, \ldots, x_k)$, and the differential is

$$U_1 \oplus \ldots \oplus U_k \ni (y_1, \ldots, y_k) \to f(y_1, x_2, \ldots, x_k)$$
$$+ f(x_1, y_2, \ldots, x_k) + \ldots + f(x_1, x_2, \ldots, x_{k-1}, y_k).$$

Another important example of a C^1 map is for two Banach spaces U and V the map f taking an invertible element $T \in L(U, V)$ to its inverse $T^{-1} \in L(V, U)$. That T^{-1} is an inverse means that

$$TT^{-1} = \mathrm{id}_V, \qquad T^{-1}T = \mathrm{id}_U.$$

If $S \in L(U, V)$ we have $(T+S)T^{-1} = \mathrm{id}_V + ST^{-1}$ so if $\|S\|\,\|T^{-1}\| < 1$ a right inverse of $T+S$ is given by

$$T^{-1}(\mathrm{id}_V + ST^{-1})^{-1} = \sum_0^\infty T^{-1}(-ST^{-1})^k.$$

In the same way we see that it is a left inverse. Thus $f(T) = T^{-1}$ is defined in an open set and

$$\|f(T+S) - f(T) + T^{-1}ST^{-1}\| \leqq \|T^{-1}\|^3 \|S\|^2 / (1 - \|S\|\,\|T^{-1}\|)$$

so f is differentiable at T and $f'(T)S = -T^{-1}ST^{-1}$. Thus f is continuous so f' is continuous.

Next we shall discuss composite functions. Let X as before be an open subset of the Banach space U, f a map from X to a Banach space V with range contained in an open set Y where another map g to a third Banach space W is defined. If f is differentiable at x and g is differentiable at $y = f(x)$ then $h = g \circ f$ is differentiable at x and

(1.1.3) $$h'(x) = g'(y)\, f'(x) \quad \text{(the chain rule).}$$

The proof is obvious. From (1.1.3) it follows that $h \in C^1$ if $g \in C^1$ and $f \in C^1$.

The differential f' may be viewed as a map

$$X \times U \ni (x, t) \xrightarrow{f'} (f(x), f'(x)\,t) \in Y \times V$$

which is linear in the second component which should be thought of as a tangent direction. Then the chain rule says that given a commutative diagram

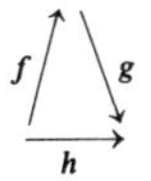

with $f, g \in C^1$ we obtain $h \in C^1$ and another commutative diagram

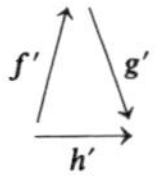

Instead of the notation f' one often writes df, particularly when f is real valued. If f is defined in an open set in $\mathbb{R}^n$ and we write $t = \sum t_j e_j$ where e_j is the jth unit vector, then

$$(df)(t) = (df)(\sum t_j e_j) = \sum t_j \, df(e_j) = \sum \partial f / \partial x_j \, t_j.$$

But $t_j = (dx_j)(t)$ so we can write this equation in the form

$$df = \sum \partial f / \partial x_j \, dx_j.$$

By the chain rule this formula remains valid if x_j are in fact functions of $z \in Z$ and both f and x_j are regarded as functions on Z so that both sides are linear functions there. This is called the invariance of the differential.

Next we prove the inverse function theorem which shows that differential calculus does accomplish its goal of reducing the study of fairly general equations to linear ones.

Theorem 1.1.7. *Let X be open in U and $f \in C^1(X, V)$, and let $x_0 \in X$, $f(x_0) = y_0$. For the existence of $g \in C^1(Y, U)$, where Y is a neighborhood of y_0, such that* a) *$f \circ g =$ identity near y_0 or* b) *$g \circ f =$ identity near x_0 or* c) *$f \circ g =$ identity near y_0 and $g \circ f =$ identity near x_0, it is necessary and sufficient that there is a linear map $A \in L(V, U)$ such that respectively*

a)′ $f'(x_0) A = \mathrm{id}_V$ b)′ $A f'(x_0) = \mathrm{id}_U$ c)′ $f'(x_0) A = \mathrm{id}_V$, $A f'(x_0) = \mathrm{id}_U$.

The condition c)′ *is equivalent to bijectivity of $f'(x_0)$ and it implies that g is uniquely determined near y_0. If $V(U)$ is finite dimensional then* a)′(b)′) *is equivalent to surjectivity (injectivity) of $f'(x_0)$.*

Proof. The necessity is an immediate consequence of the chain rule (1.1.3). In the proof of the sufficiency we first observe that if $f \circ g_1 = \mathrm{id}$ near y_0 and $g_2 \circ f = \mathrm{id}$ near x_0 then $g_1 = g_2 \circ f \circ g_1 = g_2$ near y_0 which proves uniqueness in case c) and reduces the proof to existence in cases a) and b). If we replace f by $f \circ A$ resp. $A \circ f$ we see that it suffices to study the case where $U = V$ and $f'(x_0) = \mathrm{id}$. Choose $\delta > 0$ so that

$$\|f'(x) - \mathrm{id}\| < \tfrac{1}{2} \quad \text{when } \|x - x_0\| \leqq \delta.$$

For $\|x_j - x_0\| \leqq \delta$, $j = 1, 2$, we have by (1.1.2)″

$$\|f(x_1) - f(x_2) - (x_1 - x_2)\| \leqq \|x_1 - x_2\|/2. \tag{1.1.4}$$

Hence f is injective in $\{x; \|x - x_0\| \leqq \delta\}$. To solve the equation $f(x) = y$ when $\|y - y_0\| < \delta/2$ we set

$$x_k = x_{k-1} + y - f(x_{k-1}), \quad k = 1, 2, \ldots \tag{1.1.5}$$

as long as this leads to points with $\|x_k - x_0\| < \delta$. We have

$$\|x_1 - x_0\| = \|y - y_0\| < \delta/2.$$

If $k > 1$ and $\|x_j - x_0\| < \delta$ when $j < k$ then

$$\begin{aligned} \|x_k - x_{k-1}\| &= \|x_{k-1} - f(x_{k-1}) - (x_{k-2} - f(x_{k-2}))\| \\ &\leqq \|x_{k-1} - x_{k-2}\|/2 < \delta/2^k \end{aligned}$$

by (1.1.4). Hence

$$\|x_k - x_0\| < \delta \sum_1^k 2^{-j} < \delta$$

so x_k is defined for all k and is a Cauchy sequence. For the limit x we have $\|x - x_0\| < \delta$, and when $k \to \infty$ we obtain $f(x) = y$ from (1.1.5).

To prove that the inverse $g(y) = x$, which is now defined when $\|y - y_0\| < \delta/2$, belongs to C^1 we set

$$g(y) = x, \quad g(y + k) = x + h.$$

This means that $f(x + h) = y + k$ and that $f(x) = y$. Hence

$$k = f(x + h) - f(x) = f'(x)h + o(\|h\|).$$

In view of (1.1.4) we have $\|k - h\| < \|h/2\|$, hence $\|h\|/2 < \|k\| < 2\|h\|$. Since $\|f'(x)^{-1}\| < 2$ we obtain

$$h = f'(x)^{-1} k + o(\|k\|)$$

which proves that g is differentiable and that $g'(y) = f'(g(y))^{-1}$, which is a continuous function of y.

We shall now define differentials of higher order and the space $C^k(X, V)$ where as before X is an open subset of the Banach space U but k is now an integer >1. This is done by induction so $f \in C^k(X, V)$ if $f \in C^1(X, V)$ and $f' \in C^{k-1}(X, L(U, V))$. The differential f'' of f' is called the second differential of f, and so on,

$$f^{(k)} \in C(X, L(U, L(U, \dots, L(U, V)))).$$

The vector space $L(U, L(U, \dots, L(U, V)))$ is isomorphic as a Banach space to the space $L(U, \dots, U; V) = L^k(U, V)$ of k-linear maps from U to V. In fact, we always have

$$L(U, L(U_1, \dots, U_j; V)) = L(U, U_1, \dots, U_j; V)$$

for if f is an element in the space on the left and the assertion is known when j is replaced by $j-1$, then

$$U \times U_1 \times \dots \times U_j \ni (x, x_1, \dots, x_j) \to f(x)(x_1, \dots, x_j) \in V$$

is an element in the space on the right, and all its elements can be so obtained. The correspondence is obviously linear and norm preserving.

By $L^k_s(U, V)$ we shall denote the space of symmetric k linear forms from U to V, that is, forms such that the value at $(x_1, \dots, x_k)$ is not changed by a permutation of $x_1, \dots, x_k$.

Theorem 1.1.8. *If $f \in C^k$ then $f^{(k)}$ is a symmetric multilinear form, that is, the order of differentiation is not essential.*

Proof. Set $\Delta_y F(x) = F(x+y) - F(x)$. By repeated use of (1.1.2)$''$ we obtain if L is a multilinear form

$$\begin{aligned}
&\|\Delta_{y_k} \dots \Delta_{y_1} f(x) - L(y_1, \dots, y_k)\| \\
&\quad \leqq \sup_{0<t<1} \|\Delta_{y_{k-1}} \dots \Delta_{y_1} f'(x + t y_k)\, y_k - L(y_1, \dots, y_k)\| \\
&\quad \leqq \sup_{0<t_j<1} \left\|f^{(k)}\left(x + \sum_1^k t_j y_j; y_1, \dots, y_k\right) - L(y_1, \dots, y_k)\right\|.
\end{aligned}$$

If we choose $L = f^{(k)}(x)$ it follows that

$$\|\Delta_{y_k} \dots \Delta_{y_1} f(x) - f^{(k)}(x; y_1, \dots, y_k)\| = o(\|y_1\| \dots \|y_k\|). \tag{1.1.6}$$

This determines $f^{(k)}$ completely of course, and since $\Delta_{y_k} \dots \Delta_{y_1} f(x)$ does not depend on the order of the differences it follows that $f^{(k)}(x)$ is symmetric.

From (1.1.3) it follows at once by induction that $h = g \circ f \in C^k$ if f, $g \in C^k$, for if this is proved for smaller values of k we have $g' \circ f \in C^{k-1}$

and $f' \in C^{k-1}$, and the composition of linear maps

$$L(V, W) \times L(U, V) \to L(U, W)$$

is continuous, hence infinitely differentiable. The inverse function theorem (Theorem 1.1.7) also remains valid with C^1 replaced by C^k throughout. In fact, the map

$$L(U, V) \ni T \to T^{-1} \in L(V, U)$$

is in C^k where it is defined, with k^{th} differential

$$(S_1, \dots, S_k) \to (-1)^k \sum T^{-1} S_{i_1} T^{-1} S_{i_2} T^{-1} \dots S_{i_k} T^{-1}$$

summed over all permutations of $1, \dots, k$. In the proof of Theorem 1.1.7 we therefore obtain inductively that $g \in C^k$ if $f \in C^k$, by using that $g'(y) = f'(g(y))^{-1}$.

If $f \in C^k$ in a neighborhood of the line segment $[x, x+y]$ then we have Taylor's formula

$$f(x+y) = \sum_0^{k-1} f^{(j)}(x; y, \dots, y)/j! \tag{1.1.7}$$
$$+ \int_0^1 f^{(k)}(x+ty; y, \dots, y)(1-t)^{k-1} dt/(k-1)!.$$

This follows inductively by partial integration since

$$\frac{d}{dt} f^{(k-1)}(x+ty; y, \dots, y) = f^{(k)}(x+ty; y, \dots, y).$$

In particular we obtain

$$\|f(x+y) - \sum_0^k f^{(j)}(x; y, \dots, y)/j!\| = o(\|y\|^k) \quad \text{as } y \to 0. \tag{1.1.8}$$

Let us now assume that f is defined in an open set $X \subset \mathbb{R}^n$. It follows from Theorem 1.1.6 that $f \in C^k$ if and only if all the partial derivatives

$$\frac{\partial}{\partial x_{i_1}} \dots \frac{\partial f}{\partial x_{i_j}}$$

of order $j \leq k$ are defined and continuous. By Theorem 1.1.8 we know that the order of differentiation does not matter. With the notation $\partial_j = \partial/\partial x_j$ we can therefore write these partial derivatives in the form

$$\partial_1^{\alpha_1} \dots \partial_n^{\alpha_n} f = \partial^\alpha f$$

where $\alpha = (\alpha_1, \dots, \alpha_n)$ is a *multi-index*, that is, an n-tuple of non-negative integers, and $\partial^\alpha = \partial_1^{\alpha_1} \dots \partial_n^{\alpha_n}$. We set $|\alpha| = \sum \alpha_j$, which is the order of differentiation. Note that

$$f^{(k)}(x; y, \ldots, y) = \sum_{|\alpha|=k} k!\, \partial^\alpha f(x)\, y^\alpha/\alpha!$$

where $\alpha! = \alpha_1! \ldots \alpha_n!$ and $y^\alpha = y_1^{\alpha_1} \ldots y_n^{\alpha_n}$. Hence Taylor's formula can be written

$$(1.1.7)' \qquad f(x+y) = \sum_{|\alpha|<k} \partial^\alpha f(x) y^\alpha/\alpha! + k \int_0^1 (1-t)^{k-1} \sum_{|\alpha|=k} \partial^\alpha f(x+ty) y^\alpha/\alpha!\, dt.$$

The following application is often useful:

Theorem 1.1.9. *If* $f \in C^k(B)$ *where* $B = \{x \in \mathbb{R}^n, |x| < R\}$ *and* $k \geqq 1$, *then* $f(x) - f(0) = \sum x_j f_j(x)$ *where* $f_j \in C^{k-1}(B)$, $\partial^\alpha f_j(0) = \partial^\alpha \partial_j f(0)/(1+|\alpha|)$, *and*

$$(1.1.9) \qquad \sup_B |\partial^\alpha f_j| \leqq \sup_B |\partial^\alpha \partial_j f|, \qquad |\alpha| < k.$$

Proof. Using (1.1.7)′ with $k=1$ we can take

$$f_j(x) = \int_0^1 (\partial_j f)(tx)\, dt,$$

and $f_j \in C^{k-1}$ since the integrand is in C^{k-1}. The estimate (1.1.9) is obvious.

Any linear differential operator

$$P = \sum_{|\alpha| \leqq m} a_\alpha(x) \partial^\alpha$$

can be written in the form $P(x, \partial)$ where

$$P(x, \xi) = \sum_{|\alpha| \leqq m} a_\alpha(x) \xi^\alpha$$

is a polynomial in $\xi \in \mathbb{R}^n$. By convention coefficients are put to the left of differentiations. Note that

$$P(x, \xi) = e^{-\langle x, \xi \rangle} P e^{\langle x, \xi \rangle}.$$

Leibniz' formula for the differentiation of a product takes the form

$$(1.1.10) \qquad P(x, \partial)(uv) = \sum (P^{(\alpha)}(x, \partial)\, u)\, \partial^\alpha v/\alpha!$$

where $P^{(\alpha)}(x, \xi) = \partial_\xi^\alpha P(x, \xi)$. Indeed, $u \to P(x, \partial)(uv)$ is a differential operator for fixed v, and

$$e^{-\langle x, \xi \rangle} P(x, \partial)(e^{\langle x, \xi \rangle} v) = P(x, \xi + \partial) v = \sum P^{(\alpha)}(x, \xi)\, \partial^\alpha v/\alpha!$$

by Taylor's formula. We shall refer to (1.1.10) as Leibniz' formula.

1.2. Existence of Test Functions

For an open set X in $\mathbb{R}^n$ we shall denote by $C^k(X)$ the space of k times continuously differentiable complex valued functions in X if k is a non-negative integer, and we set

$$C^\infty(X)=\bigcap C^k(X)$$

with intersection taken for all finite k. In the introduction we have seen the need for functions of the following kind:

Definition 1.2.1. By $C_0^k(X)$ we denote the space of all $u\in C^k(X)$ with compact support. The elements of $C_0^\infty(X)$ are called test functions.

Definition 1.2.2. If $u\in C(X)$ then the support of u, written supp u, is the closure in X of the set $\{x\in X;\ u(x)\neq 0\}$, that is, supp u is the smallest closed subset of X such that $u=0$ in $X\setminus \operatorname{supp} u$.

If the definition of a function $u\in C_0^k(X)$ is extended to $\mathbb{R}^n$ by letting $u=0$ in $\mathbb{R}^n\setminus X$ we obtain of course a function in $C_0^k(\mathbb{R}^n)$. Thus we may regard $C_0^k(X)$ as a subspace of $C_0^k(\mathbb{R}^n)$ which increases with X. For an arbitrary subset $M\subset\mathbb{R}^n$ we may also define $C_0^k(M)$ as the set of elements in $C_0^k(\mathbb{R}^n)$ with support contained in M. When $k=0$ we sometimes omit k in the notation.

Lemma 1.2.3. *There exists a non-negative function $\phi\in C_0^\infty(\mathbb{R}^n)$ with $\phi(0)>0$.*

Proof. Let $f(t)=\exp(-1/t)$, $t>0$ and $f(t)=0$, $t\leqq 0$. From example 1.1.3 we know that $f\in C^1(\mathbb{R})$ and repeating the argument gives $f\in C^\infty(\mathbb{R})$. Hence

$$\phi(x)=f(1-|x|^2),\qquad |x|^2=\sum_1^n x_j^2,$$

has the required properties.

By translation and change of scales we obtain the non-negative C_0^∞ function

$$x\to\phi((x-x_0)/\delta) \tag{1.2.1}$$

which is positive at x_0 and has support in the ball of radius δ with center at x_0. We can now prove a fact already alluded to in the introduction.

Theorem 1.2.4. *If $f, g \in C(X)$ and*

$$\int f\phi\, dx = \int g\phi\, dx, \quad \phi \in C_0^\infty(X) \tag{1.2.2}$$

then $f = g$.

Proof. If $h = f - g$ we have

$$\int h\phi\, dx = 0, \quad \phi \in C_0^\infty(X). \tag{1.2.3}$$

Taking real and imaginary parts we find that h may be assumed real valued provided that ϕ is taken real valued. If $h(x_0) \neq 0$ then we can take $\phi \in C_0^\infty(X)$ non-negative with $\phi(x_0) \neq 0$ and support so close to x_0 that ϕh has a constant sign which contradicts (1.2.3). Hence $h = 0$ identically as claimed.

A more general but less elementary result of the same kind is

Theorem 1.2.5. *If f, g are locally integrable functions in X and* (1.2.2) *is valid, then $f = g$ almost everywhere in X.*

Proof. Again it suffices to show that if h satisfies (1.2.3) then $h = 0$ almost everywhere. To do so we use Lebesgue's theorem stating that

$$\lim_{t \to 0} t^{-n} \int_{|x-y|<t} |h(x) - h(y)|\, dy = 0$$

for almost every x. With $\phi \in C_0^\infty$ having support in the unit ball and $\int \phi\, dx = 1$, we can write for $x \in X$ and small t

$$\begin{aligned} h(x) &= \int h(x)\phi((x-y)/t)\, dy/t^n \\ &= \int (h(x) - h(y))\, \phi((x-y)/t)\, dy/t^n + \int h(y)\, \phi((x-y)/t)\, dy/t^n. \end{aligned}$$

The last integral vanishes by hypothesis and the preceding one tends to 0 with t for almost all x, which proves that $h(x) = 0$ almost everywhere.

Remark. The theorem is also an immediate consequence of Theorem 1.3.2.

Lemma 1.2.3 is all one needs to get distribution theory started. However, a number of more subtle constructions of test functions will be needed to develop the theory, and we shall discuss them in the following two sections. As an indication of how rich the space $C_0^\infty(K)$ is we prove already now a classical theorem of Borel which will sometimes be useful.

Theorem 1.2.6. *For $j=0, 1, \dots$ let $f_j \in C_0^\infty(K)$ where K is a compact subset of $\mathbb{R}^n$, and let I be a compact neighborhood of 0 in $\mathbb{R}$. Then one can find $f \in C_0^\infty(K \times I)$ such that*

$$\partial^j f(x,t)/\partial t^j = f_j(x), \qquad t=0,\ j=0,1,\dots.$$

Proof. Choose $g \in C_0^\infty(I)$ so that $d^j(g(t)-1)/dt^j=0$ when $t=0$ for $j=0,1,\dots$. For example, if $(-\varepsilon,\varepsilon)\subset I$ we can take ϕ according to Lemma 1.2.3 with support in $(0,\varepsilon)$ and $\int \phi(t)\,dt=1$, and let g be the solution of $g'(t)=\phi(-t)-\phi(t)$ with support in I. Now

$$g_j(x,t)=g(t/\varepsilon_j)\,t^j f_j(x)/j!$$

is in $C_0^\infty(K \times I)$, and

$$|\partial^\alpha g_j(x,t)| \leqq 2^{-j} \quad \text{if } |\alpha| \leqq j-1, \tag{1.2.4}$$

provided that ε_j is sufficiently small. In fact taking t/ε_j as new variable we see that

$$|\partial^\alpha g_j(x,t)| \leqq C_{\alpha,j}\,\varepsilon_j^{j-\alpha_t}$$

where α_t is the order of differentiation with respect to t. For small ε_j we obtain (1.2.4) since $\alpha_t<j$. Hence the modified Taylor series

$$f(x,t)=\sum_0^\infty g_j(x,t)$$

is uniformly convergent. So are the series obtained by differentiation. In view of Theorem 1.1.5 it follows that $f \in C_0^\infty(K \times I)$ and that

$$\partial^j f(x,t)/\partial t^j = \sum \partial^j g_k(x,t)/\partial t^j = f_j(x) \quad \text{when } t=0.$$

1.3. Convolution

If u and v are in $C(\mathbb{R}^n)$ and either one has compact support, then the convolution $u*v$ is the continuous function defined by

$$(u*v)(x)=\int u(x-y)v(y)\,dy, \qquad x \in \mathbb{R}^n. \tag{1.3.1}$$

Thus $u*v$ is a superposition of translates of u taken with the weights $v(y)\,dy$, so we can expect $u*v$ to inherit properties of u such as differentiability. On the other hand, taking $x-y$ as a new integration variable in (1.3.1) we obtain

$$u*v=v*u, \tag{1.3.2}$$

so the properties of v should also be inherited. The reason for the commutativity (1.3.2) is perhaps more clear if we note that (1.3.1) implies

$$(1.3.1)' \qquad \int (u*v)\phi\,dx = \int\int u(x)v(y)\phi(x+y)\,dx\,dy, \qquad \phi \in C_0^0(\mathbb{R}^n),$$

which conversely implies (1.3.1) by Theorem 1.2.4. Now (1.3.1)′ shows that (1.3.2) just expresses the commutativity of addition in $\mathbb{R}^n$. Similarly the associativity leads to

$$(1.3.3) \qquad (u*v)*w = u*(v*w)$$

if all except one of the continuous functions u, v, w have compact support. The direct verification of (1.3.3) from (1.3.1) is of course an easy exercise. Taking $w=1$ we find that

$$(1.3.4) \qquad \int (u*v)\,dx = \int u\,dx \int v\,dx$$

when u and v have compact support.

If $u \in C^1$ and $v \in C^0$, either one having compact support, we can differentiate under the integral sign in (1.3.1) and obtain that $u*v \in C^1$ and

$$(1.3.5) \qquad \partial_i(u*v) = (\partial_i u)*v, \qquad i=1,\dots,n.$$

By the commutativity (1.3.2) we could differentiate on the factor v instead if $v \in C^1$. If $u \in C^j$ and $v \in C^k$ it follows therefore that $u*v \in C^{j+k}$ and that

$$(1.3.6) \qquad \partial^{\alpha+\beta}(u*v) = (\partial^\alpha u)*(\partial^\beta v) \qquad \text{if } |\alpha| \leqq j,\ |\beta| \leqq k.$$

The preceding conclusions can be strengthened in various ways. For example it is clear that $u*v$ is a continuous function if $u \in C_0$ and v is just locally integrable. If $u \in C_0^j$ we conclude for such v that $u*v \in C^j$. Summing up, we have found

Theorem 1.3.1. *If* $u \in C_0^j$ *then* $u*v \in C^j$ *if* $v \in L^1_{\text{loc}}$, and $u*v \in C^{j+k}$ *if* $v \in C^k$.

The convolution can be used to approximate functions by more differentiable ones, a technique known as *regularization:*

Theorem 1.3.2. *Let* $0 \leqq \phi \in C_0^\infty$, $\int \phi\,dx = 1$. *If* $u \in C_0^j(\mathbb{R}^n)$ *it follows that* $u_\phi = u*\phi \in C_0^\infty(\mathbb{R}^n)$. *When* $\operatorname{supp}\phi \to \{0\}$ *we have*

$$(1.3.7) \qquad \sup|\partial^\alpha u - \partial^\alpha u_\phi| \to 0 \qquad \text{if } |\alpha| \leqq j;$$

if $v \in L^p(\mathbb{R}^n)$ *then* $v_\phi \in C^\infty(\mathbb{R}^n)$ *and* $v_\phi \to v$ *in* L^p *norm if* $p < \infty$.

Proof. In view of (1.3.6) and Theorem 1.3.1 we just have to prove (1.3.7) when $\alpha=0$. If $|y|<\delta$ in $\operatorname{supp}\phi$ we obtain

$$|u(x)-u_\phi(x)|=|\int(u(x)-u(x-y))\phi(y)\,dy|\leqq\sup_{|y|<\delta}|u(x)-u(x-y)|$$

and this tends to 0 with δ since u is uniformly continuous. Since $\|v_\phi\|_{L^p}\leqq\|v\|_{L^p}$ and C_0^0 is dense in L^p the last statement follows at once.

It is sometimes useful to know more precisely how the regularizations converge, and we give a result of that kind:

Theorem 1.3.3. *Let $v\in C^j(\mathbb{R}^n)$ and $\phi\in C_0^\infty(\mathbb{R}^n)$, $\int\phi\,dx=1$. Then*

$$u(x,t)=\int v(x-ty)\phi(y)\,dy \tag{1.3.8}$$

is in C^∞ in $\{(x,t)\in\mathbb{R}^{n+1},\ t\neq0\}$ and $t^k u(x,t)$ is in $C^{k+j}(\mathbb{R}^{n+1})$ for every non-negative integer k. When $t=0$ we have

$$\partial_t^i(t^k u(x,t))=0 \quad \text{if } i<k; \qquad \partial_t^k(t^k u(x,t))=k!\,v(x). \tag{1.3.9}$$

Proof. Since $u(x,t)\to v(x)$ when $t\to0$, by Theorem 1.3.2, we obtain (1.3.9) from Taylor's formula as soon as we know that $t^k u(x,t)\in C^{k+j}$. This we shall prove for every $\phi\in C_0^\infty$. That $u\in C^j$ follows at once if we differentiate under the integral sign in (1.3.8). When $k\neq0$ we may therefore assume the statement proved already for smaller values of k. For $t\neq0$ we have

$$t^k u(x,t)=t^k|t|^{-n}\int v(y)\phi((x-y)/t)\,dy.$$

If we differentiate and change variables back again it follows that

$$\partial_i(t^k u(x,t))=t^{k-1}\int v(x-ty)\partial_i\phi(y)\,dy,$$
$$\partial_t(t^k u(x,t))=t^{k-1}\int v(x-ty)((k-n)\phi(y)-\sum y_i\partial_i\phi(y))\,dy.$$

In view of Corollary 1.1.2 this is also true when $t=0$ so the first order derivatives of $t^k u(x,t)$ are in C^{k-1+j} which proves the theorem.

Corollary 1.3.4. *Given arbitrary $u_j\in C^{k-j}(\mathbb{R}^n)$, $0\leqq j\leqq k$, one can find $u\in C^k(\mathbb{R}^{n+1})$ so that*

$$\partial_t^j u(x,t)=u_j(x) \quad \text{when } t=0,\ j=0,\dots,k. \tag{1.3.10}$$

Proof. For $r=0,\dots,k$ we shall prove inductively that one can find $u=u^r$ satisfying (1.3.10) when $0\leqq j\leqq r$. We can take $u^0(x,t)=u_0(x)$. If u^{r-1} has already been chosen we must find $u^r=u^{r-1}+U$ so that

$$\partial_t^j U=0,\ \ j<r, \qquad \partial_t^r U=v_r=u_r-\partial_t^r u^{r-1} \quad \text{when } t=0.$$

Since $v_r \in C^{k-r}$ we obtain a function U with these properties from Theorem 1.3.3.

Regularization of a function does not increase the support very much, for we have

(1.3.11) $$\operatorname{supp} u*v \subset \operatorname{supp} u + \operatorname{supp} v = \{x+y;\, x\in \operatorname{supp} u,\, y\in \operatorname{supp} v\}.$$

This is an immediate consequence of (1.3.1) or (1.3.1)′.

When applying Theorem 1.3.2 we must of course appeal to the existence of test functions proved in Lemma 1.2.3. However, test functions on $\mathbb{R}$ can also be constructed by convolutions starting from the simplest step functions, and we shall discuss this now since it leads to important quantitative information.

Set

$$H_a(x)=a^{-1} \quad \text{when } 0<x<a \quad \text{and} \quad H_a(x)=0 \text{ otherwise.}$$

If u is a continuous function then

$$u*H_a(x)=a^{-1}\int_0^a u(x-t)\,dt = a^{-1}\int_{x-a}^x u(t)\,dt$$

is in C^1 and the derivative is $(u(x)-u(x-a))/a$, so $u*H_a\in C^{k+1}$ if $u\in C^k$.

Theorem 1.3.5. *Let $a_0\geqq a_1\geqq \dots$ be a positive sequence with*

$$a=\sum_0^\infty a_j<\infty,$$

and set

$$u_k=H_{a_0}*\dots*H_{a_k}.$$

Then $u_k\in C_0^{k-1}(\mathbb{R})$ has support in $[0,a]$ and converges as $k\to\infty$ to a function $u\in C_0^\infty(\mathbb{R})$ with support in $[0,a]$ such that $\int u\,dx=1$ and

(1.3.12) $$|u^{(k)}(x)|\leqq \tfrac{1}{2}\int |u^{(k+1)}(x)|\,dx \leqq 2^k/(a_0\dots a_k), \qquad k=0,1,\dots.$$

Proof. u_1 vanishes except in $[0,a_0+a_1]$, increases with slope $1/a_0a_1$ in $[0,a_1]$, is constant in $[a_1,a_0]$ and decreases linearly to 0 in (a_0,a_0+a_1), so u_1 is continuous. Hence $u_k\in C_0^{k-1}$ by the remarks preceding the proof, and the support is in $[0,a_0+\dots+a_k]$ by (1.3.11). With the notation

$$(\tau_a u)(x)=u(x-a)$$

(which later on will be recognized as a convolution) we have

$$u_k^{(j)}=\prod_0^{j-1}\frac{1}{a_i}(1-\tau_{a_i})(H_{a_j}*\dots*H_{a_k}), \qquad \text{if } j\leqq k-1.$$

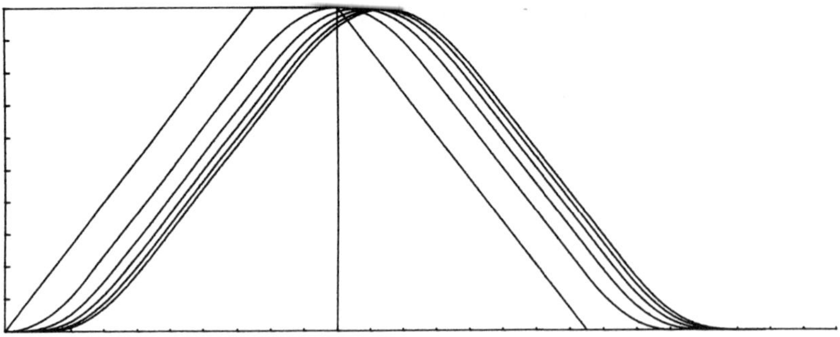

Fig. 1. u_k for $k\leqq 6$ with $a_0=1$, $a_j=1.5/(j(j+1))$

Since $\int H_a\,dx=1$ it follows from (1.3.4) that

(1.3.12)′ $\quad |u_k^{(j)}|\leqq 2^j/(a_0 \dots a_j), \qquad \int |u_k^{(j)}|\,dx \leqq 2^j/(a_0 \dots a_{j-1}),$

for a convolution $u*v$ can always be estimated by $\sup|u|\int|v|\,dx$. Now

$$\begin{aligned}|u_{k+m}-u_m| &= |u_m * H_{a_{m+1}} * \dots * H_{a_{m+k}} - u_m|\\ &\leqq (a_{m+1}+\dots+a_{m+k}) \sup|u_m'|\\ &\leqq 2a_0^{-1}a_1^{-1}(a_{m+1}+\dots+a_{m+k})\end{aligned}$$

by the proof of Theorem 1.3.2, so u_m has a uniform limit u. So has the derivative

$$u_m' = a_0^{-1}(1-\tau_{a_0})H_{a_1} * \dots * H_{a_m}$$

and so on. By Theorem 1.1.5 it follows that $u\in C^\infty$, and (1.3.12)′ gives (1.3.12). Since $\int u_k\,dx=1$ for every k by (1.3.4), we have $\int u\,dx=1$.

Only in the last step was it important that $\sum a_j$ was assumed convergent. However, if it diverges then the limit u must be identically 0. This will follow from our next lemma which shows how precise the construction in Theorem 1.3.5 really is.

Lemma 1.3.6. *If $u\in C^m((-\infty,T])$ vanishes on the negative half axis, and if a_j are positive decreasing numbers with $T\leqq a_1+\dots+a_m$, then*

(1.3.13) $\quad |u(x)| \leqq \sum_{j\in J} 2^{2j} \sup_{y<x} a_1 \dots a_j |u^{(j)}(y)| \quad$ *if* $x\leqq T$.

Here $J=\{j;\,1\leqq j\leqq m$ and $a_{j+1}<a_j$ or $j=m\}$.

Proof. The formula

$$u = H_a * au' + \tau_a u,$$

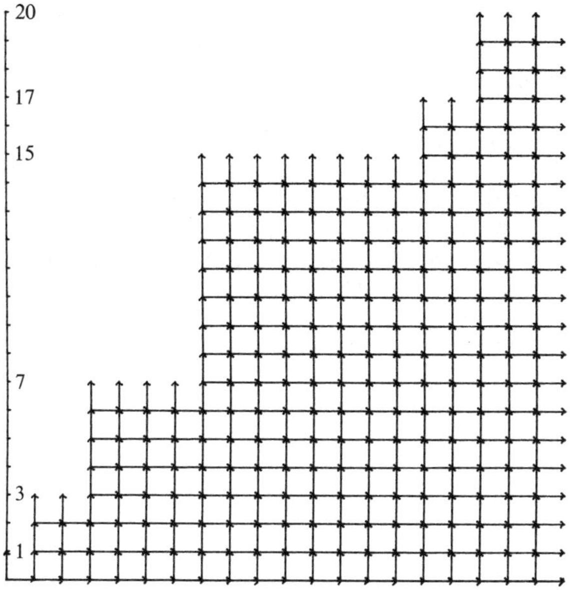

Fig. 2. Z and the moves allowed when $m=20$, $J=\{1,3,7,15,17,20\}$

used in the proof of Theorem 1.3.5, gives for $j \leqq m-1$ that

(1.3.14) $$u^{(j)} = H_{a_{j+1}} * a_{j+1} u^{(j+1)} + \tau_{a_{j+1}} u^{(j)}.$$

Starting from the case $j=0$ this allows us to write u as a sum of terms of the form

(1.3.15) $$\tau_{a_{k_1}} \dots \tau_{a_{k_i}} H_{a_1} * \dots * H_{a_j} * a_1 \dots a_j u^{(j)}; \quad i \leqq m,\ j \leqq m, \quad \text{and} \quad a_{k_\nu} \geqq a_\nu, \quad \nu \leqq i.$$

(The last condition guarantees that $\tau_{a_{k_1}} \dots \tau_{a_{k_i}}$ is a translation by at least $a_1 + \dots + a_i$.) We shall say that (1.3.15) is a legitimate term of type (i,j) and represent it by the point $(i,j) \in \mathbb{R}^2$. In any term with $(i,j) \in Z$,

$$Z = \{(i,j);\ 0 \leqq i < m,\ 0 \leqq j < m,\ a_{j+1} \geqq a_{i+1}\}$$

we apply (1.3.14) and obtain two new terms. Since $a_{j+1} \geqq a_{i+1}$ the one which comes from the second term in (1.3.14) is legitimate of type $(i+1,j)$, and $(i+1,j) \in Z$ unless $i+1=m$, for $a_{i+2} \leqq a_{i+1}$. The other term is legitimate of type $(i,j+1)$, and we have $(i,j+1) \in Z$ unless $j+1=m$, thus $j+1 \in J$, or else $a_{j+2} < a_{i+1}$; since $a_{i+1} \leqq a_{j+1}$ this implies $i+1 < j+2$ and $a_{j+2} < a_{j+1}$, hence $j+1 \in J$. After a finite number of steps we conclude that u is a sum of terms of the form (1.3.15) with either $i \geqq m$ or $j \in J$ and $i+1 \leqq j$. (In Fig. 2 (i,j) is a free arrowhead.) The number of terms of type (i,j) is at most 2^{i+j} since each term corresponds to an increasing path from $(0,0)$ to (i,j) among lattice points, so there are $i+j$ steps and at most two alternatives for each of them. Since

$$\sum_{i<j} 2^{i+j} < 2^{2j}$$

and the terms with $i=m$ vanish in $(0,T)$, we have proved (1.3.13).

If we drop the assumption $T \leqq a_1 + \ldots + a_m$ in Lemma 1.3.6 and set $c = T/(a_1 + \ldots + a_m)$, we can apply Lemma 1.3.6 to $u(cx)$ which is defined for $x \leqq a_1 + \ldots + a_m$. This gives

$$(1.3.13)' \qquad |u(x)| \leqq \sum_{j \in J} (4c)^j \sup_{y<x} a_1 \ldots a_j |u^{(j)}(y)| \quad \text{if } x \leqq T.$$

If

$$(1.3.16) \qquad |u^{(j)}(x)| \leqq K^j/(a_1 \ldots a_j) \quad \text{when } j \in J \text{ and } x \leqq T$$

then it follows that

$$(1.3.17) \qquad |u(x)| \leqq 8TK/(a_1 + \ldots + a_m) \quad \text{if } x \leqq T \leqq (a_1 + \ldots + a_m)/8K.$$

In fact,

$$\sum_1^\infty (4cK)^j = 4cK/(1-4cK) < 8cK \quad \text{if } 8cK < 1.$$

If we apply (1.3.17) to the integral of the function u in Theorem 1.3.5 taking $K=2$, we also conclude that u could not have been given support in an interval shorter than $\sum a_j/16$, for $8c/(1-8c) < 1$ if $c < 1/16$. Apart from a fixed scale factor the construction in Theorem 1.3.5 is therefore quite precise. If $\sum a_j = \infty$ and (1.3.16) is valid for all j we may also conclude that $u \equiv 0$. In fact, we shall deduce the complete Denjoy-Carleman theorem.

Let $M_0 = 1, M_1, M_2, \ldots$ be a sequence of positive numbers, and denote by $C^M([a,b])$ the set of all $u \in C^\infty([a,b])$ such that for some K

$$(1.3.18) \qquad |u^{(j)}(x)| \leqq K^{j+1} M_j \quad \text{if } a \leqq x \leqq b \text{ and } j = 0, 1, \ldots.$$

Definition 1.3.7. C^M is called quasi-analytic if $u^{(j)}(x) = 0$ for every j at some $x \in [a,b]$ implies $u = 0$ in $[a,b]$ when $u \in C^M([a,b])$.

Let

$$(1.3.19) \qquad L_j = \inf_{k \geqq j} M_k^{1/k}$$

be the largest increasing minorant of $M_k^{1/k}$ and let M_j^* be the largest logarithmically convex minorant of the sequence M_j,

(1.3.20) $\quad M_j^* = \inf\{M_j,\, M_k^{(l-j)/(l-k)} M_l^{(j-k)/(l-k)} \text{ when } k<j<l\}.$

Thus $M_j^{*2} \leqq M_{j-1}^* M_{j+1}^*$, $j>0$, which shows that $M_j^*>0$ for all j unless $M_j^*=0$ for every $j>0$. Moreover,

(1.3.21) $$a_j = M_{j-1}^*/M_j^*, \quad j=1,2,\dots$$

is a decreasing sequence. (We define $a_j=+\infty$ if $M_j^*=0$.)

Theorem 1.3.8 (*Denjoy-Carleman*). *The following conditions are equivalent:*

(i) C^M *is quasi-analytic.*

(ii) $\sum_0^\infty 1/L_j = \infty$.

(iii) $\sum_0^\infty (M_j^*)^{-1/j} = \infty$.

(iv) $\sum_1^\infty a_j = \infty$.

Proof. First we consider the rather uninteresting case where $L_j \leqq L < \infty$ for every j, that is, $\lim M_{k_j}^{1/k_j} \leqq L$ for some sequence $k_j \to \infty$. Letting l run through this sequence in (1.3.20) and taking $k=0$ there we obtain

$$L^j M_0 \geqq M_j^* = M_0/a_1 \dots a_j \geqq M_0/(a_1 \dots a_{i-1}) a_i^{i-j-1}, \quad i \leqq j,$$

and this implies $a_i^{-1} \leqq L$. Thus the conditions (ii), (iii), (iv) are fulfilled. If u satisfies (1.3.18) and $u^{(j)}(x_0)=0$ for every j, then Taylor's formula gives for every $x \in [a,b]$

$$|u(x)| \leqq K^{k_j+1} |x-x_0|^{k_j} M_{k_j}/k_j! \to 0 \quad \text{when } j\to\infty.$$

Thus (i) is also valid.

Assume now that $L_j \to \infty$ when $j\to\infty$, that is, that $M_k^{1/k}\to\infty$ as $k\to\infty$. Then the points $(k, \log M_k)$ will lie above lines with arbitrarily high slope so M_j^* is positive and $a_j \to 0$. Thus $\hat{J}=\{j; a_{j+1}<a_j\}$ is an infinite set where the graph of $\log M_k^*$ as a function of k has a corner. This implies that $M_j^*=M_j$ when $j\in\hat{J}$; analytically this follows since for $k<j<l$

$$\begin{aligned} M_k^{(l-j)/(l-k)} M_l^{(j-k)/(l-k)} &\geqq M_k^{*(l-j)/(l-k)} M_l^{*(j-k)/(l-k)} \\ &\geqq M_j^* a_j^{(l-j)(j-k)/(l-k)} a_{j+1}^{-(l-j)(j-k)/(l-k)} \geqq M_j^* (a_j/a_{j+1})^{\frac12} > M_j^*. \end{aligned}$$

Since
$$M_k^* = M_0^*/a_1 \dots a_k \leqq a_k^{-k}$$

it follows that $M_j^{1/j} \leqq 1/a_j$ when $j\in\hat{J}$. If $k<j\in\hat{J}$ and $k\leqq i<j$ implies $i\notin\hat{J}$ then $a_i=a_j$ for $k\leqq i\leqq j$ so it follows that $L_i \leqq L_j \leqq 1/a_j = 1/a_i$, hence

$L_j \leqq 1/a_j$ for every j. This proves that (iv) $\Rightarrow$ (ii). Since $\log M_j^*$ is a convex function of j and vanishes when $j=0$ the slope $j^{-1}\log M_j^*$ of the chord from (0, 0) is increasing, that is, $(M_j^*)^{1/j}$ is increasing. Hence $L_j \geqq (M_j^*)^{1/j}$ which proves that (ii) $\Rightarrow$ (iii). If (iv) is false we know from Theorem 1.3.5 that there is a function $u \neq 0$ in $C_0^\infty([a,b])$ with

$$|u^{(j)}(x)| \leqq K^{j+1}/a_1 \dots a_j = K^{j+1} M_j^* \leqq K^{j+1} M_j$$

so (i) is not valid. Thus (i) $\Rightarrow$ (iv). On the other hand, if (iv) is valid we can apply (1.3.17) with m equal to an element of $\hat{J}$. The set J in (1.3.16), defined in Lemma 1.3.6, is then the set of elements $\leqq m$ in $\hat{J}$, so (1.3.16) follows from (1.3.18) since $M_j = M_j^* = 1/a_1 \dots a_j$ when $j \in \hat{J}$. Hence (iv) $\Rightarrow$ (i). The remaining proof that (iii) $\Rightarrow$ (iv) follows from

Lemma 1.3.9 (Carleman's inequality). If $a_j > 0$ then

$$\sum_1^\infty (a_1 a_2 \dots a_n)^{1/n} \leqq e \sum_1^\infty a_n. \tag{1.3.22}$$

Proof. With c_j to be chosen later we have by the inequality between geometric and arithmetic means

$$(a_1 \dots a_n)^{1/n} = (c_1 \dots c_n)^{-1/n}(c_1 a_1 \dots c_n a_n)^{1/n} \leqq (c_1 \dots c_n)^{-1/n} n^{-1} \sum_1^n c_m a_m.$$

If we choose $c_m = (m+1)^m/m^{m-1}$ then $(c_1 \dots c_n)^{1/n} = n+1$ and we have

$$\sum (a_1 \dots a_n)^{1/n} \leqq \sum_{1 \leqq m \leqq n} c_m a_m / n(n+1) = \sum_1^\infty c_m a_m / m \leqq \sum_1^\infty e a_m$$

which proves (1.3.22).

When $M_n = n!$ then Taylor's formula gives for $f \in C^M([a,b])$ that

$$f(x) = \sum_0^\infty f^{(j)}(y)(x-y)^j/j!$$

if $x, y \in [a,b]$ and $|x-y| < r$. Thus Theorem 1.3.8 is completely elementary in this case. Note that the Taylor series is then absolutely convergent for all $x \in \mathbb{C}$ with $|x-y| < r$ and it gives an analytic continuation of f to

$$\{z \in \mathbb{C};\ |z-y| < r \text{ for some } y \in [a,b]\}.$$

Conversely, if F is an analytic function in a complex neighborhood of $[a,b]$ then the restriction of F to $[a,b]$ is in $C^M([a,b])$ by Cauchy's inequalities. Accordingly this class is called the class of *real analytic functions.* It is obvious that the preceding remarks can be applied also to functions of several variables.

1.4. Cutoff Functions and Partitions of Unity

In distribution theory one often has to replace a function by one with compact support without changing it on a large compact set. This is done by multiplication with a "cutoff function" as constructed in the following

Theorem 1.4.1. *If X is an open set in $\mathbb{R}^n$ and K is a compact subset, then one can find $\phi \in C_0^\infty(X)$ with $0 \leqq \phi \leqq 1$ so that $\phi = 1$ in a neighborhood of K.*

Proof. Choose $\varepsilon > 0$ so small that

$$|x-y| \geqq 4\varepsilon \quad \text{when } x \in K,\ y \in \complement X, \tag{1.4.1}$$

and let v be the characteristic function of

$$K_{2\varepsilon} = \{y;\ |x-y| \leqq 2\varepsilon \text{ for some } x \in K\}.$$

According to Lemma 1.2.3 we can find a non-negative function $\chi \in C_0^\infty(B)$ where B is the unit ball, such that $\int \chi\, dx = 1$. Then $\chi_\varepsilon(x) = \varepsilon^{-n} \chi(x/\varepsilon)$ has support in the ball $\{x; |x| < \varepsilon\}$ and $\int \chi_\varepsilon\, dx = 1$, so

$$\phi = v * \chi_\varepsilon \in C_0^\infty(K_{3\varepsilon})$$

by Theorem 1.3.1 and (1.3.11), and $1-\phi = (1-v) * \chi_\varepsilon$ vanishes in K_ε by (1.3.11). This proves the theorem.

For future reference we also note that

$$|\partial^\alpha \phi| \leqq \int |\partial^\alpha \chi_\varepsilon|\, dx = \varepsilon^{-|\alpha|} \int |\partial^\alpha \chi|\, dx.$$

Thus

$$|\partial^\alpha \phi| \leqq C_\alpha \varepsilon^{-|\alpha|} \tag{1.4.2}$$

where C_α only depends on α, n and the choice of the norm. Using Theorem 1.3.5 it is possible to give a still more precise result which we mention for the sake of completeness since it is sometimes useful. However, the reader can jump to Theorem 1.4.4 without loss of continuity.

Theorem 1.4.2. *Let X be an open set in the n dimensional vector space V with norm $\| \ \|$ and let K be a compact subset. If*

$$d = \inf\{\|x-y\|;\ x \in \complement X,\ y \in K\}$$

and d_j is a positive decreasing sequence with $\sum_1^\infty d_j < d$, then one can find $\phi \in C_0^\infty(X)$ with $0 \leqq \phi \leqq 1$, equal to 1 in a neighborhood of K, so that

(1.4.3) $|\phi^{(k)}(x; y_1, \dots, y_k)| \leqq C^k \|y_1\| \dots \|y_k\| / d_1 \dots d_k; \quad k = 1, 2, \dots$.

Here C depends only on the dimension n.

Proof. Assume first that $V = \mathbb{R}^n$ and that $\|x\| = \max |x_j|$. Let u be the function in Theorem 1.3.5 with $a_j = d_{j+1}$ and set $h(t) = u(t + \sum d_j/2)$. Then we have $|t| \leqq \sum |d_j|/2$ if $t \in \operatorname{supp} h$, and for every j

$$\int |h^{(j)}(t)|\, dt \leqq 2^j / d_1 \dots d_j, \qquad \int h(t)\, dt = 1.$$

We can now apply the proof of Theorem 1.4.1 with $\varepsilon = 1$ and

$$\chi(x) = h(x_1) \dots h(x_n),$$

taking for v the characteristic function of

$$\{y;\ \|x - y\| \leqq d/2 \text{ for some } x \in K\}.$$

It follows that

$$|\partial^\alpha \phi| \leqq \int |\partial^\alpha \chi|\, dx \leqq 2^{|\alpha|} / d_1 \dots d_{|\alpha|}$$

so introducing the differentials instead we have

$$|\phi^{(k)}(x; y_1, \dots, y_k)| \leqq (2n)^k \|y_1\| \dots \|y_k\| / d_1 \dots d_k,$$

which proves Theorem 1.4.2 in this case. The passage to a Euclidean norm and from there to an arbitrary norm can be made by the following lemma which gives (1.4.3) with $C = 2n^2$.

Lemma 1.4.3. *If K is a convex symmetric body in the n dimensional vector space V, then one can find an ellipsoid B with center at* 0 *and*

$$B \subset K \subset B\sqrt{n}.$$

Proof. Let B be an ellipsoid of maximal volume contained in K and choose coordinates so that B is the unit ball. We have to show that $K \subset B\sqrt{n}$. To do so we assume that K contains a point $(t, 0, \dots, 0)$ with $t > \sqrt{n}$ and prove that B cannot be maximal then. The tangent cone to B with vertex at $(t, 0, \dots, 0)$ touches B where $x_1 = 1/t$, so the part of B where $|x_1| > 1/t$ is in the interior of K because of the convexity and symmetry with respect to 0. Now consider the ellipsoid

$$(1-\varepsilon)^{n-1} x_1^2 + y^2/(1-\varepsilon) \leqq 1, \qquad y = (x_2^2 + \dots + x_n^2)^{\frac{1}{2}},$$

which has the same volume as B. The inequality may be written

$$x_1^2 + y^2 - ((n-1)\, x_1^2 - y^2)\, \varepsilon + O(\varepsilon^2) < 1,$$

that is,

$$(x_1^2 + y^2 - 1)(1 + \varepsilon) < \varepsilon (n x_1^2 - 1) + O(\varepsilon^2).$$

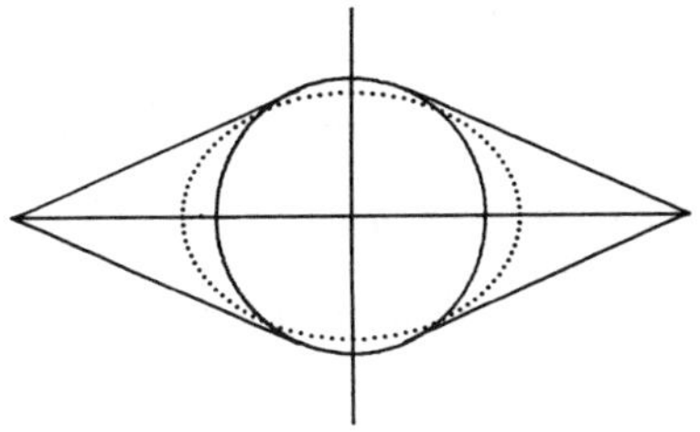

Fig. 3

For small $\varepsilon>0$ the right-hand side is negative when $|x_1|<(1/t+1/\sqrt{n})/2$ so this part of the ellipsoid is in the interior of K. But this is also true of the remaining part since $|x_1|\geqq(1/t+1/\sqrt{n})/2>1/t$. (See Fig. 3.) Hence B is not maximal.

An equivalent way of stating Lemma 1.4.3 is of course that for any norm $|\ |$ in V one can find a Euclidean norm $\|\ \|$ such that

$$|x|\leqq\|x\|\leqq|x|\sqrt{n}, \qquad x\in V.$$

In order to make conclusions about a distribution from local hypotheses, it is necessary to write arbitrary test functions as sums of test functions of small support.

Theorem 1.4.4. *Let $X_1,\dots,X_k$ be open sets in $\mathbb{R}^n$ and let $\phi\in C_0^\infty\left(\bigcup_1^k X_j\right)$. Then one can find $\phi_j\in C_0^\infty(X_j)$, $j=1,\dots,k$, such that*

$$\phi=\sum_1^k \phi_j. \tag{1.4.4}$$

If $\phi\geqq 0$ one can take all $\phi_j\geqq 0$.

Proof. We can choose compact sets $K_1,\dots,K_k$ with $K_j\subset X_j$ so that $\operatorname{supp}\phi\subset\bigcup_1^k K_j$. (In fact, every point in $\operatorname{supp}\phi$ has a compact neighborhood contained in some X_j. By the Borel-Lebesgue lemma a finite number of such neighborhoods can be chosen which cover all of $\operatorname{supp}\phi$. The union of those which belong to X_j is a compact set $K_j\subset X_j$.) Using Theorem 1.4.1 we now choose $\psi_j\in C_0^\infty(X_j)$ with $0\leqq\psi_j\leqq 1$ and $\psi_j=1$ in K_j. Then the functions

$$\phi_1=\phi\psi_1,\ \phi_2=\phi\psi_2(1-\psi_1),\dots,\phi_k=\phi\psi_k(1-\psi_1)\dots(1-\psi_{k-1})$$

have the required properties since

$$\sum_1^k\phi_j-\phi=-\phi\prod_1^k(1-\psi_j)=0$$

because either ϕ or some $1-\psi_j$ is zero at any point.

Combining Theorems 1.4.4 and 1.4.1 we obtain

Theorem 1.4.5. *Let* $X_1, \dots, X_k$ *be open sets in* $\mathbb{R}^n$ *and* K *a compact subset of* $\bigcup_1^k X_j$. *Then one can find* $\phi_j \in C_0^\infty(X_j)$ *so that* $\phi_j \geqq 0$ *and* $\sum_1^k \phi_j \leqq 1$ *with equality in a neighborhood of* K.

Remark. In Theorems 1.4.4 and 1.4.5 we could allow an infinite number of open sets X_j while in the conclusion only finitely many ϕ_j are not identically 0. In fact, by the Borel-Lebesgue lemma a finite number of the sets X_j suffice to cover $\operatorname{supp} \phi$ and K respectively.

The functions ϕ_j in Theorem 1.4.5 are said to be a *partition of unity* at K subordinate to the covering of K by the sets X_j. Infinite partitions of unity are sometimes required, and for the sake of completeness we shall give one simple and one fairly intricate example.

Theorem 1.4.6. *For every neighborhood* X *of the cube*

$$K = \{x \in \mathbb{R}^n;\ |x_j| \leqq \tfrac{1}{2}, j = 1, \dots, n\}$$

one can find a non-negative function $\phi \in C_0^\infty(X)$ *such that*

$$\sum \phi(x-g) = 1$$

where g *runs over all lattice points, that is, points with integer coordinates.*

Proof. By Theorem 1.4.1 we can choose $\psi \in C_0^\infty(X)$ so that $0 \leqq \psi \leqq 1$ and $\psi = 1$ on K. Hence

$$\Psi(x) = \sum \psi(x-g)$$

is a periodic C^∞ function with $\Psi(x) \geqq 1$ everywhere, so $\phi(x) = \psi(x)/\Psi(x)$ has the required properties.

Theorem 1.4.6 gives a partition of unity subordinate to the covering of $\mathbb{R}^n$ by the translates $X + \{g\}$ of X. We shall now construct partitions of unity corresponding to a covering by convex symmetric neighborhoods which may vary both in size and in shape. It is convenient to describe the neighborhoods in terms of the corresponding norms.

Definition 1.4.7. If X is an open set in a finite dimensional vector space V and $\| \ \|_x$ for every $x \in X$ is a norm in V, then we shall say that we have a slowly varying metric in X if

$$\text{(1.4.5)} \quad x \in X, \quad \|y-x\|_x < 1 \text{ implies } y \in X \quad \text{and} \quad \|v\|_y \leqq C\|v\|_x, \quad v \in V,$$

where $C \geqq 1$ is independent of x, y, v.

Note that if $C\|x-y\|_x<1$ then $\|x-y\|_y<1$ so

$$\|v\|_x \leqq C\|v\|_y, \quad v \in V.$$

Replacing the norm $\| \ \|_x$ by $C\| \ \|_x$ we have therefore

$$C^{-1}\|v\|_x \leqq \|v\|_y \leqq C\|v\|_x \quad \text{if } \|x-y\|_x<1 \tag{1.4.5)'}$$

which we shall assume from now on.

Example 1.4.8. Let X be an open set in V and $d(x)$ a Lipschitz continuous function, positive in X and zero in $V \setminus X$, with

$$|d(x)-d(y)| \leqq \|x-y\| \quad \text{if } x, y \in X,$$

where $\| \ \|$ is a fixed norm in V. Then

$$\|v\|_x = 2\|v\|/d(x)$$

defines a slowly varying metric in X, for $\|x-y\|_x<1$ means that $\|x-y\|<d(x)/2$ which implies $|d(x)-d(y)|<d(x)/2$ and $d(x)/2 < d(y) < 2d(x)$. For any given open set $X \neq V$ we may always take for $d(x)$ the distance to $F = \complement X$ defined by

$$d(x) = \inf_{y \in F} \|x-y\|.$$

We shall now show how a slowly varying metric gives rise first to a covering and then to a subordinate partition of unity.

Lemma 1.4.9. *Let $\varepsilon < 1/C$ and choose a maximal sequence $x_1, x_2, \ldots$ in X such that*

$$\|x_\mu - x_\nu\|_{x_\nu} \geqq \varepsilon \quad \textit{when } \nu \neq \mu. \tag{1.4.6}$$

Then the balls

$$B_\nu^R = \{x;\ \|x-x_\nu\|_{x_\nu} < R\},$$

where $\varepsilon C < R < 1$, are a covering of X, and no point belongs to more than $N = (2C^2/\varepsilon + 1)^n$ different B_ν^R.

Proof. The existence of a maximal sequence follows from the fact that a set satisfying (1.4.6) is necessarily discrete. If $x \in X$ cannot be added to the sequence then either $\|x-x_\nu\|_{x_\nu} < \varepsilon$ or else $\|x-x_\nu\|_x < \varepsilon$ for some ν, and in the latter case it follows that $\|x-x_\nu\|_{x_\nu} < C\varepsilon$ so $x \in B_\nu^R$ in any case. If $x \in B_\nu^R$ we obtain since $R<1$

$$\|x-x_\nu\|_x < C$$

and if $x \in B_\mu^R$ also then

$$\varepsilon \leqq \|x_\mu - x_\nu\|_{x_\nu} \leqq C\|x_\mu - x_\nu\|_x.$$

Hence the balls

$$\{y;\ \|y-x_\nu\|_x<\varepsilon/2C\}$$

are disjoint when $x\in B_\nu^R$, and they are all contained in the ball

$$\{y;\ \|x-y\|_x<C+\varepsilon/2C\}$$

so the number of such ν cannot exceed the ratio $(2C^2/\varepsilon+1)^n$ of the measures.

Choose $\varepsilon=1/2C$ and for every ν a non-negative function $\psi_\nu\in C_0^\infty(B_\nu^1)$ which is 1 in $B_\nu^{\frac{1}{2}}$ and for a given decreasing sequence d_j with $\sum d_j=1$ has the estimate

$$|\psi_\nu^{(k)}(x;y_1,\dots,y_k)|\leqq C_1^k\|y_1\|_{x_\nu}\dots\|y_k\|_{x_\nu}/d_1\dots d_k.$$

This is possible by Theorem 1.4.2 with C_1 equal to three times the constant in (1.4.3). Since $\|y\|_{x_\nu}\leqq C\|y\|_x$ for every $x\in\operatorname{supp}\psi_\nu$ we obtain

$$|\psi_\nu^{(k)}(x;y_1,\dots,y_k)|\leqq(CC_1)^k\|y_1\|_x\dots\|y_k\|_x/d_1\dots d_k.$$

As in the proof of Theorem 1.4.4 we now introduce

$$\phi_\nu=\psi_\nu(1-\psi_1)\dots(1-\psi_{\nu-1})$$

and obtain $\sum\phi_\nu=1$ in X. No point is in the support of more than N factors ψ_μ. Now note that if for some norm we have at x

$$|f_j^{(k)}(x,y_1,\dots,y_k)|\leqq A_j^k\|y_1\|\dots\|y_k\|/d_1\dots d_k,$$

for $k=0,1,\dots$ and $j=1,2$, then the same estimate is valid for f_1f_2 with the constant A_1+A_2 instead. This follows from the fact that d_j is decreasing and from the rules for differentiating a product. Hence

$$|\phi_\nu^{(k)}(x;y_1,\dots,y_k)|\leqq(NCC_1)^k\|y_1\|_x\dots\|y_k\|_x/d_1\dots d_k.$$

We have now proved

Theorem 1.4.10. *For any slowly varying metric in the open set X in the n dimensional vector space V one can choose a sequence $x_\nu\in X$ such that the balls*

$$B_\nu=\{x;\ \|x-x_\nu\|_{x_\nu}<1\}$$

form a covering of X for which the intersection of more than $N=(4C^3+1)^n$ balls B_ν is always empty. Moreover, for any decreasing sequence d_j with $\sum d_j=1$ one can choose non-negative $\phi_\nu\in C_0^\infty(B_\nu)$ with $\sum\phi_\nu=1$ in X so that for all k

$$|\phi_\nu^{(k)}(x;y_1,\dots,y_k)|\leqq(NCC_1)^k\|y_1\|_x\dots\|y_k\|_x/d_1\dots d_k \tag{1.4.7}$$

where C is the constant in (1.4.5) *and $C_1/3$ is that in* (1.4.3) *so it only depends on n.*

Remark. Without changing the proof one may allow d_j to be functions of x provided that they vary slowly with respect to the metric, that is,

$$d_j(y) \leqq C d_j(x) \quad \text{if } x \in X \quad \text{and} \quad \|y-x\|_x < 1. \tag{1.4.5$'$}$$

The following corollary is sometimes a useful supplement to Theorem 1.4.1.

Corollary 1.4.11. *Let F_0 and F_1 be two closed sets in $\mathbb{R}^n$. Then one can find $\phi \in C^\infty(\complement(F_0 \cap F_1))$ such that $\phi = 0$ near $F_0 \setminus (F_0 \cap F_1)$, $\phi = 1$ near $F_1 \setminus (F_0 \cap F_1)$ and, with the notation in Theorem 1.4.10,*

$$|\phi^{(k)}(x; y_1, \dots, y_k)| \leqq C_2^k \|y_1\| \dots \|y_k\| d(x)^{-k}/d_1 \dots d_k. \tag{1.4.8}$$

Here

$$d(x) = \max(d(x, F_0), d(x, F_1)), \qquad d(x, F_j) = \min_{y \in F_j} \|x-y\|.$$

The support of ϕ is bounded if F_1 is compact.

Proof. $d(x)$ is Lipschitz continuous with Lipschitz constant 1, so $\|v\|_x = 2\|v\|/d(x)$ is a slowly varying metric in $\complement(F_0 \cap F_1)$ (Example 1.4.8). A ball

$$\{y; \|x-y\|_x \leqq 1\} \tag{1.4.9}$$

cannot meet both F_0 and F_1, for if $d(x) = d(x, F_j)$ then $\|x-y\| \leqq d(x, F_j)/2$, hence $d(y, F_j) \geqq d(x, F_j)/2$ in the ball. If we form the partition of unity in Theorem 1.4.10 we can therefore take ϕ equal to the sum of all ϕ_ν with support meeting F_1. If F_1 is compact then $d(x, F_0) < 2d(x, F_1)$ when $|x|$ is large enough. Thus $d(x) < 2d(x, F_1)$ so (1.4.9) cannot meet F_1 then. Hence $\operatorname{supp} \phi$ will be bounded.

Notes

As pointed out in the summary the introductory Section 1.1 contains only classical material and we shall not discuss its history. The test functions used in Section 1.2 also have a long tradition in the calculus of variations. Thus Theorem 1.2.4 is very close to the de Bois Reymond lemma found in all introductory texts on this topic. Theorem 1.2.6 is due to Borel [1]. More refined extension theorems are given later as Corollary 1.3.4 and Theorem 2.3.6.

The construction of infinitely differentiable functions by means of repeated convolutions used in the proof of Theorem 1.3.5 has ancient roots in harmonic analysis. It was used explicitly and systematically by Mandelbrojt [1, 2] who attributes the idea to unpublished work by H.E. Bray. The method reappeared in the work by Ehrenpreis [3] on convolution operators and has been used frequently in the theory of partial differential equations since then (cf. Hörmander [21, 27, 30]). Cohen [2] observed that the proof by Bang [1] of the Denjoy-Carleman theorem can be phrased in similar terms. We use a variant of his approach combined with arguments from Mandelbrojt [2] in our proof. Quasi-analyticity will play a role in Section 8.4. Perhaps it is appropriate to mention here that the problem of characterizing quasi-analytic classes originates from the theory of partial differential equations (see Hadamard [1, p. 37]).

Continuous partitions of unity on compact sets were defined by Dieudonné [1], and the term Dieudonné decomposition was actually current for a while. The construction in the proof of Theorem 1.4.4 is taken from the third edition of the classical monograph on Riemann surfaces by Weyl [2] where the main change was actually the use of partitions of unity. However, the covering lemma 1.4.9 and the partition of unity in Theorem 1.4.10 are essentially from Whitney [1] which shows that very sophisticated partitions of unity were used several years before Dieudonné's note. Whitney's lemma occurs in the original form as Lemma 2.3.7 below. A more general version was stated by Treves [5]. The full generality given in Section 1.4 is needed in the general theory of pseudo-differential operators (Chapter XVIII). The combination of these ideas with the repeated convolutions of Section 1.3 is probably new in its full generality but special cases can be found for example in Andersson [1].

Chapter II. Definition and Basic Properties of Distributions

Summary

In the introduction we have seen how various difficulties in the theory of partial differential equations and in Fourier analysis lead one to extend the space of continuous functions to the space of distributions. In Section 2.1 we make the definition explicit and precise, using the properties of test functions proved in Chapter I. The weak topology in the space of distributions is also introduced there. The notion of support is extended to distributions in Section 2.2 and it is shown there that distributions may be defined locally provided that the local definitions are compatible. In addition it is proved that if u is a distribution then there is a unique way to define $u(\phi)$ for all $\phi \in C^\infty$ with $\operatorname{supp} u \cap \operatorname{supp} \phi$ compact. The problem of estimating $u(\phi)$ in terms of the derivatives of ϕ on the support of u only is discussed at some length in Section 2.3. The deepest result is Whitney's extension theorem (Theorem 2.3.6). We shall rarely need the results which follow from it so the reader might prefer to skip the section from Theorem 2.3.6 on.

2.1. Basic Definitions

In the introduction we were led to consider expressions such as

$$L(\phi) = \sum \int f_\alpha \partial^\alpha \phi \, dx, \quad \phi \in C_0^\infty(X) \tag{2.1.1}$$

where $f_\alpha \in C(X)$ and the sum is finite. Here X is an open set in $\mathbb{R}^n$. However, the same form L may have many different representations of this kind, so in the following definition we just keep an obvious property of the expression (2.1.1).

Definition 2.1.1. A distribution u in X is a linear form on $C_0^\infty(X)$ such that for every compact set $K \subset X$ there exist constants C and k such that

(2.1.2) $$|u(\phi)| \leqq C \sum_{|\alpha| \leqq k} \sup |\partial^\alpha \phi|, \qquad \phi \in C_0^\infty(K).$$

The set of all distributions in X is denoted by $\mathscr{D}'(X)$. If the same integer k can be used in (2.1.2) for every K we say that u is of order $\leqq k$, and we denote the set of such distributions by $\mathscr{D}'^k(X)$. Their union $\mathscr{D}'_F(X) = \bigcup \mathscr{D}'^k(X)$ is the space of distributions of finite order.

That u is a linear form on $C_0^\infty(X)$ means of course that u is a function from $C_0^\infty(X)$ to $\mathbb{C}$ such that

$$u(a\phi + b\psi) = au(\phi) + bu(\psi); \qquad a, b \in \mathbb{C};\ \phi, \psi \in C_0^\infty(X).$$

Remark. The reason for the traditional notation $\mathscr{D}'(X)$ is that Laurent Schwartz used the notation $\mathscr{D}(X)$ instead of $C_0^\infty(X)$.

The sum (2.1.1) defines a distribution of order k if $f_\alpha = 0$ when $|\alpha| > k$, and it defines a distribution whenever the sum is locally finite, that is, on every compact set there are only a finite number of functions f_α which do not vanish identically. We shall see later on that all distributions are in fact of the form (2.1.1).

Example 2.1.2. If $x_0 \in X$ then $u(\phi) = \partial^\alpha \phi(x_0)$ defines a distribution of order $|\alpha|$. That the order is not smaller follows if we choose $\psi \in C_0^\infty$ with $\psi(0) = 1$ and set $\phi_\delta(x) = (x - x_0)^\alpha \psi((x - x_0)/\delta)$, for $u(\phi_\delta) = \alpha!$ and

$$\sup |\partial^\beta \phi_\delta| \leqq C \delta^{|\alpha| - |\beta|} \to 0 \quad \text{when } \delta \to 0 \text{ if } |\beta| < |\alpha|.$$

More generally, if $x_j \in X$ is a sequence of points with no limit point in X, and if α_j are multi-indices, then

$$u(\phi) = \sum \partial^{\alpha_j} \phi(x_j)$$

is a distribution in X because a compact subset can only contain finitely many x_j. By the first part of the example we have $u \in \mathscr{D}'_F(X)$ if and only if $|\alpha_j|$ is bounded; the order is then $\max |\alpha_j|$.

The continuity condition in Definition 2.1.1 guarantees that u behaves well when acting on functions depending on parameters:

Theorem 2.1.3. *If $\phi(x, y) \in C^\infty(X \times Y)$ where Y is an open set in $\mathbb{R}^m$, and if there is a compact set $K \subset X$ such that $\phi(x, y) = 0$ when $x \notin K$, then*

$$y \to u(\phi(., y))$$

is a C^∞ function of y if $u \in \mathscr{D}'(X)$, and

$$\partial_y^\alpha u(\phi(., y)) = u(\partial_y^\alpha \phi(., y)).$$

Proof. For fixed $y \in Y$ we have by Taylor's formula

$$\phi(x, y+h) = \phi(x, y) + \sum h_j \partial\phi(x, y)/\partial y_j + \psi(x, y, h),$$
$$\sup_x |\partial_x^\alpha \psi(x, y, h)| = O(|h|^2) \quad \text{as } h \to 0, \ \forall \alpha.$$

Hence

$$u(\phi(., y+h)) = u(\phi(., y)) + \sum h_j u(\partial\phi(., y)/\partial y_j) + O(|h|^2)$$

which shows that $y \to u(\phi(., y))$ is differentiable and that

$$\frac{\partial}{\partial y_j} u(\phi(., y)) = u(\partial\phi(., y)/\partial y_j).$$

Iteration of this result proves the theorem.

The continuity condition (2.1.2) is often stated as a sequential continuity:

Theorem 2.1.4. *A linear form u on $C_0^\infty(X)$ is a distribution if and only if $u(\phi_j) \to 0$ when $j \to \infty$ for every sequence $\phi_j \in C_0^\infty(X)$ converging to 0 in the sense that* $\sup|\partial^\alpha \phi_j| \to 0$ *for every fixed α and* $\operatorname{supp} \phi_j \subset K$ *for all j and some fixed compact set $K \subset X$.*

Proof. It is obvious that (2.1.2) implies the condition in the theorem. On the other hand, suppose that there is a compact set $K \subset X$ such that (2.1.2) is not valid for any C and k. Taking $C = k = j$ we conclude that for some $\phi_j \in C_0^\infty(K)$ we have

$$|u(\phi_j)| > j \sum_{|\alpha| \leqq j} \sup |\partial^\alpha \phi_j|.$$

Since this condition is not changed if ϕ_j is multiplied by a constant factor, it is no restriction to assume that $u(\phi_j) = 1$. Then we have $|\partial^\alpha \phi_j| \leqq 1/j$ when $j \geqq |\alpha|$ so $\phi_j \to 0$ although $u(\phi_j)$ does not converge to 0. Thus the condition in the theorem is not fulfilled either, which proves the assertion.

A third equivalent form of the continuity condition is the following one:

Theorem 2.1.5. *A linear form u on $C_0^\infty(X)$ is a distribution if and only if there exist functions $\rho_\alpha \in C(X)$ such that*

$$|u(\phi)| \leqq \sum_\alpha \sup |\rho_\alpha \partial^\alpha \phi|, \quad \phi \in C_0^\infty(X), \tag{2.1.3}$$

and on each compact set in X all but a finite number of the functions ρ_α vanish identically. One can take $\rho_\alpha = 0$ when $|\alpha| > k$ if and only if u is of order $\leqq k$.

Proof. The sufficiency of (2.1.3) is obvious. On the other hand, let $u \in \mathscr{D}'(X)$. Take an increasing sequence of compact sets $K_j \subset X$ such that every compact subset of X belongs to some K_j, and choose $\chi_j \in C_0^\infty(X)$ with $\chi_j = 1$ in K_j (Theorem 1.4.1). Writing $\psi_j = \chi_j - \chi_{j-1}$ if $j > 1$ and $\psi_1 = \chi_1$ we have

$$\phi = \sum_1^\infty \psi_j \phi \quad \text{if } \phi \in C_0^\infty(X);$$

the sum is of course finite. Since the support of $\psi_j \phi$ is contained in that of ψ_j, Definition 2.1.1 gives with suitable C_j and increasing k_j

$$|u(\phi)| \leqq \sum_1^\infty |u(\psi_j \phi)| \leqq \sum_1^\infty C_j \sup_{|\alpha| \leqq k_j} \sup |\partial^\alpha (\psi_j \phi)|.$$

If we note that $\sum 2^{-j} = 1$ and differentiate the product, we obtain

$$|u(\phi)| \leqq \sup_j 2^j C_j 2^{k_j} \sup_{|\alpha+\beta| \leqq k_j} \sup |\partial^\beta \psi_j \partial^\alpha \phi|.$$

The functions

$$\rho_\alpha(x) = \sup_j \sup_{|\alpha+\beta| \leqq k_j} 2^{j+k_j} C_j |\partial^\beta \psi_j|$$

are continuous in X, for $\partial^\beta \psi_j = 0$ on K_i when $j > i$ so only a finite number of j and β must be considered. In addition $\rho_\alpha = 0$ on K_i if $|\alpha| > k_i$. Since

$$2^{j+k_j} C_j |\partial^\beta \psi_j \partial^\alpha \phi| \leqq |\rho_\alpha \partial^\alpha \phi| \quad \text{if } |\alpha+\beta| \leqq k_j,$$

we obtain $|u(\phi)| \leqq \sup_\alpha \sup |\rho_\alpha \partial^\alpha \phi|$, which implies (2.1.3).

We shall now derive a dual form of (2.1.3) which is quite close to (2.1.1). Let $\dot{C}(X)$ be the space of continuous functions in X tending to 0 at the boundary, that is, which are arbitrarily small outside compact subsets. This is a Banach space with the maximum norm, and the dual space is the space of bounded measures in X. Let B be the space of all arrays $U = \{u_\alpha\}$ with $u_\alpha \in \dot{C}(X)$ for every α and the norm

$$\|U\| = \sum \sup |u_\alpha|.$$

If L is a linear form on B with norm $\leqq 1$, then

$$L(U) = \sum \int u_\alpha \, d\mu_\alpha$$

where $d\mu_\alpha$ is a measure in X with total mass at most 1. Now (2.1.3) means that the map

$$\{\rho_\alpha \partial^\alpha \phi\} \to u(\phi), \quad \phi \in C_0^\infty(X)$$

is a linear form with norm at most 1 defined on a linear subspace of B. By the Hahn-Banach theorem it can be extended to a linear form L of norm at most 1 on all of B, which means that there are measures μ_a with $\int |d\mu_\alpha| \leqq 1$ such that

$$u(\phi) = L(\{\rho_\alpha \partial^\alpha \phi\}) = \sum_\alpha \int \rho_\alpha \partial^\alpha \phi \, d\mu_\alpha.$$

Writing $d\nu_\alpha = \rho_\alpha d\mu_\alpha$ we obtain

$$u(\phi) = \sum \int \partial^\alpha \phi \, d\nu_\alpha, \quad \phi \in C_0^\infty(X), \tag{2.1.1$'$}$$

where the supports of the measures $d\nu_\alpha$ are locally finite.

As in the introduction we identify the space of continuous functions in X with a subspace of $\mathscr{D}'(X)$ by assigning to each continuous function f the distribution

$$C_0^\infty(X) \ni \phi \to \int f \phi \, dx \tag{2.1.4}$$

which we also denote by f. This is legitimate since Theorem 1.2.4 shows that two functions defining the same distribution are identical. More generally we can make this identification when $f \in L^1_{\mathrm{loc}}(X)$, the space of functions which are integrable on compact subsets of X modulo those which vanish almost everywhere. In fact, Theorem 1.2.5 shows that functions defining the same distribution are in the same equivalence class. We can also identify arbitrary measures with distributions of order 0, for we have

Theorem 2.1.6. *If $u \in \mathscr{D}'^k(X)$ we can in a unique way extend u to a linear form on $C_0^k(X)$ such that* (2.1.2) *remains valid for all $\phi \in C_0^k(K)$ and some constant C.*

Proof. It follows from Theorem 1.3.2 that for every $\phi \in C_0^k(X)$ we can find a sequence $\phi_\nu \in C_0^\infty(X)$ with support in a fixed compact neighborhood K of $\operatorname{supp} \phi$, so that

$$\sum_{|\alpha| \leqq k} \sup |\partial^\alpha (\phi - \phi_\nu)| \to 0, \quad \nu \to \infty. \tag{2.1.5}$$

Thus we must define $u(\phi) = \lim u(\phi_\nu)$. This limit exists, for (2.1.5) implies in view of (2.1.2) that when $\nu, \mu \to \infty$ we have

$$|u(\phi_\nu) - u(\phi_\mu)| = |u(\phi_\nu - \phi_\mu)| \leqq C \sum_{|\alpha| \leqq k} \sup |\partial^\alpha (\phi_\nu - \phi_\mu)| \to 0.$$

That the limit is independent of the sequence chosen follows at once by mixing two sequences. If we apply (2.1.2) to ϕ_ν and let $\nu \to \infty$ we conclude that (2.1.2) is valid for all $\phi \in C_0^k$ with support in the interior of K, so the theorem is proved.

Since a measure can be defined to be a linear form on $C_0^0(X)$ with the continuity property (2.1.2) for $k=0$, we have now identified $\mathscr{D}'^0(X)$ with the space of measures in X. If an integrable function f is first identified with the measure $f\,dx$, as is customary in integration theory, and $f\,dx$ is then identified with a distribution, the result will of course be the same as if we identify f with a distribution directly.

A positive distribution is always a measure:

Theorem 2.1.7. *If u is a distribution in X with $u(\phi)\geqq 0$ for all non-negative $\phi\in C_0^\infty(X)$, then u is a positive measure.*

Proof. We have to show that u is of order 0. To do so we note that for any compact set $K\subset X$ Theorem 1.4.1 gives a function $\chi\in C_0^\infty(X)$ with $0\leqq\chi\leqq 1$ and $\chi=1$ on K. Then

$$\chi\sup|\phi|\pm\phi\geqq 0$$

if $\phi\in C_0^\infty(K)$ is real valued. By hypothesis it follows that

$$u(\chi)\sup|\phi|\pm u(\phi)\geqq 0$$

or equivalently that

$$|u(\phi)|\leqq u(\chi)\sup|\phi|,\qquad \phi\in C_0^\infty(K). \tag{2.1.6}$$

If we apply this to $\operatorname{Re} e^{i\theta}\phi$ when θ is real and choose θ so that $e^{i\theta}u(\phi)$ is real, we obtain (2.1.6) for complex valued ϕ also, hence $u\in\mathscr{D}'^0$.

$\mathscr{D}'(X)$ is obviously a vector space with the natural definition of addition and multiplication by complex numbers,

$$(a_1u_1+a_2u_2)(\phi)=a_1u_1(\phi)+a_2u_2(\phi);$$
$$\phi\in C_0^\infty(X),\quad u_j\in\mathscr{D}'(X),\quad a_j\in\mathbb{C}.$$

We shall always use the *weak topology* in $\mathscr{D}'(X)$ (also called the weak* topology), that is, the topology defined by the semi-norms

$$\mathscr{D}'(X)\ni u\to|u(\phi)|,$$

where ϕ is any fixed element of $C_0^\infty(X)$. Thus $u_i\to u$ means that

$$u_i(\phi)\to u(\phi)$$

for every $\phi\in C_0^\infty(X)$. Occasionally we shall need the following completeness property:

Theorem 2.1.8. *If u_j is a sequence in $\mathscr{D}'(X)$ and*

$$u(\phi)=\lim_{j\to\infty}u_j(\phi) \tag{2.1.7}$$

exists for every $\phi \in C_0^\infty(X)$, *then* $u \in \mathscr{D}'(X)$. *Thus* $u_j \to u$ *in* $\mathscr{D}'(X)$ *as* $j \to \infty$. *Moreover,* (2.1.2) *is valid for all* u_j *with constants* C *and* k *independent of* j, *and* $u_j(\phi_j) \to u(\phi)$ *if* $\phi_j \to \phi$ *in* $C_0^\infty(X)$.

Proof. When K is a compact subset of X the space $C_0^\infty(K)$ is a Fréchet space with the topology defined by the semi-norms

$$\|\phi\|_\alpha = \sup|\partial^\alpha \phi|, \qquad \phi \in C_0^\infty(K).$$

(The completeness is a consequence of Theorem 1.1.5.) (2.1.2) is valid for u_j (with constants C and k which may a priori depend on j), so u_j restricted to $C_0^\infty(K)$ is a continuous linear form on $C_0^\infty(K)$. For fixed $\phi \in C_0^\infty(K)$ it follows from (2.1.7) that the sequence $u_j(\phi)$ is bounded. Hence the principle of uniform boundedness (the Banach-Steinhaus theorem) shows that (2.1.2) is valid for all u_j with constants C and k independent of j. When $j \to \infty$ we obtain (2.1.2) for the limit u. If $\phi_j \to \phi$ in $C_0^\infty(X)$ we have $\operatorname{supp} \phi_j \subset K$ for some compact subset K of X and all j. Hence $u_j(\phi_j - \phi) \to 0$ by the uniformity of (2.1.2), which proves that $u_j(\phi_j) \to u(\phi)$.

By Cauchy's convergence principle for $\mathbb{C}$ the existence of the limit (2.1.7) means precisely that $u_j(\phi) - u_k(\phi) \to 0$ when $j, k \to \infty$. Hence an equivalent statement of Theorem 2.1.8 is that every sequence u_j in $\mathscr{D}'(X)$ such that $u_j - u_k \to 0$ when $j, k \to \infty$ must have a limit u in $\mathscr{D}'(X)$.

Theorem 2.1.9. *If* $u_j \in \mathscr{D}'(X)$, $u_j \geqq 0$ *and* $u_j \to u$ *in* $\mathscr{D}'(X)$, *then* $u \geqq 0$ *and* $u_j \to u$ *in the weak topology of measures, that is,* $u_j(\phi) \to u(\phi)$ *for every* $\phi \in C_0^0(X)$.

Proof. That $u \geqq 0$ is obvious, and Theorem 2.1.7 shows that u_j and u are measures. In addition the proof of Theorem 2.1.7 gives for every compact set $K \subset X$ a uniform bound

$$|u_j(\phi)| \leqq C_K \sup|\phi|, \qquad \phi \in C_0^0(K), \quad \forall j.$$

Since $u_j(\phi) \to u(\phi)$ for $\phi \in C_0^\infty(K)$ it follows that this remains true for all ϕ in the closure F of $C_0^\infty(K)$ in $C_0^0(K)$. By Theorem 1.3.2 F contains all continuous functions with support in the interior of K, which proves the theorem.

We shall now give some simple examples to indicate what convergence of distributions means in concrete cases. They also illustrate the fact, to be proved in Chapter IV, that every distribution is the limit of a sequence of functions in C_0^∞.

Example 2.1.10. Let $v \in C_0^\infty(\mathbb{R}^n)$ and set $u_\varepsilon(x) = \varepsilon^{-n} v(x/\varepsilon)$. Then

$$u_\varepsilon(\phi) = \int \varepsilon^{-n} v(x/\varepsilon)\phi(x)\,dx = \int v(x)\phi(\varepsilon x)\,dx \to \phi(0)\int v(x)\,dx,$$

if $\phi \in C_0^\infty(\mathbb{R}^n)$. (Compare Theorem 1.3.2.)

Example 2.1.11. Let $w \in C_0^\infty(\mathbb{R}^n)$, assume that $\int x^\alpha w(x)\,dx = 0$ when $|\alpha| < k$ and set $u_\varepsilon(x) = \varepsilon^{-n-k} w(x/\varepsilon)$. Then we have for $\phi \in C_0^\infty(\mathbb{R}^n)$

$$\begin{aligned} u_\varepsilon(\phi) &= \int \phi(\varepsilon x)\varepsilon^{-k} w(x)\,dx = \int \Big(\sum_{|\alpha| \leqq k} \varepsilon^{|\alpha|-k} x^\alpha \partial^\alpha \phi(0)/\alpha! + O(\varepsilon)\Big) w(x)\,dx \\ &\to \sum_{|\alpha|=k} \partial^\alpha \phi(0) \int x^\alpha w(x)\,dx/\alpha!. \end{aligned}$$

Example 2.1.12. If $u_t(x) = t^N e^{itx}$, $x \in \mathbb{R}$, where N is a positive integer, then

$$\begin{aligned} u_t(\phi) &= \int t^N e^{itx}\phi(x)\,dx = i\int t^{N-1} e^{itx}\phi'(x)\,dx = \dots \\ &= i^{N+1} t^{-1} \int e^{itx}\phi^{(N+1)}(x)\,dx \to 0 \quad \text{as } t \to \infty, \qquad \phi \in C_0^\infty(\mathbb{R}). \end{aligned}$$

Example 2.1.13. If $u_t(x) = t e^{itx}$, $x > 0$ and $u_t(x) = 0$, $x \leqq 0$, then

$$\begin{aligned} u_t(\phi) &= \int_0^\infty t e^{itx}\phi(x)\,dx = i\phi(0) + i\int_0^\infty e^{itx}\phi'(x)\,dx \\ &= i\phi(0) - \phi'(0)/t - \int_0^\infty e^{itx}\phi''(x)\,dx/t \to i\phi(0), \qquad \text{as } t \to \infty, \end{aligned}$$

if $\phi \in C_0^\infty(\mathbb{R})$.

Example 2.1.14. Let $u_t(x) = t^{1/k} e^{itx^k}$, $x \in \mathbb{R}$, where k is an integer > 1. To determine the limit we first examine

$$F(x) = \int_0^x e^{iy^k}\,dy.$$

When $x > 0$ we shift the integration to a segment of the line $\arg z = \pi/2k$ and a circular arc, on which $\operatorname{Im} z^k \geqq c|z|^{k-1} \operatorname{Im} z$ where $c > 0$. It follows that

$$F(x) \to e^{\pi i/2k} \int_0^\infty e^{-y^k}\,dy \qquad \text{as } x \to \infty.$$

Hence

$$F(x) \to -e^{\pi i/2k} \int_0^\infty e^{-y^k}\,dy \qquad \text{as } x \to -\infty \text{ if } k \text{ is even}$$

$$F(x) \to -e^{-\pi i/2k} \int_0^\infty e^{-y^k}\,dy \qquad \text{as } x \to -\infty \text{ if } k \text{ is odd.}$$

Now we have if $\phi \in C_0^\infty(\mathbb{R})$

$$u_t(\phi) = \int t^{1/k} F'(xt^{1/k}) \phi(x)\,dx = -\int F(xt^{1/k}) \phi'(x)\,dx$$

$$\to -F(-\infty) \int_{-\infty}^{0} \phi'(x)\,dx - F(\infty) \int_{0}^{\infty} \phi'(x)\,dx = (F(\infty) - F(-\infty))\phi(0).$$

Here

$$F(\infty) - F(-\infty) = 2\cos \pi/2k \int_0^\infty e^{-y^k} dy \qquad \text{if } k \text{ is odd}$$

$$= 2e^{\pi i/2k} \int_0^\infty e^{-y^k} dy \qquad \text{if } k \text{ is even,}$$

which is not 0. (In terms of the Γ function defined in Section 3.2 the integral is equal to $\Gamma((1+k)/k)$.) Note the contrast with the case $k=1$.

2.2. Localization

If $Y \subset X \subset \mathbb{R}^n$ and $u \in \mathscr{D}'(X)$, we can restrict u to a distribution u_Y in Y by setting

$$u_Y(\phi) = u(\phi), \qquad \phi \in C_0^\infty(Y).$$

Our next purpose is to prove the less trivial fact that a distribution is determined by the restrictions to the sets in an open covering:

Theorem 2.2.1. *If $u \in \mathscr{D}'(X)$ and every point in X has a neighborhood to which the restriction of u is 0, then $u=0$.*

Proof. If $\phi \in C_0^\infty(X)$ we can for every $x \in \operatorname{supp}\phi$ find an open neighborhood $Y \subset X$ such that the restriction of u to Y is 0. By the Borel-Lebesgue lemma we can choose a finite number of such open sets $Y_j \subset X$ which cover $\operatorname{supp}\phi$. But according to Theorem 1.4.4 we can then write $\phi = \sum \phi_j$ where $\phi_j \in C_0^\infty(Y_j)$. Thus $u(\phi_j) = 0$ which proves that $u(\phi) = \sum u(\phi_j) = 0$.

Theorem 2.2.1 makes it natural to extend Definition 1.2.2 as follows:

Definition 2.2.2. If $u \in \mathscr{D}'(X)$ then the support of u, denoted $\operatorname{supp} u$, is the set of points in X having no open neighborhood to which the restriction of u is 0.

Thus $X \setminus \operatorname{supp} u$ is the open set of points having a neighborhood in which u vanishes, so u vanishes in $X \setminus \operatorname{supp} u$ by Theorem 2.2.1, and $X \setminus \operatorname{supp} u$ contains every open set where u vanishes. Thus we have

(2.2.1) $\quad u(\phi)=0 \quad$ if $u \in \mathscr{D}'(X)$, $\phi \in C_0^\infty(X)$ and $\operatorname{supp} u \cap \operatorname{supp} \phi = \emptyset$.

Closely related to the notion of support is the notion of singular support:

Definition 2.2.3. If $u \in \mathscr{D}'(X)$, then the singular support of u, denoted $\operatorname{sing\,supp} u$, is the set of points in X having no open neighborhood to which the restriction of u is a C^∞ function.

Since every point in $X \setminus \operatorname{sing\,supp} u$ has a neighborhood where u is a C^∞ function, it follows from Theorem 2.2.1 that the restriction of u to $X \setminus \operatorname{sing\,supp} u$ is a C^∞ function. This is not true for any larger open set than $X \setminus \operatorname{sing\,supp} u$.

We shall now supplement Theorem 2.2.1 by an existence theorem:

Theorem 2.2.4. *Let X_i, $i \in I$, be an arbitrary family of open sets in $\mathbb{R}^n$, and set $X = \bigcup X_i$. If $u_i \in \mathscr{D}'(X_i)$ and $u_i = u_j$ in $X_i \cap X_j$ for all $i, j \in I$, then there exists one and only one $u \in \mathscr{D}'(X)$ such that u_i is the restriction of u to X_i for every i.*

Proof. The uniqueness is precisely Theorem 2.2.1 and the existence follows from essentially the same proof. If u is a distribution with the required properties we must have

(2.2.2) $\quad u(\phi) = \sum u_i(\phi_i) \quad$ if $\phi = \sum \phi_i, \quad \phi_i \in C_0^\infty(X_i)$

and the sum is finite. By Theorem 1.4.4 every $\phi \in C_0^\infty(X)$ can be written as such a sum, and we shall prove that $\sum u_i(\phi_i)$ is independent of how it is chosen. This will follow if we show that $\sum \phi_i = 0$ implies $\sum u_i(\phi_i) = 0$. Set $K = \bigcup \operatorname{supp} \phi_i$, which is a compact subset of X, and use Theorem 1.4.5 to choose functions $\psi_k \in C_0^\infty(X_k)$ such that $\sum \psi_k = 1$ in K and the sum is finite. Then we have $\psi_k \phi_i \in C_0^\infty(X_k \cap X_i)$ so

$$u_i(\psi_k \phi_i) = u_k(\psi_k \phi_i).$$

Hence

$$\sum u_i(\phi_i) = \sum \sum u_i(\phi_i \psi_k) = \sum \sum u_k(\phi_i \psi_k) = \sum u_k(\psi_k \sum \phi_i) = 0.$$

Having proved that (2.2.2) defines uniquely a linear form on $C_0^\infty(X)$ we must show that it has the continuity properties required of a distribution. Choose any compact set $K \subset X$ and as before functions $\psi_k \in C_0^\infty(X_k)$ with $\sum \psi_k = 1$ on K and the sum finite. If $\phi \in C_0^\infty(K)$ we have $\phi = \sum \phi \psi_k$ with $\phi \psi_k \in C_0^\infty(X_k)$ so (2.2.2) gives

$$u(\phi) = \sum u_k(\phi \psi_k).$$

Since (2.1.2) is valid for u_k and the maximum of the derivatives of $\phi \psi_k$ can be estimated in terms of those of ϕ, we conclude that

$$|u(\phi)| \leqq C \sum_{|\alpha| \leqq m} \sup |\partial^\alpha \phi|, \qquad \phi \in C_0^\infty(K),$$

which completes the proof. Note that it shows that if $u_i \in \mathscr{D}'^k$ for every i then $u \in \mathscr{D}'^k$.

Remark. If $u^\nu \in \mathscr{D}'(X)$, $\nu = 1, 2, \dots$ and u^ν restricted to X_i has a limit $u_i \in \mathscr{D}'(X_i)$ for every i, then the proof shows that u^ν has a limit in $\mathscr{D}'(X)$. In fact, if we apply (2.2.2) to each u^ν we find that $u^\nu(\phi)$ has a limit $u(\phi)$ satisfying (2.2.2). The continuity of u follows from the second part of the proof of Theorem 2.2.4 or else from Theorem 2.1.8.

If $u \in L^1_{\text{loc}}(X)$, the form

$$u(\phi) = \int u \phi \, dx$$

is of course meaningful for every $\phi \in C^\infty(X)$ such that

(2.2.3) $$\operatorname{supp} u \cap \operatorname{supp} \phi \Subset X.$$

(We write $A \Subset X$ when $\bar{A}$ is compact and contained in X.) We shall now prove that the domain of an arbitrary distribution can be extended in this way.

Theorem 2.2.5. *Let $u \in \mathscr{D}'(X)$ and let F be a relatively closed subset of X containing* $\operatorname{supp} u$. *Then there is one and only one linear form $\tilde{u}$ on $\{\phi;\, \phi \in C^\infty(X),\, F \cap \operatorname{supp} \phi \Subset X\}$ such that*

(i) $\tilde{u}(\phi) = u(\phi)$ *if* $\phi \in C_0^\infty(X)$,

(ii) $\tilde{u}(\phi) = 0$ *if* $\phi \in C^\infty(X)$ *and* $F \cap \operatorname{supp} \phi = \emptyset$.

The domain of $\tilde{u}$ is of course largest when $F = \operatorname{supp} u$, but we need the uniqueness statement also for other sets F.

Proof. a) *Uniqueness.* Let $\phi \in C^\infty(X)$ and let $F \cap \operatorname{supp} \phi = K$ be a compact subset of X. By Theorem 1.4.1 we can find $\psi \in C_0^\infty(X)$ so that $\psi = 1$ in a neighborhood of K. Then we have $\phi = \phi_0 + \phi_1$ where ϕ_0

$=\psi\phi\in C_0^\infty(X)$ and $\phi_1=(1-\psi)\phi$, so that $F\cap\operatorname{supp}\phi_1=\emptyset$. Using (i) and (ii) we obtain

$$\tilde{u}(\phi)=\tilde{u}(\phi_0)+\tilde{u}(\phi_1)=u(\phi_0),$$

which proves the uniqueness of $\tilde{u}$.

b) *Existence.* We have seen in a) that every $\phi\in C^\infty(X)$ with $F\cap\operatorname{supp}\phi$ compact can be written $\phi=\phi_0+\phi_1$ with $\phi_0\in C_0^\infty(X)$ and $F\cap\operatorname{supp}\phi_1=\emptyset$. If $\phi=\phi'_0+\phi'_1$ is another such decomposition, then $\chi=\phi_0-\phi'_0\in C_0^\infty(X)$ and $F\cap\operatorname{supp}\chi=F\cap\operatorname{supp}(\phi_1-\phi'_1)=\emptyset$ so it follows from (2.2.1) that $0=u(\chi)=u(\phi_0)-u(\phi'_0)$. Setting $\tilde{u}(\phi)=u(\phi_0)$ therefore gives a unique definition of a linear form u which obviously has the required properties.

From now on we write $u(\phi)$ instead of $\tilde{u}(\phi)$ and thus consider $u(\phi)$ as defined for all $u\in\mathscr{D}'(X)$ and all $\phi\in C^\infty(X)$ satisfying (2.2.3). In view of the symmetry of (2.2.3) we shall sometimes write $\langle u,\phi\rangle$ instead of $u(\phi)$.

2.3. Distributions with Compact Support

If $u\in\mathscr{D}'(X)$ has compact support we have seen that $u(\phi)$ can be defined for all $\phi\in C^\infty(X)$. When $\psi\in C_0^\infty(X)$ and $\psi=1$ in a neighborhood of $\operatorname{supp}u$, we have

$$u(\phi)=u(\psi\phi)+u((1-\psi)\phi)=u(\psi\phi),\qquad \phi\in C^\infty(X).$$

Hence it follows from (2.1.2) that

$$|u(\phi)|\leqq C\sum_{|\alpha|\leqq k}\sup_K|\partial^\alpha\phi|,\qquad \phi\in C^\infty(X),\tag{2.3.1}$$

where K is the support of ψ and C, k are constants. Conversely, suppose that we have a linear form v on $C^\infty(X)$ such that for some constants C and k and some compact set $L\subset X$

$$|v(\phi)|\leqq C\sum_{|\alpha|\leqq k}\sup_L|\partial^\alpha\phi|,\qquad \phi\in C^\infty(X).\tag{2.3.2}$$

Then the restriction of v to $C_0^\infty(X)$ is a distribution u with support contained in L. Since it follows from (2.3.2) that $v(\phi)=0$ if $L\cap\operatorname{supp}\phi=\emptyset$, we obtain from Theorem 2.2.5 that $v(\phi)=u(\phi)$ for every $\phi\in C^\infty(X)$. Hence we have proved

Theorem 2.3.1. *The set of distributions in X with compact support is identical with the dual space of $C^\infty(X)$ with the topology defined by the semi-norms*

$$\phi \to \sum_{|\alpha| \leqq k} \sup_K |\partial^\alpha \phi|,$$

where K ranges over all compact subsets of X and k over all integers $\geqq 0$.

Schwartz used the notation $\mathscr{E}(X)$ for the space $C^\infty(X)$ equipped with this topology. Accordingly the space of distributions with compact support in X is denoted by $\mathscr{E}'(X)$. From the proof of Theorem 2.3.1 it follows that $\mathscr{E}'(X)$ can be identified with the set of distributions in $\mathscr{E}'(\mathbb{R}^n)$ with supports contained in X. We may therefore use the notation $\mathscr{E}'(A)$ also when A is an arbitrary subset of $\mathbb{R}^n$ to denote the set of distributions in $\mathscr{E}'(\mathbb{R}^n)$ with supports contained in A. We write $\mathscr{E}'^k(A) = \mathscr{E}'(A) \cap \mathscr{D}'^k(\mathbb{R}^n)$.

The smallest k which can be used in (2.3.1) is of course the order of the distribution u. For K one can take any neighborhood of $\operatorname{supp} u$ but usually not the support itself.

Example 2.3.2. Let K be a compact set in $\mathbb{R}^n$ which is not the union of finitely many compact connected sets. Then one can find $u \in \mathscr{E}'(K)$ of order 1 so that (2.3.1) is not valid for any C and k. In fact, the hypothesis means that we can find a sequence of disjoint non-empty compact subsets K_j of K such that $K \setminus (K_1 \cup \ldots \cup K_j)$ is compact. Choose $x_j \in K_j$, let x_0 be a limit point of $\{x_j\}$ and set

$$u(\phi) = \sum m_j(\phi(x_j) - \phi(x_0))$$

where m_j is a positive sequence such that

$$\sum m_j |x_j - x_0| = 1, \qquad \sum m_j = \infty.$$

Such a sequence exists since $\liminf |x_j - x_0| = 0$. Then

$$|u(\phi)| \leqq \sup |\phi'|$$

so u is a distribution. On the other hand, if (2.3.1) is valid and we choose $\phi \in C^\infty$ equal to 1 in a neighborhood of $K_1 \cup \ldots \cup K_j$ and 0 near $K \setminus (K_1 \cup \ldots \cup K_j)$, hence at x_0, then we obtain

$$\sum_{i \leqq j} m_i \leqq C$$

which is a contradiction when $j \to \infty$.

Although (2.3.1) is not in general valid with $K = \operatorname{supp} u$ we can prove that the left-hand side must vanish then if the right-hand side does:

Theorem 2.3.3. *If* $u \in \mathscr{E}'$ *is of order* $\leqq k$ *and if* $\phi \in C^k$,

(2.3.3) $$\partial^\alpha \phi(x)=0 \quad \text{for } |\alpha| \leqq k \quad \text{and} \quad x \in \operatorname{supp} u,$$

then it follows that $u(\phi)=0$.

Recall that $u(\phi)$ was defined for all $\phi \in C_0^k$ in Theorem 2.1.6, and as in Theorem 2.2.5 we have a unique extension to all $\phi \in C^k$, for which $u(\phi)=0$ when $\operatorname{supp} u \cap \operatorname{supp} \phi = \emptyset$. The estimate (2.3.1) is valid for all $\phi \in C^k$ if K is a neighborhood of $\operatorname{supp} u$.

Proof of Theorem 2.3.3. By Theorem 1.4.1 and the remarks after it we can choose $\chi_\varepsilon \in C_0^\infty$ so that $\chi_\varepsilon = 1$ in a neighborhood of $\operatorname{supp} u$, $\chi_\varepsilon = 0$ outside

$$M_\varepsilon = \{y;\, |x-y| \leqq \varepsilon \text{ for some } x \in \operatorname{supp} u\}$$

and

$$|\partial^\alpha \chi_\varepsilon| \leqq C \varepsilon^{-|\alpha|}, \qquad |\alpha| \leqq k.$$

Since $\operatorname{supp} u$ and $\operatorname{supp}(1-\chi_\varepsilon)\phi$ do not intersect, we have

$$u(\phi) = u(\phi \chi_\varepsilon) + u(\phi(1-\chi_\varepsilon)) = u(\phi \chi_\varepsilon)$$

so using (2.3.1) we obtain

$$\begin{aligned} |u(\phi)| &\leqq C \sum_{|\alpha| \leqq k} \sup |\partial^\alpha(\phi \chi_\varepsilon)| \leqq C' \sum_{|\alpha|+|\beta| \leqq k} \sup |\partial^\alpha \phi|\, |\partial^\beta \chi_\varepsilon| \\ &\leqq C'' \sum_{|\alpha| \leqq k} \varepsilon^{|\alpha|-k} \sup_{M_\varepsilon} |\partial^\alpha \phi|. \end{aligned}$$

To show that the right-hand side tends to 0 with ε we must prove that

(2.3.4) $$\varepsilon^{|\alpha|-k} \sup_{M_\varepsilon} |\partial^\alpha \phi| \to 0 \quad \text{when } \varepsilon \to 0 \quad \text{if } |\alpha| \leqq k.$$

By the definition of M_ε we can for every $y \in M_\varepsilon$ choose $x \in \operatorname{supp} u$ so that $|x-y| \leqq \varepsilon$. This gives (2.3.4) for $|\alpha|=k$ since $\partial^\alpha \phi$ is uniformly continuous and vanishes on $\operatorname{supp} u$. If $|\alpha|<k$ and $y \in M_\varepsilon$ we obtain by Taylor's formula

$$|\partial^\alpha \phi(y)| \leqq \frac{1}{(k-|\alpha|)!} \sup_{0<t<1} |(d/dt)^{k-|\alpha|} (\partial^\alpha \phi)(x+t(y-x))|$$

for the derivatives of ϕ of order $<k$ vanish at x. If the differentiation is carried out we obtain a sum of terms involving a factor $(y-x)^\beta$, $|\beta| = k-|\alpha|$, and a derivative of ϕ of order k, which proves (2.3.4).

An important consequence of Theorem 2.3.3 is the following

Theorem 2.3.4. *If* u *is a distribution of order* k *with support equal to* $\{y\}$, *then* u *has the form*

(2.3.5) $$u(\phi)=\sum_{|\alpha|\leqq k} a_\alpha \partial^\alpha \phi(y), \qquad \phi\in C^k.$$

Proof. If we expand ϕ in a Taylor series

$$\phi(x)=\sum_{|\alpha|\leqq k} \partial^\alpha\phi(y)(x-y)^\alpha/\alpha!+\psi(x)$$

we have $\partial^\alpha\psi(y)=0$ when $|\alpha|\leqq k$ so $u(\psi)=0$ by Theorem 2.3.3. Hence (2.3.5) follows with $a_\alpha=u((.-y)^\alpha/\alpha!)$. (. denotes the variable.)

Note that Theorem 2.3.4 explains why the limits in Examples 2.1.10, 2.1.11, 2.1.13, 2.1.14 had to be of the form (2.3.5).

There is a result similar to Theorem 2.3.4 when the point is replaced by a subspace, but the proof is somewhat more complicated:

Theorem 2.3.5. *Let $x=(x', x'')$ be a splitting of the variables in $\mathbb{R}^n$ in two groups. If u is a distribution in $\mathbb{R}^n$ of order k with compact support contained in the plane $x'=0$, then*

(2.3.6) $$u(\phi)=\sum_{|\alpha|\leqq k} u_\alpha(\phi_\alpha)$$

where u_α is a distribution of compact support and order $k-|\alpha|$ in the x'' variables, $\alpha=(\alpha',0)$ and

$$\phi_\alpha(x'')=\partial^\alpha\phi(x',x'')|_{x'=0}.$$

Proof. If $\phi\in C^\infty$ and we form the Taylor expansion in x',

$$\phi(x)=\sum_{|\alpha'|\leqq k,\, \alpha''=0} \partial^\alpha\phi(0,x'')x'^\alpha/\alpha!+\Phi(x)$$

then $\partial^\alpha\Phi(x)=0$ when $x'=0$ and $|\alpha|\leqq k$ so $u(\Phi)=0$. Thus $u(\phi)$ has the form (2.3.6) with

$$u_\alpha(\psi)=u(\psi(x'')x'^\alpha/\alpha!).$$

What is not obvious is that u_α is of order $k-|\alpha|$ and not just of order k. To prove that we observe that $u_\alpha(\psi)=u(\phi)$ for all $\phi\in C^\infty$ such that

(2.3.7) $$\phi(x)=\psi(x'')x'^\alpha/\alpha!+O(|x'|^{k+1}) \quad \text{as } x'\to 0.$$

By repeated application of Corollary 1.3.4 to one x' variable at a time, we find that for any $\psi\in C^{k-|\alpha|}$ one can find $\phi\in C^k$ satisfying (2.3.7). The proof shows that if $\psi\in C^\infty$ one can find $\phi\in C^\infty$ so that for any given compact set $K\subset\mathbb{R}^n$

$$\sum_{|\gamma|\leqq k}\sup_K|\partial^\gamma\phi|\leqq C\sum_{|\beta|\leqq k-|\alpha|}\sup|\partial^\beta\psi|.$$

Hence we obtain

$$|u_\alpha(\psi)| \leqq C' \sum_{|\beta| \leqq k-|\alpha|} \sup |\partial^\beta \psi|, \qquad \psi \in C_0^\infty,$$

which completes the proof.

For the sake of completeness we shall finally discuss a more general method for extending functions than Corollary 1.3.4 and examine its implications for the structure of distributions with given support.

If $u \in C^k(\mathbb{R}^n)$ we denote its Taylor polynomial of order k at y by $u_k(x, y)$ or $u(x, y)$ for short,

$$u(x, y) = \sum_{|\alpha| \leqq k} \partial^\alpha u(y)(x-y)^\alpha/\alpha!.$$

We denote the remainder term by $R(x, y)$, thus $u(x) = u(x, y) + R(x, y)$. Note that

$$\partial_x^\alpha R(x, y) = \partial^\alpha u(x) - \sum_{|\beta| \leqq k-|\alpha|} \partial^{\alpha+\beta} u(y)(x-y)^\beta/\beta!$$

is determined by the derivatives of u at x and at y only. By Taylor's formula the quotient $|\partial_x^\alpha R(x, y)|/|x-y|^{k-|\alpha|}$ is continuous in $\mathbb{R}^n \times \mathbb{R}^n$ if it is defined as 0 on the diagonal.

Theorem 2.3.6 (Whitney's Extension Theorem). *Let K be a compact set in $\mathbb{R}^n$ and u_α, $|\alpha| \leqq k$, continuous functions on K. Set for $|\alpha| \leqq k$*

$$U_\alpha(x, y) = |u_\alpha(x) - \sum_{|\beta| \leqq k-|\alpha|} u_{\alpha+\beta}(y)(x-y)^\beta/\beta!|\,|x-y|^{|\alpha|-k}$$

when $x, y \in K$ and $x \neq y$, $U_\alpha(x, x) = 0$ when $x \in K$. If U_α is continuous on $K \times K$ when $|\alpha| \leqq k$, it is possible to find $v \in C^k(\mathbb{R}^n)$ with $\partial^\alpha v(x) = u_\alpha(x)$, $x \in K$, $|\alpha| \leqq k$. One can then choose v so that with C depending only on K

$$\sum_{|\alpha| \leqq k} \sup |\partial^\alpha v| \leqq C\Big(\sum_{|\alpha| \leqq k} \sup_{K \times K} U_\alpha + \sum_{|\alpha| \leqq k} \sup_K |u_\alpha|\Big). \tag{2.3.8}$$

We have already observed the necessity of the condition in the theorem, and the proof of sufficiency will depend on the following

Lemma 2.3.7. *There exists a partition of unity $1 = \sum \phi_j$ in $\complement K$ such that no point is in the support of more than N functions ϕ_j, the diameter of the support of ϕ_j is at most twice its distance to K, and*

$$|\partial^\alpha \phi_j(x)| \leqq C_\alpha d(x)^{-|\alpha|}, \qquad x \notin K, \tag{2.3.9}$$

where $d(x)$ is the distance from x to K.

Proof. This is a special case of Theorem 1.4.10 with the metric chosen as in Example 1.4.8 where $F=K$.

Proof of Theorem 2.3.6. Choose for every j a point $y_j \in K$ with minimal distance to $\operatorname{supp} \phi_j$. With

$$u(x,y)=\sum_{|\alpha|\leqq k} u_\alpha(y)(x-y)^\alpha/\alpha!$$

denoting the expected Taylor expansion of v at $y\in K$, we set

$$v(x)=\sum \phi_j(x)u(x,y_j), \quad x\notin K; \qquad v(x)=u_0(x), \quad x\in K.$$

Choose $x^*\in K$ with $|x-x^*|=d(x)$. If x is in the support of ϕ_j then the distance from $\operatorname{supp}\phi_j$ to K is at most $d(x)$, the diameter is at most $2d(x)$, so $|x-y_j|\leqq 3d(x)$ and $|x^*-y_j|\leqq 4d(x)$. If y is an arbitrary point in K then

$$|v(x)-u(x,y)|=o(|x-y|^k) \tag{2.3.10}$$

uniformly in y. If $x\in K$ this follows from the continuity of $U_0(x,y)$. Otherwise we use the fact that

$$|v(x)-u(x,y)|\leqq \sum \phi_j(x)\,|u(x,y_j)-u(x,y)|.$$

Here $|x-y_j|\leqq 3d(x)\leqq 3|x-y|$ when $\phi_j(x)\neq 0$. When $|\gamma|\leqq k$ we have

$$\begin{aligned}&|\partial_x^\gamma(u(x,y_1)-u(x,y_2))|\\ &\quad =o((|x-y_1|+|y_1-y_2|)^{k-|\gamma|}) \quad \text{if } y_1,y_2\in K,\end{aligned} \tag{2.3.11}$$

for $\partial_x^\gamma(u(x,y_1)-u(x,y_2))$ is a polynomial in x of degree $k-|\gamma|$ and

$$\begin{aligned}&|\partial_x^\beta\partial_x^\gamma(u(x,y_1)-u(x,y_2))_{x=y_1}|\\ &\quad =U_{\beta+\gamma}(y_1,y_2)\,|y_1-y_2|^{k-|\beta|-|\gamma|}=o(|y_1-y_2|^{k-|\beta+\gamma|}).\end{aligned}$$

(2.3.11) with $\gamma=0$ gives (2.3.10); the general case will be useful later on.

From (2.3.10) it follows that v is continuous and if $k\geqq 1$ also that v is differentiable at any point $y\in K$, with differential $d_x u(x,y)|_{x=y}$. When $x\notin K$ and $k>0$ we obtain by differentiation

$$\partial_\nu v(x)=\sum \partial_\nu\phi_j(x)u(x,y_j)+\sum \phi_j(x)\partial_\nu u(x,y_j). \tag{2.3.12}$$

Set $v_\nu(x)=\sum\phi_j(x)\partial_\nu u(x,y_j)$, $x\notin K$, and $v_\nu(x)=u_{\alpha_\nu}(x)$, $x\in K$, where $\alpha_\nu =(0,\dots,1,0,\dots)$ with 1 just in the ν-th place. If our contentions are already proved with k replaced by $k-1$ then $v_\nu\in C^{k-1}$. If we prove that the first sum in (2.3.12) and its derivatives of order $\leqq k-1$ tend to 0 when $x\to K$, it follows in view of Corollary 1.1.2 (in n variables) that $\partial_\nu v$ is continuous and in fact in C^{k-1}, and that $\partial^\alpha v = u_\alpha$ on K, $|\alpha| \leqq k$.

Now we have for $x \notin K$ when $|\beta|+|\gamma| \leqq k$ and $\beta \neq 0$

$$\sum \partial^\beta \phi_j(x) \partial_x^\gamma u(x, y_j) = \sum \partial^\beta \phi_j(x) \partial_x^\gamma (u(x, y_j) - u(x, x^*)),$$

because $\sum \partial^\beta \phi_j(x) = 0$ when $\beta \neq 0$. Recalling that $|x - y_j| \leqq 3d(x)$ we obtain from (2.3.11) and (2.3.9)

$$\sum \partial^\beta \phi_j(x) \partial_x^\gamma u(x, y_j) = o(d(x)^{-|\beta|+k-|\gamma|})$$

which proves the assertion on the first sum in (2.3.12). The preceding estimate and a trivial bound for

$$\sum \phi_j(x) \partial_x^\alpha u(x, y_j)$$

give, if we pay attention to constants also, that

$$\sum_{|\alpha| \leqq k} |\partial^\alpha v(x)| \leqq C(1+d(x))^k \Big(\sum_{|\alpha| \leqq k} \sup_{K \times K} U_\alpha + \sum_{|\alpha| \leqq k} \sup_K |u_\alpha| \Big).$$

Now if we drop from the definition of v the terms for which $\operatorname{supp} \phi_j$ has distance >1 from K, we do not change v where $d(x) < 1$ but we make $v(x) = 0$ when $d(x) > 3$, so (2.3.8) follows.

Corollary 2.3.8. *If u is a distribution of order k and compact support K, then*

$$|u(\phi)| \leqq C\Big(\sum_{|\alpha| \leqq k} \sup_{x, y \in K, x \neq y} \Big|\partial^\alpha \phi(x) - \sum_{|\beta| \leqq k - |\alpha|} \partial^{\alpha+\beta} \phi(y)(x-y)^\beta / \beta! \Big| \cdot |x-y|^{|\alpha|-k} + \sum_{|\alpha| \leqq k} \sup_K |\partial^\alpha \phi| \Big), \quad \phi \in C^\infty. \tag{2.3.13}$$

Proof. If we apply Theorem 2.3.6 with $u_\alpha = \partial^\alpha \phi|_K$ we obtain a function v with $u(v - \phi) = 0$, by Theorem 2.3.3. Hence

$$|u(\phi)| = |u(v)| \leqq C \sum_{|\alpha| \leqq k} \sup |\partial^\alpha v|$$

and (2.3.13) therefore follows from (2.3.8).

Example 2.3.2 shows that the first sum on the right-hand side cannot be omitted if K is not a finite union of compact connected sets. By Banach's theorem we also know that every continuous linear form with respect to the seminorm in the right-hand side of (2.3.13) is continuous with respect to the seminorm $\sum_{|\alpha| \leqq k} \sup_K |\partial^\alpha \phi|$ if and only if the first sum in (2.3.13) can be estimated by the second one. A necessary condition for this is given in the following

Theorem 2.3.9. *Let K be a compact connected set and assume that*

$$\sup_{x,y\in K, x\neq y} |\psi(x)-\psi(y)|/|x-y| \leqq C \sum_{|\alpha|\leqq 1} \sup_K |\partial^\alpha \psi|, \quad \psi\in C^\infty. \tag{2.3.14}$$

Then there is a constant C' such that any two points $x, y\in K$ can be joined by a rectifiable curve in K with length $\leqq C'|x-y|$.

Proof. Fix two points x_0, $y_0\in K$. If X is a connected open neighborhood of K we denote by $d(y)$ the infimum of the length of polygonal arcs from x_0 to y contained in X. If $u(y)=\min(d(y), d(y_0))$ then $u(x_0)=0$ and $u(y_0)=d(y_0)$, and we have

$$|u(x)-u(y)|\leqq|x-y| \quad \text{if } [x,y]\subset X. \tag{2.3.15}$$

Define u_ϕ according to Theorem 1.3.2 with ϕ of so small support that u_ϕ is defined in a neighborhood of K. In view of (2.3.15) we have

$$|u_\phi(x)-u_\phi(y)|\leqq|x-y|$$

in a neighborhood of K when $|x-y|$ is small, hence $|\partial_i u_\phi|\leqq 1$ on K. With $\psi=u_\phi$ we obtain from (2.3.14)

$$|u_\phi(x_0)-u_\phi(y_0)|\leqq|x_0-y_0|\, C(d(y_0)+n).$$

Letting $\operatorname{supp}\phi\to\{0\}$ we conclude that

$$d(y_0)\leqq|x_0-y_0|\, C(d(y_0)+n).$$

When $|x_0-y_0|\leqq 1/2\,C$ it follows that $d(y_0)\leqq 2n\,C|x_0-y_0|$.

For any $\varepsilon>0$ the set

$$K_\varepsilon=\{x; |x-y|<\varepsilon \text{ for some } y\in K\}$$

is a connected neighborhood of K since K is connected. Hence it contains an arc of length $<2nC|x_0-y_0|+\varepsilon$ from x_0 to y_0. Representing it as a function of the arc length we obtain when $\varepsilon\to 0$ a curve of length $\leqq 2nC|x_0-y_0|$ from x_0 to y_0 which is contained in K, provided that $|x_0-y_0|\leqq 1/2\,C$. Since K is compact and connected this proves the theorem.

The conditions given in Example 2.3.2 and Theorem 2.3.9 are also sufficient:

Theorem 2.3.10. *Let K be a compact set in $\mathbb{R}^n$ with finitely many connected components such that any two points x, y in the same component can be joined by a rectifiable curve in K of length $\leqq C|x-y|$. If u is a distribution of order k with $\operatorname{supp} u\subset K$, it follows then that*

$$|u(\phi)|\leqq C\sum_{|\alpha|\leqq k}\sup_K|\partial^\alpha\phi|, \quad \phi\in C^k(\mathbb{R}^n). \tag{2.3.16}$$

Proof. Let $s \to x(s)$ be a curve in K with $x(0)=y$ and arc length s. Then

(2.3.17) $$|F_\alpha(s)| \leqq C s^{k-|\alpha|} \sum_{|\beta|=k} \sup_K |\partial^\beta \phi|, \qquad |\alpha| \leqq k,$$

if

$$F_\alpha(s) = \partial^\alpha \phi(x(s)) - \sum_{|\beta| \leqq k-|\alpha|} \partial^{\alpha+\beta} \phi(y)(x(s)-y)^\beta/\beta!.$$

This is obvious when $|\alpha|=k$. If $|\alpha|<k$ and (2.3.17) is already proved for derivatives of higher order, we conclude that

$$|dF_\alpha(s)/ds| \leqq C n s^{k-|\alpha|-1} \sum_{|\beta|=k} \sup_K |\partial^\beta \phi|.$$

Since $F_\alpha(0)=0$ we obtain (2.3.17) with C replaced by Cn. If $d(x,y)$ is the infimum of the lengths of curves from y to x in K then (2.3.17) gives

(2.3.17)′ $$\begin{aligned} &|\partial^\alpha \phi(x) - \sum_{|\beta| \leqq k-|\alpha|} \partial^{\alpha+\beta} \phi(y)(x-y)^\beta/\beta!| \\ &\quad \leqq C d(x,y)^{k-|\alpha|} \sum_{|\beta|=k} \sup_K |\partial^\beta \phi|. \end{aligned}$$

When $d(x,y) \leqq C|x-y|$ the estimate (2.3.16) follows from (2.3.13) and (2.3.17)′

Theorem 2.3.11. *Let K be a compact set in $\mathbb{R}^n$ with finitely many connected components and assume that any two points in the same component can be joined by a rectifiable curve in K of length at most $C|x-y|^\gamma$. If u is a distribution of order k with* supp $u \subset K$ *and if m is an integer with $m\gamma \geqq k$, then*

(2.3.18) $$|u(\phi)| \leqq C \sum_{|\alpha| \leqq m} \sup_K |\partial^\alpha \phi|, \qquad \phi \in C^\infty.$$

Proof. We use (2.3.13) combined with (2.3.17)′, now with k replaced by m, noting that since $\gamma \leqq 1$

$$|x-y|^{\gamma(m-|\alpha|)} \leqq |x-y|^{k-|\alpha|} \quad \text{if } |x-y|<1.$$

Sets satisfying the hypotheses in Theorem 2.3.11 are sometimes called regular in the sense of Whitney.

Notes

As mentioned in the introduction the definition of distributions used here is that of Schwartz [1]. One of its advantages is that it suggests naturally the proof of existence theorems for differential equations by duality (the Hahn-Banach theorem). Such ideas had already led Sobolev [1] rather far towards a distribution theory.

Other spaces of test functions give different spaces of distributions. In Chapter VII we shall define a space $\mathscr{S}'$ with $\mathscr{E}'(\mathbb{R}^n)\subset\mathscr{S}'\subset\mathscr{D}'(\mathbb{R}^n)$ by using as test functions a subspace $\mathscr{S}$ of $C^\infty(\mathbb{R}^n)$ restricted by global conditions. Any non-quasi-analytic class of functions can be used as test functions for a distribution space with properties similar to those of Schwartz distributions. This has been done by Beurling [1] (see also Björck [1]). One can also use quasi-analytic classes of test functions but localization is much harder then. The typical real analytic class will be studied in Chapter IX. Also spaces of entire analytic functions can be used as test functions for special purposes (see Gelfand and Šilov [1]), but one is then rather far removed from the intuitive notion of a generalized function.

The topology in $C_0^\infty(X)$ defined by the semi-norms in the right-hand side of (2.1.3) is the inductive limit of the topology in $C_0^\infty(K)$ when the compact set K increases to X, so it is a $\mathscr{L}\mathscr{F}$ topology. (See Dieudonné-Schwartz [1].) We have avoided this terminology in order not to encourage the once common misconception that familiarity with $\mathscr{L}\mathscr{F}$ spaces is essential for the understanding of distribution theory. The convenient explicit form of the semi-norms in (2.1.3) has been adopted from Gårding and Lions [1], and it will occasionally be important later. (For an example see Section 10.7.)

The problem of estimating $u(\phi)$ by means of ϕ and its derivatives in $\operatorname{supp} u$ only was discussed in Schwartz [1] where Example 2.3.2 and Theorems 2.3.10, 2.3.11 are given in an only slightly different form. Corollary 2.3.8 and Theorem 2.3.9 were proved in Glaeser [1] (see also Hörmander [5]). The main point is of course the extension theorem 2.3.6 of Whitney [1]. His results are actually stronger and cover also the extension of C^∞ functions. In that case the extended function v does not depend linearly on the given data u_α. A linear extension of C^∞ functions from a half space to the whole space has been given by Seeley [2].

Chapter III. Differentiation and Multiplication by Functions

Summary

Our motivations for distribution theory came largely from the limitations of the classical notion of differentiability. In this chapter we shall see that differentiation of distributions is indeed always possible. In addition we shall discuss multiplication. This operation on the other hand is not always defined unless one factor is smooth.

Differentiation of distributions and multiplication by smooth functions is defined in Section 3.1. As examples we discuss differentiation of functions with simple discontinuities which leads us to the Gauss-Green formula, and to Cauchy's integral formula. As an application of the latter we digress to discuss boundary values in the distribution sense of analytic functions. As further illustration of multiplication and differentiation of distributions we discuss homogeneous distributions at some length in Section 3.2. Fundamental solutions of some classical second order differential operators are constructed in Section 3.3. In Section 3.4 finally we have collected some computations of integrals, particularly of Gaussian functions, which are needed in those constructions.

3.1. Definitions and Examples

If u is a continuous function such that $\partial_k u$ is defined everywhere and continuous, we obtain by partial integration

$$\int (\partial_k u)\,\phi\,dx = -\int u\,\partial_k \phi\,dx, \qquad \phi \in C_0^\infty(X).$$

(Compare with the weak derivatives in the introduction.) If f is a continuous function then

$$\int (fu)\,\phi\,dx = \int u(f\phi)\,dx, \qquad \phi \in C_0^\infty(X),$$

where $f\phi$ is another test function if $f \in C^\infty$. The following definition is therefore an extension of the classical one:

Definition 3.1.1. If $u\in\mathscr{D}'(X)$ we set

$$(\partial_k u)(\phi)=-u(\partial_k\phi),\qquad \phi\in C_0^\infty(X), \tag{3.1.1}$$

and if $f\in C^\infty(X)$ we define

$$(fu)(\phi)=u(f\phi),\qquad \phi\in C_0^\infty(X). \tag{3.1.2}$$

It is clear that (3.1.1) and (3.1.2) define distributions $\partial_k u$ and fu with support contained in $\operatorname{supp} u$, and that the maps $u\to\partial_k u$ and $u\to fu$ are continuous.

Remarks. 1) From the uniqueness part of Theorem 2.2.5, with $F=\operatorname{supp} u$, we conclude that (3.1.1) and (3.1.2) are valid for all $\phi\in C^\infty(X)$ with $\operatorname{supp} u\cap\operatorname{supp}\phi\Subset X$.

2) The product fu is defined for all u, $f\in\mathscr{D}'(X)$ with

$$\operatorname{sing\,supp} u\cap\operatorname{sing\,supp} f=\emptyset. \tag{3.1.3}$$

In fact, X is covered by open subsets Y such that $u\in C^\infty(Y)$ or $f\in C^\infty(Y)$, so the product is defined in Y by (3.1.2) if $f\in C^\infty(Y)$ and by

$$(fu)(\phi)=f(u\phi),\qquad \phi\in C_0^\infty(Y), \tag{3.1.2$'$}$$

if $u\in C^\infty(Y)$. In case f and u are both in $C^\infty(Y)$ the two definitions agree with the pointwise definition of the product uf, so we have a uniquely defined product $(fu)_Y\in\mathscr{D}'(Y)$. The restriction to $Z\subset Y$ is of course $(fu)_Z$. By the localization Theorem 2.2.4 it follows that there is a unique distribution $fu\in\mathscr{D}'(X)$ with restriction $(fu)_Y$ to Y for every Y.

3) If $f\in C^\infty(X)$, $u\in\mathscr{D}'(X)$ and

$$\operatorname{supp} u\cap\operatorname{supp} f\Subset X \tag{3.1.3$'$}$$

then the definitions at the end of Section 2.2 mean that

$$\langle u,f\rangle=u(f)=(fu)(1)$$

for $u(f)=u(\phi f)=(fu)(\phi)=(fu)(1)$ if $\phi\in C_0^\infty(X)$ is equal to 1 near $\operatorname{supp} u\cap\operatorname{supp} f$. In view of the preceding remark we can therefore use this formula to extend the definition of $\langle u,f\rangle$ to all u, $f\in\mathscr{D}'(X)$ satisfying (3.1.3) and (3.1.3)$'$.

4) From Theorem 2.1.6 it follows that (3.1.2) also defines a distribution $fu\in\mathscr{D}'^k(X)$ if $f\in C^k(X)$ and $u\in\mathscr{D}'^k(X)$. The preceding remarks are of course applicable also to this situation although more awkward to state.

As for smooth functions one can interchange the order of differentiation for distributions,

$$\partial_j \partial_k u = \partial_k \partial_j u, \qquad u \in \mathscr{D}'(X).$$

In fact, if $\phi \in C_0^\infty(X)$ the definition means that

$$\begin{aligned}(\partial_j \partial_k u)(\phi) &= -(\partial_k u)(\partial_j \phi) = u(\partial_k \partial_j \phi) = u(\partial_j \partial_k \phi)\\ &= (\partial_k \partial_j u)(\phi).\end{aligned}$$

We can therefore use the notation $\partial^\alpha u$ for partial derivatives as we have done in the case of functions. Iteration of (3.1.1) gives

$$(\partial^\alpha u)(\phi) = (-1)^{|\alpha|} u(\partial^\alpha \phi), \qquad \phi \in C_0^\infty(X). \tag{3.1.1$'$}$$

The usual rule for differentiation of a product remains valid,

$$\partial_k (f u) = (\partial_k f) u + f(\partial_k u); \qquad f \in C^\infty(X),\, u \in \mathscr{D}'(X); \tag{3.1.4}$$

for this means that

$$-u(f \partial_k \phi) = u((\partial_k f)\phi) - u(\partial_k (f\phi)), \qquad \phi \in C_0^\infty(X),$$

which is true since $\partial_k(f\phi) = (\partial_k f)\phi + f \partial_k \phi$.

Example 3.1.2. The function $H(x)=1$ for $x>0$, $H(x)=0$ for $x \leqq 0$, on $\mathbb{R}$ is called the Heaviside function. The derivative is by definition

$$H'(\phi) = -H(\phi') = -\int_0^\infty \phi'(x)\,dx = \phi(0).$$

One defines the Dirac measure δ_a at $a \in \mathbb{R}^n$ by

$$\delta_a(\phi) = \phi(a),$$

that is, as the unit mass at a. With this notation $H' = \delta_0$. The derivatives of the Dirac measure are

$$(\partial^\alpha \delta_a)(\phi) = (-1)^{|\alpha|} \partial^\alpha \phi(a), \qquad \phi \in C^\infty,$$

so Theorem 2.3.4 means that linear combinations of δ_a and its derivatives are the only distributions with support at a.

Theorem 3.1.3. *If u is a function in the open set $X \subset \mathbb{R}$ which is in $C^1(X \setminus \{x_0\})$ for some $x_0 \in X$, and if the function v which is equal to u' for $x \neq x_0$ is integrable in a neighborhood of x_0, then the limits*

$$u(x_0 \pm 0) = \lim_{x \to x_0 \pm 0} u(x)$$

exist and

$$u' = v + (u(x_0+0) - u(x_0-0))\,\delta_{x_0}.$$

Proof. If $x_0 < y$ and the interval $[x_0, y]$ belongs to X then

$$u(x)=u(y)-\int_x^y v(t)\,dt, \qquad x_0<x<y,$$

which shows that $u(x_0+0)$ exists. Similarly $u(x_0-0)$ exists. We have

$$u'(\phi)=-u(\phi')=\lim_{\varepsilon\to+0} -\int_{|x-x_0|>\varepsilon} u(x)\,\phi'(x)\,dx$$

$$=\lim_{\varepsilon\to+0}\Big(u(x_0+\varepsilon)\,\phi(x_0+\varepsilon)-u(x_0-\varepsilon)\,\phi(x_0-\varepsilon)+\int_{|x-x_0|>\varepsilon} v(x)\,\phi(x)\,dx\Big),$$

if $\phi\in C_0^\infty(X)$. This proves the theorem.

Note that since $u'-v=0$ in $X\setminus\{x_0\}$ it is clear that $u'-v$ must be a distribution with support at x_0, hence a linear combination of δ_{x_0} and its derivatives, which explains qualitatively the form of the conclusion.

Theorem 3.1.3 shows that by pointwise differentiation we may miss something essential at a discontinuity. However, this is not the case for the distribution derivative:

Theorem 3.1.4. *If $u\in\mathscr{D}'(X)$ where X is an open interval on $\mathbb{R}$ and if $u'=0$, then u is a constant.*

Proof. That $u'=0$ means that

$$-u(\phi')=0, \qquad \phi\in C_0^\infty(X).$$

If $\psi\in C_0^\infty(X)$ then the equation $\phi'=\psi$ has the unique solution

$$\phi(x)=\int_{-\infty}^x \psi(t)\,dt$$

vanishing to the left of the support of ψ. It is in $C_0^\infty(X)$ if and only if it vanishes to the right of it also, that is, if

$$I(\psi)=\int_{-\infty}^{\infty}\psi(x)\,dx$$

is equal to 0. Thus $u(\psi)=0$ if $I(\psi)=0$, so $u(\psi)=CI(\psi)$ for some constant C. In fact, if $\psi_0\in C_0^\infty(X)$ is chosen with $I(\psi_0)=1$, then the integral of $\psi-I(\psi)\,\psi_0$ is 0 so

$$0=u(\psi-I(\psi)\,\psi_0)=u(\psi)-I(\psi)\,u(\psi_0).$$

Hence $u=C=u(\psi_0)$ as claimed.

Corollary 3.1.5. *If $u\in\mathscr{D}'(X)$, where $X\subset\mathbb{R}$, and $u'+au=f\in C(X)$ where $a\in C^\infty(X)$, then $u\in C^1(X)$. Thus $u'+au=f$ in the classical sense.*

Proof. Assume first that $a=0$. Since f has a primitive function $v\in C^1(X)$ and $(u-v)'=u'-v'=f-f=0$, it follows from Theorem 3.1.4 that $u-v=C$, so $u\in C^1$. For a general a we let E be a solution $\neq 0$ of the equation $E'=Ea$; we can take $E=\exp\int a\,dx$, so $E\in C^\infty$. Now we have

$$\frac{d}{dx}(Eu)=Eu'+E'u=E(u'+au)=Ef\in C,$$

so $Eu\in C^1$, hence $u\in C^1$.

The corollary remains valid if $u=(u_1,\dots,u_k)$ and $f=(f_1,\dots,f_k)$ have k components and a is a $k\times k$ matrix of C^∞ functions. In the preceding proof we just have to let E be an invertible $k\times k$ matrix such that $E'=Ea$. Since ordinary differential equations of higher order can be reduced to first order systems, we can now prove

Corollary 3.1.6. *If $u\in\mathscr{D}'(X)$, $X\subset\mathbb{R}$, and if*

$$u^{(m)}+a_{m-1}u^{(m-1)}+\dots+a_0u=f\in C(X),$$

where the coefficients $a_j\in C^\infty(X)$, then $u\in C^m(X)$ so the equation is fulfilled in the classical sense.

Proof. With $u_j=u^{(j-1)}$, $1\leqq j\leqq m$, we have the equations

$$u_m'+a_{m-1}u_m+\dots+a_0u_1=f,\quad u_j'-u_{j+1}=0,\quad 1\leqq j<m,$$

so $u_j\in C^1(X)$ by the preceding remark on Corollary 3.1.5. Hence $u\in C^m$.

We shall now pass to several variables, proving first an analogue of Theorem 3.1.4.

Theorem 3.1.4′. *Let $u\in\mathscr{D}'(Y\times I)$ where Y is an open set in $\mathbb{R}^{n-1}$ and I an open interval on $\mathbb{R}$. If $\partial_n u=0$ then*

$$u(\phi)=\int u_0(\phi(.,x_n))\,dx_n,\quad \phi\in C_0^\infty(Y\times I),$$

where $u_0\in\mathscr{D}'(Y)$. Thus u is a distribution u_0 in $x'=(x_1,\dots,x_{n-1})$ independent of x_n.

Proof. Choose $\psi_0\in C_0^\infty(I)$ with $\int\psi_0(t)\,dt=1$ and define

$$u_0(\chi)=u(\chi_0);\quad \chi\in C_0^\infty(Y),\quad \chi_0(x)=\chi(x')\psi_0(x_n).$$

It is obvious that $u_0\in\mathscr{D}'(Y)$. If $\phi\in C_0^\infty(Y\times I)$ and

$$(I\phi)(x')=\int\phi(x',x_n)\,dx_n,$$

we have just as in the proof of Theorem 3.1.4

$$\phi(x)-(I\phi)(x')\,\psi_0(x_n)=\partial_n\Phi$$

where $\Phi\in C_0^\infty(Y\times I)$. Hence

$$u(\phi)=u((I\phi)_0)=u_0(I\phi)=\int u_0(\phi(.,x_n))\,dx_n$$

where the last equality follows from Theorem 2.1.3 applied to $\int\limits_{-\infty}^{x_n}\phi\,dx_n$.

We shall now prove a weak but useful analogue of Corollary 3.1.5.

Theorem 3.1.7. *If u and f are continuous functions in $X\subset\mathbb{R}^n$ and $\partial_j u=f$ in the sense of distribution theory, then $\partial_j u(x)$ exists at every $x\in X$ and is equal to $f(x)$.*

Proof. We may assume that $j=n$ and that $X=Y\times I$ as in Theorem 3.1.4′, for the assertion is local. Let T be a fixed point in I and set

$$v(x)=\int\limits_T^{x_n} f(x',t)\,dt.$$

Then $\partial_n(u-v)=f-f=0$ in the sense of distribution theory, so Theorem 3.1.4′ and its proof shows that $u(x)-v(x)=w(x')$ where w is a continuous function. Since $v(x)+w(x')$ is pointwise differentiable with respect to x_n with derivative f, the theorem is proved.

We shall now extend Example 3.1.2 by differentiating the characteristic function of an open set in $\mathbb{R}^n$ with a C^1 boundary.

Definition 3.1.8. If $Y\subset X$ are open sets in $\mathbb{R}^n$ then Y is said to have C^1 boundary in X if for every boundary point $x_0\in X$ of Y one can find a C^1 function ρ in a neighborhood X_0 of x_0 such that

$$\rho(x_0)=0,\qquad d\rho(x_0)\neq0,\qquad Y\cap X_0=\{x\in X_0;\ \rho(x)<0\}.$$

Using a partition of unity it is easy to show that there exists a function $\rho\in C^1(X)$ such that $\rho=0$, $d\rho\neq0$ on the boundary ∂Y of Y in X and $Y\cap X=\{x\in X;\ \rho(x)<0\}$. We leave the proof as an exercise for the reader. Right now it is more important for us to note that if say $\partial\rho/\partial x_1\neq0$ at x_0 then the C^1 map

$$(x_1,\dots,x_n)\to(\rho(x),x_2,\dots,x_n)$$

from a neighborhood of x_0 has a C^1 inverse. This implies that $\rho=0$ is equivalent to $x_1=\psi(x')$ where $\psi\in C^1$ and $x'=(x_2,\dots,x_n)$. If $\partial\rho/\partial x_1\lessgtr0$ say at x_0 then Y is defined by $x_1\gtrless\psi(x')$ in a neighborhood X_0' of x_0.

If χ_Y is the characteristic function of Y, then $\partial_j \chi_Y$ is supported by ∂Y. To compute $\partial_j \chi_Y$ in X'_0 we observe that if $h \in C^\infty$ vanishes in $(-\infty, 0)$ and equals 1 in $(1, \infty)$, and $Y \cap X'_0$ is defined by $x_1 > \psi(x')$ then

$$\chi_Y = \lim_{\varepsilon \to 0} h((x_1 - \psi(x'))/\varepsilon) \quad \text{in } X'_0,$$

pointwise and in the sense of distributions, so

$$\partial_j \chi_Y = \lim_{\varepsilon \to 0} v_j h'((x_1 - \psi(x'))/\varepsilon)/\varepsilon$$

in the latter sense if $v = (1, -\partial\psi/\partial x')$. Thus we have if $\phi \in C_0^\infty(X'_0)$

$$\begin{aligned}\langle \partial_j \chi_Y, \phi \rangle &= \lim_{\varepsilon \to 0} \int v_j h'((x_1 - \psi(x'))/\varepsilon)\, \varepsilon^{-1} \phi(x)\, dx \\ &= \int v_j(x')\, \phi(\psi(x'), x')\, dx'\end{aligned}$$

(cf. Theorem 1.3.2). If we observe that the Euclidean surface element dS on ∂Y is $(1 + |\psi'|^2)^{\frac{1}{2}}\, dx'$ and that $v/(1 + |\psi'|^2)^{\frac{1}{2}} = n$ is the interior unit normal of ∂Y, this means that

$$\partial_j \chi_Y = n_j dS \tag{3.1.5}$$

in X'_0 and therefore in all of X.

If $f = (f_1, \dots, f_n)$ is a vector field with components in $C_0^\infty(X)$, then (3.1.5) means that

$$\int_Y \operatorname{div} f\, dx = \langle \chi_Y, \sum \partial_j f_j \rangle = -\sum \langle \partial_j \chi_Y, f_j \rangle = -\int_{\partial Y} \langle f, n \rangle\, dS.$$

The formula

$$\int_Y \operatorname{div} f\, dx = -\int_{\partial Y} \langle f, n \rangle\, dS \tag{3.1.6}$$

is the Gauss-Green formula. By Theorem 1.3.2 it is valid for every $f \in C_0^1(X)$. It is in fact easy to verify (3.1.6) for all continuous f with compact support in X such that $\operatorname{div} f$, defined in the sense of distribution theory happens to be in $L^1(Y)$. However, we leave this as an exercise for the reader. The formulas (3.1.5) and (3.1.6) are completely equivalent, so if we had assumed (3.1.6) known we would have obtained a slightly shorter proof of (3.1.5).

Theorem 3.1.9. *Let $Y \subset X$ be open subsets of $\mathbb{R}^n$ such that Y has a C^1 boundary ∂Y in X, and let $u \in C^1(X)$. If χ_Y denotes the characteristic function of Y, dS the Euclidean surface measure on ∂Y and n the interior unit normal there, then*

$$\partial_j(u\chi_Y) = (\partial_j u)\chi_Y + u n_j dS. \tag{3.1.7}$$

Proof. If $u\in C^\infty(X)$ we obtain (3.1.7) from (3.1.5) and (3.1.4). By Theorem 1.3.2 any $u\in C_0^1(X)$ is the limit of a sequence of functions in $C_0^\infty(X)$ converging in $C_0^1(X)$, so (3.1.7) follows for all such u, hence in general since the formula has a local character.

Our proof of (3.1.5) remains valid if the function ψ is Lipschitz continuous; the normal n is just defined almost everywhere with respect to dS then. In particular this means that we can use (3.1.6) for cubes, where it is of course perfectly elementary. This is useful in the proof of the following result.

Theorem 3.1.10. *Let $P=\sum a_j(x)\partial_j+b(x)$, where $a_j\in C^1$ and $b\in C^0$, be a first order linear partial differential operator in $X\subset\mathbb{R}^n$. If u is differentiable at every point in X and there is a continuous function f such that $Pu(x)=f(x)$ for every $x\in X$, then $Pu=f$ in the sense of distribution theory, which is meaningful since $\partial_j u\in\mathscr{D}'^1$ and $a_j\in C^1$.*

Proof. Let $\phi\in C_0^\infty(X)$. We must prove that

$$\int u(b\phi-\sum\partial_j(a_j\phi))\,dx=\int f\phi\,dx.$$

For a cube I with sides parallel to the coordinate axes we set

$$F_u(I)=F(I)=\int_I(f\phi+u(\sum\partial_j(a_j\phi)-b\phi))\,dx+\int_{\partial I}u(a,n)\phi\,dS$$

where n is the interior unit normal and (a,n) the Euclidean scalar product. By (3.1.6) $F_u(I)=0$ if $u\in C^1$. For general u we first claim only that

$$\lim_{x_0\in I,\,m(I)\to 0}F(I)/m(I)=0 \tag{3.1.8}$$

for every fixed $x_0\in X$. In the proof we denote by v the first order Taylor expansion of u at x_0, thus $v(x)-u(x)=o(|x-x_0|)$. Then

$$\begin{aligned}F_u(I)&=F_u(I)-F_v(I)\\&=\int_I(f-Pv)\phi\,dx+\int_I(u-v)(\sum\partial_j(a_j\phi)-b\phi)\,dx+\int_{\partial I}(u-v)(a,n)\,\phi\,dS\end{aligned}$$

where F_v is defined with f replaced by Pv. Now

$$f(x)-Pv(x)\to 0\quad\text{as } x\to x_0,\qquad \sup_I|u-v|=o(s(I))$$

where $s(I)$ is the length of the edge of I. This proves (3.1.8).

(3.1.8) implies that $F(I)=0$ for every I. In fact, starting from any cube I_0 we can divide I_0 into 2^n cubes with sides of half the length, and for one of these cubes I_1 we must have

$$|F(I_1)| \geqq |F(I_0)|/2^n$$

because $F(I)$ is additive, that is, $F(I_0)$ is the sum of $F(I_1)$ when I_1 varies over the smaller cubes. Thus we can find a shrinking sequence of cubes I_j with $m(I_{j+1}) = m(I_j)/2^n$, such that

$$|F(I_0)|/m(I_0) \leqq |F(I_1)|/m(I_1) \leqq |F(I_2)|/m(I_2) \leqq \dots.$$

They all have a point x_0 in common so the limit is 0 by (4.1.8). Hence $F(I_0)=0$. When I_0 is taken so large that $\operatorname{supp}\phi \subset I_0$, the theorem is proved.

The proof of the preceding theorem is taken from the classical proof of the fact that if u is a differentiable function in $X \subset \mathbb{C} = \mathbb{R}^2$ and du is proportional to dz at every point, then u is an analytic function. If $z = x + iy$ we can for every differentiable function v in $\mathbb{C}$ write

$$dv = \partial v/\partial x\, dx + \partial v/\partial y\, dy = \partial v/\partial z\, dz + \partial v/\partial \bar{z}\, d\bar{z}$$

where

$$\partial/\partial z = \tfrac{1}{2}(\partial/\partial x - i\partial/\partial y), \qquad \partial/\partial \bar{z} = \tfrac{1}{2}(\partial/\partial x + i\partial/\partial y).$$

The condition on u is therefore that $\partial u/\partial \bar{z} = 0$ at every point and we have shown that this implies that $\partial u/\partial \bar{z} = 0$ in the sense of distribution theory. In Section 4.4 we shall see that such distributions are analytic functions in the usual sense, in particular that they are in C^∞. However, already now we shall make some further remarks on analytic functions and the operator $\partial/\partial \bar{z}$.

First we note that if $Y \subset X \subset \mathbb{C}$ are open sets such that Y has C^1 boundary in X, then

(3.1.9) $$2\int_Y \partial\phi/\partial\bar{z}\, dx\, dy = -i \int_{\partial Y} \phi\, d(x+iy), \qquad \phi \in C_0^1(X),$$

where the C^1 curve ∂Y is oriented so that Y is to the left of it. To prove this we observe that if s is the arc length on ∂Y then $(dx/ds, dy/ds)$ is the unit tangent and $(-dy/ds, dx/ds)$ is the interior unit normal. Taking $f = (\phi, i\phi)$ in (3.1.6) we therefore obtain

$$2\int_Y \partial\phi/\partial\bar{z}\, dx\, dy = -\int_{\partial Y} (-\phi\, dy/ds + i\phi\, dx/ds)\, ds$$

$$= -i \int_{\partial Y} \phi(dx + i\, dy).$$

The formula (3.1.9) can also be written

(3.1.10) $$\langle \partial\chi_Y/\partial\bar{z}, \phi\rangle = i/2 \int_{\partial Y} \phi(dx + i\, dy), \qquad \phi \in C_0^1(X).$$

Let $\zeta \in Y$ and apply (3.1.9) to $\phi(x,y)/(z-\zeta)$ with Y replaced by $Y \setminus D_\varepsilon$ where D_ε is a disc of radius ε with center at ζ and ε is small. Then

$$2 \int_{Y \setminus D_\varepsilon} \partial \phi(x,y)/\partial \bar{z} (z-\zeta)^{-1} dx\, dy$$
$$= -i \int_{\partial Y} \phi(x,y)(z-\zeta)^{-1} dz + i \int_{\partial D_\varepsilon} \phi(x,y)(z-\zeta)^{-1} dz,$$

where ∂D_ε is oriented as the boundary of D_ε and not as that of $Y \setminus D_\varepsilon$. Throughout we have written $z = x + iy$. Since

$$\int_{\partial D_\varepsilon} \phi(x,y)(z-\zeta)^{-1} dz = \phi(\zeta) \int_{\partial D_\varepsilon} (z-\zeta)^{-1} dz + O(\varepsilon),$$

letting $\varepsilon \to 0$ gives Cauchy's integral formula

$$\phi(\zeta) = -\pi^{-1} \int_Y \partial \phi(x,y)/\partial \bar{z} (z-\zeta)^{-1} dx\, dy \tag{3.1.11}$$
$$+ (2\pi i)^{-1} \int_{\partial Y} \phi(x,y)(z-\zeta)^{-1} dz, \quad \phi \in C_0^1(X),\ \zeta \in Y.$$

(Note that $(x,y) \to (x+iy-\zeta)^{-1}$ is locally integrable.) In particular, when $Y = X$ there is no curve integral and (3.1.11) means that for the function $E_\zeta(x,y) = \pi^{-1}(z-\zeta)^{-1}$ we have

$$\partial E_\zeta / \partial \bar{z} = \delta_\zeta. \tag{3.1.12}$$

The complete formula (3.1.11) follows from (3.1.12) and (3.1.10), for

$$\partial(\chi_Y E_\zeta)/\partial \bar{z} = \delta_\zeta + E_\zeta \partial \chi_Y / \partial \bar{z},$$

in view of (3.1.4) and the localization principles in Section 2.2 since one factor is smooth at every point.

We shall now discuss the existence of boundary values of analytic functions in the sense of distribution theory.

Theorem 3.1.11. *Let I be an open interval on $\mathbb{R}$ and let*

$$Z = \{z \in \mathbb{C};\ \operatorname{Re} z \in I,\ 0 < \operatorname{Im} z < \gamma\}$$

be a one sided complex neighborhood. If f is an analytic function in Z such that for a non-negative integer N

$$|f(z)| \leq C(\operatorname{Im} z)^{-N}, \quad z \in Z,$$

then $f(.+iy)$ has a limit $f_0 \in \mathscr{D}'^{N+1}(I)$ as $y \to 0$, that is,

$$\lim_{y \to +0} \int f(x+iy)\phi(x)\, dx = \langle f_0, \phi \rangle, \quad \phi \in C_0^{N+1}(I).$$

Proof. If $N > 0$ we choose some $z_0 \in Z$ and introduce the integral

$$F(z)=\int_{z_0}^{z} f(\zeta)\,d\zeta, \qquad z\in Z,$$

along some path in Z. The integral is an analytic function independent of the path, $F'(z)=f(z)$, and if the path from z_0 to z is taken first horizontal, then vertical, we obtain if I is bounded as we may assume

$$|F(z)|\leqq C_1(\operatorname{Im} z)^{1-N} \qquad \text{if } N>1,$$
$$|F(z)|\leqq -C\log\operatorname{Im} z+C_1 \qquad \text{if } N=1.$$

In case $N=1$ it follows that the integral of F is continuous in $\bar{Z}$ since $\log t$ is an integrable function of t near 0. In any case we obtain after $N+1$ integrations that $f(z)=G^{(N+1)}(z)$ where G is continuous in $\bar{Z}$ and analytic in Z. This proves that

$$\lim_{y\to 0} f(.+iy)=\lim_{y\to 0} d^{N+1}G(.+iy)/dx^{N+1}=d^{N+1}G(.)/dx^{N+1}$$

in $\mathscr{D}'^{N+1}(I)$ which proves the theorem.

Another proof which gives a useful formula for f_0 is obtained as follows. Set for $\phi\in C_0^{N+1}(I)$

$$\Phi(x,y)=\sum_{j\leqq N}\phi^{(j)}(x)(iy)^j/j!.$$

This is the Nth partial sum of the Taylor expansion in y which an analytic extension of ϕ would have if it did exist. Then $\Phi(x,0)=\phi(x)$ and

$$2\partial\Phi/\partial\bar{z}=(\partial/\partial x+i\partial/\partial y)\Phi=\phi^{(N+1)}(x)(iy)^N/N!.$$

Fix Y with $0<Y<\gamma$. If $0<y<\gamma-Y$ we obtain from (3.1.9) applied to $\Phi(z)f(z+iy)$ that

$$\int\Phi(x,0)f(x+iy)\,dx-\int\Phi(x,Y)f(x+iY+iy)\,dx$$
$$=2i\iint_{0<\operatorname{Im} z<Y} f(z+iy)\,\partial\Phi/\partial\bar{z}\,d\lambda(z)$$

where $d\lambda$ is the Lebesgue measure in $\mathbb{C}$. Writing $z=x+itY$, $0<t<1$ gives

(3.1.13)
$$\int\phi(x)f(x+iy)\,dx=\int\Phi(x,Y)f(x+iY+iy)\,dx$$
$$+\int\int_0^1 f(x+itY+iy)\phi^{(N+1)}(x)(iY)^{N+1}t^N/N!\,dx\,dt.$$

There is a uniform bound for the integrand in the double integral as $y\to 0$ so

$$\int \phi(x) f(x+iy)\,dx \to \int \Phi(x, Y) f(x+iY)\,dx + \int \int_0^1 f(x+itY)\phi^{(N+1)}(x)(iY)^{N+1} t^N/N!\,dx\,dt.$$

We shall often use the notation $f(x+i0)$ for the distribution limit just defined. Similarly we shall write $f(x-i0)$ for such a limit from the lower half plane.

Theorem 3.1.12. *Let I be an open interval on $\mathbb{R}$ and set*

$$Z^{\pm}=\{z\in\mathbb{C};\ \operatorname{Re} z\in I, 0<\pm\operatorname{Im} z<\gamma\}.$$

If f is an analytic function in $Z=Z^+\cup Z^-$ such that

$$|f(z)|\leqq C\,|\operatorname{Im} z|^{-N}, \qquad z\in Z,$$

then the repeated integral

$$F(\phi)=\int(\int f(x+iy)\phi(x,y)\,dx)\,dy, \qquad \phi\in C_0^N(Z\cup I), \tag{3.1.14}$$

exists and defines a distribution in $\mathscr{D}'^N(Z\cup I)$ such that

$$\langle \partial F/\partial\bar z, \phi\rangle = \frac{i}{2}\langle f(.+i0)-f(.-i0), \phi(.,0)\rangle, \qquad \phi\in C_0^{N+1}(Z\cup I). \tag{3.1.15}$$

We have $F=f$ in Z.

Proof. The first proof of Theorem 3.1.11 shows that $F=G^{(N)}$ where $|G(z)|\leqq C'\log(C'/|z|)$, hence $G\in L^1$. Partial integration gives

$$F(\phi)=(-1)^N\int\int G(x+iy)\,\partial^N\phi(x,y)/\partial x^N\,dx\,dy, \qquad \phi\in C_0^N(Z\cup I),$$

which proves the first statement. If $\phi\in C_0^{N+1}(Z\cup I)$ then

$$\begin{aligned}\langle \partial F/\partial\bar z, \phi\rangle &= -\langle F, \partial\phi/\partial\bar z\rangle = -\lim_{\varepsilon\to 0}\int\int_{|y|>\varepsilon} f(x+iy)\,\partial\phi(x,y)/\partial\bar z\,dx\,dy\\ &=\lim_{\varepsilon\to 0}\frac{i}{2}(\int f(x+i\varepsilon)\phi(x,\varepsilon)\,dx-\int f(x-i\varepsilon)\,\phi(x,-\varepsilon)\,dx)\\ &\to \frac{i}{2}\langle f(.+i0)-f(.-i0), \phi(.,0)\rangle,\end{aligned}$$

which proves (3.1.15).

Example 3.1.13. $(x+i0)^{-1}-(x-i0)^{-1}=-2\pi i\delta_0$ by (3.1.15) since $\partial(1/z)/\partial\bar z=\pi\delta_{0,0}$ by (3.1.12). (Here $\delta_{0,0}$ is the Dirac measure in $\mathbb{C}$ at 0.)

When $f(.+i0)=f(.-i0)$ we have $\partial F/\partial\bar z=0$. As already mentioned, we shall show in Section 4.4 that this implies that F is defined by an analytic function. Thus f is the restriction to Z of an analytic function in $Z \cup I$ if and only if the two limits coincide.

The hypothesis in Theorem 3.1.11 cannot be relaxed very much:

Theorem 3.1.14. *If f is analytic in the rectangle Z defined in Theorem 3.1.11 and $\lim_{y\to 0} f(.+iy)$ exists in $\mathscr{D}'^k(I)$, then*

$$|f(z)| \leqq C'(\operatorname{Im} z)^{-k-1}, \quad z \in Z',$$

if Z' is the product of an interval $J \Subset I$ and $(0, \gamma/2)$, say.

Proof. Choose $\phi \in C_0^\infty(\mathbb{R}^2)$ equal to 1 near $\overline{Z'}$ and with support in $\{z; \operatorname{Re} z \in I, |\operatorname{Im} z| < \gamma\}$. Cauchy's integral formula applied to $f\phi$ in the set $\operatorname{Im} z > \operatorname{Im} \zeta/2$ gives if $\zeta = \xi + i\eta \in Z'$

$$f(\zeta) = -\pi^{-1} \iint_{y>\eta/2} f(x+iy)\, \partial\phi(x,y)/\partial\bar{z}\,(z-\zeta)^{-1}\, dx\, dy$$
$$+ (2\pi i)^{-1} \int \phi(x, \eta/2)(x-\xi-i\eta/2)^{-1} f(x+i\eta/2)\, dx.$$

The hypothesis implies a uniform bound for $f(.+iy)$ in $\mathscr{D}'^k$ when $0 < y < \gamma/2$, so the last integral can be estimated by

$$C_1 \sum_{|\alpha| \leqq k} \sup |\partial_x^\alpha (\phi(x, \eta/2)(x-\xi-i\eta/2)^{-1})| \leqq C_2 |\eta|^{-k-1}.$$

The double integral is even bounded. This proves the theorem.

Theorem 3.1.11 has an analogue for analytic functions of several variables:

Theorem 3.1.15. *Let X be an open set in $\mathbb{R}^n$, Γ an open convex cone in $\mathbb{R}^n$, and set for some $\gamma > 0$*

$$Z = \{z \in \mathbb{C}^n;\ \operatorname{Re} z \in X,\ \operatorname{Im} z \in \Gamma,\ |\operatorname{Im} z| < \gamma\}.$$

If f is an analytic function in Z such that

$$|f(z)| \leqq C |\operatorname{Im} z|^{-N}, \quad z \in Z, \tag{3.1.16}$$

then $f(.+iy)$ has a limit $f_0 \in \mathscr{D}'^{N+1}(X)$ as $\Gamma \ni y \to 0$. If $f_0 = 0$ then $f = 0$.

Proof. Choose a fixed $Y \in \Gamma$ with $|Y| < \gamma$. We may assume that $0 \notin \Gamma$ since the theorem is trivial otherwise, and then we can choose C so that

$$t \leqq C|y + tY| \quad \text{if } y \in \Gamma \text{ and } t > 0. \tag{3.1.17}$$

It is sufficient to prove this when $t = 1$ and then we just have to note that $-Y \notin \bar{\Gamma}$ since 0 would otherwise be in Γ. Now set for $\phi \in C_0^{N+1}(X)$

$$\Phi(x,y)=\sum_{|\alpha|\leqq N}\partial^\alpha\phi(x)(iy)^\alpha/\alpha!. \tag{3.1.18}$$

Then we have if $y\in\Gamma$ and $|y|+|Y|<\gamma$

$$\begin{aligned}\int\phi(x)f(x+iy)\,dx=&\int\Phi(x,Y)f(x+iy+iY)\,dx\\ &+(N+1)\iint_{0<t<1} f(x+iy+itY)\sum_{|\alpha|=N+1}\partial^\alpha\phi(x)(iY)^\alpha/\alpha!\,t^N\,dx\,dt.\end{aligned} \tag{3.1.19}$$

In fact, the formula is clearly invariant under linear changes of variables so we may assume that Y lies along the positive x_1 axis. Then the formula (3.1.19) is obtained by applying (3.1.13) for fixed $x_2,\dots,x_n$ and integrating afterwards with respect to these variables. By (3.1.16) and (3.1.17) we have $t^N|f(x+iy+itY)|\leqq C'$ so the integrand in the double integral has an integrable majorant. Hence $f(.+iy)$ converges in $\mathscr{D}'^{N+1}$ to f_0 where

$$\begin{aligned}\langle f_0,\phi\rangle=&\int\Phi(x,Y)f(x+iY)\,dx\\ &+(N+1)\iint_{0<t<1} f(x+itY)\sum_{|\alpha|=N+1}\partial^\alpha\phi(x)(iY)^\alpha/\alpha!\,t^N\,dx\,dt.\end{aligned} \tag{3.1.20}$$

Assume now that $f_0=0$. Take $y\in\Gamma$ and $\phi\in C_0^\infty(X)$ and form

$$F(w)=\int\phi(x)f(x+wy)\,dx=\int\phi(x-\operatorname{Re}wy)f(x+i\operatorname{Im}wy)\,dx.$$

This is an analytic function of w when $0<\operatorname{Im}w|y|<\gamma$ and $|\operatorname{Re}w|\,|y|<d$, the distance from $\operatorname{supp}\phi$ to ∂X. When $\operatorname{Im}w\to 0$ we know that F and all its derivatives tend to 0 since $f_0=0$. Hence F remains analytic when $|\operatorname{Re}w|\,|y|<d$ if we define $F=0$ when $\operatorname{Im}w<0$. By the uniqueness of analytic continuation it follows that $F=0$ identically. Hence $f=0$ in Z, which completes the proof.

Remark. If $f_0=0$ in an open non-void subset X_0 of X we still obtain $f=0$ in Z when $\operatorname{Re}x\in X_0$. Hence $f=0$ in Z and $f_0=0$ in X if X is connected.

Finally we shall prove an analogue of Theorem 3.1.4 for multiplication.

Theorem 3.1.16. *If $u\in\mathscr{D}'(X)$ and $x_ju=0$, $j=1,\dots,n$, then $u=c\delta_0$ for some constant c.*

Proof. The support of u is at 0. If $\phi\in C^\infty(X)$ we can by Theorem 1.1.9 write

$$\phi(x)=\phi(0)+\sum x_j\phi_j(x),\qquad \phi_j\in C^\infty,$$

and obtain

$$u(\phi)=\phi(0)\,u(1)+\sum\langle x_ju,\phi_j\rangle=c\,\phi(0).$$

3.2. Homogeneous Distributions

If a is a complex number with $\operatorname{Re} a > -1$ then the function on $\mathbb{R}$

$$x_+^a = x^a \quad \text{if } x>0, \qquad x_+^a = 0 \quad \text{if } x \leqq 0,$$

is locally integrable so it defines a distribution. (We define $\log x$ to be real when $x>0$ and this defines x^a uniquely when $x>0$.) It is clear that

$$x x_+^a = x_+^{a+1} \quad \text{if } \operatorname{Re} a > -1, \tag{3.2.1}$$

and by Theorem 3.1.3 we have

$$\frac{d}{dx} x_+^a = a x_+^{a-1} \quad \text{if } \operatorname{Re} a > 0. \tag{3.2.2}$$

We want to extend the definition of x_+^a to all $a \in \mathbb{C}$, as a distribution, so that these properties are preserved as far as possible. That some exception must occur is clear though for if $a=0$ then the left-hand side of (3.2.2) is $H'(x)=\delta_0$ (Example 3.1.2) and the right-hand side must be 0.

For $\phi \in C_0^\infty(\mathbb{R})$ the function

$$a \to I_a(\phi) = \langle x_+^a, \phi \rangle = \int_0^\infty x^a \phi(x)\, dx$$

is analytic when $\operatorname{Re} a > -1$ for the differential is

$$da \int_0^\infty x^a \log x\, \phi(x)\, dx.$$

Now (3.2.2) means that

$$I_a(\phi') = -a I_{a-1}(\phi) \quad \text{if } \operatorname{Re} a > 0, \qquad \phi \in C_0^\infty(\mathbb{R}), \tag{3.2.2$'$}$$

so for $\operatorname{Re} a > -1$ and any integer $k>0$ we have

$$I_a(\phi) = (-1)^k I_{a+k}(\phi^{(k)})/((a+1)\dots(a+k)). \tag{3.2.3}$$

The right-hand side is analytic for $\operatorname{Re} a > -k-1$ except for simple poles at $-1, -2, \dots, -k$. If a is not a negative integer we can thus define $I_a(\phi)$ by analytic continuation with respect to a, or equivalently by (3.2.3) with any $k > -1 - \operatorname{Re} a$. By (3.2.3) I_a then defines a distribution of order $\leqq k$. We shall denote it by x_+^a. At $a=-k$ the residue of the function $a \to I_a(\phi)$ is

$$\lim_{a \to -k} (a+k) I_a(\phi) = (-1)^k I_0(\phi^{(k)})/(1-k)\dots(-1) = \phi^{(k-1)}(0)/(k-1)!$$

so

$$(a+k) x_+^a \to (-1)^{k-1} \delta_0^{(k-1)}/(k-1)! \quad \text{as } a \to -k. \tag{3.2.4}$$

Subtracting the singular part we obtain as $a+k=\varepsilon\to 0$

$$\begin{aligned}&I_a(\phi)-\phi^{(k-1)}(0)/((k-1)!\,\varepsilon)\\&\quad=(-1)^k\int_0^\infty (x^\varepsilon-1)\phi^{(k)}(x)/((\varepsilon+1-k)\dots\varepsilon)\,dx\\&\qquad+\phi^{(k-1)}(0)(1/((k-1-\varepsilon)\dots(1-\varepsilon))-1/(k-1)!)/\varepsilon\to\\&\quad-\int_0^\infty (\log x)\phi^{(k)}(x)\,dx/(k-1)!+\phi^{(k-1)}(0)\left(\sum_1^{k-1}1/j\right)/(k-1)!.\end{aligned}$$

Thus we define

$$x_+^{-k}(\phi)=-\int_0^\infty (\log x)\phi^{(k)}(x)\,dx/(k-1)!+\phi^{(k-1)}(0)\left(\sum_1^{k-1}1/j\right)/(k-1)!. \tag{3.2.5}$$

The relation (3.2.1) or equivalently

$$\langle x_+^a, x\phi\rangle=\langle x_+^{a+1},\phi\rangle \tag{3.2.1$'$}$$

follows for all $a\in\mathbb{C}$ since it is valid when $\operatorname{Re}a>-1$, hence by analytic continuation when a is not a negative integer, and the value of both sides when $a=-k$ is obtained by letting $a\to -k$ after subtracting a term $C/(a+k)$, which must be the same on both sides. Also (3.2.2) follows if a is not a negative integer or 0. If k is a non-negative integer then (3.2.4) gives

$$\begin{aligned}\lim_{a\to -k}\left(\frac{d}{dx}x_+^a+kx_+^{a-1}\right)&=\lim_{a\to -k}(a+k)x_+^{a-1}\\&=\lim_{a\to -k-1}(a+k+1)x_+^a=(-1)^k\delta_0^{(k)}/k!,\end{aligned}$$

so dropping terms of the form $C\delta_0^{(k)}/(a+k)$ which must cancel, we obtain

$$\frac{d}{dx}x_+^{-k}=-kx_+^{-k-1}+(-1)^k\delta_0^{(k)}/k!. \tag{3.2.2$''$}$$

This follows also directly from (3.2.5) by an easy calculation.

The preceding argument is essentially due to Marcel Riesz. There is an older method for defining the distribution x_+^a, due to Hadamard, which consists of omitting first a neighborhood of the singularity at 0. Thus one forms, now for any $a\in\mathbb{C}$

$$H_{a,\varepsilon}(\phi)=\int_\varepsilon^\infty x^a\phi(x)\,dx,\qquad \phi\in C_0^\infty(\mathbb{R}).$$

Assume that a is not a negative integer and let k be the smallest integer $\geqq 0$ such that $k+\operatorname{Re}a>-1$. If we integrate by parts k times and use Taylor's formula to express $\phi^{(j)}(\varepsilon)$ in terms of derivatives of ϕ at 0, we obtain an identity of the form

$$H_{a,\varepsilon}(\phi)=(-1)^k \int_0^\infty x^{a+k}\phi^{(k)}(x)/((a+1)\dots(a+k))\,dx \qquad (3.2.6)$$
$$+\sum_0^{k-1} A_j\phi^{(j)}(0)\varepsilon^{a+1+j}+o(1), \qquad \varepsilon\to 0.$$

There can be no other decomposition of the form

$$H_{a,\varepsilon}(\phi)=B_0+\sum B_j\varepsilon^{-\lambda_j}+o(1), \qquad \varepsilon\to 0,$$

where the sum is finite and $\operatorname{Re}\lambda_j\geqq 0$, $\lambda_j\neq 0$. In fact, we have

Lemma 3.2.1. *If $C_0, \dots, C_k$ and $\lambda_1, \dots, \lambda_k$ are different complex numbers with $\operatorname{Re}\lambda_j\geqq 0$ and $\lambda_j\neq 0$, then*

$$C_0+\sum_1^k C_j\varepsilon^{-\lambda_j}\to 0, \qquad \varepsilon\to 0,$$

implies that $C_0=\dots=C_k=0$.

Proof. Assume first that all λ_j are purely imaginary. Replace ε by εe^{-t} and let $\varepsilon\to 0$ through a sequence such that $\varepsilon^{-\lambda_j}$ has a limit γ_j for every j. Then $|\gamma_j|=1$ and

$$C_0+\sum C_j\gamma_j e^{-\lambda_j t}=0$$

for all real t, hence for all complex t. When $t\to\infty$ on the imaginary axis one term dominates so this is not possible unless all $C_j=0$. If $\max\operatorname{Re}\lambda_j=\sigma>0$ in the general case, we have

$$\sum_{\operatorname{Re}\lambda_j=\sigma} C_j\varepsilon^{(\sigma-\lambda_j)}\to 0, \qquad \varepsilon\to 0,$$

so all the coefficients here must vanish which completes the proof.

By Lemma 3.2.1 the terms in the expansion (3.2.6) with $\operatorname{Re}a+1+j\leqq 0$ are therefore uniquely determined, so it is legitimate to discard the singular terms and define the *finite part* of the integral $\int_0^\infty x^a\phi(x)\,dx$ to be

$$(-1)^k \int_0^\infty x^{a+k}\phi^{(k)}(x)/((a+1)\dots(a+k))\,dx.$$

But this agrees with our previous definition of $\langle x_+^a, \phi\rangle$ by (3.2.3). If $a=-k$ is a negative integer the procedure is somewhat different,

$$
\begin{aligned}
H_{-k,\varepsilon}(\phi) &= \frac{1}{(k-1)!}\int_{\varepsilon}^{\infty} x^{-1}\phi^{(k-1)}(x)\,dx + \sum_{0}^{k-2}\phi^{(j)}(\varepsilon)(k-j-2)!\,\varepsilon^{j+1-k}/(k-1)! \\
&= -\frac{1}{(k-1)!}\int_{0}^{\infty}(\log x)\phi^{(k)}(x)\,dx + \phi^{(k-1)}(0)\left(\sum_{1}^{k-1} 1/j\right)\Big/(k-1)! \\
&\quad + \sum_{0}^{k-2} A_j\phi^{(j)}(0)\varepsilon^{j+1-k} - \frac{\log\varepsilon}{(k-1)!}\phi^{(k-1)}(0) + o(1).
\end{aligned}
$$

To define the finite part we must discard not only linear combinations of powers $\varepsilon^{-\lambda}$ with $\operatorname{Re}\lambda \geqq 0$, $\lambda \neq 0$, but also a multiple of $\log\varepsilon$. This can be justified by an analogue of Lemma 3.2.1 and gives (3.2.5) again. However, the notion of finite part is now more delicate for if we replace ε by 2ε say it will change.

The function x_+^a is homogeneous of degree a for $\operatorname{Re} a > -1$. This means that for $t > 0$

$$
\langle x_+^a, \phi\rangle = \int_0^\infty x^a\phi(x)\,dx = t^a\int_0^\infty x^a\phi(tx)\,t\,dx = t^a\langle x_+^a, \phi_t\rangle
$$

where $\phi_t(x) = t\phi(tx)$. The analytic continuation of the two sides must agree, so

$$
\langle x_+^a, \phi\rangle = t^a\langle x_+^a, \phi_t\rangle \quad \text{if } \phi \in C_0^\infty(\mathbb{R}), \tag{3.2.7}
$$

a not a negative integer. If $a = -k$ we obtain from (3.2.5)

$$
\begin{aligned}
t^{-k}\langle x_+^{-k}, \phi_t\rangle &= -\int_0^\infty \log x\,\phi^{(k)}(tx)\,d(tx)/(k-1)! \\
&\quad + \phi^{(k-1)}(0)\left(\sum_1^{k-1} 1/j\right)/(k-1)! \\
&= \langle x_+^{-k}, \phi\rangle + \log t\int_0^\infty \phi^{(k)}(x)\frac{dx}{(k-1)!}.
\end{aligned}
$$

Hence

$$
\langle x_+^{-k}, \phi\rangle = t^{-k}\langle x_+^{-k}, \phi_t\rangle + \log t\,\phi^{(k-1)}(0)/(k-1)! \tag{3.2.8}
$$

so the homogeneity is partly lost.

In addition to x_+^a we shall also have to use the function

$$
x_-^a = 0 \quad \text{if } x > 0, \qquad x_-^a = |x|^a \quad \text{if } x < 0,
$$

where $a > -1$. This is the reflection of x_+^a with respect to the origin,

$$
\langle x_-^a, \phi\rangle = \langle x_+^a, \check{\phi}\rangle \qquad \text{where } \check{\phi}(x) = \phi(-x),
$$

so it is clear that all we have said about the definition of x_+^a for arbitrary complex a remains true for x_-^a.

By Theorem 3.1.11 the function z^a, defined in $\mathbb{C}\setminus\mathbb{R}_-$ as $e^{a\log z}$ where $\log z$ is real if $z\in\mathbb{R}_+$, has boundary values $(x\pm i0)^a$ on the real axis from the upper and lower half planes. When $\operatorname{Re} a>0$ it is clear that

$$(x\pm i0)^a = x_+^a + e^{\pm\pi i a} x_-^a. \tag{3.2.9}$$

Now $\langle (x\pm i0)^a, \phi\rangle$ is for every test function ϕ an entire analytic function of a since it is the limit of entire analytic functions. Hence (3.2.9) remains valid when a is not a negative integer. When $a\to -k$, where k is a positive integer, we have by (3.2.4) and the derivation of (3.2.5)

$$x_+^a - (-1)^{k-1}\delta_0^{(k-1)}/((k-1)!\,(a+k)) \to x_+^{-k}$$

and since

$$e^{\mp\pi i a} = (-1)^k(1\mp\pi i(a+k)+O(a+k)^2) \quad \text{as } a\to -k,$$

we obtain when $a\to -k$

$$e^{\mp\pi i a}x_+^a + \delta_0^{(k-1)}/((k-1)!(a+k)) \mp \pi i\,\delta_0^{(k-1)}/(k-1)! \to (-1)^k x_+^{-k}.$$

When $a\to -k$ in (3.2.9) we must have cancellation of the singular terms and it follows that

$$(x\pm i0)^{-k} = x_+^{-k} + (-1)^k x_-^{-k} \pm \pi i(-1)^k\,\delta_0^{(k-1)}/(k-1)!. \tag{3.2.10}$$

In particular

$$(x+i0)^{-k} - (x-i0)^{-k} = 2\pi i(-1)^k\,\delta_0^{(k-1)}/(k-1)! \tag{3.2.11}$$

which agrees with Example 3.1.13 if $k=1$. In fact (3.2.11) follows for any k from the case $k=1$ by differentiation since

$$\frac{d}{dx}(x\pm i0)^a = a(x\pm i0)^{a-1}. \tag{3.2.12}$$

The average of the two distributions in (3.2.10) is sometimes denoted by $\underline{x}^{-k}$, thus

$$\underline{x}^{-k} = ((x+i0)^{-k} + (x-i0)^{-k})/2 = x_+^{-k} + (-1)^k x_-^{-k}. \tag{3.2.10$'$}$$

From (3.2.12) and the obvious fact that $x(x\pm i0)^a = (x\pm i0)^{a+1}$ we obtain

$$\frac{d}{dx}\underline{x}^{-k} = -k\underline{x}^{-k-1}, \quad x\underline{x}^{-k} = \underline{x}^{1-k}. \tag{3.2.12$'$}$$

By (3.2.5) we have

$$\underline{x}^{-1}(\phi) = x_+^{-1}(\phi-\check{\phi}) = -\int_0^\infty \log x(\phi'(x)+\phi'(-x))\,dx$$
$$= -\int_{-\infty}^\infty (\log|x|)\phi'(x)\,dx.$$

Hence

$$\underline{x}^{-1}=\frac{d}{dx}\log|x|. \tag{3.2.13}$$

Since $\phi(x)-\phi(-x)=O(x)$ as $x\to 0$, an integration by parts also gives

$$\underline{x}^{-1}(\phi)=\int_0^\infty (\phi(x)-\phi(-x))\,dx/x=\lim_{\varepsilon\to 0}\int_{|x|>\varepsilon}\phi(x)\,dx/x.$$

The last integral, where a symmetric neighborhood of the singularity tending to 0 has been removed, is called a *principal value*. Thus

$$\langle \underline{x}^{-1},\phi\rangle=\lim_{\varepsilon\to 0}\int_{|x|>\varepsilon}\phi(x)\,dx/x=PV\int\phi(x)\,dx/x,\qquad \phi\in C_0^1. \tag{3.2.14}$$

The problems we have encountered in the discussion of x_+^a when a is a negative integer were caused by the factor a in (3.2.2). By a change of normalizations they can be made to disappear. First note that (3.2.2)′ assumes a particularly simple form if $\phi'=-\phi$, that is, $\phi(x)=e^{-x}$. This is not a function of compact support but it decreases so fast at $+\infty$ that the proof of (3.2.2)′ is valid for it. Set

$$\Gamma(a)=\int_0^\infty x^{a-1}e^{-x}\,dx,\qquad \operatorname{Re}a>0, \tag{3.2.15}$$

which in our old notation is $I_{a-1}(e^{-\cdot})$. Then (3.2.2)′ means that

$$\Gamma(a+1)=a\Gamma(a)\qquad \text{if } \operatorname{Re}a>0. \tag{3.2.16}$$

Using (3.2.16) we can extend $\Gamma(a)$ analytically to a meromorphic function in $\mathbb{C}$ with simple poles at the integers $\leqq 0$, and (3.2.16) remains valid outside the poles. The residue at an integer $-k\leqq 0$ is

$$\lim_{a\to -k}(a+k)\Gamma(a)=\lim_{a\to -k}\Gamma(a+k+1)/a(a+1)\dots(a+k-1)$$
$$=\Gamma(1)/(-k)\dots(-1)=(-1)^k/k!$$

which is of course just (3.2.4) with k replaced by $k+1$, acting on e^{-x}. In Section 3.4 we shall prove that

$$\Gamma(a)\Gamma(1-a)=\pi/\sin(\pi a).$$

This implies that the Γ-function has no zeros, so the quotient defined by

$$\chi_+^a=x_+^a/\Gamma(a+1),\qquad \operatorname{Re}a>-1 \tag{3.2.17}$$

is analytic when $\operatorname{Re}a>-1$. Since (3.2.2)′ gives, when combined with (3.2.16),

$$\chi_+^a(\phi')=-\chi_+^{a-1}(\phi) \tag{3.2.2$'''$}$$

it is now clear that χ_+^a can be continued analytically to all $a\in\mathbb{C}$ so that $d\chi_+^a/dx=\chi_+^{a-1}$. Noting that $\chi_+^0=H$ we obtain

$$\chi_+^{-k}=\delta_0^{(k-1)}, \quad k=1,2,\ldots. \tag{3.2.17$'$}$$

We shall now carry some of the preceding results over to $\mathbb{R}^n$. First note that if $u\in L^1_{\text{loc}}(\mathbb{R}^n\setminus 0)$ is homogeneous of degree a, that is, $u(tx)=t^a u(x)$ when $x\neq 0$ and $t>0$, then

$$\langle u,\phi\rangle=t^a\langle u,\phi_t\rangle \quad \text{if } \phi\in C_0^\infty(\mathbb{R}^n\setminus 0), \quad \phi_t(x)=t^n\phi(tx), \quad t>0, \tag{3.2.18}$$

and conversely this implies that u is homogeneous. If $\operatorname{Re} a>-n$ then u is integrable in a neighborhood of 0 because with polar coordinates $x=r\omega$, $|\omega|=1$, we have

$$|u(r\omega)|=r^{\operatorname{Re} a}|u(\omega)| \quad \text{and} \quad dx=r^{n-1}\,dr\,d\omega$$

where $d\omega$ is the surface measure on the unit sphere. In that case u defines a distribution in $\mathbb{R}^n$ and (3.2.18) is valid when $\phi\in C_0^\infty(\mathbb{R}^n)$.

Definition 3.2.2. A distribution u in $\mathbb{R}^n\setminus 0$ is called homogeneous of degree a if (3.2.18) is valid. If u is a distribution in $\mathbb{R}^n$ and (3.2.18) is valid for all $\phi\in C_0^\infty(\mathbb{R}^n)$ then u is said to be homogeneous of degree a in $\mathbb{R}^n$.

The problem which we shall discuss is the extension of homogeneous distributions from $\mathbb{R}^n\setminus 0$ to $\mathbb{R}^n$, which as we know from the case $n=1$ is not always possible. However, we shall first rephrase (3.2.18). If we differentiate with respect to t using Theorem 2.1.3 and put $t=1$ it follows that

$$(a+n)\langle u,\phi\rangle+\langle u,\lambda\phi\rangle=0, \quad \phi\in C_0^\infty(\mathbb{R}^n\setminus 0) \tag{3.2.19}$$

where $\lambda=\sum x_j\partial_j$ is the radial vector field. We shall prove that (3.2.19) implies

$$u(\psi)=0 \quad \text{if } \psi\in C_0^\infty(\mathbb{R}^n\setminus 0) \text{ and } \int_0^\infty r^{a+n-1}\psi(rx)\,dr\equiv 0; \tag{3.2.20}$$

since the integral is a homogeneous function of x of degree $-a-n$ we may take $|x|=1$ here. (3.2.20) follows from (3.2.19) if we show that the equation

$$(a+n)\phi+\lambda\phi=\psi$$

has a solution $\phi\in C_0^\infty(\mathbb{R}^n\setminus 0)$. With polar coordinates this equation can be written

$$\frac{\partial}{\partial r}(r^{a+n}\phi(r\omega))=\psi(r\omega)\,r^{a+n-1}$$

so the solution which vanishes for small r is also zero for large r. This proves (3.2.20), and since

$$\int_0^\infty r^{a+n-1}(\phi(rx)-t^{a+n}\phi(rtx))\,dr=0, \qquad \phi\in C_0^\infty(\mathbb{R}^n\setminus 0),$$

we see that (3.2.18), (3.2.19) and (3.2.20) are equivalent. If we note that

$$\langle u,\lambda\phi\rangle=\sum\langle x_j u,\partial_j\phi\rangle=-\sum\langle\partial_j x_j u,\phi\rangle=-\langle\lambda u,\phi\rangle-n\langle u,\phi\rangle$$

we can also write (3.2.19) in the form

(3.2.19)′ $$\lambda u=au$$

(Euler's identity for homogeneous "functions"). In particular we find using Corollary 3.1.5 that homogeneous distributions when $n=1$ are just multiples of $|x|^a$ on each half axis. It is also clear that if ψ is a homogeneous C^∞ function in $\mathbb{R}^n\setminus 0$ of degree b then ψu is homogeneous of degree $a+b$. Since

$$\lambda\partial_j u=\partial_j\lambda u-\partial_j u=(a-1)\,\partial_j u$$

differentiation lowers the degree of homogeneity by one unit.

Theorem 3.2.3. *If $u\in\mathscr{D}'(\mathbb{R}^n\setminus 0)$ is homogeneous of degree a, and a is not an integer $\leqq -n$, then u has a unique extension $\dot u\in\mathscr{D}'(\mathbb{R}^n)$ which is homogeneous of degree a. If P is a homogeneous polynomial then $(Pu)^{\cdot}=P\dot u$, and if $a\neq 1-n$ then $(\partial_j u)^{\cdot}=\partial_j\dot u$. The map*

$$\mathscr{D}'(\mathbb{R}^n\setminus 0)\ni u\to\dot u\in\mathscr{D}'(\mathbb{R}^n)$$

is continuous.

Proof. a) Uniqueness. The difference between two homogeneous extensions is supported at 0, hence a linear combination of derivatives of δ_0. Now $\partial^\alpha\delta_0$ is homogeneous of degree $-n-|\alpha|$ since

$$\langle\partial^\alpha\delta_0,\phi_t\rangle=t^{n+|\alpha|}\langle\partial^\alpha\delta_0,\phi\rangle, \qquad \phi\in C_0^\infty(\mathbb{R}^n),$$

so a homogeneous distribution with support at 0 must be homogeneous of integer degree $\leqq -n$.

b) Existence. If u is a function and $\phi\in C_0^\infty(\mathbb{R}^n\setminus 0)$ then

$$\langle u,\phi\rangle=\int_{|\omega|=1}\int_0^\infty u(\omega)\,r^{a+n-1}\phi(r\omega)\,dr\,d\omega.$$

This suggests that for arbitrary $\phi\in C_0^\infty(\mathbb{R}^n)$ we should introduce

(3.2.21) $$(R_a\phi)(x)=\langle t_+^{a+n-1},\phi(tx)\rangle, \qquad 0\neq x\in\mathbb{R}^n.$$

It follows from (3.2.7) that $R_a\phi$ is a homogeneous function of degree $-n-a$. By Theorem 2.1.3 R_a is a continuous map from $C_0^\infty(K)$ to $C^\infty(\mathbb{R}^n \setminus 0)$ for every compact set $K \subset \mathbb{R}^n$.

Choose a fixed function $\psi \in C_0^\infty(\mathbb{R}^n \setminus 0)$ such that

$$\int_0^\infty \psi(tx)\,dt/t = 1, \qquad x \neq 0. \tag{3.2.22}$$

It suffices to take $\psi(x)$ as a function of $|x|$ so that (3.2.22) is valid for one x. Then $\psi R_a\phi \in C_0^\infty(\mathbb{R}^n \setminus 0)$ and

$$\begin{aligned} R_a(\psi R_a\phi)(x) &= \int_0^\infty t^{a+n-1}\psi(tx)(R_a\phi)(tx)\,dt \\ &= (R_a\phi)(x)\int_0^\infty \psi(tx)\,dt/t = (R_a\phi)(x). \end{aligned}$$

Hence it follows from (3.2.20) that $u(\psi R_a\phi)$ is always independent of the choice of ψ and that $u(\psi R_a\phi) = u(\phi)$ if $\phi \in C_0^\infty(\mathbb{R}^n \setminus 0)$. Thus

$$\langle \dot{u}, \phi\rangle = \langle u, \psi R_a\phi\rangle, \qquad \phi \in C_0^\infty(\mathbb{R}^n), \tag{3.2.23}$$

defines a distribution $\dot{u}$ in $\mathbb{R}^n$ which extends u. The map $u \to \dot{u}$ is obviously continuous. Since

$$(R_a\phi_t)(x) = \langle r_+^{a+n-1}, t^n\phi(rtx)\rangle = t^{n-1-(a+n-1)}R_a\phi(x) = t^{-a}R_a\phi(x)$$

by (3.2.7) again, it follows that $\langle \dot{u}, \phi_t\rangle = t^{-a}\langle \dot{u}, \phi\rangle$, so $\dot{u}$ is homogeneous.

Finally we note that $(Pu)^{\cdot} - P\dot{u}$ and $(\partial_j u)^{\cdot} - \partial_j\dot{u}$ are homogeneous distributions of degree $a+$degree P and $a-1$ respectively and supported by 0, so they must be zero by our hypotheses. This completes the proof of the theorem.

(3.2.23) still defines an extension $\dot{u}$ of u if a is an integer $-n-k \leq -n$, but it may depend on the choice of ψ then. $\dot{u}$ may also fail to be homogeneous, for (3.2.8) gives

$$\begin{aligned} (R_{-n-k}\phi_t)(x) &= \langle r_+^{-k-1}, t^n\phi(trx)\rangle \\ &= t^{n-1+k+1}\left(\langle r_+^{-k-1}, \phi(rx)\rangle - \log t\,\frac{\partial^k}{\partial r^k}\phi(rx)/k!|_{r=0}\right) \\ &= t^{k+n}(R_{-n-k}\phi(x) - \log t\,\Phi_k(x)) \end{aligned}$$

where Φ_k is the homogeneous part of degree k in the Taylor expansion of ϕ. Thus

$$\langle \dot{u}, \phi\rangle = t^{-k-n}\langle \dot{u}, \phi_t\rangle + \log t \sum_{|\alpha|=k}\langle u, x^\alpha\psi\rangle\,\partial^\alpha\phi(0)/\alpha!. \tag{3.2.24}$$

Any other extension of u is of the form $U=\dot{u}+\sum a_\alpha \partial^\alpha \delta_0$ where the sum is finite. Replacing $\dot{u}$ by U does not change the logarithmic term in (3.2.24) but introduces a new term

$$\sum (1-t^{|\alpha|-k})\, a_\alpha \partial^\alpha \delta_0$$

which can only be 0 if $a_\alpha=0$ when $|\alpha| \neq k$. Thus the weakened homogeneity property (3.2.24) cannot be improved. In particular there exists a homogeneous extension if and only if

$$\langle u, x^\alpha \psi\rangle = 0 \quad \text{when } |\alpha| = k = -n-a. \tag{3.2.25}$$

That (3.2.25) is independent of the choice of ψ is clear from the preceding result but can also be seen as follows. If v is homogeneous of degree $-n$ in $\mathbb{R}^n \setminus 0$ then $v(\psi)$ is independent of the choice of ψ satisfying (3.2.22), in virtue of (3.2.20), so we may write

$$S(v) = \langle v, \psi\rangle$$

for such ψ. If v is a continuous function then introducing polar coordinates gives

$$S(v) = \iint_{|\omega|=1} v(\omega)\, \psi(\omega r)\, d\omega\, dr/r = \int_{|\omega|=1} v(\omega)\, d\omega$$

so $S(v)$ is the integral of v over the unit sphere. (This has nothing to do with the Euclidean metric really, we could equally well integrate the Kronecker form

$$\sum (-1)^{j-1} v(x)\, x_j dx_1 \wedge \ldots \wedge dx_{j-1} \wedge dx_{j+1} \wedge \ldots \wedge dx_n$$

over the C^1 boundary of any neighborhood of 0.) In (3.2.25) $x^\alpha u$ is homogeneous of degree $-n$, so (3.2.25) can be written

$$S(x^\alpha u) = 0 \quad \text{when } |\alpha| + \text{degree}\, u = -n. \tag{3.2.25$'$}$$

We can also rewrite (3.2.24) as follows

$$\langle \dot{u}, \phi\rangle = t^{-k-n} \langle \dot{u}, \phi_t\rangle + \log t \sum_{|\alpha|=k} S(x^\alpha u)\, \partial^\alpha \phi(0)/\alpha!. \tag{3.2.24$'$}$$

Note that if u is homogeneous of degree a and a is not an integer which is $\leqq -n$ then (3.2.23) can be written in the form

$$\langle \dot{u}, \phi\rangle = S(u R_a \phi) = S(u \langle t_+^{a+n-1}, \phi(t.)\rangle), \qquad \phi \in C_0^\infty(\mathbb{R}^n), \tag{3.2.23$'$}$$

for $uR_a\phi$ is then homogeneous of degree $a-n-a=-n$. However, when $a=-n-k$ is an integer $\leqq -n$ then (3.2.23) depends on the choice of ψ. Our choice of $\dot{u}$ with a fixed function ψ guarantees that

$$(Pu)^{\cdot} = P\dot{u} \tag{3.2.26}$$

if P is a homogeneous polynomial. In fact, if the degree is m then

$$\langle t_+^{-k-1}, P(tx)\,\phi(tx)\rangle = P(x)\langle t_+^{m-k-1}, \phi(tx)\rangle$$

by (3.2.1)′, hence

$$\begin{aligned}\langle P\dot u, \phi\rangle &= \langle \dot u, P\phi\rangle = \langle u, \psi(R_{-n-k}P\phi)\rangle = \langle u, \psi P R_{-n+m-k}\phi\rangle \\ &= \langle Pu, \psi R_{-n+m-k}\phi\rangle = \langle (Pu)^{\cdot}, \phi\rangle.\end{aligned}$$

(3.2.25) is automatically fulfilled if u is homogeneous of degree $a = -n-k$ where k is an integer $\geqq 0$ and u has parity opposite to k. By this we mean that (3.2.18) is strengthened to

(3.2.18)′ $$\langle u, \phi\rangle = \operatorname{sgn} t\, t^a \langle u, \phi_t\rangle \quad \text{if } \phi \in C_0^\infty(\mathbb{R}^n \setminus 0),$$
$$\phi_t(x) = t^n \phi(tx),$$

for every $t \neq 0$. It is of course sufficient to assume (3.2.18)′ for $t=-1$ in addition to (3.2.18). If u is a function then (3.2.18)′ means for $t=-1$ that

$$\int u(x)\,\phi(x)\,dx = (-1)^{1+a+n}\int u(x)\,\phi(-x)\,dx = (-1)^{1+k}\int u(-x)\,\phi(x)\,dx,$$

that is, $u(x) = (-1)^{k+1} u(-x)$.

(3.2.18)′ always implies (3.2.25). In fact, if ψ is even and satisfies (3.2.22), and if $\phi(x) = x^\alpha \psi(x)$, $|\alpha| = n$, then $\phi_{-1} = (-1)^{n+k}\phi = (-1)^a \phi$, hence (3.2.18)′ gives $\langle u, \phi\rangle = -\langle u, \phi\rangle$, that is, $\langle u, \phi\rangle = 0$. Thus u has a homogeneous extension. We claim that there is a unique extension $\dot u$ satisfying (3.2.18)′ for all $\phi \in C_0^\infty(\mathbb{R}^n)$ and that it is given by

(3.2.23)″ $$\langle \dot u, \phi\rangle = S(u\langle \underline{t}^{a+n-1}, \phi(t.)\rangle/2), \qquad \phi \in C_0^\infty(\mathbb{R}^n).$$

Here $\underline{t}^{a+n-1}$ is defined by (3.2.10)′. The uniqueness is obvious, for if $|\alpha| = k$ then

$$\langle \partial^\alpha \delta, \phi\rangle = t^{-k-n}\langle \partial^\alpha \delta, \phi_t\rangle$$

so the usually undetermined part of the extension has the wrong parity. The second part of (3.2.10)′ gives (recall that $a+n=-k$)

$$\langle \underline{t}^{-k-1}, \phi(t.)\rangle = \langle t_+^{-k-1}, \phi(t.) + (-1)^{k-1}\phi(-t.)\rangle.$$

If U is the extension of u defined by (3.2.23) then (3.2.23)″ means that

$$2\langle \dot u, \phi\rangle = \langle U, \phi\rangle + (-1)^{k+n-1}\langle U, \phi_{-1}\rangle, \qquad \phi \in C_0^\infty(\mathbb{R}^n).$$

Hence (3.2.23)″ does define a distribution. If $\phi \in C_0^\infty(\mathbb{R}^n \setminus 0)$ then the right-hand side is equal to $2\langle U, \phi\rangle$ by (3.2.18)′ so $\dot u = U$ in $\mathbb{R}^n \setminus 0$. Finally we obtain (3.2.18)′ with u replaced by $\dot u$ for all $\phi \in C_0^\infty(\mathbb{R}^n)$ and $t=-1$, since

$$2\langle \dot u, \phi_{-1}\rangle = \langle U, \phi_{-1}\rangle + (-1)^{k+n-1}\langle U, \phi\rangle = 2(-1)^{k+n-1}\langle \dot u, \phi\rangle.$$

Summing up, we have now proved

Theorem 3.2.4. *If* $u \in \mathscr{D}'(\mathbb{R}^n \setminus 0)$ *is homogeneous of integer degree* $a = -n-k \leqq -n$, *then u has an extension* $\dot{u} \in \mathscr{D}'(\mathbb{R}^n)$ *satisfying* (3.2.24)′. *This determines u apart from a linear combination of derivatives of order k of* δ_0. *A consistent choice of extension can be made so that* (3.2.26) *is fulfilled for every homogeneous polynomial P. A homogeneous extension exists if and only if* (3.2.25)′ *is valid. If u satisfies* (3.2.18)′ *then there is a unique extension* $\dot{u}$ *with the same property for every* $\phi \in C_0^\infty(\mathbb{R}^n)$. *It is given by* (3.2.23)″.

Remark. If u is homogeneous of integer order $a = -n-k > -n$ and satisfies (3.2.18)′ then we also have (3.2.23)″ for the unique homogeneous extension.

We shall refrain from discussing the difference $\partial_j \dot{u} - (\partial_j u)^{\cdot}$ in general because it depends on the choice of ψ. However, one useful case where ψ does not matter is the following one.

Theorem 3.2.5. *Let* $u_1, \dots, u_n \in \mathscr{D}'(\mathbb{R}^n \setminus 0)$ *all be homogeneous of degree* $1-n$ *in* $\mathbb{R}^n \setminus 0$ *and let* $\sum \partial_j u_j = 0$ *there. Then it follows that*

$$\sum \partial_j \dot{u}_j = c\delta_0, \qquad c = \sum S(u_j \psi_j)$$

where $\psi_j(x) = x_j/|x|^2$, $|x|$ *denoting the Euclidean metric.*

Proof. We know that $\sum \partial_j \dot{u}_j$ is homogeneous of degree $-n$ and supported by 0, so $\sum \partial_j \dot{u}_j = c\delta_0$ for some c. If $\phi \in C_0^\infty(\mathbb{R}^n)$ and $\phi(0) = 1$ then

$$c = \sum \langle \partial_j \dot{u}_j, \phi \rangle = -\sum \langle \dot{u}_j, \partial_j \phi \rangle.$$

Choose $\phi(x) = \chi(|x|)$ where $\chi \in C_0^\infty(\mathbb{R})$ and $\chi = 1$ near 0. Then

$$-\partial_j \phi = \psi_j(x)\,\psi(x), \qquad \psi(x) = -\chi'(|x|)|x|,$$

and since

$$\int_0^\infty \psi(tx)\,dt/t = -\int_0^\infty \chi'(t|x|)\,|x|\,dt = 1, \qquad x \neq 0,$$

the condition (3.2.22) is fulfilled so $-\langle u_j, \partial_j \phi \rangle = S(\psi_j u_j)$ by the definition of S.

3.3. Some Fundamental Solutions

In Section 3.1 we saw that Cauchy's integral formula is closely related to the fact that $\partial E/\partial \bar{z} = \delta_0$ if $E = 1/\pi z$. Integration of a function on $\mathbb{R}$ means convolving it with the Heaviside function H, and $dH/dx = \delta_0$. These are two examples of fundamental solutions:

Definition 3.3.1. A distribution $E\in\mathscr{D}'(\mathbb{R}^n)$ is called a fundamental solution of the differential operator $P=\sum a_\alpha\partial^\alpha$ with constant (complex) coefficients if $PE=\delta_0$.

Fundamental solutions are very important in the study of existence and regularity of solutions of differential equations. Such applications will be discussed in Section 4.4 after we have studied convolution of distributions. It will be convenient then to have available the examples of fundamental solutions which we shall give now.

Theorem 3.3.2. *Put* $E(x)=(2\pi)^{-1}\log|x|$ *if* $x\in\mathbb{R}^2\setminus 0$ *and for* $n>2$

$$E(x)=-|x|^{2-n}/(n-2)\,c_n, \quad x\in\mathbb{R}^n\setminus 0,$$

where $|x|$ *is the Euclidean norm and* c_n *the area of the unit sphere. Then* $\partial_j E$ *is defined by the locally integrable function* $x_j|x|^{-n}/c_n$ *and*

$$\Delta E=\sum_1^n \partial_j^2 E=\delta_0. \tag{3.3.1}$$

Proof. If $\phi\in C_0^\infty(\mathbb{R}^n)$ then

$$\langle\partial_j E,\phi\rangle=-\langle E,\partial_j\phi\rangle=-\lim_{\varepsilon\to 0}\int_{|x|>\varepsilon} E(x)\,\partial_j\phi(x)\,dx$$
$$=\int\phi(x)\,\partial_j E(x)\,dx+\lim_{\varepsilon\to 0}\int_{|x|=\varepsilon} E(x)\,\phi(x)\,x_j/|x|\,dS$$

by Gauss' formula. The surface integral is $O(\varepsilon)$ if $n>2$ and $O(\varepsilon\log 1/\varepsilon)$ if $n=2$ so the limit is 0. Thus $\partial_j E$ is defined by the locally integrable function $\partial_j E(x)$ which for $n>2$ also follows from Theorem 3.2.3. For $x\neq 0$ we have

$$\Delta E=(n|x|^{-n}-\sum n x_j^2|x|^{-n-2})/c_n=0$$

so Theorem 3.2.5 and the fact that $S(\sum x_j^2/|x|^{n+2}c_n)=1$ gives

$$\Delta E=\sum\partial_j\partial_j E=\delta_0.$$

We could also make this conclusion without appealing to Theorem 3.2.5:

$$\langle\Delta E,\phi\rangle=\langle E,\Delta\phi\rangle=\lim_{\varepsilon\to 0}\int_{|x|>\varepsilon}(E\Delta\phi-\phi\Delta E)\,dx$$
$$=\lim_{\varepsilon\to 0}\int_{|x|>\varepsilon}\operatorname{div}(E\operatorname{grad}\phi-\phi\operatorname{grad}E)\,dx$$
$$=\lim_{\varepsilon\to 0}\int_{|x|=\varepsilon}\langle\phi\operatorname{grad}E-E\operatorname{grad}\phi,x/|x|\rangle\,dS=\phi(0).$$

Note that when $n=2$ we have $\Delta=4\partial^2/\partial z\partial\bar{z}$ and

$$4\partial E/\partial z=\bar{z}/|z|^2\pi=1/\pi z$$

so we get back our old result that $\partial(1/\pi z)/\partial\bar{z}=\delta_0$.

Next we consider the heat equation in $\mathbb{R}^{n+1}$:

Theorem 3.3.3. *Denote the variables in* $\mathbb{R}^{n+1}$ *by* $(x,t)\in\mathbb{R}^n\times\mathbb{R}$ *and set*

$$E(x,t)=(4\pi t)^{-n/2}\exp(-|x|^2/4t),\quad t>0,\quad E(x,t)=0,\quad t\leqq 0.$$

Then E is locally integrable in $\mathbb{R}^{n+1}$, $E\in C^\infty(\mathbb{R}^{n+1}\setminus 0)$, *and*

$$(\partial/\partial t-\Delta_x)E=\delta_0. \tag{3.3.2}$$

Proof. That E is C^∞ in $\mathbb{R}^{n+1}\setminus 0$ follows from Corollary 1.1.2 as in the closely related Example 1.1.3. By (3.4.1)″ below the integral of $E(x,t)$ with respect to x is equal to 1 when $t>0$, so E is locally integrable and defines a distribution. When $t>0$ we have

$$\partial E/\partial x_j=-x_jE/2t,\quad \Delta_xE=-nE/2t+|x|^2E/4t^2=\partial E/\partial t$$

so $(\partial/\partial t-\Delta_x)E$ is supported by 0. When $\phi\in C_0^\infty$ we have

$$\begin{aligned}\langle(\partial/\partial t-\Delta_x)E,\phi\rangle&=-\langle E,\partial\phi/\partial t+\Delta_x\phi\rangle\\ &=\lim_{\varepsilon\to 0}\int_{t>\varepsilon}-E(x,t)(\partial\phi/\partial t+\Delta_x\phi)\,dx\,dt=\lim_{\varepsilon\to 0}\int E(x,\varepsilon)\,\phi(x,\varepsilon)\,dx\\ &=\lim_{\varepsilon\to 0}\int E(x,1)\,\phi(\sqrt{\varepsilon}\,x,\varepsilon)\,dx=\phi(0)\end{aligned}$$

by bounded convergence. The theorem is proved.

We shall now consider the closely related Schrödinger operator $i\partial/\partial t+\Delta_x$ or more generally operators of the form

$$L=\partial/\partial t-\sum_{j,k=1}^{n}A_{jk}\partial_j\partial_k \tag{3.3.2'}$$

where the symmetric matrix $A=(A_{jk})$ is constant and $\det A\neq 0$. In analogy to Theorem 3.3.3 we try to find a fundamental solution of the form

$$E(x,t)=ct^{-n/2}\exp(-\langle Bx,x\rangle/t)$$

for $t>0$ where B is another symmetric matrix. Then

$$\begin{aligned}&\partial_jE=-2E(Bx)_j/t,\qquad \partial_j\partial_kE=-2EB_{jk}/t+4E(Bx)_j(Bx)_k/t^2,\\ &LE=(2\,Tr\,BA-n/2)E/t+(\langle Bx,x\rangle-4\langle ABx,Bx\rangle)E/t^2.\end{aligned}$$

To make this vanish we must take B so that $4BAB=B$, that is, $B=A^{-1}/4$. Since E increases exponentially as $t\to +0$ unless $\operatorname{Re} B\geqq 0$ we must make this assumption. Now $\operatorname{Re}\langle Bx,\bar{x}\rangle=\operatorname{Re}\langle Ay,\bar{y}\rangle$ if $x=2Ay$ so this is equivalent to assuming that $\operatorname{Re} A\geqq 0$. However, even then we do not necessarily get a locally integrable function if $n>1$, so we need the following

Theorem 3.3.4. *If B is a symmetric non-singular matrix with* $\operatorname{Re} B\geqq 0$ *then*

$$(3.3.3)\qquad (\pi t)^{-n/2}(\det B)^{\frac{1}{2}}\int e^{-\langle Bx,x\rangle/t}\phi(x)\,dx\to\phi(0)\qquad \text{as } t\to +0,$$

if $\phi\in C_0^\infty(\mathbb{R}^n)$. *For any even integer* $k\geqq n$ *we have*

$$(3.3.4)\qquad t^{-n/2}\left|\int e^{-\langle Bx,x\rangle/t}\phi(x)\,dx\right|\leqq C_k\sum_{|\alpha|\leqq k}\left(\int|\partial^\alpha\phi|\,dx+\sup|\partial^\alpha\phi|\right).$$

Here $(\det B)^{\frac{1}{2}}$ *is defined as explained in Section 3.4.*

We postpone the proof a moment and give an immediate consequence:

Theorem 3.3.5. *Let A be a symmetric $n\times n$ matrix with* $\operatorname{Re} A\geqq 0$ *and* $\det A\neq 0$. *Then*

$$(3.3.5)\qquad \langle E,\phi\rangle=\int_0^\infty(4\pi t)^{-n/2}(\det A)^{-\frac{1}{2}}\,dt\left(\int e^{-\langle A^{-1}x,x\rangle/4t}\phi(x,t)\,dx\right),$$

$\phi\in C_0^\infty(\mathbb{R}^{n+1})$, *defines a distribution in* $\mathbb{R}^{n+1}$ *of order* $\leqq n+1$, *and*

$$(3.3.6)\qquad LE=\delta_0$$

if L is defined by (3.3.2)$'$.

The proof is an obvious modification of that of Theorem 3.3.3 when Theorem 3.3.4 is available. We therefore leave the proof for the reader and pass to the main step in the proof of Theorem 3.3.4.

Lemma 3.3.6. *If B satisfies the hypotheses of Theorem 3.3.4 then*

$$(3.3.7)\qquad \left|\int e^{-\langle Bx,x\rangle/t}\phi(x)\,dx\right|\leqq C_j t^j\sum_{|\alpha|\leqq 2j}N(\partial^\alpha\phi),\qquad \phi\in C_0^{2j}(\mathbb{R}^n),$$

provided that $\partial^\alpha\phi(0)=0$ *when* $|\alpha|<2j$. *Here* $j=0,1,\dots,0<t<1$ *and*

$$N(\psi)=\sup|\psi|+\int|\psi|\,dx.$$

Proof. (3.3.7) is obvious when $j=0$ so we assume $j>0$ and that (3.3.7) has already been proved for smaller values of j. We can write

$$\phi(x)=\sum_1^n x_k\phi_k(x)$$

where $\phi_k\in C_0^{2j-1}$, $\partial^\alpha\phi_k(0)=0$ if $|\alpha|+1<2j$ and

$$\sum_k \sum_{|\alpha|\leqq 2j-1} N(\partial^\alpha\phi_k)\leqq C\sum_{|\alpha|\leqq 2j} N(\partial^\alpha\phi).$$

In fact, when $|x|>\frac{1}{2}$ we could take $\phi_k^1=x_k\phi/|x|^2$, and when $|x|<1$ we could use the functions ϕ_k^0 given by Theorem 1.1.9. If $\chi\in C_0^\infty$ has support in the unit ball and $\chi(x)=1$ when $|x|<\frac{1}{2}$, then

$$\phi_k(x)=\chi(x)\,\phi_k^0(x)+(1-\chi(x))\,\phi_k^1(x)$$

will have the required properties. Writing $C=B^{-1}/2$ we have

$$x_k=\sum C_{ki}\partial\langle Bx,x\rangle/\partial x_i,$$

so an integration by parts gives

$$\int e^{-\langle Bx,x\rangle/t}x_k\phi_k(x)\,dx=t\sum C_{ki}\int e^{-\langle Bx,x\rangle/t}\,\partial\phi_k/\partial x_i\,dx.$$

Since $\partial\phi_k/\partial x_i$ satisfies the hypotheses of the lemma with j replaced by $j-1$, the estimate (3.3.7) now follows by the inductive hypothesis.

Proof of Theorem 3.3.4. First note that the proof of (3.3.7) is valid for all $\phi\in C^{2j}$ with derivatives tending fast to 0 at ∞ but not necessarily of compact support. In particular it is valid for functions such as $\exp(-\langle\lambda x,x\rangle)$ where λ is a diagonal matrix with positive diagonal elements λ_j. By (3.4.1.)″ we have

$$\int e^{-\langle Bx,x\rangle/t}e^{-\langle\lambda x,x\rangle}dx=(\pi t)^{n/2}(\det(B+\lambda t))^{-\frac{1}{2}}.$$

If we differentiate with respect to λ_j and put $\lambda_j=1$ afterwards, it follows that

$$\int e^{-\langle Bx,x\rangle/t}x^{2\alpha}e^{-|x|^2}dx=O(t^{n/2+|\alpha|}),\qquad t\to 0.$$
$$(\pi t)^{-n/2}\int e^{-\langle Bx,x\rangle/t}e^{-|x|^2}dx\to(\det B)^{-\frac{1}{2}},\qquad t\to 0.$$

In addition

$$\int e^{-\langle Bx,x\rangle/t}x^\alpha e^{-|x|^2}dx=0\qquad\text{if some }\alpha_j\text{ is odd.}$$

Now we write for the function ϕ in Theorem 3.3.4

$$\phi(x)=T(x)\,e^{-|x|^2}+\psi(x)$$

where T is a polynomial of degree $k-1$ and ψ vanishes of order k at 0. This means just that T is the Taylor polynomial of $\phi(x)\,e^{|x|^2}$. Since

$$\sum_{|\alpha|\leqq k} N(\partial^\alpha\psi)\leqq C\sum_{|\alpha|\leqq k} N(\partial^\alpha\phi)$$

and the coefficients of T can be estimated by the derivatives of ϕ at 0, the estimate (3.3.4) follows if we use (3.3.7) with $2j=k$ and ϕ replaced by ψ. Using Taylor polynomials of order $k+2$ instead, we obtain (3.3.3).

In Chapter VII we shall see that the preceding results are more easily accessible by means of the Fourier transformation.

3.4. Evaluation of Some Integrals

To avoid interruption of the main argument we have postponed some elementary computations to this section. The main point is the study of the integral

$$\int_{\mathbb{R}^n} e^{-|x|^2}\,dx=\Big(\int_{-\infty}^{\infty} e^{-t^2}\,dt\Big)^n$$

where $|x|^2=x_1^2+\ldots+x_n^2$. If c_n is the area of the unit sphere $S^{n-1}\subset\mathbb{R}^n$ then introduction of polar coordinates gives

$$\Big(\int_{-\infty}^{\infty} e^{-t^2}\,dt\Big)^n=\int_0^{\infty} e^{-r^2} r^{n-1} c_n\,dr=c_n/2\int_0^{\infty} t^{n/2-1}e^{-t}\,dt=c_n\Gamma(n/2)/2.$$

When $n=2$ this is equal to π so

$$\int_{-\infty}^{\infty} e^{-t^2}\,dt=\pi^{\frac{1}{2}} \tag{3.4.1}$$

and it follows that in general

$$c_n=2\pi^{n/2}/\Gamma(n/2). \tag{3.4.2}$$

When $n=3$ this gives $4\pi=2\pi^{\frac{3}{2}}/\Gamma(\frac{3}{2})$ so $2\,\Gamma(\frac{3}{2})=\pi^{\frac{1}{2}}$, or

$$\Gamma(\tfrac{1}{2})=\pi^{\frac{1}{2}}. \tag{3.4.3}$$

Thus

$$c_{2n}=2\pi^n/(n-1)!,\qquad c_{2n+1}=2^{n+1}\pi^n/((2n-1)\ldots 3.1). \tag{3.4.4}$$

The volume C_n of the unit ball in $\mathbb{R}^n$ is c_n/n, which follows for example from Gauss' formula applied to the radial vector field x. Thus

$$C_{2n}=\pi^n/n!,\qquad C_{2n+1}=2^{n+1}\pi^n/((2n+1)(2n-1)\ldots 3.1). \tag{3.4.5}$$

From (3.4.1) we obtain by a change of variables

$$\int_{-\infty}^{\infty} e^{-at^2}\,dt=(\pi/a)^{\frac{1}{2}},\qquad a>0. \tag{3.4.1$'$}$$

More generally, if A is a symmetric positive definite $n\times n$ matrix, then

$$\int_{\mathbb{R}^n} e^{-\langle Ax,x\rangle}\,dx=\pi^{n/2}(\det A)^{-\frac{1}{2}}. \tag{3.4.1$''$}$$

This is an immediate consequence of (3.4.1)$'$ if A has diagonal form, and we can always give A diagonal form by an orthogonal transformation. Now the set H of symmetric matrices A with $\operatorname{Re} A$ positive definite is an open convex set in the $n(n+1)/2$ dimensional complex vector space of symmetric $n\times n$ matrices. If $A\in H$ then $\det A\neq 0$ since $Ax=0$, $x\in\mathbb{C}^n$ implies

$$0=\operatorname{Re}\langle Ax,\bar{x}\rangle=\langle(\operatorname{Re} A)x,\bar{x}\rangle, \qquad \text{hence } x=0.$$

Since H is convex it follows that there is a unique analytic branch of $H\ni A\to(\det A)^{\frac{1}{2}}$ such that $(\det A)^{\frac{1}{2}}>0$ when A is real. Both sides of (3.4.1)$''$ are analytic when $A\in H$ so it follows that (3.4.1)$''$ is valid for all $A\in H$.

$(\det A)^{\frac{1}{2}}$ is also uniquely defined by continuity when A is in the closure of H, for if $\det A\neq 0$ we have two analytic branches of the square root in a neighborhood of A, only one of which agrees with the definition chosen in H. We shall compute $(\det A)^{\frac{1}{2}}$ when $A=iB$ is purely imaginary and non-singular. To do so we may assume that B has diagonal form for a real orthogonal transformation does not change $(\det A)^{\frac{1}{2}}$ when $A\in\overline{H}$ since it does not when A is real. Thus we assume that $\langle Bx,x\rangle=\sum b_j x_j^2$ with all $b_j\neq 0$. Then

$$\det(iB+\varepsilon I)=\prod(\varepsilon+ib_j)$$

and the square root which is positive when all b_j vanish has the argument

$$\tfrac{1}{2}\sum\arg(\varepsilon+ib_j)$$

where each term lies between $-\pi/2$ and $\pi/2$. When $\varepsilon\to 0$ this converges to

$$\tfrac{1}{2}\sum\frac{\pi}{2}\operatorname{sgn} b_j.$$

Now $\sum\operatorname{sgn} b_j$ is the *signature* $\operatorname{sgn} B$ of B, so we have proved that

$$(\det(iB))^{\frac{1}{2}}=|\det B|^{\frac{1}{2}}\exp(\pi i(\operatorname{sgn} B)/4). \tag{3.4.6}$$

Occasionally we shall also need some properties of the Γ function. First we observe that if $a>0$ and $b>0$ then

$$x_+^{a-1}*x_+^{b-1}=B(a,b)\,x_+^{a+b-1} \tag{3.4.7}$$

where

$$B(a,b)=\int_0^1 t^{a-1}(1-t)^{b-1}\,dt \tag{3.4.8}$$

is called the *beta function.* Taking the scalar product with e^{-t} in both sides of (3.4.7) we obtain by the definition of the Γ function

$$\Gamma(a)\Gamma(b)=B(a,b)\Gamma(a+b),$$

that is,

$$B(a,b)=\Gamma(a)\Gamma(b)/\Gamma(a+b). \tag{3.4.9}$$

Hence (3.4.7) may be written, with the notation (3.2.17),

$$\chi_+^{a-1}*\chi_+^{b-1}=\chi_+^{a+b-1}, \tag{3.4.10}$$

which will follow by analytic continuation for all $a,b\in\mathbb{C}$ when convolution has been defined for distributions in Section 4.2.

Taking $t=1/(1+s)$ as new variable we obtain when $0<a<1$

$$B(a,1-a)=\int_0^1 t^{a-1}(1-t)^{-a}dt=\int_0^\infty s^{-a}(1+s)^{-1}ds.$$

Integrating $s^{-a}/(1+s)$ in $\mathbb{C}$ slit along $\mathbb{R}_+$, first from $R-i0$ to 0, then to $R+i0$ and finally along the circle $|s|=R$ to $R-i0$, we obtain when $R\to\infty$

$$B(a,1-a)(1-e^{-2\pi ia})=2\pi ie^{-\pi ia},$$

which can be written in the form

$$\Gamma(a)\Gamma(1-a)=B(a,1-a)=\pi/\sin(\pi a). \tag{3.4.11}$$

By analytic continuation this remains valid when a is not an integer.

Notes

In Section 3.1 we just rewrote some basic real and complex analysis in the language of distribution theory. The discussion of boundary values of analytic functions will be continued in Chapter VIII where it plays an important role. The discussion of x_+^a in Section 3.2 goes back to Hadamard [1] and to Marcel Riesz [1]. Homogeneous distributions in several variables were also considered by them, and they were studied at great length by Gelfand and Šilov [2]. The reader can find more information and examples there. Sections 3.3 and 3.4 are also entirely classical in contents. For a proof that a product with reasonable algebraic properties cannot always be defined we refer to Schwartz [3].

Chapter IV. Convolution

Summary

In Section 1.3 we defined the convolution $u_1 * u_2$ of two continuous functions u_1 and u_2, one of which has compact support. The definition can be applied without change if $u_1 \in \mathscr{D}'$ (resp. $\mathscr{E}'$) and $u_2 \in C_0^\infty$ (resp. C^∞); we have $u_1 * u_2 \in C^\infty$ then. Section 4.1 is devoted to such convolutions. As in the case of functions this is an efficient method to approximate distributions by C^∞ functions. It can often be used to extend statements concerning smooth functions to distributions, particularly when translation invariant questions are concerned. As examples of this we give in Section 4.1 a discussion of convex, subharmonic and plurisubharmonic functions.

Convolution of two distributions u_1, u_2 one of which has compact support is defined in Section 4.2 so that the associativity

$$(u_1 * u_2) * \phi = u_1 * (u_2 * \phi), \qquad \phi \in C_0^\infty,$$

is preserved. It is then elementary to see that

$$\operatorname{supp}(u_1 * u_2) \subset \operatorname{supp} u_1 + \operatorname{supp} u_2.$$

Section 4.3 is devoted to the proof of the theorem of supports which states that when u_1 and u_2 both have compact supports, then there is equality if one takes convex hulls of the supports. The standard proofs of this depend on analytic function theory, and we shall return to them later on. The reader might therefore prefer to wait for Section 7.3 rather than studying the end of the proof of Theorem 4.3.3.

Section 4.4 is intended to present the basic methods used to derive results on existence and smoothness of solutions of constant coefficient partial differential equations from the properties of a fundamental solution. This is an important application of the convolution. The final Section 4.5 is then devoted to L^p estimates for convolutions. In addition to estimates related to Hölder's inequality we prove potential estimates of the Hardy-Littlewood-Sobolev type and derive from them relations between the L^p (or Hölder) classes of a function and its derivatives. These are basic particularly in the study of elliptic differential equations. They will be supplemented in Section 7.9 when we have Fourier analysis at our disposal.

4.1. Convolution with a Smooth Function

The convolution $u*\phi$ of a distribution $u\in\mathscr{D}'(\mathbb{R}^n)$ and a function $\phi\in C_0^\infty(\mathbb{R}^n)$ is defined by

$$(u*\phi)(x)=u(\phi(x-.))$$

where the right-hand side denotes u acting on $\phi(x-y)$ as a function of y. If u is a continuous function this agrees of course with the previous definition (1.3.1), and the properties of the convolution proved in Section 1.3 remain valid for distributions:

Theorem 4.1.1. *If* $u\in\mathscr{D}'(\mathbb{R}^n)$ *and* $\phi\in C_0^\infty(\mathbb{R}^n)$, *then* $u*\phi\in C^\infty(\mathbb{R}^n)$ *and*

$$\operatorname{supp}(u*\phi)\subset\operatorname{supp}u+\operatorname{supp}\phi. \tag{4.1.1}$$

For any multi-index α *we have*

$$\partial^\alpha(u*\phi)=(\partial^\alpha u)*\phi=u*(\partial^\alpha\phi). \tag{4.1.2}$$

Proof. It follows from Theorem 2.1.3 that $u*\phi\in C^\infty$ and that

$$\partial^\alpha(u*\phi)=u*\partial^\alpha\phi.$$

This proves (4.1.2) for the second equality in (4.1.2) follows at once from the definition of $\partial^\alpha u$. To prove (4.1.1) we note that $u*\phi(x)=0$ unless $x-y\in\operatorname{supp}\phi$ for some $y\in\operatorname{supp}u$, which means that $x\in\operatorname{supp}u+\operatorname{supp}\phi$. This is a closed set since $\operatorname{supp}\phi$ is compact. The theorem is proved.

Commutativity of the convolution would not make sense in the present asymmetric setup, but we shall prove the important associativity:

Theorem 4.1.2. *If* $u\in\mathscr{D}'(\mathbb{R}^n)$ *and* $\phi,\psi\in C_0^\infty(\mathbb{R}^n)$, *then*

$$(u*\phi)*\psi=u*(\phi*\psi).$$

The proof is an easy consequence of the following

Lemma 4.1.3. *If $\phi \in C_0^j(\mathbb{R}^n)$ and $\psi \in C_0^0(\mathbb{R}^n)$, then the Riemann sum*

$$\sum_{k \in \mathbb{Z}^n} \phi(x-kh)\, h^n \psi(kh) \tag{4.1.3}$$

converges to $\phi * \psi(x)$ in C_0^j when $h \to 0$.

Proof. The support of the sum (4.1.3) is contained in the compact set $\operatorname{supp}\phi + \operatorname{supp}\psi$. Since the function $(x,y) \to \phi(x-y)\psi(y)$ is uniformly continuous, the sum (4.1.3) converges uniformly to $\phi * \psi(x)$ when $h \to 0$. Differentiating the sum at most j times we obtain the same conclusion for the derivatives since $\partial^\alpha(\phi * \psi) = (\partial^\alpha \phi) * \psi$ when $|\alpha| \leqq j$. This proves the lemma.

Proof of Theorem 4.1.2. With the assumptions in the theorem we obtain from Lemma 4.1.3

$$\begin{aligned} u * (\phi * \psi)(x) &= \lim_{h \to 0} u\big(\textstyle\sum \phi(x - . - kh)\, h^n \psi(kh)\big) \\ &= \lim_{h \to 0} \textstyle\sum (u * \phi)(x-kh)\, h^n \psi(kh) = \int (u * \phi)(x-y)\psi(y)\,dy \end{aligned}$$

which proves the statement even for any $\psi \in C_0^0$.

We can now prove an analogue of Theorem 1.3.2 for distributions:

Theorem 4.1.4. *Let $0 \leqq \phi \in C_0^\infty$, $\int \phi\, dx = 1$. If $u \in \mathscr{D}'(\mathbb{R}^n)$ it follows that $u_\phi = u * \phi \in C^\infty(\mathbb{R}^n)$ and that $u_\phi \to u$ in $\mathscr{D}'(\mathbb{R}^n)$ as* $\operatorname{supp}\phi \to \{0\}$.

Proof. To clarify the computations we note that $u(\psi) = u * \check{\psi}(0)$ if $\psi \in C_0^\infty(\mathbb{R}^n)$ and $\check{\psi}(x) = \psi(-x)$. This gives

$$u_\phi(\psi) = u_\phi * \check{\psi}(0) = u * \phi * \check{\psi}(0) = u(\check{\phi} * \psi).$$

In view of Theorem 1.3.2 we have $\check{\phi} * \psi \to \psi$ in C_0^∞ as $\operatorname{supp}\phi \to \{0\}$ so it follows that $u_\phi(\psi) \to u(\psi)$ as claimed.

Theorem 4.1.4 shows that $\mathscr{D}'(\mathbb{R}^n)$ could have been defined by completion of $C^0(\mathbb{R}^n)$ or even of $C^\infty(\mathbb{R}^n)$ in the manner suggested by examples from physics given at the end of the introduction. This is also true for $\mathscr{D}'(X)$ if X is any open set in $\mathbb{R}^n$:

Theorem 4.1.5. *If $u \in \mathscr{D}'(X)$ there is a sequence $u_j \in C_0^\infty(X)$ such that $u_j \to u$ in $\mathscr{D}'(X)$.*

Proof. Choose a sequence $\chi_j \in C_0^\infty(X)$ such that on any compact subset of X we have $\chi_j = 1$ for all large j. Then choose $\phi_j \in C_0^\infty(\mathbb{R}^n)$ satisfying the hypothesis of Theorem 4.1.4 with so small support that

$$\operatorname{supp}\phi_j+\operatorname{supp}\chi_j\subset X \tag{4.1.4}$$

and $|x|<1/j$ if $x\in\operatorname{supp}\phi_j$. Since $\chi_j u\in\mathscr{E}'(X)\subset\mathscr{E}'(\mathbb{R}^n)$ we can form

$$u_j=(\chi_j u)*\phi_j$$

and obtain a function in $C_0^\infty(X)$ by (4.1.4) and (4.1.1). If $\psi\in C_0^\infty(X)$ we have as in the proof of Theorem 4.1.4

$$u_j(\psi)=(\chi_j u)(\check{\phi}_j*\psi)=u(\chi_j(\check{\phi}_j*\psi)).$$

Since $\operatorname{supp}\check{\phi}_j*\psi$ belongs to any neighborhood of $\operatorname{supp}\psi$ for large j, we have $\chi_j(\check{\phi}_j*\psi)=\check{\phi}_j*\psi$ then, and it follows that $u_j(\psi)\to u(\psi)$ as stated.

Remark. That $C_0^\infty(X)$ is dense in $\mathscr{D}'(X)$ follows also from the Hahn-Banach theorem since the dual space of $\mathscr{D}'(X)$ (with the weak topology) is $C_0^\infty(X)$ by an elementary fact concerning weak topologies. Also note that formal rules of computation such as (3.1.4) follow for distributions by means of Theorem 4.1.5 when they are known for C^∞ functions.

If $u\in\mathscr{D}'(X)$ and $\phi\in C_0^\infty(\mathbb{R}^n)$, the convolution $u*\phi$ is defined in

$$\{x;\, x-y\in X \text{ if } y\in\operatorname{supp}\phi\} \tag{4.1.5}$$

which is close to X when ϕ has small support. With obvious modifications all properties proved above when $X=\mathbb{R}^n$ remain valid.

Regularization by convolution can often be used to reduce questions concerning distributions to smooth functions. We shall give some important examples.

Theorem 4.1.6. *If u, $v\in\mathscr{D}'(X)$ where X is an open interval on $\mathbb{R}$ then $u'\geqq 0$ if and only if u is defined by an increasing function, and $v''\geqq 0$ if and only if v is defined by a convex function, that is, a continuous function with*

$$v(tx+(1-t)y)\leqq tv(x)+(1-t)v(y);\quad 0<t<1;\ x,y\in X. \tag{4.1.6}$$

Proof. a) Assume $u, v\in C^\infty$. Then $u'\geqq 0$ implies

$$u(x+y)-u(x)=y\int_0^1 u'(x+ty)\,dt\geqq 0 \quad \text{if } y\geqq 0,$$

and since v' is therefore increasing we conclude that

$$(v(x+y)-v(x))/y=\int_0^1 v'(x+ty)\,dt$$

is an increasing function of y if $y \geqq 0$. Hence

$$v(x+ty)-v(x) \leqq t(v(x+y)-v(x)) \quad \text{if } 0 \leqq t \leqq 1 \text{ and } x, x+y \in X$$

which is equivalent to convexity. The converse of these conclusions is obvious. Also note that if $0 \leqq \psi^+ \in C_0^\infty(\mathbb{R}^+)$, $\int \psi^+ \, dx = 1$, then

$$u * \psi_\varepsilon^+(x) = \int u(x-\varepsilon y) \psi^+(y) \, dy$$

is an increasing function of x and a decreasing function of ε (where it is defined), while if $0 \leqq \psi^e \in C_0^\infty(\mathbb{R})$, $\int \psi^e dx = 1$ and ψ^e is even we have that

$$v * \psi_\varepsilon^e(x) = \int v(x-\varepsilon y) \psi^e(y) \, dy = \int_{y>0} (v(x-\varepsilon y) + v(x+\varepsilon y)) \psi^e(y) \, dy$$

is a convex function of x and an increasing function of ε when $\varepsilon > 0$ since

$$\frac{d}{dy}(v(x+y)+v(x-y)) = v'(x+y) - v'(x-y) \geqq 0 \quad \text{if } y \geqq 0.$$

b) In general we choose ϕ as in Theorem 4.1.4 and form the regularizations $u_\phi = u * \phi$, $v_\phi = v * \phi$. Assume $u' \geqq 0$, $v'' \geqq 0$. Then $u'_\phi = u' * \phi \geqq 0$, $v''_\phi = v'' * \phi \geqq 0$ so $u_\phi * \psi_\varepsilon^+(x)$ is an increasing function of x and a decreasing function of ε while $v_\phi * \psi_\varepsilon^e(x)$ is a convex function of x which increases with ε. Letting $\operatorname{supp} \phi \to \{0\}$ we conclude that $u * \psi_\varepsilon^+$ and $v * \psi_\varepsilon^e$ have the same properties, so when $\varepsilon \downarrow 0$ we obtain

$$u * \psi_\varepsilon^+ \uparrow u_0, \quad v * \psi_\varepsilon^e \downarrow v_0$$

where u_0 is increasing, v_0 satisfies (4.1.6), and

$$\langle u, \chi \rangle = \int u_0 \chi \, dx, \quad \langle v, \chi \rangle = \int v_0 \chi \, dx, \quad 0 \leqq \chi \in C_0^\infty(X).$$

It follows that u_0 is finite everywhere because it would otherwise be $+\infty$ in an interval, and v_0 is finite since by (4.1.6) it would otherwise be $-\infty$ in an interval. Now (4.1.6) implies continuity for

$$v(x) - v(x-y) \leqq (v(x+hy) - v(x))/h \leqq v(x+y) - v(x), \quad |h| < 1,$$

so u_0 and v_0 have the stated properties. Conversely, if u and v are defined by increasing and convex functions respectively, then so are u_ϕ and v_ϕ. Thus $u'_\phi \geqq 0$ and $v''_\phi \geqq 0$ which gives $u' \geqq 0$ and $v'' \geqq 0$ when $\operatorname{supp} \phi \to \{0\}$.

Theorem 4.1.7. *If $v \in \mathscr{D}'(X)$ where X is open in $\mathbb{R}^n$ and if*

(4.1.7) $$\sum \sum y_j y_k \partial_j \partial_k v \geqq 0 \quad \textit{for all } y \in \mathbb{R}^n,$$

then v is defined by a continuous function satisfying (4.1.6) *on every line segment in X and conversely. One calls v a convex function.*

Proof. We may assume that X is convex. If $v \in C^\infty$ then (4.1.7) means precisely that

$$\frac{d^2}{dt^2} v(x+ty) \geqq 0 \quad \text{when } x+ty \in X,$$

so the statement follows from Theorem 4.1.6. If $0 \leqq \psi \in C_0^\infty$ is an even function with $\int \psi\, dx = 1$, then

$$v * \psi_\varepsilon(x) = \int v(x-\varepsilon y)\psi(y)\, dy$$

is also a convex function and it increases with ε. If v is just known to be in $\mathscr{D}'(X)$ we can now argue exactly as in the proof of Theorem 4.1.6. v_ϕ satisfies (4.1.7) so $v_\phi * \psi_\varepsilon$ is a convex function which increases with ε, hence $v * \psi_\varepsilon$ is convex and increases with ε. The decreasing limit v_0 as $\varepsilon \downarrow 0$ defines v and satisfies (4.1.6) so v_0 is finite everywhere and upper semicontinuous. This implies continuity since for sufficiently small $|y|$

$$v(x+hy) - v(x) \geqq h(v(x) - v(x-y)) \geqq -Ch, \qquad 0 < h < 1.$$

The converse is obvious.

We shall now prove an analogue of the second part of Theorem 4.1.6 for several variables.

Theorem 4.1.8. *Let X be an open set in $\mathbb{R}^n$. If $u \in \mathscr{D}'(X)$ is real and $\Delta u \geqq 0$ it follows that u is defined by a subharmonic function u_0, that is, an upper semi-continuous function with values in $[-\infty, \infty)$ such that*

$$M(x,r) = \int_{|\omega|=1} u_0(x+r\omega)\, d\omega / c_n, \qquad c_n = \int_{|\omega|=1} d\omega,$$

is an increasing function of r for $x \in X$ and $0 \leqq r < d(x, \complement X)$, the distance from x to ∂X. Conversely, if u_0 is an upper semi-continuous function with values in $[-\infty, \infty)$ which is not identically $-\infty$ in any component of X and if $u_0(x) \leqq M(x,r)$ when $r < d(x, \complement X)$, then $u_0 \in L^1_{\text{loc}}(X)$ and $\Delta u \geqq 0$ for the distribution u defined by u_0. The function u_0 is uniquely determined by u at every point. If K is a compact subset of X then the maximum principle is valid,

$$\sup_{\partial K} u_0 = \sup_K u_0.$$

Proof. Assume first that $u \in C^\infty$ and that $\Delta u \geqq 0$. Let $0 < r < R$ and set

$$v(x) = 0. \quad |x| > R; \qquad v(x) = e(R) - E(x), \quad r < |x| < R;$$
$$v(x) = e(R) - e(r), \quad |x| < r;$$

where E is defined in Theorem 3.3.2, $E(x)=e(|x|)$. Then v is continuous so grad v is the function which is

$$\operatorname{grad} v = -\operatorname{grad} E, \quad r<|x|<R$$

and 0 otherwise. Again by Theorem 3.1.9

$$\begin{aligned}\Delta v = \operatorname{div} \operatorname{grad} v &= \langle -\operatorname{grad} E, -x/|x|\rangle dS_R + \langle -\operatorname{grad} E, x/|x|\rangle dS_r \\ &= dS_R/c_n R^{n-1} - dS_r/c_n r^{n-1}\end{aligned}$$

where dS_r and dS_R are the Euclidean surface measures on the spheres $|x|=r$ and $|x|=R$. When $d(x, \complement X)>R$ we have since $\Delta u \geqq 0$ and $v \geqq 0$

$$0 \leqq (\Delta u) * v(x) = u * \Delta v(x) = M(x,R) - M(x,r)$$

which proves that $M(x,r)$ is increasing for $r>0$, hence for $r \geqq 0$ by the continuity. Note that if $0 \leqq \psi \in C_0^\infty$, $\int \psi\, dx = 1$, and ψ is a function of $|x|$ then

$$u*\psi_\varepsilon(x) = \int u(x-\varepsilon y)\psi(y)\,dy$$

is an increasing function of ε. This follows if we introduce polar coordinates.

On the other hand, we have if $u \in C^\infty$

$$\begin{aligned}\int u(x+r\omega)\,d\omega &= \int \Big(u(x) + r\sum \omega_j \partial_j u(x) + \frac{r^2}{2}\sum \omega_j\omega_k \partial_j\partial_k u(x) + O(r^3)\Big)\,d\omega \\ &= c_n(u(x) + r^2 \Delta u(x)/2n + O(r^3)).\end{aligned}$$

In fact, $\int \omega_j \omega_k\, d\omega = 0$ if $j \neq k$, and

$$\int \omega_j^2\, d\omega = n^{-1} \int \sum \omega_j^2\, d\omega = c_n/n.$$

Hence

$$\Delta u(x) = \lim_{r\to 0} 2n(M(x,r)-u(x))/r^2 \geqq 0 \qquad \text{if } u(x) \leqq M(x,r).$$

Now let $u \in \mathscr{D}'(X)$ and $\Delta u \geqq 0$. If ϕ satisfies the conditions in Theorem 4.1.4 then $u_\phi = u*\phi \in C^\infty$ and $\Delta u_\phi \geqq 0$ where u_ϕ is defined. Hence

$$u_\phi * \psi_\varepsilon$$

is an increasing function of ε so letting $\operatorname{supp} \phi \to \{0\}$ we conclude that $u*\psi_\varepsilon$ is an increasing function of ε, and

$$\int (u*\psi_\varepsilon)(x+r\omega)\,d\omega$$

is an increasing function of r in the interval $[0,a)$ where it is defined. Hence

$$u*\psi_\varepsilon \downarrow u_0 \quad \text{as } \varepsilon \downarrow 0,$$

where u_0 is upper semicontinuous, $M(x,r)$ is an increasing function of r when $0 \leqq r < d(x, \complement X)$, and

$$\langle u, \chi\rangle = \int u_0(x)\chi(x)\,dx \qquad \text{if } 0 \leqq \chi \in C_0^\infty(X).$$

Hence $u_0 \in L^1_{\text{loc}}(X)$.

Assume now that u_0 is an upper semicontinuous function in a connected open set X such that $u_0(x) \leqq M(x,r)$ for $0 < r < d(x, \complement X)$ and $u_0 \not\equiv -\infty$. If $u_0(x) > -\infty$ it follows that u_0 is integrable in the ball with center x and any radius $r < d(x, \complement X)$. The open set $X_0 \subset X$ of points such that u_0 is integrable in a neighborhood is therefore closed, for if $x \in X$ is a limit point of X_0 then we can find $y \in X_0$ with $u_0(y) > -\infty$ and $|x-y| < d(y, \complement X)$. Hence $X_0 = X$ so $u_0 \in L^1_{\text{loc}}$. If ψ is chosen as above then

$$u_0(x) \leqq u_0 * \psi_\varepsilon(x)$$

by the mean value property, and

$$\overline{\lim_{\varepsilon\to 0}}\, u_0 * \psi_\varepsilon(x) \leqq u_0(x)$$

since u_0 is upper semicontinuous. Hence

$$u_0 * \psi_\varepsilon(x) \to u_0(x) \qquad \text{as } \varepsilon \to 0$$

so u_0 is determined by the corresponding distribution. Since $u_0 * \psi_\varepsilon$ inherits the mean value property from u_0, we have

$$0 \leqq \Delta(u_0 * \psi_\varepsilon) = (\Delta u_0) * \psi_\varepsilon \to \Delta u_0 \qquad \text{as } \varepsilon \to 0.$$

When proving the maximum principle we may assume that $\sup_K u_0 = 0$. Then $u_0(x) = 0$ for some $x \in K$ since u_0 is upper semicontinuous. If r is the distance from x to ∂K then

$$0 = u_0(x) \leqq \int_{|\omega|=1} u_0(x + r\omega)\,d\omega \Big/ \int_{|\omega|=1} d\omega.$$

For some ω_0 we have $x + r\omega_0 \in \partial K$ by the definition of r. If $u_0(x + r\omega_0) < 0$ then $u_0 < 0$ in a neighborhood of $x + r\omega_0$, and since $u \leqq 0$ in K it follows that $\int u_0(x + r\omega)\,d\omega < 0$. This is a contradiction proving that $u_0(x + r\omega_0) = 0$, hence $\sup_{\partial K} u_0 = 0$. The proof is complete.

For later reference we also give a property of subharmonic functions with a closely related proof:

Theorem 4.1.9. *Let v_j be a sequence of subharmonic functions in a connected open set $X \subset \mathbb{R}^n$, which have a uniform upper bound on any compact set. Then*

a) *if* v_j *does not converge to* $-\infty$ *uniformly on every compact set in* X *then there is a subsequence* v_{j_k} *which is convergent in* $L^1_{\text{loc}}(X)$.

b) *if* v *is a subharmonic function and* $v_j \to v$ *in* $\mathscr{D}'(X)$, *then* $v_j \to v$ *in* $L^1_{\text{loc}}(X)$,

$$\varlimsup_{j\to\infty} v_j(x) \leqq v(x), \qquad x \in X, \tag{4.1.8}$$

with the two sides equal and finite almost everywhere. More generally,

$$\varlimsup_{j\to\infty} \sup_K (v_j - f) \leqq \sup_K (v-f) \tag{4.1.9}$$

for every compact set $K \subset X$ *and every continuous function* f *on* K.

Proof. a) By hypothesis one can find j_k and x_k such that all x_k belong to a compact subset of X and $v_{j_k}(x_k)$ is bounded. We may assume that $x_k \to x_0 \in X$ and to simplify notation that $j_k = k$. If B is a closed ball $\subset X$ with center x_0 it follows that $\int_B v_k\,dx$ is bounded from below. In fact, for large k there is a closed ball B_k with center at x_k such that $B \subset B_k \subset X$ and $m(B_k) \to m(B)$. Hence

$$\int_B v_k\,dx = \int_{B_k} v_k\,dx - \int_{B_k \setminus B} v_k\,dx \geqq m(B_k)\,v_k(x_k) - \int_{B_k \setminus B} v_k\,dx$$

is bounded from below, which implies that $\int_B |v_k|\,dx$ is bounded since v_k has a uniform upper bound on B. If $x \in B$ then the mean value of v_k over a ball with center at x and radius r is an increasing function of r for which we have a bound when r is small. Hence we obtain a bound for the L^1 norm of v_k over any such ball contained in X. The argument used in the proof of Theorem 4.1.8 to show that $u_0 \in L^1_{\text{loc}}(X)$ now gives that v_k is bounded in $L^1_{\text{loc}}(X)$. Hence there is a subsequence v_{j_k} converging weakly as a measure to a limit v with $\Delta v = \lim \Delta v_{j_k} \geqq 0$ (Theorem 2.1.9). By Theorem 4.1.8 v is a subharmonic function, so b) will prove that $v_{j_k} \to v$ in L^1_{loc}.

b) Choose ψ_δ as in the proof of Theorem 4.1.8. Then

$$v_j(x) \leqq v_j * \psi_\delta(x) \to v * \psi_\delta(x)$$

uniformly on compact sets in X as $j \to \infty$, if δ is small enough. If $0 \leqq \chi \in C_0^\infty$ then

$$\begin{aligned}&\int (v*\psi_\delta(x) + \varepsilon - v_j(x))\,\chi(x)\,dx\\ &\quad\to \int (v*\psi_\delta(x) + \varepsilon - v(x))\,\chi(x)\,dx, \qquad j\to\infty,\end{aligned}$$

and if $\varepsilon > 0$ the integrand is positive for large j. Hence

$$\varlimsup_{j\to\infty} \int |v - v_j|\,\chi\,dx \leqq 2\int |v*\psi_\delta + \varepsilon - v|\,\chi\,dx.$$

Since ε and δ are arbitrary it follows that $v \to v_j$ in L^1_{loc}. By Dini's theorem

$$\sup_K (v_j - f) \leqq \sup_K (v_j * \psi_\delta - f)$$
$$\to \sup_K (v * \psi_\delta - f) \leqq \sup_K (v - f) + \varepsilon, \qquad \delta < \delta_\varepsilon,$$

which proves (4.1.8) and (4.1.9). If $0 \leqq \chi \in C_0^\infty(X)$ we obtain by Fatou's lemma

$$\int \overline{\lim}\, v_j \chi\, dx \geqq \overline{\lim} \int v_j \chi\, dx = \int v \chi\, dx$$

so using (4.1.8) we conclude that $\overline{\lim}\, v_j(x) = v(x)$ almost everywhere when $\chi(x) > 0$. The proof is complete.

Example 4.1.10. If f is an analytic function in $X \subset \mathbb{C}$ then $\log|f|$ is subharmonic and in any component where $f \not\equiv 0$ we have

$$\Delta \log|f| = 2\pi \sum m_j \delta_{z_j}.$$

where z_j are the zeros and m_j the multiplicities.

Proof. Near a point where $f \neq 0$ there is an analytic branch g of $\log f$, hence $\log|f| = \operatorname{Re} g$ is harmonic. In a neighborhood of z_j we can write

$$f(z) = (z - z_j)^{m_j} g(z), \qquad \log|f(z)| = m_j \log|z - z_j| + \log|g(z)|$$

where $g(z_j) \neq 0$. By Theorem 3.3.2 it follows that $\Delta \log|f| = 2\pi m_j \delta_{z_j}$ there.

If N is a positive integer then $\log|f|^{1/N} = N^{-1} \log|f|$ is of course also subharmonic with Laplacean $2\pi \sum m_j/N\, \delta_{z_j}$. We can approximate any measure by such measures so it is quite plausible that subharmonic functions can be approximated by functions of the form $\log|f|^{1/N}$ in $X \subset \mathbb{C}$. We shall have more to say about this topic in Section 15.1.

Theorem 4.1.11. *Let X be an open set in $\mathbb{C}^n$. Every real $u \in \mathscr{D}'(X)$ such that*

$$\sum w_j \bar{w}_k \partial^2 u / \partial z_j \partial \bar{z}_k \geqq 0 \text{ in } X, \quad \text{if } w \in \mathbb{C}^n, \tag{4.1.10}$$

can be defined by a plurisubharmonic function u_0, that is, an upper semicontinuous function such that $\mathbb{C} \ni t \to u(z + tw)$ is subharmonic where it is defined, for arbitrary z, $w \in \mathbb{C}^n$. Conversely, every such function u_0 which is not identically $-\infty$ in a component of X is in L^1_{loc} and defines a distribution satisfying (4.1.10). *The function u_0 is uniquely determined by u.*

Proof. The statement is obvious if $u \in C^\infty$. Note that (4.1.10) gives

$$\Delta u \geqq 0$$

if we choose w equal to any basis vector and add. The approximation by regularization used in the proof of Theorem 4.1.8 is therefore applicable without change to prove the theorem in general. This is left as an exercise for the reader.

As an example we shall now discuss $u=\log|f|$ when f is analytic $\neq 0$ in an open set $X\subset\mathbb{C}^n$. Near a point where $f\neq 0$ we can choose an analytic branch g of $\log f$, so $\log|f|=\operatorname{Re} g$ and (4.1.10) is then zero. Thus the sum in (4.1.10) is supported by the zero set. Let 0 be a zero, of order k say. In the Taylor expansion

$$f=f_k+f_{k+1}+\dots$$

the homogeneous part f_k of order k is not identically 0 then, and we shall compute (4.1.10) near 0 when $f_k(w)\neq 0$. This is sufficient since a quadratic form (also with distribution coefficients) is determined by a finite number of its values. Let w be the z_1 axis and set $z'=(z_2,\dots,z_n)$. For fixed z'

$$\frac{\partial^2}{\partial z_1\partial\bar z_1}\log|f(z_1,z')|=\frac{\pi}{2}\sum m_j(z')\delta_{\zeta_j}$$

where $\zeta_j(z')$ are the small zeros of $f(\zeta,z')=0$ and m_j the multiplicities, which are $\leqq k$ since $\partial^k f/\partial z_1^k\neq 0$. The "counting" measure on the right depends continuously on z' since $\log|f(z_1,z')|$ does (in the distribution topology by Theorem 4.1.9), and for $\phi\in C_0^\infty$ with support close to 0 we have

$$\langle\partial^2\log|f|/\partial z_1\partial\bar z_1,\phi\rangle=\frac{\pi}{2}\int\sum m_j(z')\phi(\zeta_j(z'),z')\,d\lambda(z') \tag{4.1.11}$$

where $d\lambda$ is the Lebesgue measure in $\mathbb{C}^{n-1}$.

Now we distinguish two cases,

(i) $m_1\equiv k$, $\zeta_1(z')$ is analytic and $f(z)=(z_1-\zeta_1(z'))^k g(z)$, $g\neq 0$.

(ii) $m_1<k$ except for z' in a null set.

One of these must occur. In fact, a zero of order k is a zero of

$$\partial^{k-1}f(z)/\partial z_1^{k-1}$$

so the implicit function theorem shows that $z_1=\zeta(z')$ for some analytic function ζ. Now either

$$f=\partial f/\partial z_1=\dots=\partial^{k-1}f/\partial z_1^{k-1}\equiv 0\quad\text{when } z_1=\zeta(z'),$$

and then we have case (i), or else one of these functions is not identically zero and then we have case (ii). In case (ii) we omit from the integration in (4.1.11) the set where $m_1=k$. Repeating the argument a finite number of times, until the order of the zero cannot be lowered any longer because one comes across case (i), we obtain

Theorem 4.1.12. *Let f be an analytic function $\not\equiv 0$ in an open set $X \subset \mathbb{C}^n$. Let Z be the set of zeros z_0 of f such that in a neighborhood of z_0 we can write*

$$f(z) = g(z)\,h(z)^m$$

where $g(z_0) \neq 0$, $h(z_0) = 0$ and $\partial h(z_0)/\partial z \neq 0$, and let dS be the Euclidean surface measure on Z. Then dS is locally integrable in X and

$$\sum w_j \bar{w}_k \partial^2 \log|f|/\partial z_j \partial \bar{z}_k = \frac{\pi}{2}\mu(w)\,dS \tag{4.1.12}$$

where near z_0

$$\mu(w) = m|\langle w, h'\rangle|^2/|h'|^2.$$

In particular $\Delta \log|f| = 2\pi m\,dS$.

Proof. It only remains to note that (4.1.12) follows from (4.1.11) in case (i) above since $dS = (1 + |\partial\zeta_1/\partial z'|^2)\,d\lambda(z')$.

Every plurisubharmonic function in $X \subset \mathbb{C}^n$ is of course subharmonic as a function in $X \subset \mathbb{R}^{2n}$, but it has additional properties:

Theorem 4.1.13. *If u is plurisubharmonic in $X \subset \mathbb{C}^n$ and μ is the positive measure Δu, then*

$$r^{2-2n} \int_{|z-\zeta|<r} d\mu(z)$$

is an increasing function of r for $0 < r < d(\zeta, \complement X)$.

Proof. We may assume that $\zeta = 0$. a) Assume first that $u \in C^\infty$ and that u is a function of $|z|$ only. Write $u(z) = F(|z|^2)$. Then $F \in C^\infty$ and the plurisubharmonicity of u means that

$$F'(|z|^2)|w|^2 + F''(|z|^2)|\langle w, \bar{z}\rangle|^2 \geqq 0,$$

that is, $F'(s) \geqq 0$ and $F'(s) + sF''(s) \geqq 0$. Thus $sF'(s)$ is a positive increasing function of s. (Note that this condition is independent of n.) Now

$$\Delta u = 4(nF'(s) + sF''(s)), \qquad s = |z|^2,$$

$$\iint_{|z|<r} \Delta u\,dx\,dy = 4\int_0^{r^2} (nF'(s) + sF''(s))\,d(C_{2n}s^n)$$

where C_{2n} is the volume of the unit ball in $\mathbb{R}^{2n}$. Hence

$$r^{2-2n} \iint_{|z|<r} \Delta u\,dx\,dy = 4nC_{2n}r^2F'(r^2) = 4\pi C_{2n-2}r^2F'(r^2)$$

where the last equality follows from (3.4.5). This is an increasing function. We shall use later on that

$$\Big(\iint_{|z|<r} \Delta u/2\pi\,dx\,dy\Big)/(r^{2n-2}C_{2n-2}) = 2r^2F'(r^2). \tag{4.1.13}$$

b) Assume still that $u \in C^\infty$ but not that u is a function of $|z|$ only. If U is a unitary transformation, then $u(Uz)$ is also plurisubharmonic and the Laplacean is $(\Delta u)(Uz)$. Hence

$$v(z) = \int u(Uz)\,dU,$$

where dU is the Haar measure on the unitary group, is a C^∞ function of $|z|$ only, and

$$\int_{|z|<r} \Delta v\,dx\,dy = \int_{|z|<r} \Delta u\,dx\,dy$$

so the assertion follows from the case a) already studied. If $\psi \in C_0^\infty(\mathbb{R})$ is a decreasing function of r^2 it follows that

$$r^{2-2n} \int \Delta u \psi(|x|/r)\,dx = \int_{R>0} -d\psi(R) r^{2-2n} \int_{|x|<Rr} \Delta u\,dx$$

is an increasing function of r.

c) For a general u we form the regularizations u_ϕ according to Theorem 4.1.4. Then $u_\phi \to u$ in $\mathscr{D}'$ so $\Delta u_\phi \to \Delta u$ in $\mathscr{D}'$. Hence

$$r^{2-2n} \int \psi(|x|/r)\,d\mu(x)$$

is an increasing function of r. Letting ψ increase to the characteristic function of $(-1, 1)$ we have proved the theorem.

It is customary and convenient to use the normalization

$$\Theta(u, r, \zeta) = (r^{2n-2} C_{2n-2})^{-1} \int_{|z-\zeta|<r} \Delta u/2\pi \tag{4.1.14}$$

which by Theorem 4.1.13 is an increasing function of r.

Proposition 4.1.14. *If u is a plurisubharmonic function such that e^u is homogeneous of degree $k \geqq 0$, then $\Theta(u, r, 0) = k$.*

Proof. By averaging over the unitary group as in part b) of the proof of Theorem 4.1.13 we reduce the proof to the case where u is a function of $|z|$, thus $u(z) = k \log|z| + C$. Then $F(s) = (k/2)\log s + C$ so $\Theta(u, r, 0) = k$ by (4.1.13).

Theorem 4.1.15. *If f is an analytic function in $X \subset \mathbb{C}^n$, and not identically 0 in any component, then*

$$\Theta(\log|f|, r, \zeta) \to k \quad \text{when } r \to 0, \qquad \text{if } \zeta \in X, \tag{4.1.15}$$

where k is the order of the zero at ζ. If f is a polynomial then the limit as $r \to \infty$ is the degree of the polynomial.

Proof. We may assume that $\zeta = 0$. The definition of k means that

$$f(z) = f_k(z) + O(|z|^{k+1}), \qquad z \to 0,$$

where f_k is a homogeneous polynomial of degree 0 which is not identically 0. Put

$$F_r(z)=f(rz)/r^k.$$

Then F_r converges locally uniformly to f_k as $r\to 0$, so $\log|F_r|\to\log|f_k|$ in $\mathscr{D}'$ in view of Theorem 4.1.9. Hence the same is true for the Laplaceans and we conclude as in part c) of the proof of Theorem 4.1.13 that when $r\to 0$

$$\Theta(\log|F_r|,1,0)\to\Theta(\log|f_k|,1,0)=k$$

where the last equality is a consequence of Proposition 4.1.14. Now

$$\Theta(\log|F_r|,1,0)=\Theta(\log|f|,r,0)$$

which follows immediately if $\log|f|$ is replaced by a smooth approximation. This proves (4.1.15). If f is a polynomial we can let $r\to\infty$ instead and obtain as limit of F_r the homogeneous part of highest degree, which proves the last statement.

4.2. Convolution of Distributions

To define the convolution of two distributions we shall use the properties of the convolution

$$C_0^\infty(\mathbb{R}^n)\ni\phi\to u*\phi\in C^\infty(\mathbb{R}^n)$$

defined in Section 4.1 when $u\in\mathscr{D}'(\mathbb{R}^n)$. It is obvious in view of (4.1.2) that $u*\phi_j\to 0$ in $C^\infty(\mathbb{R}^n)$ if $\phi_j\to 0$ in $C_0^\infty(\mathbb{R}^n)$. (See Theorems 2.3.1 and 2.1.4 for the definition of convergence.) If $h\in\mathbb{R}^n$ we define the translation operator τ_h by $(\tau_h\phi)(x)=\phi(x-h)$ (which is convolution by δ_h) and obtain

$$u*(\tau_h\phi)=\tau_h(u*\phi).$$

Thus $u*$ commutes with translations. Conversely, we have

Theorem 4.2.1. *If U is a linear map from $C_0^\infty(\mathbb{R}^n)$ to $C^\infty(\mathbb{R}^n)$ which is continuous in the sense that $U\phi_j\to 0$ in $C(\mathbb{R}^n)$ when $\phi_j\to 0$ in $C_0^\infty(\mathbb{R}^n)$, and if U commutes with all translations, then there exists a unique distribution u such that $U\phi=u*\phi$, $\phi\in C_0^\infty(\mathbb{R}^n)$.*

Proof. If such a distribution exists we must have $u(\check{\phi})=U\phi(0)$. (We recall the notation $\check{\phi}(x)=\phi(-x)$.) Now the linear form

$$C_0^\infty\ni\phi\to(U\check{\phi})(0)$$

is by hypothesis a distribution u. From the fact that $(U\phi)(0)=(u*\phi)(0)$ we obtain by replacing ϕ by $\tau_h\phi$ and using the commutativity with τ_h that

$$(U\phi)(-h)=(\tau_h U\phi)(0)=(U\tau_h\phi)(0)=(u*\tau_h\phi)(0)=(u*\phi)(-h)$$

which proves that $U\phi=u*\phi$, $\phi\in C_0^\infty(\mathbb{R}^n)$. The proof is complete.

If $u\in\mathscr{E}'(\mathbb{R}^n)$ it follows from (4.1.1) that $\phi\to u*\phi$ is a continuous map from $C_0^\infty(\mathbb{R}^n)$ to $C_0^\infty(\mathbb{R}^n)$, that is, sequences converging to 0 are mapped to other such sequences. The convolution $u*\phi$ is also defined for arbitrary $\phi\in C^\infty(\mathbb{R}^n)$ then and gives a continuous map from $C^\infty(\mathbb{R}^n)$ to $C^\infty(\mathbb{R}^n)$.

There is a unique way to define the convolution of two distributions u_1 and u_2, one of which has compact support, so that the associativity

$$(u_1*u_2)*\phi=u_1*(u_2*\phi)$$

remains valid for $\phi\in C_0^\infty(\mathbb{R}^n)$. In fact, the mapping

$$C_0^\infty(\mathbb{R}^n)\ni\phi\to u_1*(u_2*\phi)$$

is linear, translation invariant and continuous because it is the composition of two such mappings. Hence there is a unique $u\in\mathscr{D}'(\mathbb{R}^n)$ such that

$$u_1*(u_2*\phi)=u*\phi,\qquad \phi\in C_0^\infty(\mathbb{R}^n). \tag{4.2.1}$$

Definition 4.2.2. The convolution u_1*u_2 of two distributions u_1 and u_2 one of which has compact support is defined to be the unique distribution u such that (4.2.1) is valid.

By Theorem 4.1.2 the definition is consistent with our original one when $u_2\in C_0^\infty$, and a simple modification of Theorem 4.1.2 shows that it is also consistent with our earlier definition when $u_1\in\mathscr{E}'(\mathbb{R}^n)$ and $u_2\in C^\infty(\mathbb{R}^n)$. Somewhat more generally we have

Theorem 4.2.3. *If $u_1\in\mathscr{D}'^k(\mathbb{R}^n)$, $u_2\in C_0^k(\mathbb{R}^n)$ (or $u_1\in\mathscr{E}'^k(\mathbb{R}^n)$, $u_2\in C^k(\mathbb{R}^n)$) then u_1*u_2 is the continuous function $x\to u_1(u_2(x-.))$.*

Proof. If this function is denoted by u then the proof of Theorem 4.1.2 shows without change that when $\psi\in C_0^\infty$, $u_1\in\mathscr{D}'^k(\mathbb{R}^n)$, $u_2\in C_0^k(\mathbb{R}^n)$ then

$$u*\psi=u_1*(u_2*\psi).$$

This proves the first part of the statement and the other follows in the same way.

By its definition the convolution is associative, that is,

$$u_1*(u_2*u_3)=(u_1*u_2)*u_3$$

if all the distributions u_j except at most one have compact support.

Theorem 4.2.4. *The convolution is commutative, that is,*

$$u_1*u_2=u_2*u_1$$

if one of the distributions u_1, u_2 has compact support. We have

$$\operatorname{supp}(u_1*u_2)\subset\operatorname{supp}u_1+\operatorname{supp}u_2. \tag{4.2.2}$$

Proof. To prove that two distributions v_1 and v_2 are equal it suffices to show that

$$v_1*(\phi*\psi)=v_2*(\phi*\psi) \quad \text{when } \phi,\psi\in C_0^\infty.$$

For then we have $(v_1*\phi)*\psi=(v_2*\phi)*\psi$ by Theorem 4.1.2, hence $v_1*\phi=v_2*\phi$ and so $v_1=v_2$. Now we have

$$\begin{aligned}(u_1*u_2)*(\phi*\psi)&=u_1*(u_2*(\phi*\psi))=u_1*((u_2*\phi)*\psi)\\&=u_1*(\psi*(u_2*\phi))=(u_1*\psi)*(u_2*\phi)\end{aligned}$$

where in addition to Theorem 4.1.2 we have used the commutativity of convolution of functions. In the same way we obtain

$$\begin{aligned}(u_2*u_1)*(\phi*\psi)&=(u_2*u_1)*(\psi*\phi)=(u_2*\phi)*(u_1*\psi)\\&=(u_1*u_2)*(\phi*\psi)\end{aligned}$$

which proves that $u_1*u_2=u_2*u_1$. To prove the last statement we choose ϕ as in Theorem 4.1.4 and note that since $(u_1*u_2)*\phi=u_1*(u_2*\phi)$ we have

$$\operatorname{supp}((u_1*u_2)*\phi)\subset\operatorname{supp}u_1+\operatorname{supp}u_2+\operatorname{supp}\phi$$

by (4.1.1). When $\operatorname{supp}\phi\to\{0\}$ it follows that (4.2.2) is valid. The theorem is proved.

If u_2 has compact support it follows from (4.2.2) that u_1*u_2 is determined in a neighborhood of x if u_1 is known only in a neighborhood of $\{x\}-\operatorname{supp}u_2$. Hence the convolution u_1*u_2 is defined in

$$\{x;\, x-y\in X \text{ for all } y\in\operatorname{supp}u_2\}$$

if $u_1\in\mathscr{D}'(X)$.

Theorem 4.2.5. *If u_1 and u_2 are distributions in $\mathbb{R}^n$, one of which has compact support, then*

$$\operatorname{sing\,supp}(u_1*u_2)\subset\operatorname{sing\,supp}u_1+\operatorname{sing\,supp}u_2. \tag{4.2.3}$$

Proof. Assume $u_2 \in \mathscr{E}'$, choose $\psi \in C_0^\infty$ equal to 1 near $\operatorname{sing\,supp} u_2$, and set $u_2 = v_2 + w_2$ where $v_2 = \psi u_2$ and $w_2 = (1-\psi) u_2 \in C_0^\infty$. Then $u_1 * w_2 \in C^\infty$ and $u_1 * v_2$ is a C^∞ function in

$$\{x;\, \{x\} - \operatorname{supp} v_2 \subset \complement \operatorname{sing\,supp} u_1\}.$$

This means that

$$\operatorname{sing\,supp} u_1 * u_2 = \operatorname{sing\,supp} u_1 * v_2 \subset \operatorname{sing\,supp} u_1 + \operatorname{supp} v_2$$

and since $\operatorname{supp} v_2 \subset \operatorname{supp} \psi$ can be taken as close to $\operatorname{sing\,supp} u_2$ as we wish, we obtain (4.2.3).

Differentiation can be interpreted as a convolution, for we have

$$\partial^\alpha u = (\partial^\alpha \delta_0) * u, \qquad u \in \mathscr{D}'(\mathbb{R}^n). \tag{4.2.4}$$

In fact, using (4.1.2) we obtain for $\phi \in C_0^\infty(\mathbb{R}^n)$

$$(\partial^\alpha u) * \phi = u * (\partial^\alpha \phi) = u * (\delta_0 * (\partial^\alpha \phi)) = u * (\partial^\alpha \delta_0) * \phi$$

which proves (4.2.4). Note in particular that convolution with δ_0 is the identity operator. If u_1 and u_2 are two distributions, one of which has compact support, it follows that

$$\partial^\alpha (u_1 * u_2) = (\partial^\alpha u_1) * u_2 = u_1 * \partial^\alpha u_2. \tag{4.2.5}$$

In fact, if the differentiations are rewritten as convolutions with $\partial^\alpha \delta_0$ this follows from the associativity and commutativity of the convolution. More generally, if

$$P = \sum a_\alpha \partial^\alpha,$$

where $a_\alpha \in \mathbb{C}$ and the sum is finite, is a partial differential operator with constant coefficients, then

$$Pu = (P\delta_0) * u, \qquad u \in \mathscr{D}'(\mathbb{R}^n), \tag{4.2.4$'$}$$

$$P(u_1 * u_2) = (Pu_1) * u_2 = u_1 * (Pu_2), \qquad u_1 \in \mathscr{D}'(\mathbb{R}^n), \quad u_2 \in \mathscr{E}'(\mathbb{R}^n). \tag{4.2.5$'$}$$

We shall now prove a local form of Theorem 4.2.3 containing Theorem 4.2.5.

Theorem 4.2.6. *Let $u_1 \in \mathscr{D}'(\mathbb{R}^n)$ and $u_2 \in \mathscr{E}'(\mathbb{R}^n)$. Then $u_1 * u_2$ is in C^k in a neighborhood of x if for every $y \in \operatorname{supp} u_2$ one can find an integer $j \geqq 0$ and an open neighborhood V_y of y such that $u_1 \in \mathscr{D}'^j(\{x\} - V_y)$ and $u_2 \in C^{k+j}(V_y)$ or $u_1 \in C^{k+j}(\{x\} - V_y)$ and $u_2 \in \mathscr{D}'^j(V_y)$.*

Proof. We may assume in the proof that $x = 0$ for otherwise we just have to make a translation of u_1. If $\phi, \psi \in C_0^\infty(V_y)$ it follows from Theorem 4.2.3 and (4.2.5) that $(\check{\phi} u_1) * (\psi u_2) \in C_0^k(\mathbb{R}^n)$. Choose an open covering $W_1, \dots, W_N$ of $\operatorname{supp} u_2$ so fine that $W_\nu \cap W_\mu \neq \emptyset$ implies $W_\nu \cup W_\mu \subset V_y$ for some y. Let $\phi_\nu \in C_0^\infty(W_\nu)$ and $\sum \phi_\nu = 1$ in a neighborhood of $\operatorname{supp} u_2$. Then $u_2 = \sum \phi_\nu u_2$ and

$$u_1 * u_2 = ((1 - \sum \check{\phi}_\mu) u_1) * u_2 + \sum_{\nu, \mu} (\check{\phi}_\mu u_1) * (\phi_\nu u_2).$$

The terms in the sum where $\operatorname{supp} \phi_\nu \cap \operatorname{supp} \phi_\mu \neq \emptyset$ are in C^k since $\operatorname{supp} \phi_\nu \cup \operatorname{supp} \phi_\mu \subset V_y$ for some y. The other terms vanish in a neighborhood of 0 by (4.2.2). This proves the theorem.

The convolution $u_1 * u_2$ can be defined in many cases where neither u_1 nor u_2 has compact support. What one needs in just that the map

(4.2.6) $$\operatorname{supp} u_1 \times \operatorname{supp} u_2 \ni (x, y) \to x + y \in \mathbb{R}^n$$

is *proper*, that is, that the inverse image of each compact set is compact. This condition is very natural if we look at the convolution of functions defined by (1.3.1)′. Assume now that u_j are distributions and that (4.2.6) is proper. If X is a bounded open set in $\mathbb{R}^n$ it follows that for some constant C

(4.2.7) $$x \in \operatorname{supp} u_1, \quad y \in \operatorname{supp} u_2, \quad x + y \in X \Rightarrow |x| \leqq C \quad \text{and} \quad |y| \leqq C.$$

Then the restriction of $(\phi_1 u_1) * (\phi_2 u_2)$ to X is independent of the choice of $\phi_j \in C_0^\infty(\mathbb{R}^n)$ provided that $\phi_j = 1$ in a neighborhood of $\{x; |x| \leqq C\}$. In fact, suppose that we change ϕ_1 by adding a function ψ_1 vanishing near this ball. By (4.2.2)

$$\operatorname{supp} (\psi_1 u_1) * (\phi_2 u_2) \subset \operatorname{supp} (\psi_1 u_1) + \operatorname{supp} (\phi_2 u_2)$$

which contains no point in X by (4.2.7), so

$$((\phi_1 + \psi_1) u_1) * (\phi_2 u_2) = (\phi_1 u_1) * (\phi_2 u_2) \quad \text{in } X.$$

We take this as definition of $u_1 * u_2$ in X and note that Theorem 2.2.4 shows that these local definitions together define $u_1 * u_2$ in $\mathscr{D}'(\mathbb{R}^n)$. More generally, $u_1 * u_2$ is defined in X if X is the largest open set such that (4.2.6) is proper if $\mathbb{R}^n$ is replaced by X; and (4.2.2) remains valid.

As an example we observe that if $u_j \in \mathscr{D}'(\mathbb{R})$, $j = 1, 2$, and $\operatorname{supp} u_j \subset \overline{\mathbb{R}}_+$, then $u_1 * u_2$ is defined and $\operatorname{supp}(u_1 * u_2) \subset \overline{\mathbb{R}}_+$. More generally, let $\Gamma \subset \mathbb{R}^n$ be a closed convex cone which is *proper* in the sense that it does not contain any straight line. Then

$$\{(x, y) \in \Gamma \times \Gamma, \ |x + y| \leqq C\}$$

is bounded, for if $(x_j, y_j) \in \Gamma \times \Gamma$ and $|x_j + y_j| \leqq C$, $|x_j| \to \infty$, we can pass to a subsequence such that $x_j / |x_j| \to x \in \Gamma$, hence $y_j / |x_j| \to -x \in \Gamma$. Then the straight line $\mathbb{R}x$ lies in Γ which is a contradiction. For every such cone the convolution of distributions in

$$\{u \in \mathscr{D}'(\mathbb{R}^n),\ \operatorname{supp} u \subset \Gamma\}$$

is therefore always defined and makes this set an algebra.

4.3. The Theorem of Supports

In this section we shall prove an inclusion opposite to (4.2.2) for the support of a convolution. This is not possible without restrictions, for if we take $u_1 = 1$ and u_2 with $\int u_2\, dx = 0$ then $u_1 * u_2 = 0$. We shall therefore have to assume that both u_1 and u_2 have compact supports. Even so, if u_1 is the characteristic function of a bounded open set X, and $\operatorname{supp} u_2 \subset B$ for some ball B, then

$$\operatorname{supp} u_1 * u_2 \subset (\partial X) + B,$$

which does not contain $X + \operatorname{supp} u_2$ if B is small enough. This forces us to take convex hulls of the supports, and we digress to discuss this concept before returning to the main topic.

Definition 4.3.1. If E is a compact set in $\mathbb{R}^n$ then $ch E$ is the closed convex hull of E, that is, the intersection of all closed convex sets containing E.

Equivalently, $ch E$ is the closure of the convex set of centers of gravity

$$\{\sum \lambda_j x_j;\ 0 \leqq \lambda_j, \sum \lambda_j = 1 \text{ and } x_j \in E\}.$$

By the Hahn-Banach theorem, if $y \notin ch E$ then one can separate y from $ch E$ by a hyperplane $\langle x, \xi\rangle = c$, so that $\langle x, \xi\rangle < c$ if $x \in ch E$ but $\langle y, \xi\rangle > c$. Set

$$H_E(\xi) = \sup_{x \in E} \langle x, \xi\rangle = \sup_{x \in ch E} \langle x, \xi\rangle, \qquad \xi \in \mathbb{R}^n. \tag{4.3.1}$$

We have seen that $y \notin ch E$ implies $\langle y, \xi\rangle > H_E(\xi)$ for some ξ, so $ch E$ is the set of all $x \in \mathbb{R}^n$ such that

$$\langle x, \xi\rangle \leqq H_E(\xi), \qquad \xi \in \mathbb{R}^n. \tag{4.3.2}$$

One calls H_E the *supporting function* of E. Since it is the supremum of linear functions we have

$$H_E(\xi + \eta) \leqq H_E(\xi) + H_E(\eta), \qquad H_E(t\xi) = t H_E(\xi); \quad t \geqq 0,\ \xi, \eta \in \mathbb{R}^n. \tag{4.3.3}$$

$H_E(\xi)<\infty$ for every $\xi\in\mathbb{R}^n$ since E is bounded, and it follows from (4.3.3) that H_E is convex.

Theorem 4.3.2. *For every convex positively homogeneous function H in $\mathbb{R}^n$ there is precisely one convex compact set K such that $H=H_K$, in fact,*

$$K=\{x;\ \langle x,\xi\rangle\leqq H(\xi),\ \xi\in\mathbb{R}^n\}. \tag{4.3.4}$$

Proof. If K is convex and $H=H_K$ we have already proved that K must be given by (4.3.4). All that remains is therefore to prove that $H_K=H$ if K is defined by (4.3.4). By the definition of K we have $H_K\leqq H$. Set

$$G=\{(\xi,\tau)\in\mathbb{R}^{n+1};\ \tau\geqq H(\xi)\}.$$

G is a closed convex set since H is convex and therefore continuous. If $\eta\in\mathbb{R}^n$ it follows that there is a half space containing G with $(\eta, H(\eta))$ on its boundary. Thus we have for some $(y,t)\in\mathbb{R}^{n+1}\setminus 0$ and $a\in\mathbb{R}$

$$\langle y,\xi\rangle+t\tau\geqq a \quad \text{if } (\xi,\tau)\in G; \qquad \langle y,\eta\rangle+tH(\eta)=a.$$

Here τ can be replaced by any larger number so $t\geqq 0$. If $t=0$ we would obtain $\langle y,\xi\rangle\geqq a$ for every $\xi\in\mathbb{R}^n$ which is impossible since $y\neq 0$ then. Hence $t>0$. Since H is positively homogeneous, the set G is invariant under multiplication by positive scalars which implies that $a\leqq 0$ and that

$$\langle y/t,\xi\rangle+H(\xi)\geqq 0, \qquad \xi\in\mathbb{R}^n.$$

Hence $x=-y/t\in K$ and

$$H_K(\eta)\geqq\langle x,\eta\rangle=H(\eta)-a/t\geqq H(\eta).$$

This completes the proof.

If K_1 and K_2 are compact sets in $\mathbb{R}^n$, the supporting function of the sum

$$K_1+K_2=\{x_1+x_2;\ x_j\in K_j\}$$

is obviously given by

$$H_{K_1+K_2}=H_{K_1}+H_{K_2}.$$

If K is a compact set and $t\in\mathbb{R}$, we set

$$tK=\{tx;\ x\in K\}$$

and obtain $H_{tK}(\xi)=H_K(t\xi)$, that is,

$$H_{tK}(\xi)=tH_K(\xi) \quad \text{if } t>0, \qquad H_{tK}(\xi)=-tH_K(-\xi) \quad \text{if } t<0.$$

Finally, if K_α is any family of compact sets contained in a fixed compact set and K is the closure of the union, then

$$H_K = \sup_\alpha H_{K_\alpha}.$$

Conversely, if the right-hand side is finite for every ξ, it is the supporting function of a convex compact set K' containing K_α for every α, and the equality is therefore valid. Thus there is an easy way to translate operations on sets to operations on supporting functions.

Now we state the main result in this section, the theorem of supports:

Theorem 4.3.3. *If* $u_1, u_2 \in \mathscr{E}'(\mathbb{R}^n)$, *then*

$$ch\,\operatorname{supp} u_1 * u_2 = ch\,\operatorname{supp} u_1 + ch\,\operatorname{supp} u_2. \tag{4.3.5}$$

Proof. That the left-hand side is contained in the right-hand side follows from (4.2.2) so it suffices to show that

$$\operatorname{supp} u_1 + \operatorname{supp} u_2 \subset ch\,\operatorname{supp} u_1 * u_2. \tag{4.3.6}$$

In doing so we may assume that $u_j \in C_0^\infty$. In fact, if ϕ is chosen as in Theorem 4.1.4, and (4.3.6) is known for the convolution of smooth functions, then

$$\begin{aligned}\operatorname{supp}(u_1 * \phi) + \operatorname{supp}(u_2 * \phi) &\subset ch\,\operatorname{supp}(u_1 * \phi * u_2 * \phi)\\ &\subset ch\,\operatorname{supp}(u_1 * u_2) + ch\,\operatorname{supp} \phi * \phi,\end{aligned}$$

and as $\operatorname{supp}\phi \to \{0\}$ we obtain (4.3.6). We shall now prove that (4.3.6) follows for $u_j \in C_0^\infty$ if we assume the special case

$$2\operatorname{supp} u \subset ch\,\operatorname{supp} u * u, \quad u \in C_0^\infty(\mathbb{R}^n) \tag{4.3.7}$$

which will be proved afterwards.

Set for fixed $u_1, u_2 \in C_0^\infty(\mathbb{R}^n)$ and $j = 0, 1, \dots$

$$K_j = ch \bigcup_{\deg p_1 p_2 \leqq j} \operatorname{supp}(p_1 u_1) * (p_2 u_2)$$

where p_1, p_2 are polynomials. Then $K_0 = ch\,\operatorname{supp} u_1 * u_2$, and (4.3.6) will follow if we prove that $K_j = K_0$ for every j. In fact, this means that

$$\operatorname{supp}(p_1 u_1) * (p_2 u_2) \subset K_0$$

for arbitrary polynomials p_j, hence for arbitrary entire analytic functions p_j since they are limits of their Taylor polynomials. Now take

$$p_j(x) = E(x - x_j, t)$$

where E is defined as in Theorem 3.3.3 and x_j are points with $u_j(x_j)\neq 0$. When $t\searrow 0$ it follows then that

$$\operatorname{supp} u_1(x_1)\,u_2(x_2)\,\delta_{x_1}*\delta_{x_2}\subset K_0,$$

hence $\operatorname{supp} u_1+\operatorname{supp} u_2\subset K_0$.

To prove that $K_j=K_0$ for every j it suffices to show that

$$2K_j\subset K_{j-1}+K_{j+1},\quad j>0. \tag{4.3.8}$$

In fact, if H_j is the supporting function of K_j, we obtain from (4.3.8)

$$H_j-H_{j-1}\leqq H_{j+1}-H_j,\quad j>0,$$

hence $H_{j+k}\geqq H_{j-1}+(k+1)(H_j-H_{j-1})$. Since H_{j+k} is bounded by the sum of the supporting functions of $\operatorname{supp} u_1$ and of $\operatorname{supp} u_2$, it follows that $H_j-H_{j-1}\leqq 0$. But $H_{j-1}\leqq H_j$ since $K_{j-1}\subset K_j$, so $H_j=H_{j-1}$ for every $j>0$ as claimed.

(4.3.8) means that if $\deg p_1+\deg p_2\leqq j$ then

$$2\operatorname{supp}(p_1u_1)*(p_2u_2)\subset K_{j-1}+K_{j+1}$$

or by (4.3.7) that

$$\operatorname{supp}(p_1u_1)*(p_2u_2)*(p_1u_1)*(p_2u_2)\subset K_{j-1}+K_{j+1}. \tag{4.3.9}$$

When proving (4.3.9) we may assume that $p_1(x)=x_\nu q_1(x)$. If we write $q_2(x)=x_\nu p_2(x)$ then

$$x_\nu((q_1u_1)*(p_2u_2))=(p_1u_1)*(p_2u_2)+(q_1u_1)*(q_2u_2)$$

has support in K_{j-1}, so the convolution with $(p_1u_1)*(p_2u_2)$ has support in $K_{j-1}+K_j\subset K_{j-1}+K_{j+1}$. Moreover,

$$(q_1u_1)*(q_2u_2)*(p_1u_1)*(p_2u_2)=((q_1u_1)*(p_2u_2))*((p_1u_1)*(q_2u_2))$$

has support in $K_{j-1}+K_{j+1}$, so this completes the proof of (4.3.9).

The remaining proof of (4.3.7) is based on the following

Lemma 4.3.4. *If* $u\in C_0^\infty(\mathbb{R}^n)$ *and* $\tilde{u}(x)=\overline{u(-x)}$ *then*

$$\|u*\tilde{u}\|_{L^2}^2=\|u*u\|_{L^2}^2. \tag{4.3.10}$$

Proof. For any $g\in C_0^\infty$ we have

$$\|g\|_{L^2}^2=g*\tilde{g}(0) \tag{4.3.11}$$

so both sides of (4.3.10) are equal to $u*\tilde{u}*u*\tilde{u}(0)$.

Lemma 4.3.5. *If* K *is a compact set in* $\mathbb{R}^n$ *then*

$$\sup|u|\leqq C\|\partial_1^2\dots\partial_n^2 u\|_{L^2},\quad u\in C_0^\infty(K). \tag{4.3.12}$$

Proof. By Taylor's formula applied to each variable we have

$$u(x)=\int_{y_j<x_j}(x_1-y_1)\dots(x_n-y_n)\,\partial_1^2\dots\partial_n^2 u(y)\,dy$$

which proves (4.3.12).

End of proof of Theorem 4.3.3. Combination of (4.3.10) and (4.3.12) gives in view of (4.3.11)

$$\begin{aligned}\|u\|_{L^2}^2 &\leqq \sup|u*\tilde{u}| \leqq C\|(\partial_1\dots\partial_n u)*(\partial_1\dots\partial_n\tilde{u})\|_{L^2}\\ &= C\|\partial_1^2\dots\partial_n^2 u*u\|_{L^2}, \qquad u\in C_0^\infty(K).\end{aligned}$$

Now replace u by u_ξ,

$$u_\xi(x)=e^{\langle x,\xi\rangle}u(x)$$

and note that

$$u_\xi*u_\xi(x)=e^{\langle x,\xi\rangle}u*u(x),$$
$$\partial^\alpha(u_\xi*u_\xi)(x)=e^{\langle x,\xi\rangle}(\xi+\partial)^\alpha(u*u)(x).$$

Then we obtain

$$\begin{aligned}\int e^{2\langle x,\xi\rangle}|u(x)|^2dx &\leqq C'\sum_{|\alpha|\leqq 2n}|\xi|^{2n-|\alpha|}\left(\int e^{2\langle x,\xi\rangle}|\partial^\alpha u*u|^2dx\right)^{\frac{1}{2}}\\ &\leqq C''(1+|\xi|)^{2n}e^{H(\xi)}\end{aligned}$$

where H is the supporting function of $\operatorname{supp} u*u$. If we replace ξ by $t\xi$ and let $t\to+\infty$, it follows that

$$2\langle x,\xi\rangle\leqq H(\xi) \quad \text{if } u(x)\neq 0.$$

Hence $2\langle x,\xi\rangle\leqq H(\xi)$ when $x\in\operatorname{supp} u$, which proves (4.3.7) and Theorem 4.3.3.

4.4. The Role of Fundamental Solutions

Let us first recall that a differential operator with constant coefficients in $\mathbb{R}^n$ is a finite sum

$$P=\sum a_\alpha\partial^\alpha$$

with $a_\alpha\in\mathbb{C}$. According to Definition 3.3.1 a distribution $E\in\mathscr{D}'(\mathbb{R}^n)$ is called a fundamental solution of P if

$$PE=\delta_0. \tag{4.4.1}$$

Fundamental solutions were constructed for some operators in Section 3.3. We shall prove later (Theorem 7.3.10) that every operator has a fundamental solution.

The importance of fundamental solutions is due to the following two consequences of (4.2.5)′:

$$E*(Pu)=u, \quad u\in\mathscr{E}'(\mathbb{R}^n), \tag{4.4.2}$$

$$P(E*f)=f, \quad f\in\mathscr{E}'(\mathbb{R}^n). \tag{4.4.3}$$

Thus convolution with E is both a left and a right inverse of P. From (4.4.3) it follows that the equation $Pu=f$ has a solution for every $f\in\mathscr{E}'(\mathbb{R}^n)$, and (4.4.2) makes it possible to obtain information on say the singularities of u from those of Pu.

Theorem 4.4.1. *If P has a fundamental solution with* $\operatorname{sing\,supp} E=\{0\}$ *and X is any open set in $\mathbb{R}^n$, then*

$$\operatorname{sing\,supp} u=\operatorname{sing\,supp} Pu, \quad u\in\mathscr{D}'(X). \tag{4.4.4}$$

Proof. It is always true that $\operatorname{sing\,supp} Pu\subset\operatorname{sing\,supp} u$. If $u\in\mathscr{E}'$ we obtain from (4.4.2) and Theorem 4.2.5 that

$$\operatorname{sing\,supp} u=\operatorname{sing\,supp} E*(Pu)\subset\operatorname{sing\,supp} Pu$$

so the assertion is valid when $u\in\mathscr{E}'$. If $\psi\in C_0^\infty(X)$ is equal to 1 in an open subset Y, it follows that

$$\begin{aligned} Y\cap\operatorname{sing\,supp} Pu&=Y\cap\operatorname{sing\,supp} P(\psi u)\\ &=Y\cap\operatorname{sing\,supp}\psi u=Y\cap\operatorname{sing\,supp} u\end{aligned}$$

which proves (4.4.4).

Examples of operators for which Theorem 4.4.1 is applicable are the Cauchy-Riemann, Laplace and heat operators for which fundamental solutions with singularity only at 0 were given in Section 3.3. Note that if we take $X=\mathbb{R}^n$ and u equal to a fundamental solution E, we see that (4.4.4) implies that $\operatorname{sing\,supp} E=\{0\}$. In Section 11.1 we shall characterize the operators to which Theorem 4.4.1 is applicable; they are called hypoelliptic. By repeating the proof of Theorem 4.4.1 we shall now show that an analogue of the Stieltjes-Vitali theorem is valid for all of them:

Theorem 4.4.2. *Let P satisfy the hypothesis in Theorem 4.4.1. If $u_j\in C^\infty(X)$, $Pu_j=0$, and $u_j\to u$ in $\mathscr{D}'(X)$ when $j\to\infty$, then $u\in C^\infty(X)$ and $u_j\to u$ in $C^\infty(X)$.*

Proof. Since $Pu=\lim Pu_j=0$ we have $u\in C^\infty(X)$ by Theorem 4.4.1. We may therefore assume that $u=0$. If ψ is chosen as in the proof of Theorem 4.4.1 then $f_j=P(\psi u_j)\to 0$ in $\mathscr{E}'$ and $\operatorname{supp} f_j\subset\operatorname{supp} d\psi\subset X\smallsetminus Y$.

On a compact set $K \subset Y$ we have

$$\partial^\alpha u_j(x) = \partial^\alpha E * f_j(x) = f_j(\partial^\alpha E(x - .)) \to 0 \quad \text{uniformly as } j \to \infty,$$

which proves the theorem.

For the Cauchy-Riemann and the Laplace operators we have a fundamental solution E which is real analytic in $\mathbb{R}^n \setminus \{0\}$, that is, analytic in a neighborhood in $\mathbb{C}^n$. One can then improve Theorem 4.4.2:

Theorem 4.4.3. *Assume that P has a fundamental solution E which is real analytic in $\mathbb{R}^n \setminus 0$. For every open $X \subset \mathbb{R}^n$ one can then find an open neighborhood $Z \subset \mathbb{C}^n$ such that every solution in X of the equation $Pu = 0$ can be extended to an analytic function in Z. If $Pu_j = 0$ in X and $u_j \to u$ in $\mathscr{D}'(X)$, then the extension of $u_j - u$ to Z converges uniformly to 0 on compact sets in Z.*

Proof. Let Z_0 be an open set in $\mathbb{C}^n$ such that $Z_0 \cap \mathbb{R}^n = \mathbb{R}^n \setminus \{0\}$ and E is analytic in Z_0. Choose Y, ψ as in the proof of Theorem 4.4.1 and set $f = P(\psi u)$. Then

$$u(x) = E * f(x) = \int E(x - y) f(y)\, dy, \qquad x \in Y,$$

and the right-hand side is defined and analytic in

$$\{z;\ \operatorname{Re} z \in Y \text{ and } z - y \in Z_0 \text{ if } y \in \operatorname{supp} d\psi\} \tag{4.4.5}$$

which is a neighborhood of Y in $\mathbb{C}^n$. We can choose Z_0 so that

$$x + iy \in Z_0 \Rightarrow x + ity \in Z_0, \qquad 0 \leqq t \leqq 1,$$

which implies the same property for the set (4.4.5). An analytic function in the set (4.4.5) is then uniquely determined by its restriction to real arguments, so letting ψ and Y vary we obtain an analytic continuation of u to the union Z of all the open sets (4.4.5). This proves the first statement and the second one follows in the same way.

In Chapters VII and VIII we shall show that Theorem 4.4.3 is applicable if and only if P is elliptic, that is, m denoting the order of P

$$\sum_{|\alpha| = m} a_\alpha \xi^\alpha \neq 0, \qquad 0 \neq \xi \in \mathbb{R}^n.$$

Corollary 4.4.4. *Assume that P satisfies the hypothesis in Theorem 4.4.3 and that X is a connected open set in $\mathbb{R}^n$. If $u \in \mathscr{D}'(X)$ and $Pu = 0$ it follows then that $u = 0$ if u vanishes in an open non-empty subset Y of X.*

Proof. This follows by the uniqueness of analytic continuation.

We can now prove an extension of the classical approximation theorem of Runge:

Theorem 4.4.5. *Assume that P satisfies the hypothesis in Theorem 4.4.3, and let $Y \subset X \subset \mathbb{R}^n$ be open sets such that $X \setminus Y$ is not a disjoint union $F \cup K$ where K is compact and non-empty, and F is closed in X. Every solution $u \in C^\infty(Y)$ of the equation $Pu=0$ is then a limit in $C^\infty(Y)$ of restrictions to Y of solutions of the same equation in X.*

Proof. By the Hahn-Banach theorem we must show that if $w \in \mathscr{E}'(Y)$ is orthogonal to all solutions of the equation $Pu=0$ in X, then w is orthogonal to all solutions in Y. Set

$$\langle \check{E}, \phi \rangle = \langle E, \check{\phi} \rangle$$

where $\check{\phi}(x) = \phi(-x)$. Thus $\check{E}(x) = E(-x)$ for $x \neq 0$ so $\check{E}$ is analytic then. We have ${}^tP\check{E} = \delta$ where ${}^tP = \sum a_\alpha(-\partial)^\alpha$ is the transpose of P, that is,

$$\langle {}^tPv, u \rangle = \langle v, Pu \rangle \quad \text{if } v \in \mathscr{E}'(Y), \quad u \in C^\infty(Y).$$

Now set

$$v = \check{E} * w.$$

Then ${}^tPv = w$. If we show that $v \in \mathscr{E}'(Y)$, then

$$\langle w, u \rangle = \langle {}^tPv, u \rangle = \langle v, Pu \rangle = 0$$

if $Pu=0$ in Y so the theorem will be proved.

v is an analytic function outside $M = \operatorname{supp} w$. If $x \notin X$ then

$$\partial^\alpha v(x) = \langle \partial_x^\alpha E(.-x), w \rangle = 0$$

for every α since

$$P_y \partial_x^\alpha E(y-x) = 0 \quad \text{if } y \in X.$$

Hence $v=0$ in every component O of $\complement M$ which contains some point in $\complement X$. If O is a component of $\complement M$ which is contained in X and is bounded, then $K = O \cap (X \setminus Y) \subset X \setminus Y$ is bounded and closed, because $\partial O \subset \partial M \subset Y \subset X$. Hence K is compact and $F = \complement O \cap (X \setminus Y)$ is closed in X so $K = \emptyset$, hence $O \subset Y$. It follows that $v=0$ in every component of $\complement M$ which is not contained in Y, with the exception of the unbounded component when it is a subset of X. The theorem is therefore proved unless X contains a neighborhood of infinity.

By repeated use of the part of the theorem proved now we conclude that every solution of $Pu=0$ in a ball B is the limit of solutions u_j in $\mathbb{R}^n$. In fact, if $B = B_0 \subset B_1 \subset B_2 \subset \dots$ is a sequence of concentric balls with radius $\to\infty$, then we can find u_j^1 with $Pu_j^1=0$ in B_1 so that $u_j^1\to u$ in B, and successively solutions u_j^k in B_k such that

$$|u_j^{k+1}-u_j^k|<2^{-j-k} \quad \text{in } B_{k-1}.$$

It follows then that

$$u_j=\lim_{k\to\infty} u_j^k$$

exists in $\mathbb{R}^n$ and that $Pu_j=0$. Moreover,

$$|u_j-u_j^1|\leqq 2^{-j} \quad \text{in } B_0$$

which proves the assertion.

Returning to the proof above we find that if $|y|<R$ in $\operatorname{supp} w$ then

$$v(x)=\langle E(.-x), w\rangle=0, \qquad |x|<R,$$

because $y\to E(y-x)$ is the limit in $C^\infty(\{y;|y|<R\})$ of solutions in $\mathbb{R}^n$, which are orthogonal to w by assumption. Hence $v=0$ in the unbounded component also, which completes the proof.

Remark. The approximation theorem is not valid if the hypothesis on $X\smallsetminus Y$ in Theorem 4.4.5 is not fulfilled, for a sequence of solutions u_j in X which converges in Y must also converge in the open set $Y\cup K$ $=X\smallsetminus F$ then. The reason is that we can take $\phi\in C_0^\infty(X\smallsetminus F)$ equal to 1 in a neighborhood O of K and obtain in O

$$u_j=\phi u_j=E*f_j$$

where $f_j=P(\phi u_j)$ has support in $\operatorname{supp} d\phi\subset X\smallsetminus(F\cup K)=Y$, so u_j converges to a solution in $X\cup O$. However, if $x_0\in K$ the solution $u(y)=E(y-x_0)$ of the equation $Pu=0$ in Y cannot have an extension U to $Y\cup K$, for $w=u-U$ would then vanish in $Y\cup K\smallsetminus\{x_0\}$ which is absurd since $Pw=\delta_{x_0}$ in $Y\cup K$. (We assume that P is not a constant.)

Finally we shall prove an existence theorem which exploits both (4.4.2) and (4.4.3):

Theorem 4.4.6. *Let P satisfy the hypothesis in Theorem 4.4.3, let X be an open set in $\mathbb{R}^n$ and $f\in\mathscr{D}'(X)$. Then one can find $u\in\mathscr{D}'(X)$ so that $Pu=f$.*

Proof. Let X_j be the set of points $x\in X$ with $|x|<j$ and distance $>1/j$ to $\complement X$. Then the hypothesis in Theorem 4.4.5 is fulfilled for $Y=X_j$. In fact, for a point $x\in K$ the distance to a point $y\in\complement X$ must exceed $1/j$ since the segment between y and x contains points in X_j, and $|x|<j$ since x lies on an interval with end points in X_j. Choose $\phi_j\in C_0^\infty(X)$ equal to 1 in X_j and set $v_j=E*(\phi_j f)$. Then (4.4.3) gives

$$Pv_j=\phi_j f=f \quad \text{in } X_j.$$

We must now correct v_j to obtain a limit as $j\to\infty$. To do so we observe that $v_{j+1}-v_j$ satisfies the equation $P(v_{j+1}-v_j)=0$ in X_j. By Theorem 4.4.5 we can therefore find $w_j\in C^\infty(X)$ with $Pw_j=0$ so that

$$|v_{j+1}-v_j-w_j|<2^{-j} \quad \text{in } X_{j-1}.$$

Then

$$u=\lim_{j\to\infty}(v_j-\sum_{i<j} w_i)$$

exists in $\mathscr{D}'(X)$ and satisfies the equation $Pu=f$. In fact, on X_k we have when $j>k+1$

$$v_j-\sum_{i<j} w_i=\sum_{k+1}^{j-1}(v_{i+1}-v_i-w_i)+v_{k+1}-\sum_1^k w_i.$$

The sum converges uniformly on X_k to a solution v of the equation $Pv=0$. This completes the proof, for $Pv_{k+1}=f$ in X_k.

It is now easy to prove that all distributions are of the form (2.1.1):

Theorem 4.4.7. *For every $f\in\mathscr{D}'(X)$, $X\subset\mathbb{R}^n$, one can find $f_\alpha\in C(X)$ such that*

$$f(\phi)=\sum\int f_\alpha\partial^\alpha\phi\,dx, \qquad \phi\in C_0^\infty(X),$$

and the sets $\operatorname{supp} f_\alpha$ *are locally finite. If $f\in\mathscr{D}'_F$ the sum can be taken finite.*

Proof. Since $x_+^m/m!$ is a fundamental solution of $(d/dx)^{m+1}$ on $\mathbb{R}$, if m is a non-negative integer, the product

$$E(x)=x_{1+}^m\dots x_{n+}^m/(m!)^n$$

is a fundamental solution of $P=(\partial_1\dots\partial_n)^{m+1}$ in $\mathbb{R}^n$, and $E\in C^k$ if $m>k$. If $f\in\mathscr{E}'^k(\mathbb{R}^n)$ it follows that $u=E*f\in C(\mathbb{R}^n)$ satisfies the equation $Pu=f$. To prove the theorem we proceed as in the proof of Theorem 2.1.5, choosing first a partition of unity $1=\sum\psi_j$ in X. Let $\chi_j\in C_0^\infty(X)$ be equal to 1 near the support of ψ_j. If m_j is larger than the order of $\chi_j f$ we have just seen that one can find $u_j\in C(\mathbb{R}^n)$ such that $(\partial_1\dots\partial_n)^{m_j+1}u_j=\chi_j f$. It follows that

$$\langle f,\phi\rangle=\sum\langle\chi_j f,\psi_j\phi\rangle=\sum\langle u_j,\ \pm(\partial_1\dots\partial_n)^{m_j+1}(\psi_j\phi)\rangle.$$

If we carry out the differentiations we obtain the desired representation.

As an application of Theorem 4.4.7 we shall prove an extension of Corollary 3.1.6:

Theorem 4.4.8. *Let $X=Y\times I\subset\mathbb{R}^n$ where Y is an open set in $\mathbb{R}^{n-1}$ and I is an open interval on $\mathbb{R}$, and assume that $u\in\mathscr{D}'(X)$ satisfies a differential equation of the form*

$$\partial_n^m u+a_{m-1}\partial_n^{m-1}u+\ldots+a_0u=f$$

where a_j is a differential operator in $x'=(x_1,\ldots,x_{n-1})$ with coefficients in $C^\infty(X)$ and f is a continuous function of $x_n\in I$ with values in $\mathscr{D}'(Y)$. Then it follows that u is a C^m function of $x_n\in I$ with values in $\mathscr{D}'(Y)$.

Proof. If we allow u to be vector valued and the coefficients to be square matrices the proof reduces to the case $m=1$ just as that of Corollary 3.1.6. We assume this from now on. Shrinking Y and I if necessary we can by Theorem 4.4.7 write u as a finite sum

$$u=\sum_{|\alpha|\leqq\mu}\partial^\alpha u_\alpha \tag{4.4.6}$$

where $u_\alpha\in C(X)$, and the proof of Theorem 4.4.7 also shows that

$$f=\sum_{|\alpha|\leqq\mu',\alpha_n=0}\partial^\alpha f_\alpha,\qquad f_\alpha\in C(X).$$

(We just regard x_n as a parameter and apply the proof in the other variables.) If $u_\alpha=0$ when $\alpha_n>0$ then u and $\partial_n u=f-a_0u$ are continuous with values in $\mathscr{D}'(Y)$. Otherwise let ν be the smallest integer such that $u_\alpha=0$ when $|\alpha_n|>\nu$. Then we have

$$\partial_n u=f-a_0u=\sum\partial^\alpha f_\alpha-\sum a_0\partial^\alpha u_\alpha=\sum\partial^\alpha v_\alpha=\partial_n\sum\partial^\alpha V_\alpha$$

where v_α and V_α are continuous functions and $v_\alpha=0$ when $\alpha_n>\nu$, $V_\alpha=0$ when $\alpha_n>\nu-1$. In fact, we obtain the expression involving v_α by commuting the coefficients of a_0 through the differentiations successively; recall that a_0 only involves x' derivatives. To pass to the representation with V_α we just take out a factor ∂_n if $\alpha_n>0$ and take a primitive function of v_α with respect to x_n if $\alpha_n=0$. If $w=u-\sum\partial^\alpha V_\alpha$ we have $\partial_n w=0$. Hence

$$w=\sum\partial^\alpha w_\alpha(x')$$

by Theorem 3.1.4′ and Theorem 4.4.7. Now we obtain

$$u=\sum\partial^\alpha w_\alpha(x')+\sum\partial^\alpha V_\alpha$$

where $\alpha_n\leqq\nu-1$ in the sum. Iterating the argument ν times we obtain a similar representation where $\alpha_n=0$ in every term, and the theorem is proved.

If there exists a representation (4.4.6) of u with u_α continuous in $Y\times\bar{I}$ the proof gives more:

Theorem 4.4.8′. *Assume in addition to the hypotheses in Theorem 4.4.8 that u can be extended to a distribution in $Y \times J$ where J is an open neighborhood of $\bar{I}$, and that f is continuous in $\bar{I}$ with values in $\mathscr{D}'(Y)$. Then u is in C^m in $\bar{I}$ with values in $\mathscr{D}'(Y)$.*

Note that Theorem 3.1.11 is essentially only the special case of solutions of the Cauchy-Riemann equation. The importance of Theorem 4.4.8′ is that it allows us to interpret boundary conditions such as

$$u = u_0 \quad \text{when } x_n = x_0$$

if x_0 is an end point of I and $u_0 \in \mathscr{D}'(Y)$. The theorem also gives us a unique way of extending u to $Y \times \mathbb{R}$ by defining u as a function of x_n to be 0 outside I.

4.5. Basic L^p Estimates for Convolutions

In this section we shall use the notation

$$\|u\|_p = (\int |u(x)|^p \, dx)^{1/p}$$

if $u \in L^p(\mathbb{R}^n)$, $1 \leqq p < \infty$. To avoid convergence questions we shall usually assume $u \in C_0$. We write $\|u\|_\infty = \sup |u|$ then. The following is essentially Hölder's inequality, which is really the special case $k=2$.

Theorem 4.5.1. *If $u_1, \dots, u_k \in C_0$ then*

$$|u_1 * u_2 * \dots * u_k(0)| \leqq \|u_1\|_{p_1} \dots \|u_k\|_{p_k}, \tag{4.5.1}$$

if

$$1/p_1 + \dots + 1/p_k = k-1 \quad \textit{and} \quad 1 \leqq p_j \leqq \infty. \tag{4.5.2}$$

Proof. Let us first consider the exceptional case when $p_j = \infty$ for some j, say $p_1 = \infty$. Then $p_j = 1$ when $j \neq 1$ and (4.5.1) is obvious. Writing $1/p_j = t_j$ and $v_j = |u_j|^{p_j}$ it is otherwise clear that (4.5.1) is equivalent to the inequality

$$v_1^{t_1} * \dots * v_k^{t_k}(0) \leqq 1 \quad \text{when} \quad 0 \leqq v_j \in C_0, \quad \int v_j \, dx = 1, \tag{4.5.3}$$

where $0 \leqq t_j \leqq 1$ and $\sum t_j = k-1$. The left-hand side is a convex function of t since $\prod v_j(x_j)^{t_j} = \exp(\sum t_j \log v_j(x_j))$ is convex. We have

$$(t_1, \dots, t_k) = \sum (1-t_j)\tilde{e}_j$$

where $\tilde{e}_j$ is the vector with j^{th} coordinate 0 and the others equal to 1. Since $0 \leqq 1-t_j \leqq 1$ and $\sum(1-t_j)=1$, the inequality follows from the cases where $t=\tilde{e}_j$ which were discussed at the beginning of the proof.

Corollary 4.5.2. *If* $1 \leqq p_j \leqq \infty$, $j=1, \ldots, k$, *and*

$$1/p_1 + \ldots + 1/p_k = k-1+1/q, \qquad 1 \leqq q \leqq \infty,$$

then

$$\|u_1 * \ldots * u_k\|_q \leqq \|u_1\|_{p_1} \ldots \|u_k\|_{p_k}. \tag{4.5.4}$$

Proof. If u is the convolution on the left and $v \in C_0$, then

$$|u * v(0)| \leqq \|u_1\|_{p_1} \ldots \|u_k\|_{p_k} \|v\|_{q'}, \qquad 1/q+1/q'=1,$$

by (4.5.1) with $k+1$ factors now. The converse of Hölder's inequality gives (4.5.4).

In particular, we have

$$\|u * k\|_q \leqq C \|u\|_p \tag{4.5.5}$$

if $1 \leqq p \leqq q \leqq \infty$ and $k \in L^r$, $1/r = 1+1/q-1/p$. If $q=\infty$ then the converse of Hölder's inequality shows that (4.5.5) cannot be valid unless $k \in L^r$. The same conclusion follows for $p=1$ if we observe that by Hölder's inequality and its converse (4.5.5) is equivalent to

$$|u * v * k(0)| \leqq C \|u\|_p \|v\|_{q'}$$

which means that (4.5.5) remains valid with q and p replaced by p' and q' if $1/p+1/p'=1$, $1/q+1/q'=1$. This means in particular that we cannot apply our result (4.5.5) to any homogeneous k when $r<\infty$, for if say

$$k(y) = |y|^{-n/a}, \qquad y \in \mathbb{R}^n, \tag{4.5.6}$$

then $\int k(y)^r dy$ diverges at ∞ if $r/a \leqq 1$ and at 0 if $r/a \geqq 1$. However, we shall now prove the Hardy-Littlewood-Sobolev inequality which states that (4.5.5) remains valid for this k as if k were in L^a, except for the extreme cases where we know that it must fail:

Theorem 4.5.3. *If* $1<a<\infty$ *and* $1<p<q<\infty$,

$$1/p+1/a = 1+1/q, \tag{4.5.7}$$

then, with k_a *defined by* (4.5.6),

$$\|k_a * u\|_q \leqq C_{p,a} \|u\|_p, \qquad u \in C_0. \tag{4.5.5$'$}$$

The proof will be based on a few lemmas. We write $1/a+1/a'=1$.

Lemma 4.5.4. *If* $1 \leqq p < a'$ *then*

(4.5.8) $$\|k_a * u\|_\infty \leqq C_{p,a} \|u\|_p^{p/a'} \|u\|_\infty^{1-p/a'}, \qquad u \in L^p \cap L^\infty.$$

Proof. For every $R > 0$ we have

$$|k_a * u(x)| \leqq \int_{|y|<R} |y|^{-n/a} |u(x-y)|\,dy + \int_{|y|>R} |y|^{-n/a} |u(x-y)|\,dy$$
$$\leqq C(R^{n-n/a} \|u\|_\infty + R^{n/p'-n/a} \|u\|_p),$$

for

$$\int_{|y|>R} |y|^{-np'/a}\,dy = C_1 R^{n-np'/a}$$

for a finite C_1 since $p' > a$. If we balance the terms by choosing R so that $R^{n/p} = \|u\|_p / \|u\|_\infty$ then $R^{n/a'} = \|u\|_p^{p/a'} \|u\|_\infty^{-p/a'}$ and (4.5.8) follows.

Next we need a fundamental covering lemma of Calderón and Zygmund:

Lemma 4.5.5. *Let* $u \in L^1(\mathbb{R}^n)$ *and let s be a number* >0. *Then we can write*

(4.5.9) $$u = v + \sum_1^\infty w_k$$

where all terms are in L^1,

(4.5.10) $$\|v\|_1 + \sum_1^\infty \|w_k\|_1 \leqq 3\,\|u\|_1,$$

(4.5.11) $$|v(x)| \leqq 2^n s \quad \textit{almost everywhere},$$

and for certain disjoint cubes I_k

(4.5.12) $$w_k(x) = 0 \quad \textit{if} \quad x \notin I_k, \qquad \int w_k\,dx = 0,$$

(4.5.13) $$s \sum_1^\infty m(I_k) \leqq \|u\|_1.$$

If u has compact support, the supports of v and all w_k *are contained in a fixed compact set.*

Proof. Divide the whole space $\mathbb{R}^n$ into a mesh of cubes of volume $> s^{-1} \int |u|\,dx$. The mean value of $|u|$ over every cube is thus $< s$. Divide each cube into 2^n equal cubes, and let $I_{11}, I_{12}, I_{13}, \ldots$ be those (open) cubes so obtained over which the mean value of $|u|$ is $\geqq s$. We have

(4.5.14) $$s\,m(I_{1k}) \leqq \int_{I_{1k}} |u|\,dx < 2^n s\,m(I_{1k}).$$

For if I_{1k} was obtained by subdivision of the cube I, the construction gives

$$sm(I_{1k}) \leqq \int_{I_{1k}} |u|\,dx \leqq \int_I |u|\,dx < sm(I) = 2^n sm(I_{1k}).$$

We set

$$v(x) = \int_{I_{1k}} u\,dy/m(I_{1k}), \qquad x \in I_{1k}; \tag{4.5.15}$$

$$w_{1k}(x) = u(x) - v(x), \qquad x \in I_{1k}$$
$$= 0, \qquad x \notin I_{1k}.$$

Next we make a new subdivision of the cubes which are not among the cubes I_{1k}, select those new cubes $I_{21}, I_{22}, \dots$ over which the mean value of $|u|$ is $\geqq s$ and extend the definitions (4.5.15) to these cubes. Continuing in this way we obtain disjoint cubes I_{jk} and functions w_{jk}, which we rearrange as a sequence. If the definition of v is completed by setting $v(x) = u(x)$ when $x \notin O = \bigcup I_k$, it is clear that (4.5.9) holds. To prove (4.5.10) we first note that

$$\int_{I_k} (|v| + |w_k|)\,dx \leqq 3 \int_{I_k} |u|\,dx.$$

Since the cubes are disjoint, w_k vanishes outside I_k and $v = u$ in $\complement O$, we immediately get (4.5.10). Further, (4.5.11) follows from (4.5.14) if $x \in O$. On the other hand, if $x \notin O$, there are arbitrarily small cubes containing x over which the mean value of $|u|$ is $< s$. Hence $|u(x)| \leqq s$ at every Lebesgue point in $\complement O$, that is, almost everywhere. (4.5.12) follows from the construction, and adding the inequalities (4.5.14) we obtain (4.5.13) since the cubes are disjoint. The proof is complete.

The reason for wanting w_k to have integral 0 is the following lemma:

Lemma 4.5.6. *If $w \in L^1$ has support in a cube I, $\int w\,dx = 0$, and if I^* is the doubled cube, with the same center and twice the side, then*

$$\Big(\int_{\complement I^*} |k_a * w|^a\,dx\Big)^{1/a} \leqq C_a \|w\|_1. \tag{4.5.16}$$

Proof. We may assume that the center of I is at 0, and we denote the side by L. By the mean value theorem

$$|k_a * w(x)| = \Big|\int (k_a(x-y) - k_a(x))\,w(y)\,dy\Big| \leqq CL|x|^{-1-n/a}\|w\|_1, \quad x \notin I^*,$$

and this proves (4.5.16) since

$$\Big(\int_{|x|>L} |x|^{-a-n}\,dx\Big)^{1/a} \leqq C/L.$$

Lemma 4.5.7. *The operator k_a* is of weak type* 1, *a in the sense that*

$$m\{x;\,|k_a*u(x)|>t\}\,t^a \leqq C_a\,\|u\|_1^a, \qquad u\in L^1. \tag{4.5.17}$$

Proof. Assume that $\|u\|_1=1$ and decompose u by means of Lemma 4.5.5. Then we have by (4.5.8), with $p=1$ now,

$$|k_a*v|\leqq C s^{1/a}.$$

Define s so that $Cs^{1/a}=t/2$. Then $|k_a*u(x)|>t$ implies $\sum|k_a*w_k|>t/2$. If $O=\bigcup I_k^*$ we have by (4.5.13), (4.5.10) and (4.5.16)

$$m(O)\leqq 2^n/s, \qquad \int_{\complement O} (\sum|k_a*w_k|)^a\,dx\leqq C,$$

which means that

$$m\{x;\,\sum|k_a*w_k|>t/2\}\leqq 2^n/s+C(t/2)^{-a}<C't^{-a}$$

so (4.5.17) is proved.

(4.5.17) is a substitute for (4.5.5)′ which remains true when $p=1$. Using an argument of Marcinkiewicz we shall now prove that together with Lemma 4.5.4 it proves Theorem 4.5.3:

Proof of Theorem 4.5.3. Assume that $\|u\|_p=1$. If

$$m(t)=m\{x;\,|k_a*u(x)|>t\}$$

then

$$\|k_a*u\|_q^q=-\int_0^\infty t^q\,dm(t)=q\int_0^\infty t^{q-1}m(t)\,dt$$

so we must estimate $m(t)$. With a number s to be chosen later we split u,

$$u=v+w, \qquad \text{where } v=u \text{ when } |u|\leqq s,\quad w=u \text{ when } |u|>s.$$

Then we have by (4.5.8) and (4.5.7)

$$\|k_a*v\|_\infty\leqq Cs^{1-p/a'}=Cs^{p/q},$$

and we choose s so that

$$Cs^{p/q}=t/2.$$

Then $|k_a*w(x)|>t/2$ if $|k_a*u(x)|>t$, so (4.5.17) gives

$$m(t)\leqq m\{x;\,|k_a*w(x)|>t/2\}\leqq C't^{-a}\|w\|_1^a,$$

and we obtain by Minkowski's inequality

$$\|k_a*u\|_q^q\leqq C''\int t^{q-1-a}\Big(\int_{|u|>s}|u|\,dx\Big)^a dt$$
$$\leqq C''\Big(\int\Big(\int_{s<|u(x)|}t^{q-1-a}\,dt\Big)^{1/a}|u(x)|\,dx\Big)^a.$$

The integral with respect to t is proportional to t^{q-a} since $q>a$, and when $s=|u(x)|$ this is proportional to

$$|u(x)|^{(q-a)p/q}=|u(x)|^{ap/p'}.$$

Altogether we have therefore

$$\|k_a*u\|_q^q \leqq C_3(\int |u(x)|^{1+p/p'}\,dx)^a = C_3(\int |u(x)|^p\,dx)^a = C_3,$$

which completes the proof of (4.5.5)′.

As an application we shall now prove the Sobolev embedding theorems, for which Theorem 4.5.3 was in fact originally intended. First we give a local form.

Theorem 4.5.8. *Let $u\in\mathscr{D}'(X)$ where X is an open set in $\mathbb{R}^n$, and assume that $\partial_j u\in L^p_{\text{loc}}(X)$, $j=1,\dots,n$, where $1<p<n$. Then it follows that $u\in L^q_{\text{loc}}(X)$ if*

$$1/p=1/q+1/n. \tag{4.5.18}$$

Proof. Let E be the fundamental solution of the Laplacean given in Theorem 3.3.2 and set $E_j=\partial_j E=x_j|x|^{-n}/c_n$. Then

$$|E_j(x)|\leqq |x|^{-n/a}/c_n, \qquad 1/a=1-1/n,$$

so Theorem 4.5.3 gives, C_0 being dense in L^p,

$$\|E_j*v\|_q \leqq C\,\|v\|_p, \qquad v\in L^p\cap\mathscr{E}'. \tag{4.5.19}$$

Now choose $\chi\in C_0^\infty(X)$ equal to 1 in a large open subset Y of X. Then

$$\begin{aligned}\chi u = E*\Delta(\chi u) &= \sum E_j*\partial_j(\chi u)\\ &= \sum E_j*(\chi(\partial_j u)) + \sum E_j*(u\partial_j\chi).\end{aligned}$$

Here $E_j*(\chi(\partial_j u))\in L^q$ by (4.5.19) and $E_j*(u\partial_j\chi)\in C^\infty(Y)$ by Theorem 4.2.5, so $u\in L^q_{\text{loc}}(Y)$, which proves the theorem.

Next we give a global version of Theorem 4.5.8:

Theorem 4.5.9. *Let $u\in\mathscr{D}'(\mathbb{R}^n)$ and assume that $\partial_j u\in L^p(\mathbb{R}^n)$, $j=1,\dots,n$ where $1<p<n$. Then there is a constant C such that $u-C\in L^q(\mathbb{R}^n)$ where q is determined by* (4.5.18).

Proof. With the notation used in the proof of Theorem 4.5.8 we set

$$v=\sum E_j*\partial_j u\in L^q(\mathbb{R}^n)$$

and must prove that $\partial_k v = \partial_k u$, $k=1,\dots,n$, which implies that $v-u$ is a constant. Here $L^p \ni f \to E_j * f \in L^q$ denotes the continuous extension of the convolution map defined on C_0. Choose $\chi \in C_0^\infty(\mathbb{R}^n)$ with $0 \leqq \chi \leqq 1$ and $\chi = 1$ in a neighborhood of 0. With $E_j^\varepsilon(x) = \chi(\varepsilon x) E_j(x)$ we have if $w \in L^p$

$$E_j^\varepsilon * w \to E_j * w \quad \text{in } L^q(\mathbb{R}^n) \quad \text{when } \varepsilon \to 0.$$

Since $\|E_j^\varepsilon * w\|_q \leqq C\|w\|_p$ with C independent of ε it suffices to prove this when $w \in C_0$. Then we have $E_j^\varepsilon * w = E_j * w$ on any compact set when ε is small, and $|E_j^\varepsilon * w| \leqq |E_j| * |w| \in L^q$ so the statement follows by dominated convergence. Hence we have with convergence in $\mathscr{D}'$

$$v = \lim_{\varepsilon \to 0} \sum E_j^\varepsilon * \partial_j u, \qquad \partial_k v = \lim_{\varepsilon \to 0} \sum E_j^\varepsilon * \partial_k \partial_j u = \lim_{\varepsilon \to 0} \sum \partial_j E_j^\varepsilon * \partial_k u.$$

Here

$$\sum \partial_j E_j^\varepsilon = \chi(\varepsilon x) \Delta E + \varepsilon \sum \chi_j(\varepsilon x) E_j = \delta_0 + \varepsilon \sum \chi_j(\varepsilon x) E_j$$

where $\chi_j = \partial_j \chi$. The L^q norm of $(\chi_j(\varepsilon x) E_j) * \partial_k u$ has a uniform bound as $\varepsilon \to 0$, so $\partial_k v = \partial_k u$ as claimed.

When p increases to n then the exponent q increases to ∞, and the preceding result breaks down in the limiting case. However, when $p > n$ we have a substitute result based on a supplement to Theorem 4.5.3.

Theorem 4.5.10. *Assume that $k \in C^1(\mathbb{R}^n \setminus 0)$ is homogeneous of degree $-n/a$, let $1 \leqq p \leqq \infty$ and assume that*

$$0 < \gamma = n(1 - 1/a - 1/p) < 1. \tag{4.5.20}$$

Then we have

$$\sup_{x \neq y} |k * u(x) - k * u(y)|\, |x-y|^{-\gamma} \leqq C \|u\|_{L^p}, \qquad u \in L^p(\mathbb{R}^n) \cap \mathscr{E}'(\mathbb{R}^n). \tag{4.5.21}$$

Proof. The convolution is a continuous function for

$$\Big(\int_{|y|<R} |k(y)|^{p'} dy\Big)^{1/p'} \leqq C R^{(n - np'/a)/p'} = C R^\gamma. \tag{4.5.22}$$

In proving (4.5.21) we assume $y = 0$ and set $h = |x|$. Then

$$k * u(x) - k * u(0) = \int (k(x-y) - k(-y)) u(y)\, dy$$

and we split the integral in the part with $|y| < 2h$ and that with $|y| > 2h$. The first part is $< C' h^\gamma \|u\|_p$ by (4.5.22). In the second part we use that $|k(x+y) - k(y)| \leqq C h |y|^{-1-n/a}$ by the mean value theorem (see also the proof of Lemma 4.5.6) which is in $L^{p'}$ at ∞ since

$$n - p'(1 + n/a) = p' n(1/p' - 1/a) - p' = p'(\gamma - 1) < 0.$$

Hence

$$\Big(\int_{|y|>2h} |k(x-y)-k(-y)|^{p'}\,dy\Big)^{1/p'} \leqq C h^{\gamma}$$

which proves (4.5.21).

The following is an analogue of Theorem 4.5.9.

Theorem 4.5.11. *Let $u\in\mathscr{D}'(\mathbb{R}^n)$ and assume that $\partial_j u\in L^p(\mathbb{R}^n)$, $j=1,\dots,n$, where $p>n$. Then u is a continuous function and with $\gamma=1-n/p$ we have*

$$\sup_{x\neq y} |u(x)-u(y)|/|x-y|^{\gamma} \leqq C\sum \|\partial_j u\|_p. \tag{4.5.23}$$

Proof. The modified convolution

$$v(x)=\int\sum (E_j(x-y)-E_j(-y))\,\partial_j u(y)\,dy$$

defines a continuous function by the proof of Theorem 4.5.10, and the estimate (4.5.23) is valid with v instead of u in the left hand side. Here the notations are the same as in the proof of Theorem 4.5.9. We can repeat the proof there to show that $\partial_k v=\partial_k u$ for every k if we note that the $L_{p'}$ norm of $\varepsilon\chi_j(\varepsilon x)E_j$ is $O(\varepsilon^{1-\gamma})$ by (4.5.22). This proves the theorem.

The local version of Theorem 4.5.11 which follows is proved just as Theorem 4.5.8, with reference to Theorem 4.5.10 instead of Theorem 4.5.3.

Theorem 4.5.12. *Let $u\in\mathscr{D}'(X)$ where X is an open set in $\mathbb{R}^n$, and assume that $\partial_j u\in L^p_{\text{loc}}(X)$, $j=1,\dots,n$, where $p>n$. Then it follows that u is Hölder continuous of order $\gamma=1-n/p$, that is,*

$$\sup_{x\neq y;\, x,y\in K} |u(x)-u(y)|/|x-y|^{\gamma}<\infty \quad \text{if } K\Subset X. \tag{4.5.24}$$

By repeated use of the preceding theorems we obtain the full Sobolev embedding theorems:

Theorem 4.5.13. *Let $u\in\mathscr{D}'(X)$ where X is an open set in $\mathbb{R}^n$ and assume that $\partial^\alpha u\in L^p_{\text{loc}}(X)$ when $|\alpha|=m$. Here m is a positive integer and $1<p<\infty$. If $|\alpha|<m$ then*

(i) *$\partial^\alpha u\in L^q_{\text{loc}}(X)$ if $q<\infty$ and $1/p\leqq 1/q+(m-|\alpha|)/n$*

(ii) *$\partial^\alpha u$ is Hölder continuous of order γ if $0<\gamma<1$ and*

$$1/p\leqq (m-|\alpha|-\gamma)/n.$$

Proof. (i) follows from Theorem 4.5.8 by induction for decreasing $|\alpha|$. To prove (ii) it is by Theorem 4.5.12 sufficient to show that $\partial_j \partial^\alpha u \in L^q_{\mathrm{loc}}$ when $j=1, \dots, n$ if $n/q = 1-\gamma$. But then we have

$$1/p \leqq 1/q + (m - |\alpha| - 1)/n$$

so this follows from (i) or the hypothesis.

The global result is parallel but with equality in the conditions (i), (ii), and we omit the statement.

Notes

The applications of regularization to convex, subharmonic and plurisubharmonic functions in Section 4.1 are mainly intended to illustrate the use of regularization, but some of the results are required in Chapters XV and XVI. For a more thorough discussion of convex functions and sets the reader might consult the classical text by Bonnesen and Fenchel [1]. We shall continue the study of subharmonic and plurisubharmonic functions in Chapter XVI using potential representation formulas. There is a recent monograph by Hayman and Kennedy [1] on subharmonic functions, and the reader could consult Lelong [1] for the theory of plurisubharmonic functions. Theorem 4.1.12 actually goes back to Poincaré [1], and Theorem 4.1.15 is due to Lelong [2].

In Section 4.2 we followed Gelfand and Shilov [1] in the definition of the convolution. The definition of Schwartz will be given in Section 5.1. The theorem of supports in Section 4.3 is due to Titchmarsh [1] in one dimension. The simple extension to n variables was given by Lions [1]. Most proofs depend more or less on analytic function theory (see Section 16.3). Mikusiński [1] gave an argument reproduced here which reduces the proof to the case of a convolution with equal factors. The theorem is then a consequence of the Paley-Wiener theorem (Theorem 7.3.1). We give a direct elementary proof using only convolutions. A similar argument occurs in Mikusiński [2].

The observations in Section 4.4 on the use of fundamental solutions will be developed very systematically in Chapters X and XI so we refrain from discussing the results here. The estimates in Section 4.5 are due to Hardy and Littlewood [1] when $n=1$. Sobolev [2] gave a rather difficult reduction of the n dimensional case to the one dimensional case by means of spherical symmetrization. Later on

DuPlessis [1] has observed that there is a very simple reduction by means of the inequality between geometric and arithmetic means. Here we have chosen proofs in the spirit of Zygmund [1] which will be applicable also in a related context in Section 7.9. The main point is a covering lemma due to Calderón and Zygmund [1], which is actually due to F. Riesz [1] in the one dimensional case. The reader should consult Nirenberg [4] for a thorough discussion of results related to the Sobolev embedding theorems. A very simple derivation from the Hardy-Littlewood maximal theorem can be found in Hedberg [1].

Chapter V. Distributions in Product Spaces

Summary

We were not able to define the product of arbitrary distributions in Chapter III. However, as we shall now see this can always be done when they depend on different sets of variables. Thus to arbitrary distributions $u_j \in \mathscr{D}'(X_j)$, X_j open in $\mathbb{R}^{n_j}$ $(j=1,2)$, we define in Section 5.1 a product $u_1 \otimes u_2 \in \mathscr{D}'(X_1 \times X_2)$ in $X_1 \times X_2 \subset \mathbb{R}^{n_1+n_2}$. In case u_j are functions this is the function $X_1 \times X_2 \ni (x_1, x_2) \to u_1(x_1) u_2(x_2)$.

On the other hand, a function $K \in C(X_1 \times X_2)$ can be viewed as the kernel of an integral operator $\mathscr{K}$,

$$(\mathscr{K} u)(x_1) = \int K(x_1, x_2) u(x_2) dx_2,$$

mapping $C_0(X_2)$ to $C(X_1)$ say. It is not easy to characterize the operators having such a kernel. However, the analogue in the theory of distributions is very satisfactory. It is called the Schwartz kernel theorem and states that the distributions $K \in \mathscr{D}'(X_1 \times X_2)$ can be identified with the continuous linear maps $\mathscr{K}$ from $C_0^\infty(X_2)$ to $\mathscr{D}'(X_1)$ which they define. This will be proved in Section 5.2. We shall return to this topic in Section 8.2. A rather precise classification of singularities will then allow us to discuss the regularity of $\mathscr{K} u$ and its definition when u is not smooth.

5.1. Tensor Products

If X_j is an open set in $\mathbb{R}^{n_j}$, $j=1,2$, and if $u_j \in C(X_j)$, then the function $u_1 \otimes u_2$ in $X_1 \times X_2 \subset \mathbb{R}^{n_1+n_2}$ defined by

$$(u_1 \otimes u_2)(x_1, x_2) = u_1(x_1) u_2(x_2), \qquad x_j \in X_j,$$

is called the direct (or tensor) product of u_1 and u_2. To extend the definition to distributions we observe that $u_1 \otimes u_2 \in C(X_1 \times X_2)$ and that

$$\iint (u_1 \otimes u_2)(\phi_1 \otimes \phi_2)\, dx_1\, dx_2 = \int u_1\, \phi_1\, dx_1 \int u_2\, \phi_2\, dx_2, \qquad \phi_j \in C_0^\infty(X_j).$$

Theorem 5.1.1. *If $u_j \in \mathscr{D}'(X_j)$, $j=1,2$, then there is a unique distribution $u \in \mathscr{D}'(X_1 \times X_2)$ such that*

$$u(\phi_1 \otimes \phi_2) = u_1(\phi_1) u_2(\phi_2), \qquad \phi_j \in C_0^\infty(X_j). \tag{5.1.1}$$

We have

$$u(\phi) = u_1[u_2(\phi(x_1, x_2))] = u_2[u_1(\phi(x_1, x_2))], \qquad \phi \in C_0^\infty(X_1 \times X_2), \tag{5.1.2}$$

where u_j acts on the following function as a function of x_j only. If $u_j \in \mathscr{E}'$, $j=1, 2$, the same formula is valid for $\phi \in C^\infty$. The distribution u is called the tensor product and one writes $u = u_1 \otimes u_2$.

Proof. a) Uniqueness. We must show that if $u \in \mathscr{D}'(X_1 \times X_2)$ and if

$$u(\phi_1 \otimes \phi_2) = 0 \qquad \text{for } \phi_j \in C_0^\infty(X_j),$$

then $u=0$. To do so we take $\psi_j \in C_0^\infty(\mathbb{R}^{n_j})$ with $\psi_j \geqq 0$, $\int \psi_j dx_j = 1$, and $|x_j| \leqq 1$ if $x_j \in \operatorname{supp} \psi_j$. With

$$\Psi_\varepsilon(x_1, x_2) = \varepsilon^{-n_1 - n_2} \psi_1(x_1/\varepsilon) \psi_2(x_2/\varepsilon)$$

we know that $u * \Psi_\varepsilon \to u$ in $\mathscr{D}'(Y)$ if $Y \Subset X_1 \times X_2$ (Theorem 4.1.4). However, $u * \Psi_\varepsilon = 0$ in Y for small ε since $\Psi_\varepsilon(x_1 - y_1, x_2 - y_2)$ is the product of a function of y_1 and one of y_2. Hence $u=0$ in Y and therefore in X.

b) Existence of u and (5.1.2). Let K_j be a compact subset of X_j. Then

$$|u_j(\phi_j)| \leqq C_j \sum_{|\alpha| \leqq k_j} \sup |\partial^\alpha \phi_j|, \qquad \phi_j \in C_0^\infty(K_j).$$

If $\phi \in C_0^\infty(K_1 \times K_2)$ then

$$I_\phi(x_1) = u_2(\phi(x_1, .))$$

is in $C_0^\infty(K_1)$ by Theorem 2.1.3, and

$$\partial_{x_1}^\alpha I_\phi(x_1) = u_2(\partial_{x_1}^\alpha \phi(x_1, .)).$$

Hence

$$\sup |\partial_{x_1}^\alpha I_\phi(x_1)| \leqq C_2 \sum_{|\beta| \leqq k_2} \sup |\partial_{x_1}^\alpha \partial_{x_2}^\beta \phi(x_1, x_2)|$$

so $u_1(I_\phi)$ is defined and

$$|u_1(I_\phi)| \leqq C_1 C_2 \sum_{|\alpha_j| \leqq k_j} \sup |\partial_{x_1}^{\alpha_1} \partial_{x_2}^{\alpha_2} \phi(x_1, x_2)|.$$

Writing $u(\phi)=u_1(I_\phi)$ we obtain a distribution satisfying (5.1.1) and the first part of (5.1.2). In the same way we obtain a distribution satisfying (5.1.1) and with the second property in (5.1.2). By the uniqueness both conditions (5.1.2) must therefore be valid. The remaining statement follows in the same way; note that

$$\operatorname{supp} u_1 \otimes u_2 = \operatorname{supp} u_1 \times \operatorname{supp} u_2. \tag{5.1.3}$$

Example 5.1.2. If δ_{a_j} is the Dirac measure at $a_j \in X_j$, then $\delta_{a_1} \otimes \delta_{a_2}$ is the Dirac measure δ_a at $a=(a_1, a_2) \in X_1 \times X_2$. Theorem 2.3.5 can now be stated as follows: If $u \in \mathscr{D}'(X_1 \times X_2)$ is of order k and if $\operatorname{supp} u \subset X_1 \times \{a_2\}$, $a_2 \in X_2$, then

$$u = \sum_{|\alpha| \leqq k} u_\alpha \otimes \partial^\alpha \delta_{a_2}$$

where $u_\alpha \in \mathscr{D}'^{k-|\alpha|}(X_1)$ and α is a multi-index corresponding to the X_2 variables.

The direct product allows us to justify the definition (1.3.1)′ of the convolution in general. In fact, if $u_1, u_2 \in \mathscr{D}'(\mathbb{R}^n)$ and either one has compact support, then

$$(u_1 \otimes u_2)(\phi(x_1+x_2)) = (u_1 * u_2)(\phi), \qquad \phi \in C_0^\infty(\mathbb{R}^n), \tag{5.1.4}$$

for if $\check{\phi}(x)=\phi(-x)$ we have

$$u_2(\phi(x_1+x_2)) = (u_2 * \check{\phi})(-x_1)$$

so the left-hand side is $u_1 * (u_2 * \check{\phi})(0) = (u_1 * u_2) * \check{\phi}(0)$. (5.1.4) could also have been taken as definition of convolution. However, it was convenient to have convolution available in the proof of Theorem 5.1.1.

5.2. The Kernel Theorem

Every function $K \in C(X_1 \times X_2)$ defines an integral operator $\mathscr{K}$ from $C_0(X_2)$ to $C(X_1)$ by the formula

$$(\mathscr{K}\phi)(x_1) = \int K(x_1, x_2)\, \phi(x_2)\, dx_2, \qquad \phi \in C_0(X_2),\ x_1 \in X_1.$$

We shall now show that the definition can be extended to arbitrary distributions K if ϕ is restricted to C_0^∞ and $\mathscr{K}\phi$ is allowed to be a distribution. To do so we start from the observation that when $K \in C(X_1 \times X_2)$

$$\langle \mathscr{K}\phi, \psi \rangle = K(\psi \otimes \phi); \qquad \psi \in C_0^\infty(X_1),\ \phi \in C_0^\infty(X_2). \tag{5.2.1}$$

Theorem 5.2.1 (The Schwartz kernel theorem). *Every $K \in \mathscr{D}'(X_1 \times X_2)$ defines according to (5.2.1) a linear map $\mathscr{K}$ from $C_0^\infty(X_2)$ to $\mathscr{D}'(X_1)$*

which is continuous in the sense that $\mathscr{K}\phi_j \to 0$ *in* $\mathscr{D}'(X_1)$ *if* $\phi_j \to 0$ *in* $C_0^\infty(X_2)$. *Conversely, to every such linear map* $\mathscr{K}$ *there is one and only one distribution* K *such that* (5.2.1) *is valid. One calls* K *the kernel of* $\mathscr{K}$.

Proof. If $K \in \mathscr{D}'(X_1 \times X_2)$ then (5.2.1) defines a distribution $\mathscr{K}\phi$ since $\psi \to K(\psi \otimes \phi)$ is continuous; $\mathscr{K}$ is continuous since $\phi \to K(\psi \otimes \phi)$ is continuous. To prove the converse we first note that the uniqueness is identical to the uniqueness in Theorem 5.1.1. To prove the existence we observe that for any compact sets $K_j \subset X_j$ there are constants C, N_j such that

$$(5.2.2) \qquad |\langle \mathscr{K}\phi, \psi\rangle| \leqq C \sum_{|\alpha| \leqq N_1} \sup |\partial^\alpha \psi| \sum_{|\beta| \leqq N_2} \sup |\partial^\beta \phi|;$$

$$\psi \in C_0^\infty(K_1),\ \phi \in C_0^\infty(K_2).$$

In fact, by hypothesis the bilinear form

$$C_0^\infty(K_1) \times C_0^\infty(K_2) \ni (\psi, \phi) \to \langle \mathscr{K}\phi, \psi\rangle$$

is continuous with respect to ϕ (resp. ψ) for fixed ψ (resp. ϕ), and every separately continuous bilinear form in a product of Fréchet spaces is continuous.

Let $Y_j \Subset X_j$, choose the compact sets K_j as neighborhoods of $\bar{Y}_j$ and set for $(x_1, x_2) \in Y_1 \times Y_2$ and small $\varepsilon > 0$

$$(5.2.3) \qquad K_\varepsilon(x_1, x_2) = \varepsilon^{-n_1-n_2} \langle \mathscr{K}\psi_2((x_2 - .)/\varepsilon), \psi_1((x_1 - .)/\varepsilon)\rangle$$

where ψ_j are chosen as in the proof of Theorem 5.1.1. Note that if we already knew that there is a distribution K satisfying (5.2.1) then K_ε would be $K * \Psi_\varepsilon$ and therefore converge to K as $\varepsilon \to 0$. Our program is now to show that K_ε does have a limit in $\mathscr{D}'(Y_1 \times Y_2)$ when $\varepsilon \to 0$ and then to show that (5.2.1) is fulfilled for the limit.

(5.2.3) is well defined when ε is smaller than the distance from Y_j to $\complement K_j$, and by (5.2.2) we have with $\mu = N_1 + N_2 + n_1 + n_2$

$$(5.2.4) \qquad |K_\varepsilon(x_1, x_2)| \leqq C\varepsilon^{-\mu} \quad \text{if } x_j \in Y_j,\ j = 1, 2.$$

We shall prove that K_ε has a limit in $\mathscr{D}'^{\mu+1}(Y_1 \times Y_2)$ as $\varepsilon \to 0$ by using an argument which is very close to the proof of Theorem 3.1.11. Note that if $\psi \in C^\infty(\mathbb{R}^n)$ then

$$(5.2.5) \qquad \frac{\partial}{\partial \varepsilon}(\varepsilon^{-n}\psi(x/\varepsilon)) = \sum \frac{\partial}{\partial x_j}(\varepsilon^{-n}\psi_j(x/\varepsilon)), \qquad \psi_j(x) = -x_j\psi(x).$$

In fact, by the homogeneity

$$\varepsilon \frac{\partial}{\partial \varepsilon}(\varepsilon^{-n}\psi(x/\varepsilon)) + \sum x_j \frac{\partial}{\partial x_j}(\varepsilon^{-n}\psi(x/\varepsilon)) = -n\varepsilon^{-n}\psi(x/\varepsilon)$$

which implies (5.2.5). Now it follows from the continuity (5.2.2) that we may differentiate with respect to ε or x_j in (5.2.3), and by (5.2.5) this gives

$$\partial K_\varepsilon(x_1, x_2)/\partial\varepsilon = \sum_\nu \partial L_\varepsilon^\nu(x_1, x_2)/\partial x_\nu$$

where x_ν runs over all coordinates of (x_1, x_2). Here L_ε^ν is defined by replacing ψ_1 or ψ_2 by the product with $-x_\nu$, so (5.2.4) is valid for L_ε^ν. Repeating this process we conclude that

$$K_\varepsilon^{(j)}(x_1, x_2) = \partial^j K_\varepsilon(x_1, x_2)/\partial\varepsilon^j$$

is a sum of derivatives of order j of functions having a bound of the form (5.2.4), so $\varepsilon^\mu K_\varepsilon^{(j)}$ is bounded in $\mathscr{D}'^j(Y_1 \times Y_2)$ for every j. With a fixed small δ and $\varepsilon \to 0$ we now use Taylor's formula

$$K_\varepsilon = \sum_0^\mu (\varepsilon-\delta)^j K_\delta^{(j)}/j! + (\varepsilon-\delta)^{\mu+1} \int_0^1 K_{\delta+t(\varepsilon-\delta)}^{(\mu+1)} (1-t)^\mu/\mu!\, dt.$$

Since

$$(1-t)^\mu/(\delta+t(\varepsilon-\delta))^\mu \leqq \delta^{-\mu}$$

it follows for $\Phi \in C_0^{\mu+1}(Y_1 \times Y_2)$ that when $\varepsilon \to 0$

$$\langle K_\varepsilon, \Phi\rangle \to \langle K_0, \Phi\rangle$$
$$= \sum_0^\mu (-\delta)^j \langle K_\delta^{(j)}/j!, \Phi\rangle + (-\delta)^{\mu+1} \int_0^1 \langle K_{\delta(1-t)}^{(\mu+1)}, \Phi\rangle (1-t)^\mu/\mu!\, dt,$$

where $K_0 \in \mathscr{D}'^{\mu+1}(Y_1 \times Y_2)$.

Let $\phi_j \in C_0^\infty(Y_j)$ and form

$$\langle K_\varepsilon, \phi_1 \otimes \phi_2\rangle = \iint K_\varepsilon(x_1, x_2)\phi_1(x_1)\phi_2(x_2)\, dx_1\, dx_2.$$

With the notation $\check{\psi}_{j,\varepsilon}(x_j) = \varepsilon^{-n_j}\psi_j(-x_j/\varepsilon)$ we have

$$\iint K_\varepsilon(x_1, x_2)\phi_1(x_1)\phi_2(x_2)\, dx_1\, dx_2$$
$$= \iint \langle \mathscr{K}\check{\psi}_{2,\varepsilon}(\cdot - x_2)\phi_2(x_2), \check{\psi}_{1,\varepsilon}(\cdot - x_1)\phi_1(x_1)\rangle\, dx_1\, dx_2.$$

Replacing the integral by a Riemann sum first we conclude as in the proof of Lemma 4.1.3 that the integration can performed "under the sign", hence

$$\langle K_\varepsilon, \phi_1 \otimes \phi_2\rangle = \langle \mathscr{K}(\phi_2 * \check{\psi}_{2,\varepsilon}), \phi_1 * \check{\psi}_{1,\varepsilon}\rangle.$$

Since $\phi_j * \check{\psi}_{j,\varepsilon} \to \phi_j$ in $C_0^\infty(Y_j)$ when $\varepsilon \to 0$, it follows from (5.2.2) that the right-hand side converges to $\langle \mathscr{K}\phi_2, \phi_1\rangle$ when $\varepsilon \to 0$. Thus

$$\langle K_0, \phi_1 \otimes \phi_2\rangle = \langle \mathscr{K}\phi_2, \phi_1\rangle \quad \text{if } \phi_j \in C_0^\infty(Y_j),$$

and since Y_j are arbitrary relatively compact subsets of X_j, this completes the proof.

Example 5.2.2. The kernel of the identity map $\mathscr{K}: C_0^\infty(X) \to C_0^\infty(X)$, where X is an open set in $\mathbb{R}^n$, is the distribution

$$\langle K, \Phi \rangle = \int \Phi(x, x)\, dx, \qquad \Phi \in C_0^\infty(X \times X),$$

with support in the diagonal $\{(x, x), x \in X\}$.

Theorem 5.2.3. *The kernel of a continuous map* $\mathscr{K}: C_0^\infty(X) \to \mathscr{D}'(X)$ *is supported by the diagonal if and only if*

$$\mathscr{K}\phi = \sum a_\alpha \partial^\alpha \phi \tag{5.2.6}$$

where $a_\alpha \in \mathscr{D}'(X)$ *and the sum is locally finite.*

Proof. For the operator (5.2.6) we have

$$\langle \mathscr{K}\phi, \psi \rangle = \sum \langle a_\alpha, (\partial^\alpha \phi)\psi \rangle$$

so the kernel is given by

$$\langle K, \Phi \rangle = \sum \langle a_\alpha, \partial_y^\alpha \Phi(x, y)|_{x=y} \rangle$$

which is obviously supported by the diagonal. Conversely, if the kernel K of $\mathscr{K}$ is supported by the diagonal, it follows from Theorem 2.3.5 that K has the preceding form, which proves the theorem.

The preceding operators preserve supports; more generally, we have

Theorem 5.2.4. *If* $K \in \mathscr{D}'(X_1 \times X_2)$ *and* $\mathscr{K}$ *is the corresponding operator, then*

$$\operatorname{supp} \mathscr{K}u \subset \operatorname{supp} K \circ \operatorname{supp} u, \qquad u \in C_0^\infty(X_2). \tag{5.2.7}$$

Here $\operatorname{supp} K \subset X_1 \times X_2$ is considered as a relation acting on $\operatorname{supp} u \subset X_2$. Thus

$$\operatorname{supp} K \circ M = \{x_1 \in X_1;\ \exists x_2 \in M, (x_1, x_2) \in \operatorname{supp} K\}.$$

This is a closed set when M is compact, for $\operatorname{supp} K$ is closed.

Proof. Assume that $x_1 \notin \operatorname{supp} K \circ \operatorname{supp} u$. Then there is a neighborhood V of x_1 such that $V \cap (\operatorname{supp} K \circ \operatorname{supp} u) = \emptyset$. If $v \in C_0^\infty(V)$ then

$$(\operatorname{supp} v \otimes u) \cap \operatorname{supp} K = \emptyset$$

which proves that $\langle \mathscr{K}u, v \rangle = 0$, hence $\mathscr{K}u = 0$ in V.

Example 5.2.5. If $f: X_1 \to X_2$ is a continuous map and $\mathscr{K}\phi = \phi \circ f$, $\phi \in C_0^\infty(X_2)$, then the kernel is given by

$$\langle K, \Phi \rangle = \int \Phi(x, f(x))\, dx, \qquad \Phi \in C_0^\infty(X_1 \times X_2),$$

so the support is in the graph of f.

The operator (5.2.6) has a natural extension to all $\phi \in \mathscr{E}'$ if the coefficients $a_\alpha \in C^\infty$. General sufficient smoothness conditions for the existence of such extensions will be given in Chapter VIII, but we give an elementary example now:

Theorem 5.2.6. *If* $K\in C^\infty(X_1\times X_2)$ *then the map* $\mathscr{K}$ *defined by* (5.2.1) *has a continuous extension from* $\mathscr{E}'(X_2)$ *to* $C^\infty(X_1)$,

$$\mathscr{K}u(x_1)=u(K(x_1,.)),\qquad u\in\mathscr{E}'(X_2),\ x_1\in X_1. \tag{5.2.8}$$

Conversely, every continuous linear map $\mathscr{K}$ *from* $\mathscr{E}'(X_2)$ *to* $C^\infty(X_1)$ *is defined in this way by a kernel* $K\in C^\infty(X_1\times X_2)$.

Proof. If $K\in C^\infty$ it follows from Theorem 2.1.3 that (5.2.8) defines a map $\mathscr{E}'(X_2)\to C^\infty(X_1)$, and the continuity is a consequence of Theorem 2.1.8. Conversely, if we are given a continuous map $\mathscr{K}$: $\mathscr{E}'(X_2)\to C^\infty(X_1)$ then

$$K(.,x_2)=\mathscr{K}\delta_{x_2},\qquad x_2\in X_2,$$

is a continuous function of x_2 with values in $C^\infty(X_1)$. Taking difference quotients we find that K is continuously differentiable in x_2,

$$\langle y,\partial_{x_2}\rangle K(.,x_2)=-\mathscr{K}\langle y,\partial\rangle\delta_{x_2}.$$

Repeating the argument gives $K\in C^\infty(X_1\times X_2)$. We obtain (5.2.8) since this is true for finite linear combinations of Dirac measures and they are dense in $\mathscr{E}'$.

Notes

The tensor product was defined in Schwartz [1], and the kernel theorem was announced shortly afterwards in Schwartz [2]. In both cases the main point is the decomposition of test functions in $X_1\times X_2$ into sums of tensor products of test functions in X_1 and in X_2. Thus the topological tensor product of $C_0^\infty(X_1)$ and $C_0^\infty(X_2)$ is involved, and Schwartz [4] gave a proof emphasizing this aspect. Ehrenpreis [4] published a more elementary proof where the decomposition was made by Fourier series expansion (see also Gask [1]). Here we have used instead the fact that a regularization of any test function in $X_1\times X_2$ by a product of two test functions in the corresponding spaces $\mathbb{R}^{n_1}$ and $\mathbb{R}^{n_2}$ can be considered as a superposition of tensor products of test functions in X_1 and in X_2. (To be able to use this argument we had to define convolution before the tensor product.)

There is an interesting addendum to Theorem 5.2.3 due to Peetre [2]: If $\mathscr{K}$ is any linear map $C_0^\infty(X)\to C^\infty(X)$ with $\operatorname{supp}\mathscr{K}u\subset\operatorname{supp}u$, $u\in C_0^\infty(X)$, then $\mathscr{K}$ is a differential operator with C^∞ coefficients, that is, (5.2.6) is valid with $a_\alpha\in C^\infty$. Note that no continuity is assumed; it follows from the restriction on the supports.

Chapter VI. Composition with Smooth Maps

Summary

If f is a map $\mathbb{R}^n \to \mathbb{R}^m$ then a function u in $\mathbb{R}^m$ can be pulled back to a function $u \circ f$ in $\mathbb{R}^n$, the composition. In Section 6.1 we show that this operation can be defined for all distributions u if $f \in C^\infty$ and the differential is surjective. (In Section 8.2 we shall find that the composition can be defined for more general maps f when the location of the singularities of u is known in a rather precise sense.) As an example we discuss in Section 6.2 how powers of real quadratic forms can be used to construct fundamental solutions for homogeneous second order differential operators with real coefficients. In Section 6.3 we use the fact that distributions can be composed with diffeomorphisms to define distributions on C^∞ manifolds simply as distributions in the local coordinates which behave right when the coordinates are changed. In Section 6.4 we continue the discussion of manifolds by giving a short review of the calculus of differential forms on a manifold, ending up with the Hamilton-Jacobi integration theory for first order differential equations. These results will not be used until Chapter VIII, and the geometrical notions related to the Hamilton-Jacobi theory will be discussed in much greater depth in Chapter XXI.

6.1. Definitions

Let X_j be an open set in $\mathbb{R}^{n_j}$, $j=1,2$, and let $f: X_1 \to X_2$ be a C^∞ map. We wish to extend the definition of the composition

$$C^0(X_2) \ni u \to u \circ f \in C^0(X_1)$$

to distributions u so that the map

$$\mathscr{D}'(X_2) \ni u \to u \circ f \in \mathscr{D}'(X_1)$$

is continuous. If this is possible it follows from Theorem 4.1.5 that it can only be done in one way. However, we must put conditions on f.

Theorem 6.1.1. *If $u_j \circ f \to 0$ in $\mathscr{D}'(X_1)$ for every sequence $u_j \in C_0^\infty(X_2)$ such that $u_j \to 0$ in $\mathscr{D}'(X_2)$, then f is open, that is, $f(V)$ is open in X_2 if $V \subset X_1$ is open. If $u \in C_0^\infty(X_2)$ implies $u \circ f \in C_0^\infty(X_1)$, then f is proper.*

Proof. The second statement is obvious, for if K is a compact subset of X_2 we can choose u equal to 1 on K and obtain $u \circ f = 1$ on $f^{-1}(K)$ which must therefore be compact if $u \circ f \in C_0^\infty(X_1)$. To prove the first statement we assume that f is not open. Assume for example that $0 \in X_1$, $f(0)=0$, and that V is a compact neighborhood of 0 in X_1 such that $f(V)$ is not a neighborhood of 0 in X_2. Choose a sequence $y_j \notin f(V)$ so that $y_j \to 0$, and set $|y_j| = \varepsilon_j$. With $0 \leqq u \in C_0^\infty(\mathbb{R}^{n_2})$ equal to 1 in the unit ball, we set $u_\varepsilon(y) = \varepsilon^{-n_1} u(y/\varepsilon)$ and obtain $u_\varepsilon \circ f(x) \geqq \varepsilon^{-n_1}$ when $C|x| < \varepsilon$. Since $u_\varepsilon \circ f \geqq 0$ it follows in view of Theorem 2.1.9 that $u_\varepsilon \circ f$ does not converge to 0 in $\mathscr{D}'(X_1)$ as $\varepsilon \to 0$. Now

$$v_j = u_{\varepsilon_j} - \sum_{|\alpha| \leqq \mu} a_{j,\alpha} \partial^\alpha \delta_{y_j}$$

tends to 0 in $\mathscr{D}'(X_2)$ when $j \to \infty$ if $n_2 - n_1 + \mu \geqq 0$ and $a_{j,\alpha}$ are suitably chosen. In fact,

$$\begin{aligned}\langle v_j, \phi \rangle &= \varepsilon_j^{n_2-n_1} \int \phi(\varepsilon_j y) u(y) dy - \sum_{|\alpha| \leqq \mu} a_{j,\alpha} (-\partial)^\alpha \phi(y_j) \\ &= \varepsilon_j^{n_2-n_1} \int \sum_{|\alpha| \leqq \mu} \partial^\alpha \phi(y_j)(\varepsilon_j y - y_j)^\alpha/\alpha!\, u(y) dy + O(\varepsilon_j^{n_2-n_1+\mu+1}) \\ &\quad - \sum_{|\alpha| \leqq \mu} a_{j,\alpha} (-\partial)^\alpha \phi(y_j)\end{aligned}$$

so we just have to take

$$a_{j,\alpha} = (-1)^{|\alpha|} \varepsilon_j^{n_2-n_1} \int (\varepsilon_j y - y_j)^\alpha u(y) dy/\alpha!.$$

If we replace $\partial^\alpha \delta_{y_j}$ by a smooth approximation with support in $\complement f(V)$ we obtain a sequence $v_j' \to 0$ in $\mathscr{D}'(X_2)$ such that $v_j' \circ f = u_{\varepsilon_j} \circ f$ in V. This does not converge to 0 in $\mathscr{D}'(X_1)$ so the theorem is proved.

If f is open then $f'(x)$ is surjective for all x in an open dense subset of X_1. On the other hand, by the inverse function theorem f is open if $f'(x)$ is surjective for every x. We shall now prove that one can then define the composition with f.

Theorem 6.1.2. *Let $X_j \subset \mathbb{R}^{n_j}$, $j=1,2$, be open sets, and $f: X_1 \to X_2$ a C^∞ map such that $f'(x)$ is surjective for every $x \in X_1$. Then there is a unique continuous linear map $f^*: \mathscr{D}'(X_2) \to \mathscr{D}'(X_1)$ such that $f^* u = u \circ f$ when*

$u \in C^0(X_2)$. It maps $\mathscr{D}'^k(X_2)$ into $\mathscr{D}'^k(X_1)$ for every k. One calls f^*u the pullback of u by f.

Proof. As already observed, the uniqueness follows from Theorem 4.1.5. To prove the existence we choose for any fixed $x_0 \in X_1$ a C^∞ map $g: X_1 \to \mathbb{R}^{n_1-n_2}$, for example a linear map, such that the direct sum $f \oplus g$,

$$X_1 \ni x \to (f(x), g(x)) \in \mathbb{R}^{n_1} = \mathbb{R}^{n_2} \oplus \mathbb{R}^{n_1-n_2}$$

has a bijective differential at x_0. By the inverse function theorem there is an open neighborhood $Y_1 \subset X_1$ of x_0 such that the restriction of $f \oplus g$ to Y_1 is a diffeomorphism on an open neighborhood Y_2 of $(f(x_0), g(x_0))$. We denote the inverse by h. If $u \in C^0(X_2)$ and $\phi \in C_0^\infty(Y_1)$ then a change of variables gives

$$\int (f^*u)\phi\, dx = \int u(f(x))\phi(x)\, dx = \int u(y')\phi(h(y))\,|\det h'(y)|\, dy$$

where we have written $y = (y', y'') \in \mathbb{R}^{n_2} \oplus \mathbb{R}^{n_1-n_2}$. Hence

$$(6.1.1) \qquad (f^*u)(\phi) = (u \otimes 1)(\Phi), \qquad \Phi(y) = \phi(h(y))\,|\det h'(y)|.$$

Here 1 is the function 1 in $\mathbb{R}^{n_1-n_2}$. If $u \in \mathscr{D}'(X_2)$ and we choose $u_j \in C_0^\infty(X_2)$ so that $u_j \to u$ in $\mathscr{D}'(X_2)$, it follows in view of the remark after Theorem 2.2.4 that f^*u_j converges in $\mathscr{D}'(X)$ to a distribution f^*u defined by (6.1.1) in Y_1. Thus (6.1.1) gives a local definition of f^*u, and the continuity of the map $u \to f^*u$ follows at once from (6.1.1). The theorem is proved.

Remark. The proof shows that if $f \in C^{k+1}$ only, then f^* is well defined and continuous in $\mathscr{D}'^k$. In fact, $\phi \to \Phi$ is continuous from C_0^k to C_0^k. (We need an extra derivative for f since $\det h'$ involves one derivative of f.)

Since we have defined f^*u by continuous extension from the case of functions u, it is clear that the usual rules of computation remain valid:

$$(6.1.2) \qquad \partial_j f^*u = \sum \partial_j f_k f^* \partial_k u, \qquad u \in \mathscr{D}'(X_2) \quad \text{(the chain rule)},$$

$$(6.1.3) \qquad f^*(au) = (f^*a)(f^*u); \qquad a \in C^\infty(X_2),\ u \in \mathscr{D}'(X_2).$$

Here f is assumed to satisfy the hypotheses of Theorem 6.1.2. If in addition we have a C^∞ map $g: X_2 \to X_3$ with surjective differential, then

$$(6.1.4) \qquad (g \circ f)^* u = f^* g^* u, \qquad u \in \mathscr{D}'(X_3),$$

In practice it is often convenient to use the notation $u(f)$, $u \circ f$ or even $u(f(x))$ instead of f^*u since (6.1.2)–(6.1.4) look more familiar then. However, one must always keep in mind then that one is referring to an extension of the pointwise definition given by (6.1.1).

Example 6.1.3. If f is a diffeomorphism $X_1 \to X_2$ between open sets in $\mathbb{R}^n$ then $f^*\delta_y = |\det f'(x)|^{-1}\delta_x$ where $f(x)=y$. This follows from (6.1.1) with $h=f^{-1}$.

Example 6.1.4. If $M_t x = tx$, $x \in \mathbb{R}^n$, $t>0$, then

$$(M_t^* u)(\phi) = u(\phi(./t)/t^n).$$

Thus (3.2.18) with t replaced by $1/t$ means that

$$M_t^* u = t^a u \quad \text{in } \mathbb{R}^n \setminus 0,$$

which is just the usual definition of homogeneity of degree a.

Theorem 6.1.5. *If ρ is a real valued function in $C^\infty(X)$, $X \subset \mathbb{R}^n$, and if $|\rho'| = (\sum |\partial\rho/\partial x_j|^2)^{\frac{1}{2}} \neq 0$ when $\rho=0$, then $\rho^*\delta_0 = dS/|\rho'|$ where dS is the Euclidean surface measure on the surface $\{x;\, \rho(x)=0\}$.*

Proof. Let $\rho(x_0)=0$ and assume for example that $\partial\rho(x_0)/\partial x_1 \neq 0$. Then we can apply (6.1.1) in a neighborhood with

$$h^{-1}(x) = (\rho(x), x_2, \dots, x_n).$$

Then $h(0, y_2, \dots, y_n) = (\psi(y_2, \dots, y_n), y_2, \dots, y_n)$ lies on the surface $\rho=0$, and we have for $\phi \in C_0^\infty(Y)$ if Y is a small neighborhood of x_0

$$\langle \rho^*\delta_0, \phi\rangle = \int (\phi/|\partial_1 \rho|) \circ h(0, y_2, \dots, y_n)\, dy_2 \dots dy_n.$$

Since $\rho(\psi, y_2, \dots, y_n)=0$ we have for $j=2, \dots, n$

$$\partial_1\rho\, \partial\psi/\partial y_j + \partial_j \rho = 0.$$

Hence

$$|\rho'| = |\partial_1 \rho| M, \qquad M = \Big(1 + \sum_2^n (\partial\psi/\partial y_j)^2\Big)^{\frac{1}{2}}.$$

Since $dS = M\, dy_2 \dots dy_n$ with the parameters $y_2 \dots y_n$, this proves the theorem.

From (6.1.2) it follows that if H is the Heaviside function then

$$\partial_j \rho^* H = (\partial_j \rho)\rho^*\delta_0 = (\partial_j\rho)/|\rho'|\, dS,$$

which means that we have given another proof of the Gauss-Green formula (3.1.5). One calls $\rho^*\delta_0$ a *simple layer* on the surface $\rho=0$, and its derivatives are called *multiple layers*. They are essentially the pull-backs by ρ of the derivatives of δ_0. In fact, let

$$L = \sum a_\alpha(x)\partial^\alpha$$

be a differential operator with C^∞ coefficients and define L^k so that $L^0 = L$ and $L^{k+1} = [L^k, \rho]$, that is,

$$L^{k+1} u = L^k \rho u - \rho L^k u.$$

It is clear that L^k are differential operators of decreasing order. Then we have

$$L(u\rho^* \delta_0) = \sum_k (L^k u)\, \rho^* \delta_0^{(k)}/k!, \qquad u \in C^\infty, \tag{6.1.5}$$

where the sum is locally finite. By a change of variables we reduce the proof to the case where $\rho(x) = x_1$, and we may assume that $L = \partial_1^m$ then. Since $L^k = \partial_1^{m-k} m!/(m-k)!$ the formula (6.1.5) becomes Leibniz' formula for the differentiation of a product then.

6.2. Some Fundamental Solutions

Let A be a real non-singular quadratic form in $\mathbb{R}^n$, thus $\partial A/\partial x \neq 0$ when $x \neq 0$. Then $A^* f$ is defined in $\mathbb{R}^n \setminus 0$ if $f \in \mathscr{D}'(\mathbb{R})$. When f is homogeneous of degree a then $f(A) = A^* f$ is homogeneous of degree $2a$, for if $M_t x = tx$ for $x \in \mathbb{R}^n$, $t > 0$, we have

$$M_t^* A^* f = (A M_t)^* f = (t^2 A)^* f = A^* m_{t^2}^* f = t^{2a} A^* f$$

where m_s denotes multiplication by s on $\mathbb{R}$. Unless $2a$ is an integer $\leqq -n$ it follows from Theorem 3.2.3 that $f(A)$ has a unique extension to $\mathbb{R}^n$ which is homogeneous of degree $2a$, and we shall use the notation $f(A)$ for the extension also, when no ambiguity can arise.

We can write

$$A(x) = \sum a_{jk} x_j x_k$$

with $a_{jk} = a_{kj}$, and introduce the differential operator

$$B(\partial) = \sum b_{jk} \partial_j \partial_k$$

where (b_{jk}) is the inverse of (a_{jk}). We shall now compute $B(\partial) A^* f$ in $\mathbb{R}^n \setminus 0$ when f is homogeneous of degree a on $\mathbb{R}$. Since

$$\partial_k f(A) = \partial A/\partial x_k f'(A),$$
$$\partial_j \partial_k f(A) = 2a_{jk} f'(A) + \partial A/\partial x_j \partial A/\partial x_k f''(A)$$

and $\sum b_{jk} \partial A/\partial x_j = 2x_k$, we have

$$B(\partial) f(A) = 2n f'(A) + 4A f''(A) = g(A)$$

where by (3.2.19)′

$$g=2nf'+4tf''=(2n+4(a-1))f'.$$

This is 0 if $a=(2-n)/2$, so $B(\partial)f(A)$ vanishes in $\mathbb{R}^n \setminus 0$ then. It follows that $B(\partial)f(A)$ is homogeneous of degree $-n$ and supported by $\{0\}$, hence it is a multiple of δ_0 which will now be determined.

Theorem 6.2.1. *Let the signature of A be (n_+, n_-), that is, $n_+ + n_- = n$ and A is positive (negative) definite in some n_+ (n_-) dimensional plane. If c_n is the area of the unit sphere in $\mathbb{R}^n$ and $n>2$, then*

$$B(\partial)(A \pm i0)^{(2-n)/2} = (2-n)c_n |\det A|^{-\frac{1}{2}} e^{\mp \pi i n_-/2} \delta_0, \tag{6.2.1}$$

$$B(\partial)A^* \chi_{\pm}^{(2-n)/2} = \pm 4\pi^{(n-2)/2} \sin(\pi n_\pm/2) |\det A|^{-\frac{1}{2}} \delta_0. \tag{6.2.1$'$}$$

Proof. It is sufficient to verify that

$$B(\partial)(A+i0)^{(2-n)/2} = (2-n)c_n |\det A|^{-\frac{1}{2}} e^{-\pi i n_-/2} \delta_0. \tag{6.2.1$^+$}$$

In fact, complex conjugation gives the other case of (6.2.1). The two cases of (6.2.1)′ are interchanged if A is replaced by $-A$. By (3.2.9) we have

$$x_+^a (e^{\pi i a} - e^{-\pi i a}) = (x-i0)^a e^{\pi i a} - (x+i0)^a e^{-\pi i a}, \qquad \operatorname{Re} a > 0.$$

Since $\Gamma(a+1)\Gamma(-a) = -\pi/\sin(\pi a)$ and $\chi_+^a = x_+^a/\Gamma(a+1)$, it follows that

$$\chi_+^a = i\Gamma(-a)/2\pi((x-i0)^a e^{\pi i a} - (x+i0)^a e^{-\pi i a}), \qquad \operatorname{Re} a > 0,\ a \notin \mathbb{Z}_+,$$

and by analytic continuation this is extended to all $a \notin \mathbb{Z}_+$. If we note that

$$(2-n)c_n \Gamma(n/2-1)/\pi = -4\pi^{(n-2)/2}$$

and

$$\pi i n_-/2 + \pi i(1-n/2) = \pi i(1-n_+/2),$$

then (6.2.1)′ is a consequence of (6.2.1).

First we shall prove that

$$B(\partial)A^{(2-n)/2} = (2-n)c_n (\det A)^{-\frac{1}{2}} \delta_0 \tag{6.2.1$''$}$$

when $\operatorname{Re} A$ is positive definite and $(\det A)^{-\frac{1}{2}}$ is defined as in Section 3.4. Here $A^{(2-n)/2}$ is of course the homogeneous extension of $A(x)^{(2-n)/2}$ where $-\pi/2 < \arg A(x) < \pi/2$. By the uniqueness of analytic continuation it suffices to prove (6.2.1)″ when A is positive definite. We can then choose a linear bijection T in $\mathbb{R}^n$ such that the pullback T^*A of A by T is the Euclidean metric form. (6.2.1)″ is then a consequence of Theorem 3.3.2, for

$$|\det A|^{-\frac{1}{2}} T^* \delta_0 = |\det A|^{-\frac{1}{2}} |\det T|^{-1} \delta_0 = \delta_0$$

so the two sides have the same pullback under T by Theorem 3.3.2.

Assume now that A is real and non-degenerate and apply (6.2.1)″ to $A_\varepsilon(x) = -iA(x) + \varepsilon|x|^2$, $\varepsilon > 0$. Then

$$(\det A_\varepsilon)^{-\frac{1}{2}} \to |\det A|^{-\frac{1}{2}} e^{\pi i(\operatorname{sgn} A)/4}$$

(cf. (3.4.6)), $B_\varepsilon \to iB$, and

$$A_\varepsilon^{(2-n)/2} = (e^{-\pi i/2}(A + \varepsilon i|x|^2))^{(2-n)/2} \to i^{-1} e^{\pi i n/4}(A + i0)^{(2-n)/2}$$

in $\mathscr{D}'(\mathbb{R}^n \setminus 0)$ by Lemma 6.2.2 below. From Theorem 3.2.3 it follows that we have convergence in $\mathscr{D}'(\mathbb{R}^n)$ for the homogeneous extensions, so (6.2.1)$^+$ follows from (6.2.1)″.

Lemma 6.2.2. *Let F be a C^∞ function in $X \times J$ where $X \subset \mathbb{R}^n$ is an open set and J is a neighborhood of 0 in $\mathbb{R}$. Let f, I and Z be as in Theorem 3.1.11. If $F(x,0) \in I$ and $\partial F(x,0)/\partial x \neq 0$ when $x \in X$, and $F(x,\varepsilon) \in Z$ when $x \in X$, $0 < \varepsilon \in J$, it follows that $f(F(.,\varepsilon)) \to F(.,0)^* f(. + i0)$ as $\varepsilon \to +0$.*

Proof. The statement is local and invariant under coordinate changes so we may assume that $F(x,0) = x_1$. From the proof of Theorem 3.1.11 we know that $f = G^{(N+1)}$ where G is analytic in Z and continuous in $\bar{Z}$. If $\varphi \in C_0^\infty(X)$ it follows by partial integrations with respect to x_1 that as $\varepsilon \to 0$

$$\begin{aligned}\int \varphi(x) f(F(x,\varepsilon))\,dx &= \int \varphi(x)\, G^{(N+1)}(F(x,\varepsilon))\,dx\\ &= \int G(F(x,\varepsilon))\,(\partial_1(-\partial F(x,\varepsilon)/\partial x_1)^{-1})^{N+1}\, \varphi(x)\,dx\\ &\to \int G(F(x,0))\,(-\partial_1)^{N+1}\, \varphi(x)\,dx\\ &= \int G(x_1)\,(-\partial_1)^{N+1}\, \varphi(x)\,dx\\ &= \int \langle f(. + i0), \varphi(., x')\rangle\, dx'.\end{aligned}$$

This completes the proof.

If we divide (6.2.1) by the constant in the right-hand side of (6.2.1) we obtain a fundamental solution E of $B(\partial)$ which is homogeneous of degree $2-n$. (When $n_+ = n_- = 1$ it is easy to see that (6.2.1)′ remains valid, so it gives a fundamental solution also in this case.) Let us now consider the special case of the wave operator in $\mathbb{R}^{n+1}$

$$\square = c^{-2}\partial^2/\partial t^2 - \Delta$$

where $t \in \mathbb{R}$, c is the speed of light, and Δ is the Laplacean in $x \in \mathbb{R}^n$. In our earlier notation this is the operator $B(\partial)$ if

$$A = c^2 t^2 - |x|^2.$$

By (6.2.1)′ a fundamental solution is given by ($n \geqq 1$)

$$E = \pi^{(1-n)/2} c/4 A^* \chi_+^{(1-n)/2}.$$

It has support in the double cone where $A \geqq 0$. Let

$$E_+ = 2E \quad \text{when } t > 0, \qquad E_+ = 0 \quad \text{when } ct < |x|.$$

The two definitions agree in the overlap of the two regions and define a distribution which is homogeneous of degree $1-n$ in $\mathbb{R}^{n+1} \setminus 0$. We use the same notation for the homogeneous extension to the whole space. Then

$$E = (E_+ + E_-)/2$$

where $E_- = s^* E_+$, s denoting reflection with respect to the origin. Now $\Box E_+ = \Box E_- = \delta_0$, so both are fundamental solutions supported respectively by the forward cone where $t \geqq 0$ and the backward cone where $t \leqq 0$. When n is odd and $\neq 1$ the support is in the boundary of the cone by (3.2.17)′.

We can compute $A^* \chi_+^{(1-n)/2}$ when $t > 0$ by using (6.1.1) with h equal to the inverse of

$$(t, x) \to (A, x).$$

Thus

$$h(s, x) = ((s + |x|^2)^{\frac{1}{2}}/c, x), \qquad |\det h'| = (2c)^{-1}(s + |x|^2)^{-\frac{1}{2}}.$$

It follows that

(6.2.4) $$\langle E_+, \phi \rangle = \pi^{(1-n)/2} 4^{-1} \langle \chi_+^{(1-n)/2}, \Phi \rangle,$$

(6.2.5) $$\Phi(s) = \int \phi((s + |x|^2)^{\frac{1}{2}}/c, x)(s + |x|^2)^{-\frac{1}{2}} dx,$$

if $t > 0$ in supp ϕ. Set

(6.2.6) $$\tilde{\phi}(t, r) = r^{n-2} \int_{|\omega|=1} \phi(t, r\omega) d\omega.$$

Then introduction of polar coordinates in (6.2.5) gives

$$\begin{aligned} \Phi(s) &= \int \tilde{\phi}((s + r^2)^{\frac{1}{2}}/c, r)\, r\, dr (s + r^2)^{-\frac{1}{2}} \\ &= c \int_{s < c^2 t^2} \tilde{\phi}(t, (c^2 t^2 - s)^{\frac{1}{2}})\, dt. \end{aligned}$$

As stated by Theorem 4.4.8 E_+ is thus a continuous function of $t > 0$ with values in $\mathscr{E}'(\mathbb{R}^n)$,

(6.2.4)′ $$\langle E_+(t), \psi \rangle = \pi^{(1-n)/2} c/4 \langle \chi_+^{(1-n)/2}, \tilde{\psi}((c^2 t^2 - .)^{\frac{1}{2}}) \rangle, \quad \psi \in C^\infty(\mathbb{R}^n),\ t > 0,$$

(6.2.6)′ $$\tilde{\psi}(r) = r^{n-2} \int_{|\omega|=1} \psi(r\omega) d\omega; \qquad \tilde{\psi}(ir) = 0, \ r \in \mathbb{R}.$$

Since $\chi_+^0 = H$, $\chi_+^{-\frac{1}{2}} = x_+^{-\frac{1}{2}} \pi^{-\frac{1}{2}}$, $\chi_+^{(-\nu)} = \delta_0^{(\nu-1)}$ for integer $\nu > 0$, we have

$$\begin{aligned}
&\langle E_+(t), \psi\rangle = c/2 \int_{|x|<ct} \psi(x)\,dx, \qquad n=1,\\
(6.2.7)\qquad &\langle E_+(t), \psi\rangle = c/2\pi \int_{|x|<ct} \psi(x)(c^2t^2-|x|^2)^{-\frac{1}{2}}\,dx, \qquad n=2,\\
&\langle E_+(t), \psi\rangle = \pi^{-\nu} c/4(d/ds)^{\nu-1} \tilde{\psi}(s^{\frac{1}{2}})_{s=c^2t^2}, \qquad n=2\nu+1.
\end{aligned}$$

By Theorem 4.4.8′ we know that $E_+(t)$ and $E'_+(t) = \partial_t E_+(t)$ have limits in $\mathscr{D}'$ when $t \to +0$, and

$$\langle E_+, \phi\rangle = \int_0^\infty \langle E_+(t), \phi(t,.)\rangle\,dt, \qquad \phi \in C_0^\infty(\mathbb{R}^{n+1}),$$

for the right-hand side is homogeneous of degree $1-n$ and the equality is valid when $\phi \in C_0^\infty(\mathbb{R}^{n+1} \setminus 0)$. Since E_+ is a fundamental solution it follows that $E_+(+0) = 0$, $E'_+(+0) = c^2\delta_0$.

Theorem 6.2.3. *The wave operator has a unique fundamental solution* E_+ (resp. E_-) *with support in the forward cone where* $ct \geqq |x|$ (resp. *the backward cone where* $ct \leqq -|x|$). *When n is odd the support is in the boundary of the cone if* $n \neq 1$.

Proof. Only the uniqueness remains to be proved. We shall prove more, that there is no fundamental solution other than E_+ with support in the half space $t \geqq 0$. Any other fundamental solution must be of the form $E_+ + u$ where $\Box u = 0$ and $t \geqq 0$ in supp u. But then

$$u = \delta * u = (\Box E_+) * u = E_+ * \Box u = 0$$

where the computations are legitimate since

$$\operatorname{supp} E_+ \times \operatorname{supp} u \ni ((x,t),(y,s)) \to (x+y, t+s)$$

is proper. (See the end of Section 4.2.) In fact, a bound for $t+s$ implies a bound for t and s since $t \geqq 0$ and $s \geqq 0$, and since $|x| \leqq ct$ a bound for $x+y$ gives a bound for x and y then. The proof is complete.

The fundamental solutions in Theorem 6.2.3 are called the advanced and retarded fundamental solutions respectively, while those given by normalization of (6.2.1) are called the Feynman fundamental solutions. The fundamental solutions lead quickly to the solution of the *Cauchy problem* for the wave equation:

Theorem 6.2.4. *For arbitrary* $\phi_0, \phi_1 \in C^\infty(\mathbb{R}^n)$ *and* $f \in C^\infty(\mathbb{R}_+^{n+1})$, $\mathbb{R}_+^{n+1} = \{(t,x);\ t \geqq 0,\ x \in \mathbb{R}^n\}$, *the Cauchy problem*

$$\Box u = f \quad \text{in } \mathbb{R}_+^{n+1}, \tag{6.2.8}$$

and

$$u=\phi_0, \quad \partial u/\partial t=\phi_1 \quad \textit{when } t=0,$$

has a unique solution $u\in C^\infty(\mathbb{R}^{n+1}_+)$, *and it is given by*

(6.2.9) $$u(t,.)=c^{-2}E_+(t)*\phi_1+c^{-2}E'_+(t)*\phi_0+\int_0^t E_+(t-s)*f(s,.)\,ds.$$

Proof. If $f=0$, $\phi_0=\phi_1=0$ then (6.2.8) implies $\Box u_0=0$ in $\mathbb{R}^{n+1}$ if $u_0=u$ in $\mathbb{R}^{n+1}_+$ and $u_0=0$ in $\mathbb{R}^{n+1}\setminus\mathbb{R}^{n+1}_+$. The proof of Theorem 6.2.3 shows that u_0 must then be equal to 0, so the uniqueness is proved. Now (6.2.9) defines a solution in $C^\infty(\mathbb{R}^{n+1}_+)$ of (6.2.8). Indeed, that $u\in C^\infty$ follows from the fact that $E(t)$ and all its t derivatives are continuous with values in $\mathscr{E}'$ when $t\geqq 0$. Since $E_+(+0)=0$, $E'_+(+0)=c^2\delta_0$ we have

$$\partial/\partial t\int_0^t E_+(t-s)*f(s,.)\,ds=\int_0^t E'_+(t-s)*f(s,.)\,ds,$$

$$\partial^2/\partial t^2\int_0^t E_+(t-s)*f(s,.)\,ds=\int_0^t E''_+(t-s)*f(s,.)\,ds+c^2\delta_0*f(t,.).$$

The equation $\Box u=f$ follows now since $\Box E_+=0$, $t>0$. The boundary conditions in (6.2.8) are obtained if we also note that $E''_+(0)=c^2\Delta E_+(+0)=0$. The proof is complete.

We shall return to the Cauchy problem in Chapter XII.

6.3. Distributions on a Manifold

The definition of composition of distributions with diffeomorphisms allows us to define distributions on arbitrary C^∞ manifolds. First we recall the definition of manifolds.

Definition 6.3.1. An n-dimensional manifold is a Hausdorff space with countable basis in which each point has a neighborhood homeomorphic to some open set in $\mathbb{R}^n$. A C^∞ structure on a manifold X is a family $\mathscr{F}$ of homeomorphisms κ, called local coordinate systems, of open set $X_\kappa\subset X$ on open sets $\tilde{X}_\kappa\subset\mathbb{R}^n$ such that

i) If κ, $\kappa'\in\mathscr{F}$, then the map

(6.3.1) $$\kappa'\kappa^{-1}\colon\ \kappa(X_\kappa\cap X_{\kappa'})\to\kappa'(X_\kappa\cap X_{\kappa'})$$

(between open sets in $\mathbb{R}^n$) is infinitely differentiable. (This is then true of the inverse map also.)

ii)
$$\bigcup X_\kappa = X.$$

iii) If κ_0 is a homeomorphism of an open set $X_0 \subset X$ on an open set in $\mathbb{R}^n$ and the map

$$\kappa\kappa_0^{-1}\colon\ \kappa_0(X_0 \cap X_\kappa) \to \kappa(X_0 \cap X_\kappa)$$

as well as its inverse is infinitely differentiable for every $\kappa \in \mathscr{F}$, it follows that $\kappa_0 \in \mathscr{F}$.

A manifold with a C^∞ structure is called a C^∞ manifold. The sets X_κ are called coordinate patches and the cartesian coordinates of $\kappa(x)$, $x \in X_\kappa$, are called local coordinates in X_κ.

The condition iii) in Definition 6.3.1 is in a way superfluous. For if $\mathscr{F}$ satisfies i) and ii) we can extend $\mathscr{F}$ in one and only one way to a family $\mathscr{F}'$ satisfying i), ii) and iii). In fact, the only such family $\mathscr{F}'$ is the set of all homeomorphisms κ' of open subsets $X_{\kappa'}$ of X on open subsets of $\mathbb{R}^n$ such that (6.3.1) and its inverse are infinitely differentiable for every $\kappa \in \mathscr{F}$. The simple verification is left for the reader. (Clearly every extension of $\mathscr{F}$ satisfying i) is contained in this family $\mathscr{F}'$. That $\mathscr{F}'$ satisfies i) and ii) and contains $\mathscr{F}$ follows from the fact that $\mathscr{F}$ satisfies i) and ii).) A C^∞ structure can thus be defined by an arbitrary family $\mathscr{F}$ satisfying i) and ii) only, but if the condition iii) is dropped there are many families defining the same structure. Such a family is called a C^∞ atlas, and two atlases are called equivalent if they define the same C^∞ structure.

Definition 6.3.2. Let X be a C^∞ manifold. A function u defined in X will be said to be in $C^k(X)$ or in $L^p_{\text{loc}}(X)$ if for every coordinate system the composite function $(\kappa^{-1})^* u$ defined by

$$(\kappa^{-1})^* u(x) = u(\kappa^{-1}(x)), \qquad x \in \tilde{X}_\kappa,$$

is in $C^k(\tilde{X}_\kappa)$ or $L^p_{\text{loc}}(\tilde{X}_\kappa)$.

We leave as an exercise for the reader to verify that it is sufficient to require that $u \circ \kappa^{-1}$ be in $C^k(\tilde{X}_\kappa)$ or in $L^p_{\text{loc}}(\tilde{X}_\kappa)$ for every κ in an atlas. Also note that if v is a function with compact support in $\tilde{X}_\kappa$ and we set

$$u = v \circ \kappa \quad \text{in } X_\kappa, \qquad u = 0 \text{ elsewhere},$$

it follows that $u \in C^k(X)$ if and only if $v \in C^k(\tilde{X}_\kappa)$; the analogous statement is valid for L^p_{loc}. We shall somewhat incorrectly denote u by $v \circ \kappa$.

To motivate our definition of a distribution in X we shall now give a slightly different description of $C^k(X)$. Thus let $u \in C^k(X)$ and set

$$u_\kappa = u \circ \kappa^{-1}$$

where κ is an arbitrary coordinate system. u_κ is then in $C^k(\tilde{X}_\kappa)$, and since for any other coordinate system we have $u = u_\kappa \circ \kappa = u_{\kappa'} \circ \kappa'$ in $X_\kappa \cap X_{\kappa'}$, it follows that

$$u_{\kappa'} = u_\kappa \circ (\kappa \circ \kappa'^{-1}) \quad \text{in } \kappa'(X_\kappa \cap X_{\kappa'}). \tag{6.3.2}$$

Conversely, if to every coordinate system κ we are given a function u_κ in $\tilde{X}_\kappa$ in such a way that (6.3.2) is valid for any two coordinate systems κ and κ', then there exists one and only one function u in X such that $u_\kappa = u \circ \kappa^{-1}$ for every κ, and $u \in C^k(X)$ if and only if $u_\kappa \in C^k(\tilde{X}_\kappa)$ for every κ. In analogy with this description of a function in $C^k(X)$ as a system of functions $u_\kappa \in C^k(\tilde{X}_\kappa)$ satisfying (6.3.2), we can introduce distributions on a manifold:

Definition 6.3.3. Let X be a C^∞ manifold. If to every coordinate system κ in X we are given a distribution $u_\kappa \in \mathscr{D}'(\tilde{X}_\kappa)$ such that

$$u_{\kappa'} = (\kappa \circ \kappa'^{-1})^* u_\kappa \quad \text{in } \kappa'(X_\kappa \cap X_{\kappa'}), \tag{6.3.3}$$

we call the system u_κ a distribution u in X. The set of all distributions in X is denoted by $\mathscr{D}'(X)$. Similarly we define $\mathscr{D}'^k(X)$.

It is convenient to use the notation $u_\kappa = u \circ \kappa^{-1}$ as in the case discussed above where u was a continuous function on X. Thus $\mathscr{D}'(X)$ now appears as an extension of $C^0(X)$ if we identify a function $u \in C^0(X)$ with the system $u_\kappa = u \circ \kappa^{-1}$. The following theorem shows in particular that Definition 6.3.3 coincides with our previous one if X is an open subset of $\mathbb{R}^n$.

Theorem 6.3.4. *Let $\mathscr{F}$ be an atlas for X. If for every $\kappa \in \mathscr{F}$ we have a distribution $u_\kappa \in \mathscr{D}'(\tilde{X}_\kappa)$, and* (6.3.3) *is valid when κ and κ' belong to $\mathscr{F}$, it follows that there exists one and only one distribution $u \in \mathscr{D}'(X)$ such that $u \circ \kappa^{-1} = u_\kappa$ for every $\kappa \in \mathscr{F}$.*

Proof. Let ψ be an arbitrary C^∞ coordinate system in X. We first note that there exists one and only one distribution U_ψ in $\mathscr{D}'(\tilde{X}_\psi)$ such that for every κ

$$U_\psi = (\kappa \circ \psi^{-1})^* u_\kappa \quad \text{in } \psi(X_\psi \cap X_\kappa) \subset \tilde{X}_\psi.$$

In fact, this follows from Theorem 2.2.4 in view of the hypotheses in Theorem 6.3.4. In particular, if $\psi \in \mathscr{F}$ we have $U_\psi = u_\psi$ since we may choose $\kappa=\psi$. Furthermore, it is immediately verified that the distributions U_ψ satisfy (6.3.3) for any two coordinate systems κ and κ'. Hence they define a distribution u with the required properties. This proves the theorem, for the uniqueness of u is a trivial consequence of the proof.

The reader may have asked himself why we did not define $\mathscr{D}'(X)$ as the space of continuous linear forms on $C_0^\infty(X)$. The reason for this is that if $f\in C(X)$ and $\phi\in C_0^\infty(X)$ we have no invariant way of integrating $f\phi$ in order to identify f with such a linear form. However, we would have obtained something rather close to $\mathscr{D}'(X)$. In fact, let u be a continuous linear form on $C_0^\infty(X)$. Then u defines a distribution $u_\kappa\in\mathscr{D}'(\tilde{X}_\kappa)$ by

$$u_\kappa(\phi)=u(\phi\circ\kappa), \qquad \phi\in C_0^\infty(\tilde{X}_\kappa).$$

(We define $\phi\circ\kappa$ as 0 outside X_κ.) If $\phi\in C_0^\infty(\kappa'(X_\kappa\cap X_{\kappa'}))$ then

$$u_{\kappa'}(\phi)=u(\phi\circ\kappa')=u(\phi\circ\psi_1\circ\kappa)=u_\kappa(\phi\circ\psi_1)$$

where $\psi_1=\kappa'\circ\kappa^{-1}$. In view of (6.1.1) this means that

$$u_{\kappa'}=|\det\psi'|\psi^* u_\kappa \quad \text{in } \kappa'(X_\kappa\cap X_{\kappa'}), \tag{6.3.4}$$

where $\psi=\kappa\circ\kappa'^{-1}$. Conversely, assume given distributions u_κ in $\tilde{X}_\kappa$ satisfying (6.3.4) for all κ in an atlas $\mathscr{F}$. Choose a partition of unity $1=\sum\chi_j$ with $\chi_j\in C_0^\infty(X_{\kappa_j})$ for some $\kappa_j\in\mathscr{F}$. Then

$$U(\phi)=\sum\langle u_{\kappa_j},(\chi_j\phi)\circ\kappa_j^{-1}\rangle, \qquad \phi\in C_0^\infty(X),$$

is a continuous linear form on $C_0^\infty(X)$. We leave as an exercise for the reader to verify by means of Theorem 6.1.2 that $U_\kappa=u_\kappa$ for every $\kappa\in\mathscr{F}$. The continuous linear forms u on $C_0^\infty(X)$ can thus be identified with the systems $u_\kappa\in\mathscr{D}'(\tilde{X}_\kappa)$ satisfying (6.3.4). They are called distribution *densities*. If all u_κ are continuous (C^∞) we say that u is continuous (C^∞). When u is a distribution and ϕ is a C^∞ density it follows from (6.3.3) and (6.3.4) that ϕu is a distribution density, so

$$\langle u,\phi\rangle=\langle\phi u,1\rangle$$

is defined if ϕ in addition has compact support. Again we leave as an exercise for the reader to verify that this identifies $\mathscr{D}'(X)$ with the space of continuous linear forms on C_0^∞ densities.

As soon as a strictly positive C^∞ density a is chosen in X, the map

$$u\to au$$

identifies distributions with distribution densities and functions with densities. This is the case if $X\subset\mathbb{R}^n$ where the Lebesgue measure is used to give such an identification. Another important case is any Riemannian manifold.

6.4. The Tangent and Cotangent Bundles

Having introduced the notion of C^∞ manifold we shall now discuss some basic facts on differential calculus on manifolds. These will not be used until Chapter VIII. Throughout we denote by X a C^∞ manifold.

Definition 6.4.1. The vector space $T_x(X)$ of tangent vectors to X at x is the space of real distribution densities t in X of order one with support at x and $t(1)=0$.

To justify the definition we observe that if $X \subset \mathbb{R}^n$ then Theorem 2.3.4 gives for some $t_1, \dots, t_n \in \mathbb{R}$

$$t(\phi)=\sum_1^n t_j \partial\phi(x)/\partial x_j$$

so $t(\phi)$ is the derivative of ϕ at x in the direction t, which it is natural to consider as a tangent vector at x. If X is a general manifold and x is in the coordinate patch X_κ then $t^\kappa(\phi)=t(\phi\circ\kappa)$ has the form

$$t^\kappa(\phi)=\sum t_j^\kappa \partial_j\phi(\kappa x), \quad \phi\in C_0^\infty(\tilde{X}_\kappa).$$

Thus $T_x(X)$ is always a vector space of dimension $n=\dim X$, and we have identified $\bigcup_{x\in X_\kappa} T_x(X)$ with $\tilde{X}_\kappa\times\mathbb{R}^n$. If $x\in X_\kappa\cap X_{\kappa'}$ then

$$\langle t^{\kappa'},\phi\rangle=\langle t^\kappa,\phi\circ f\rangle=\sum t_k^\kappa \partial_k(\phi\circ f)(\kappa x), \quad \phi\in C_0^\infty(\kappa'(X_\kappa\cap X_{\kappa'}))$$

where $f=\kappa'\circ\kappa^{-1}$: $\kappa(X_\kappa\cap X_{\kappa'})\to\kappa'(X_\kappa\cap X_{\kappa'})$. Thus

$$t_j^{\kappa'}=\sum \partial_k f_j t_k^\kappa, \quad \text{that is,} \quad t^{\kappa'}=f'(\kappa x)\,t^\kappa.$$

This is a C^∞ map so the atlas consisting of the maps

$$\bigcup_{x\in X_\kappa} T_x(X)\ni t\to(\kappa(x), t^\kappa)\in\tilde{X}_\kappa\times\mathbb{R}^n$$

makes $T(X)$ a C^∞ manifold. These observations are summed up by saying that $T(X)$ is a vector bundle over X with fiber dimension n:

Definition 6.4.2. A C^∞ real vector bundle over X with fiber dimension N is a C^∞ manifold V with

(i) a C^∞ map π: $V\to X$ called the projection,
(ii) a vector space structure in each fiber $V_x=\pi^{-1}(x)$,
(iii) local isomorphisms between V and the product of open subsets of X and $\mathbb{R}^N$.

Explicitly condition (iii) means that for each $x \in X$ there is an open neighborhood Y and a C^∞ map ψ of $V_Y = \pi^{-1}(Y)$ onto $Y \times \mathbb{R}^N$ such that ψ^{-1} is also in C^∞, $\psi(V_x) = \{x\} \times \mathbb{R}^N$ for every $x \in Y$, and the composed map

$$V_x \to \{x\} \times \mathbb{R}^N \to \mathbb{R}^N$$

is a linear isomorphism. In the case of $T(X)$ we have defined isomorphisms

$$\pi^{-1} X_\kappa \to \tilde{X}_\kappa \times \mathbb{R}^n \to X_\kappa \times \mathbb{R}^n$$

with these properties. One calls $T(X)$ the *tangent bundle* of X.

Let V be any vector bundle over X and choose an open covering $\{X_i\}_{i \in I}$ of X such that for each i there is a C^∞ map ψ_i of $\pi^{-1}(X_i)$ onto $X_i \times \mathbb{R}^N$ with the properties listed above. Then

$$g_{ij} = \psi_i \circ \psi_j^{-1}$$

can be regarded as a C^∞ map from $X_i \cap X_j$ to the group $GL(N, \mathbb{R})$ of invertible $N \times N$ matrices with real entries, and we have

$$\begin{aligned} &g_{ij} g_{ji} = \text{identity in } X_i \cap X_j; \\ &g_{ij} g_{jk} g_{ki} = \text{identity in } X_i \cap X_j \cap X_k. \end{aligned} \tag{6.4.1}$$

A system of such $N \times N$ matrices g_{ij} with C^∞ coefficients is called a system of *transition matrices.* One can recover the bundle V from them by forming the set V' of all $(i, x, t) \in I \times X \times \mathbb{R}^N$ such that $x \in X_i$, and defining (i, x, t) to be equivalent to (i', x', t') if $x = x'$ and $t' = g_{i'i} t$. It follows from (6.4.1) that this is an equivalence relation, and it is easily proved that the quotient of V' by this equivalence relation is a vector bundle. It is isomorphic to V if g_{ij} were obtained from local trivializations of the vector bundle V as explained above. We shall sometimes find it convenient to look at a vector bundle in this way which is directly suitable for calculations.

A vector bundle is thus a family of vector spaces V_x, $x \in X$, varying smoothly with x. If $Y \subset X$ then a *section* u of V over Y is a map $Y \ni y \to u(y) \in V_y$, that is, u is a map from Y to V with $\pi \circ u = \text{identity}$. Since V is a C^∞ manifold the set $C^k(Y, V)$ of C^k sections is well defined for $k = 0, 1, \ldots$ or $k = \infty$. If we have a covering $X = \bigcup X_i$ as above with local trivializations ψ_i of $V_{|X_i}$ then $u_i = \psi_i \circ u \in C^k(Y \cap X_i, \mathbb{R}^N)$ and

$$u_i = g_{ij} u_j \quad \text{in } Y \cap X_i \cap X_j. \tag{6.4.2}$$

Conversely, any system $u_i \in C^k(Y \cap X_i, \mathbb{R}^N)$ satisfying (6.4.2) defines a section of the vector bundle. We can therefore also define say the space of distribution sections $\mathscr{D}'(Y, V)$ (if Y is open) as the space of all systems $u_i \in \mathscr{D}'(Y \cap X_i, \mathbb{R}^N)$ satisfying (6.4.2). We could also allow the distributions u_i to be complex valued which strictly speaking means that we complexify the bundle V. (The definition of complex vector bundles is obtained by substituting $\mathbb{C}$ for $\mathbb{R}$ in Definition 6.4.2.)

Examples. 1) A line bundle Ω over X is defined by taking

$$g_{\kappa'\kappa} = |\det(\kappa \circ \kappa'^{-1})'| \circ \kappa' \quad \text{in } X_\kappa \cap X_{\kappa'}$$

for arbitrary coordinate patches X_κ and $X_{\kappa'}$ in X. The sections of Ω are then the densities introduced in Section 6.3.

2) A C^∞ section u of $T(X)$ over Y is called a C^∞ vector field in Y. If $\phi \in C^1(Y)$, then $\phi \to u(\phi)$ is a first order differential operator in Y with no constant term, that is, u annihilates constants.

The map from $T(X)_{|X_\kappa}$ to $\tilde{X}_\kappa \times \mathbb{R}^n$ above is a special case of a very general construction: If X_1, X_2 are C^∞ manifolds and $f\colon X_1 \to X_2$ is a C^∞ map, then we have a map $f_*\colon T(X_1) \to T(X_2)$ with $f_*\, T_x(X_1) \to T_{f(x)}(X_2)$ defined by

$$(f_* t)(\phi) = t(\phi \circ f), \quad \phi \in C_0^\infty(X_2),\ t \in T(X_1).$$

If local coordinates are introduced in X_1 and in X_2, then the matrix of f_* is just the Jacobian matrix with respect to these coordinates, so we shall sometimes use the notation f' instead of f_*. In particular, if $f\colon \mathbb{R} \to X$ is a curve this defines the tangent vector $f' \in T_f(X)$ as the image of d/dt on $\mathbb{R}$.

The dual $V' = \bigcup\limits_x V'_x$ of a vector bundle V is of course also a vector bundle since any trivialization of V defines a trivialization of V'. The transition matrices of V' are ${}^t g_{ij}^{-1}$. In particular, the dual $T^*(X)$ of $T(X)$ is a vector bundle called the *cotangent* bundle. For any $\phi \in C^k(X)$, $k > 0$, the differential

$$T_x(X) \ni t \to t(\phi)$$

at x is an element $d\phi(x) \in T_x^*(X)$, so ϕ defines a C^{k-1} section $d\phi$ (or ϕ') of $T^*(X)$. An arbitrary section of $T^*(X)$ is called a *one form.*

We shall now look at $T^*(X)$ in local coordinates. If

$$X_\kappa \ni x \to \kappa(x) = (x_1, \ldots, x_n) \in \tilde{X}_\kappa \subset \mathbb{R}^n$$

is a local coordinate system, we have identified $T(X)_{|X_\kappa}$ with $\tilde{X}_\kappa \times \mathbb{R}^n$ so that (x, t) is the tangent vector

$$C_0^\infty(\tilde{X}_\kappa) \ni \phi \to (\sum t_j \partial/\partial x_j)\, \phi(x).$$

Now $\langle t, \xi\rangle = \sum t_j \partial/\partial x_j \, (\sum \xi_j x_j)$, so in the corresponding trivialization of $T^*(X)$ the form $\sum \xi_j dx_j$ at x corresponds to (x, ξ), $\xi \in \mathbb{R}^n$. If x_j and ξ_j are regarded as functions on $\tilde{X}_\kappa \times \mathbb{R}^n$ then $\sum \xi_j dx_j$ can also be considered as a differential form there. Its value on the tangent vector t to $\tilde{X}_\kappa \times \mathbb{R}^n$ is $\langle \pi_* t, \xi\rangle$ if π is the projection $\tilde{X}_\kappa \times \mathbb{R}^n \to \tilde{X}_\kappa$. Thus a one form ω on $T^*(X)$ is invariantly defined by letting its value on the tangent vector t at $\gamma \in T^*(X)$ be

$$\langle t, \omega\rangle = \langle \pi_* t, \gamma\rangle.$$

In the standard local coordinates in $T^*(X)$ we have $\omega = \sum \xi_j dx_j$. One calls ω the *canonical one form* on $T^*(X)$.

If $f\colon X_1 \to X_2$ is a C^k map, we obtain a map

$$f^*\colon C^{k-1}(T^*(X_2)) \to C^{k-1}(T^*(X_1)),$$

called the *pullback* of the one form by f, if we define for a one form u in X_2 and a tangent vector $t \in T(X)$

$$\langle t, f^* u\rangle = \langle f_* t, u\rangle.$$

For example, if $\phi \in C^1(X)$ then ϕ' is a map $X \to T^*(X)$ which we can use to pull back the canonical one form ω from $T^*(X)$ to X. As is obvious in the local coordinate representation of ω we obtain

$$(\phi')^* \omega = d\phi. \tag{6.4.3}$$

This is in fact an alternative characterization of the one form ω.

If Y is a C^∞ submanifold of X then the conormal bundle $N(Y)$ of Y is defined as

$$\{\gamma \in T^*(X), \pi_* \gamma = y \in Y \text{ and } \gamma \text{ vanishes on } T_y(Y)\}.$$

If we introduce local coordinates $x = (x_1, \dots, x_n)$ in X such that Y is defined by $x_1 = \dots = x_k = 0$ then $N(Y)$ is defined by $x_1 = \dots = x_k = \xi_{k+1} = \dots = \xi_n = 0$ in the corresponding coordinates in $T^*(X)$. Thus $N(Y)$ is a vector bundle of total dimension n. The definition of ω or the coordinate representation of ω both show at once that the restriction of ω to $N(Y)$ is equal to 0. $N(Y)$ is the dual of the bundle on Y with fiber $T_y(X)/T_y(Y)$ which is called the normal bundle. We shall hardly ever use it and will therefore often drop the prefix in conormal when no confusion seems likely.

Another important vector bundle over X is the bundle $\Lambda^k T^*(X)$ for which the fiber at x consists of the alternating multilinear forms on $T_x(X)$. If $\xi_1, \dots, \xi_k \in T^*(X)_x$ then

$$T_x(X)^k \ni (t_1, \dots, t_k) \to \det \langle t_i, \xi_j\rangle_{i,j=1}^k$$

is an element in $\Lambda^k T^*(X)_x$ denoted by $\xi_1 \wedge \ldots \wedge \xi_k$, and the whole fiber is spanned by such elements. A bilinear product

$$\Lambda^k T^*(X)_x \times \Lambda^l T^*(X)_x \to \Lambda^{k+l} T^*(X)_x,$$

also denoted by $\wedge$, can be uniquely defined so that the product of $\xi_1 \wedge \ldots \wedge \xi_k$ and $\eta_1 \wedge \ldots \wedge \eta_l$ is $\xi_1 \wedge \ldots \wedge \xi_k \wedge \eta_1 \wedge \ldots \wedge \eta_l$. A C^∞ section of $\Lambda^k T^*(X)$ is called a k *form*. It can always be written as a linear combination of forms of the special type

$$f_0 \, df_1 \wedge \ldots \wedge df_k; \quad f_j \in C^\infty(X), \tag{6.4.4}$$

even with $f_1, \ldots, f_k$ chosen among local coordinates. There is a unique first order differential operator d from k forms to $k+1$ forms such that

$$d(f_0 \, df_1 \wedge \ldots \wedge df_k) = df_0 \wedge df_1 \wedge \ldots \wedge df_k;$$

we have $d^2 = 0$, and for any k form f with $df = 0$, $k > 0$, one can locally find a $k-1$ form u such that $f = du$. (C^∞ functions are considered as 0 forms.) If $\psi: Y \to X$ is a C^∞ map and f is a form on X, then $\psi^* f$ can be defined just as in the case of 1 forms, and we have

$$d\psi^* f = \psi^* df$$

since this is true for 0 forms (the invariance of the differential) and consequently also for forms of the type (6.4.4).

One calls X *oriented* if an atlas is given such that for any κ, κ' in the atlas the Jacobian of $\kappa \circ \kappa'^{-1}$ is positive where it is defined. If X is of dimension n and f is an n form in X, we have

$$(\kappa^{-1})^* f = f_\kappa \, dx_1 \wedge \ldots \wedge dx_n \quad \text{in } \tilde{X}_\kappa$$

where $\kappa(x) = (x_1, \ldots, x_n)$. Since

$$f_{\kappa'} = (\det \psi') \psi^* f_\kappa \quad \text{in } \kappa'(X_\kappa \cap X_{\kappa'}), \qquad \psi = \kappa \circ \kappa'^{-1}$$

for arbitrary local coordinates we obtain the transformation law (6.3.4) when κ and κ' are in the atlas defining the orientation, for the Jacobian is then positive. (Note that on the other hand every n form f which is different from 0 at every point defines an orientation where the atlas consists of all κ such that $f_\kappa > 0$. One says that X is oriented by $f > 0$.) Hence f defines a distribution density, which as linear form on C_0^∞ is denoted by

$$\phi \to \int_X \phi f, \quad \phi \in C_0^\infty(X).$$

If we have instead a k form f and an oriented k dimensional submanifold Y of X then a distribution density with support in Y is defined by

$$\phi \to \int_Y \phi f, \quad \phi \in C_0^\infty(X).$$

The right-hand side is interpreted by pulling ϕ and f back to Y by means of the inclusion map $Y \hookrightarrow X$. If f is a $k-1$ form then

$$0=\int_Y d(\phi f)=\int_Y \phi\, df+\int_Y d\phi \wedge f$$

which is clear if ϕ has support in a coordinate patch. Suppose now that $M \subset Y$ is an open subset with smooth boundary ∂M and that $\bar{M}$ is compact. Letting ϕ approach the characteristic function of M we then obtain as in Section 3.1

(6.4.5) $$\int_M df=\int_{\partial M} f. \quad \text{(Stokes' formula.)}$$

Here ∂M is oriented by taking a local coordinate system $y_1, \dots, y_k$ for Y with $y_1=0$ on ∂M and $y_1<0$ in M; if this is in the positive atlas for M then $y_2, \dots, y_k$ is in the positive atlas for ∂M. This ends our brief review of the differential calculus on manifolds; the reader not previously familiar with this topic should consult a detailed exposition or supply the missing details. However, we shall make some additional comments on how the cotangent bundle occurs in the study of differential operators.

If X is a manifold and E, F two (complex) vector bundles on X, then a linear differential operator from sections of E to sections of F is a linear map $C^\infty(X,E) \to C^\infty(X,F)$ which is a differential operator in the usual sense in terms of local coordinates in X and local frames for E and F. Thus let $(x_1, \dots, x_n)$ be local coordinates in $X_\kappa \subset X$ and let $e_1, \dots, e_N$ (resp. $f_1, \dots, f_M$) be sections of E and F forming bases at every point in X_κ. Then the condition is that in $\tilde{X}_\kappa$

$$(Pu)_j=\sum P_{jk} u_k$$

if $u=\sum u_k e_k$ is a section of E and $Pu=\sum (Pu)_j f_j$; here

$$P_{jk}=\sum_\alpha P_{jk\alpha}(x)\,\partial^\alpha$$

is a differential operator in the local coordinates. This condition is obviously independent of the choice of local coordinates and of local frames for E and F. We say that P is of order $\leqq m$ if $|\alpha| \leqq m$ in the sum. Then the *principal symbol* $p(\gamma)$ of P at $\gamma \in T^*(X)$ with local coordinates x, ξ is defined as the linear map from E_x to F_x, $x=\pi\gamma$, which in terms of the bases e_k, f_j is given by

(6.4.6) $$(p(\gamma)u)_j=\sum_{|\alpha|=m} P_{jk\alpha}(x)\,\xi^\alpha u_k.$$

The definition is in fact independent of the choice of local coordinates and bases in the bundle, for it means that if $\phi \in C^\infty(X)$ and $(x, \phi'(x))$

$=\gamma$, then for any section u of E we have at x

(6.4.6)′ $$p(\gamma)u=\lim_{t\to\infty} t^{-m} e^{-t\phi} P(e^{t\phi} u),$$

and the right hand side is invariantly defined. (Later on we shall modify p slightly by a factor i^m but this is not important here.) To prove the equivalence of (6.4.6) and the coordinate free definition (6.4.6)′ we just note that when (6.4.6)′ is evaluated in terms of local coordinates then we must let m derivatives fall on the exponential to get a non-zero contribution.

If P is a scalar differential operator, that is, a differential operator from $C^\infty(X)$ to $C^\infty(X)$, then p is a function defined on $T^*(X)$. The zeros of p (outside the zero section 0 of the cotangent bundle) are called *characteristics* of P. A surface in X defined by $\phi=c$ is called characteristic (at x) if $p(x,\phi'(x))=0$ when $\phi(x)=c$ (at x), that is, the conormal bundle (at x) is in the characteristic set. We shall now discuss the problem of integrating the characteristic equation $p(x,\phi'(x))=0$, that is, the construction of functions with all level surfaces characteristic. In this discussion we allow p to be any real valued C^∞ function on $T^*(X)$, thus we drop the property of the principal symbol that it is a homogeneous polynomial in each fiber.

Denote by ξ the section ϕ' of $T^*(X)$. We must satisfy two conditions:

(i) the section must lie in the zero set of p,

(ii) ξ must be of the form ϕ'.

Now we know from (6.4.3) that (ii) implies

$$d\phi=\xi^*\omega.$$

Hence

(6.4.7) $$d\xi^*\omega=\xi^* d\omega=0.$$

Conversely, if (6.4.7) is fulfilled then $\xi^*\omega$ is locally of the form $d\phi$ and we obtain locally a solution of the characteristic equation $p(x,\phi'(x))=0$. The differential form

(6.4.8) $$\sigma=d\omega$$

is called the *symplectic form* of $T^*(X)$ and plays a fundamental role in what follows. (Later on we shall devote the entire Chapter XXI to a study of the geometry to which it leads.) In the standard local coordinates x,ξ in $T^*(X)$ we have

(6.4.8)′ $$\sigma=\sum d\xi_j\wedge dx_j$$

which means that for two tangent vectors to $T^*(X)$ with the coordinates (t',τ') and (t'',τ'') the symplectic form is

$$(6.4.8)'' \qquad \sum \begin{vmatrix} \tau_j' & t_j' \\ \tau_j'' & t_j'' \end{vmatrix} = \sum (\tau_j' t_j'' - t_j' \tau_j'') = \langle t'', \tau' \rangle - \langle t', \tau'' \rangle.$$

This is a non-degenerate bilinear form, that is, it vanishes for all (t'', τ'') if and only if $(t', \tau') = 0$. Since $\sigma = d\omega$ we have of course $d\sigma = 0$.

We can now reinterpret (6.4.7). Let $S = \{(x, \xi(x))\} \subset T^*(X)$ be the graph of the section. It is an n dimensional manifold parametrized by x, and (6.4.7) means precisely that the restriction of σ to S is equal to 0, that is, the pullback of σ to S by the inclusion $S \hookrightarrow T^*(X)$ is equal to 0. In other words, the tangent plane of S is at every point γ orthogonal to itself with respect to the bilinear form σ. Since σ is non-degenerate the annihilator of a k-dimensional subspace of $T_\gamma(T^*(X))$ with respect to σ is necessarily of dimension $2n-k$, so it follows that the tangent plane of S must be its own annihilator.

Summing up, it follows from (i) and (ii) that the n dimensional manifold $S = \{(x, \xi(x))\} \subset T^*(X)$ satisfies the conditions

(i)′ $p = 0$ on S

(ii)′ the restriction of σ to S is equal to 0, that is, the tangent plane of S is its own annihilator with respect to σ at every point in S.

Conversely, if S satisfies (i)′, (ii)′ and the restriction of the projection π to S is a diffeomorphism on X, then S is a section satisfying (i) and (ii). Thus (i)′, (ii)′ represent a slightly generalized, geometrical form of our problem.

That $p = 0$ on S implies that $dp = 0$ on the tangent planes of S. To exploit this we define a vector field H_p on $T^*(X)$ by

$$(6.4.9) \qquad \langle t, dp \rangle = \sigma(t, H_p), \qquad t \in T(T^*(X)),$$

which is possible since σ is non-degenerate. In terms of local coordinates (6.4.9) means if $t = (t_x, t_\xi)$ and $H_p = (h_x, h_\xi)$ that

$$\langle t_x, \partial p/\partial x \rangle + \langle t_\xi, \partial p/\partial \xi \rangle = \langle t_\xi, h_x \rangle - \langle t_x, h_\xi \rangle,$$

that is, $h_x = \partial p/\partial \xi$, $h_\xi = -\partial p/\partial x$. Thus

$$(6.4.10) \qquad H_p = \sum (\partial p/\partial \xi_j \, \partial/\partial x_j - \partial p/\partial x_j \, \partial/\partial \xi_j)$$

in terms of local coordinates. One calls H_p the *Hamilton vector field* of p. Now we obtain from (i)′

$$\sigma(t, H_p) = 0 \quad \text{if } \gamma \in S \text{ and } t \in T_\gamma(S),$$

so H_p must be in the tangent plane of S in view of (ii)′. Hence S is generated by integral curves of H_p, that is, solutions of the Hamilton equations

$$(6.4.11) \qquad dx_j/dt = \partial p(x, \xi)/\partial \xi_j, \qquad d\xi_j/dt = -\partial p(x, \xi)/\partial x_j$$

in local coordinates. Note that (6.4.11) implies

$$dp/dt = \sum \partial p/\partial x_j \, dx_j/dt + \sum \partial p/\partial \xi_j \, d\xi_j/dt = 0$$

so p is constant on such a curve. If p vanishes at one point of the curve it follows that p vanishes identically on it; the curve is then called a *bicharacteristic* of p.

Theorem 6.4.3. *Assume that S_0 is an $n-1$ dimensional manifold $\subset T^*(X)$ such that*

(a) *$p=0$ on S_0,*

(b) *the restriction of σ to S_0 is equal to* 0, *that is, the tangent space of S_0 is self orthogonal with respect to σ,*

(c) *H_p is not a tangent to S_0 at any point of S_0.*

Then the union of the bicharacteristics of p starting on S_0 defines an n dimensional manifold satisfying (i)′ *and* (ii)′ *in some neighborhood of S_0.*

Proof. Let $\Phi(t)\gamma$ be the solution of the Hamilton equations (6.4.11) at time t when $\Phi(0)\gamma = \gamma$. This map is defined for small t. Then

$$S_0 \times \mathbb{R} \ni (\gamma, t) \to \Phi(t)\gamma \in T^*(X)$$

is well defined in a neighborhood of $S_0 \times \{0\}$, and the differential is injective there by hypothesis for its image contains both H_p and $T(S_0)$ so the dimension is n by condition (c). The image of a suitably small neighborhood U is therefore an n dimensional manifold S. Since $\sigma(t, H_p) = \langle t, dp\rangle = 0$ for $t \in T(S_0)$, and since $\sigma(H_p, H_p) = 0$, it follows from (b) that (ii)′ is satisfied at S_0, and (i)′ is true since S is a union of bicharacteristics. It remains to prove that (ii)′ is valid in S. To do so it suffices to show that

$$\Phi(t)^* \sigma = \sigma \tag{6.4.12}$$

for the fact that σ restricted to S vanishes at S_0 then shows that this is also true at $\Phi(t)S_0$, hence in S.

To verify (6.4.12) we first observe that

$$\Phi(s+t) = \Phi(s)\Phi(t).$$

If we verify that the derivative of $\Phi(t)^*\sigma$ with respect to t is 0 when $t = 0$ it will therefore follow that this is true for any t, hence that (6.4.12) is valid. In evaluating the derivative when $t=0$ we use local coordinates. Since

$$\Phi(t)(x, \xi) = (x + t\partial p(x, \xi)/\partial \xi, \ \xi - t\partial p(x, \xi)/\partial x) + O(t^2),$$

we have

$$\begin{aligned}\Phi(t)^*\sigma &= \sum (d\xi_j - t\,d\,\partial p/\partial x_j)\wedge(dx_j + t\,d\,\partial p/\partial \xi_j) + O(t^2)\\ &= \sigma + t\left(\sum d\xi_j \wedge d\,\partial p/\partial \xi_j - d\,\partial p/\partial x_j \wedge dx_j\right) + O(t^2)\\ &= \sigma - t\,d^2 p + O(t^2) = \sigma + O(t^2).\end{aligned}$$

This completes the proof of (6.4.12) and of Theorem 6.4.3. At the same time we have proved that the solution operators $\Phi(t)$ of the Hamiltonian equations are canonical transformations:

Definition 6.4.4. A C^∞ map Φ from $\Omega_1 \subset T^*(X)$ to $\Omega_2 \subset T^*(X)$ is called canonical (or sympletic) if $\Phi^*\sigma = \sigma$.

Note that $\Phi^*\omega^n = \omega^n$ and that ω^n is a $2n$ form on $T^*(X)$ which is nowhere 0 since $\omega^n = n!\, d\xi_1 \wedge dx_1 \wedge \ldots \wedge d\xi_n \wedge dx_n$ in local coordinates. Thus ω^n can be used to orient $T^*(X)$, and we conclude that the differential of a canonical transformation is always bijective.

Theorem 6.4.3 solves a geometrical form of the initial value problem for the equation $p(x, \phi'(x)) = 0$. The following is a more conventional analytical version of the result:

Theorem 6.4.5. *Let p be a real valued C^∞ function in a neighborhood of $(0, \eta) \in T^*(\mathbb{R}^n)$ such that*

(6.4.13) $$p(0, \eta) = 0, \qquad \partial p(0, \eta)/\partial \eta_n \neq 0$$

and let ψ be a real valued C^∞ function in $\mathbb{R}^{n-1}$ such that

$$\partial \psi(0)/\partial x_j = \eta_j, \qquad j < n.$$

Then there is in a neighborhood of $0 \in \mathbb{R}^n$ a unique real valued solution ϕ of the equation

(6.4.14) $$p(x, \phi'(x)) = 0$$

satisfying the boundary condition

(6.4.15) $$\phi(x', 0) = \psi(x'), \qquad \partial \phi(0)/\partial x = \eta.$$

Here $x' = (x_1, \ldots, x_{n-1})$.

Proof. By the implicit function theorem the equation

$$p(x', 0, \partial \psi(x')/\partial x', \xi_n) = 0$$

has a unique solution $\xi_n = \xi_n(x')$ with $\xi_n(0) = \eta_n$, provided that $|x'|$ is small. Then

$$S_0 = \{(x', 0, \partial \psi/\partial x', \xi_n(x'))\}$$

is an $n-1$ dimensional manifold satisfying (a) and (b) in Theorem 6.4.3. Condition (c) is fulfilled since the x_n component $\partial p/\partial \xi_n$ of H_p is not 0. For the surface S given by Theorem 6.4.3 the projection π restricted to S has therefore a surjective differential at $(0, \eta)$, so S is a section of $T^*(\mathbb{R}^n)$ near 0. Hence we obtain a solution ϕ of (6.4.14) near 0 such that $(x, \phi'(x)) \in S$, in particular

$$\partial\phi(x',0)/\partial x' = \partial\psi/\partial x', \quad \partial\phi(0)/\partial x = \eta.$$

Thus $\phi(x',0) = \psi(x') + C$ so $\phi - C$ has the required properties. The uniqueness follows from the discussion preceding Theorem 6.4.3 of the various ways of stating (6.4.14), so this completes the proof.

Sometimes one wants to solve two equations

$$p(x, \phi'(x)) = 0, \quad q(x, \phi'(x)) = 0$$

at the same time. Since the integral curves of the Hamilton field H_p must lie in the surface S defined by ϕ' and since along these curves

$$dq/dt = \sum (\partial p/\partial \xi_j \partial q/\partial x_j - \partial p/\partial x_j \partial q/\partial \xi_j) \tag{6.4.16}$$

it follows that the *Poisson bracket*

$$\{p, q\} = \sum (\partial p/\partial \xi_j \partial q/\partial x_j - \partial p/\partial x_j \partial q/\partial \xi_j) \tag{6.4.17}$$

must also vanish in S. The argument can be repeated to show that the repeated Poisson brackets $\{p, \{p, q\}\}, \ldots$ must vanish in S. When $\{p, q\}$ vanishes identically on $\{(x, \xi);\ p(x, \xi) = q(x, \xi) = 0\}$ and dp, dq are linearly independent it is not hard to modify the proof of Theorem 6.4.3 to show that one can find S satisfying (i)′ for p and q as well as condition (ii)′, so that S passes through a given manifold S_0 of dimension $n-2$ satisfying (a), (b) and a suitable form of (c). We shall not pursue this further since a thorough discussion will be given in Chapter XXI. However, the notion of Poisson bracket will be important in Chapter VIII, and it should be kept in mind that by (6.4.16) the Poisson bracket of two functions on $T^*(X)$ is invariantly defined.

Notes

As indicated at the end of Section 6.3 one can define $\mathscr{D}'(X)$ when X is a manifold as the dual of the space of C^∞ densities of compact support. This is a special case of the theory of currents of de Rham [1], which also contains a study of distribution valued differential forms of arbitrary degree. The results in Section 6.1 are thus essentially contained in de Rham's theory, for composition with a map f having surjective differential can locally be split into a tensor product with the function 1 in some new variables and a change of variables in the domain of f. The formula (6.1.5) is from John [5] though.

Just as we recalled differential calculus in Section 1.1 we have summed up in Sections 6.3 and 6.4 the notions of manifold, tangent bundle, cotangent bundle as well as general vector bundle, and the calculus of differential forms. This should suffice to define our notation but the reader previously unfamiliar with these topics will at least have to look elsewhere for a detailed presentation of differential forms. Besides the book of de Rham [1] he might consult for example Warner [1] or Sternberg [1].

In Section 6.4 we also discussed the Hamilton-Jacobi integration theory for first order non-linear differential equations. We chose a geometrical presentation which leads to the notions of Hamilton vector field, bicharacteristic and Poisson bracket which will all be required for some results in Section 8.5. They form the basis for the symplectic geometry which pervades a large part of the modern theory of linear partial differential operators with variable coefficients. We shall have to devote the entire Chapter XXI to developing all the symplectic geometry required in our program.

The reader can very well postpone the study of Sections 6.3 and 6.4 until he reaches part III of this book. It is largely to make this possible that we chose to define composition of distributions with maps without reference to distributions on manifolds although the reversed order would have been preferable conceptually.

Chapter VII. The Fourier Transformation

Summary

The Fourier transformation of a function $u \in L^1$ is defined by

$$\hat{u}(\xi) = \int e^{-i\langle x, \xi \rangle} u(x)\, dx.$$

In Section 7.1 we extend the definition to all $u \in \mathscr{S}'$, the space of temperate distributions, which is the smallest subspace of $\mathscr{D}'$ containing L^1 which is invariant under differentiation and multiplication by polynomials. That this is possible is not surprising since the Fourier transformation exchanges differentiation and multiplication by coordinates. (See also the introduction.) It is technically preferable though to define $\mathscr{S}'$ as the dual of the space $\mathscr{S}$ of rapidly decreasing test functions. After proving the Fourier inversion formula and basic rules of computation, we study in Section 7.1 the Fourier transforms of L^2 functions, distributions of compact support, homogeneous distributions and densities on submanifolds. As an application fundamental solutions of elliptic equations are discussed. Section 7.2 is devoted to Poisson's summation formula and Fourier series expansions. We return to the Fourier-Laplace transform of distributions with compact support in Section 7.3. After proving the Paley-Wiener-Schwartz theorem we give applications such as the existence of fundamental solutions for arbitrary differential operators with constant coefficients, Asgeirsson's mean value theorem and Kirchoff's formulas for solutions of the wave equation. The Fourier-Laplace transform of distributions which do not necessarily have compact support is studied in Section 7.4. In particular we compute the Fourier-Laplace transform of the advanced fundamental solution of the wave equation. The Fourier transformation gives a convenient method for approximating C^∞ functions by analytic functions. This is used in Section 7.5 to prove the Malgrange preparation theorem after we have recalled the classical analytical counterpart of Weierstrass.

Section 7.6 is devoted to the Fourier transform of Gaussian functions and the convolution operators which they define. This prepares for a rather detailed discussion in Section 7.7 of the method of stationary phase, which is a fundamental tool in the study of pseudo-differential and Fourier integral operators in Chapters XVIII and XXV. The Malgrange preparation theorem plays an essential role in many of the proofs. As an application of the simplest form of the method of stationary phase we introduce in Section 7.8 the notion of oscillatory integral. This gives a precise meaning to equations such as

$$\delta(\xi)=(2\pi)^{-n}\int e^{i\langle x,\xi\rangle}dx$$

and will simplify notation later on. In Section 7.9 finally we continue the proof of L^p estimates for convolution operators started in Section 4.5. Applications are given concerning the regularity of solutions of elliptic differential equations with constant coefficients. Although the results are very important in the study of non-linear elliptic differential equations they will not be essential in this book so the reader can skip Section 7.9 without any loss of continuity.

7.1. The Fourier Transformation in $\mathscr{S}$ and in $\mathscr{S}'$

The purpose of Fourier analysis in $\mathbb{R}^n$ is to decompose arbitrary functions into usually continuous sums of characters. By a character one means an eigenfunction for the translations, that is, a function f such that for every $y\in\mathbb{R}^n$

$$f(x+y)=f(x)c(y), \qquad x\in\mathbb{R}^n,$$

for some $c(y)$. If $f(0)=0$ we conclude that f vanishes identically. Excluding this uninteresting case we can normalize f so that $f(0)=1$. Then $x=0$ gives $f(y)=c(y)$, hence

$$f(x+y)=f(x)f(y), \qquad f(0)=1. \tag{7.1.1}$$

Assuming that f is continuous we obtain if $g\in C_0^\infty$ and $\int fg\,dy=1$

$$f(x)=\int f(x+y)g(y)\,dy\in C^\infty$$

(Theorem 1.3.1). Differentiation of (7.1.1) with respect to y gives when $y=0$

$$\partial_j f=a_j f, \qquad a_j=\partial_j f(0)$$

and since $f(0)=1$ it follows that

$$f(x)=\exp\langle x,a\rangle, \qquad \langle x,a\rangle=\sum_1^n x_j a_j. \tag{7.1.2}$$

Conversely, the exponential (7.1.2) satisfies (7.1.1) so we have determined all continuous characters.

Which characters are needed for the expansion of a given function u depends on the properties of u. We shall mainly consider functions and distributions which are fairly well behaved at ∞ and shall only use bounded characters then, that is, take a purely imaginary in (7.1.2).

Definition 7.1.1. If $f \in L^1(\mathbb{R}^n)$ then the Fourier transform $\hat{f}$ is the bounded continuous function in $\mathbb{R}^n$ defined by

$$\hat{f}(\xi) = \int e^{-i\langle x, \xi\rangle} f(x)\, dx, \qquad \xi \in \mathbb{R}^n. \tag{7.1.3}$$

If $\hat{f}$ also happens to be integrable one can express f in terms of $\hat{f}$ by Fourier's inversion formula

$$f(x) = (2\pi)^{-n} \int e^{i\langle x, \xi\rangle} \hat{f}(\xi)\, d\xi, \tag{7.1.4}$$

so $\hat{f}(\xi)$ is the density of the character $e^{i\langle x, \xi\rangle}$ in the harmonic decomposition of f. To study the Fourier transform and in particular to prove (7.1.4) we shall first consider functions in a subset of C^∞ containing C_0^∞.

Definition 7.1.2. By $\mathscr{S}$ or $\mathscr{S}(\mathbb{R}^n)$ we denote the set of all $\phi \in C^\infty(\mathbb{R}^n)$ such that

$$\sup_x |x^\beta \partial^\alpha \phi(x)| < \infty \tag{7.1.5}$$

for all multi-indices α and β. The topology in $\mathscr{S}$ defined by the semi-norms in the left-hand side of (7.1.5) makes $\mathscr{S}$ a Fréchet space.

The importance of the class $\mathscr{S}$ is due to the following result, where we use the notation $D_j = -i\partial_j$ which is much more convenient as soon as the Fourier transformation is involved. Note that $D_j \mathscr{S} \subset \mathscr{S}$, $x_j \mathscr{S} \subset \mathscr{S}$, and that $\mathscr{S} \subset L^1$.

Lemma 7.1.3. *The Fourier transformation $\phi \to \hat{\phi}$ maps $\mathscr{S}$ continuously into $\mathscr{S}$. The Fourier transform of $D_j\phi$ is $\xi_j \hat{\phi}(\xi)$, and the Fourier transform of $x_j\phi$ is $-D_j\hat{\phi}$.*

Proof. Differentiation of (7.1.3) gives

$$D^\alpha \hat{\phi}(\xi) = \int e^{-i\langle x, \xi\rangle} (-x)^\alpha \phi(x)\, dx$$

and is legitimate since the integral obtained is uniformly convergent. Hence $\hat{\phi} \in C^\infty$ and $D^\alpha \hat{\phi}$ is the Fourier transform of $(-x)^\alpha \phi$. Integrating by parts we also obtain

(7.1.6) $$\xi^\beta D^\alpha \hat{\phi}(\xi) = \int e^{-i\langle x,\xi\rangle} D^\beta((-x)^\alpha \phi(x))\,dx.$$

These operations are legitimate since $\phi \in \mathscr{S}$. Hence

$$\sup |\xi^\beta D^\alpha \hat{\phi}(\xi)| \leqq C \sup_x (1+|x|)^{n+1} |D^\beta((-x)^\alpha \phi(x))|$$

where $C = \int (1+|x|)^{-n-1}\,dx$, so the Fourier transformation maps $\mathscr{S}$ continuously into $\mathscr{S}$. When $\alpha = 0$ we obtain from (7.1.6) that $\xi^\beta \hat{\phi}$ is the Fourier transform of $D^\beta \phi$, which completes the proof.

Lemma 7.1.4. *If T: $\mathscr{S} \to \mathscr{S}$ is a linear map such that*

$$T D_j \phi = D_j T\phi, \quad T x_j \phi = x_j T\phi, \quad j = 1, \ldots, n, \quad \phi \in \mathscr{S},$$

then $T\phi = c\phi$ for some constant c.

Proof. If $\phi \in \mathscr{S}$ and $\phi(y) = 0$ then we can write

$$\phi(x) = \sum (x_j - y_j)\phi_j(x)$$

where $\phi_j \in \mathscr{S}$. In fact, Theorem 1.1.9 shows that we can do so with $\phi_j \in C^\infty$, and for $x \neq y$ we could take $\phi_j(x) = \phi(x)(x_j - y_j)|x-y|^{-2}$ which behaves at ∞ as a function in $\mathscr{S}$. Combining these two choices by a partition of unity we obtain ϕ_j with the desired properties. Hence

$$T\phi(x) = \sum (x_j - y_j) T\phi_j(x) = 0 \quad \text{if } x = y.$$

It follows that for all $\phi \in \mathscr{S}$

$$T\phi(x) = c(x)\phi(x)$$

where c is independent of ϕ. Taking $\phi \neq 0$ everywhere we obtain $c \in C^\infty$. Now

$$0 = D_j T\phi - T D_j \phi = (D_j c)\phi, \quad \phi \in \mathscr{S},$$

so c must be a constant.

Theorem 7.1.5. *The Fourier transformation F: $\phi \to \hat{\phi}$ is an isomorphism of $\mathscr{S}$ with inverse given by Fourier's inversion formula* (7.1.4).

Proof. By Lemma 7.1.3 F^2 maps $\mathscr{S}$ into $\mathscr{S}$ and anticommutes with D_j and x_j. With the notation $R\phi(x) = \phi(-x)$ we conclude from Lemma 7.1.4 applied to $T = RF^2$ that $RF^2 = c$. To determine c we can take $\phi(x) = \exp(-|x|^2/2)$, which is a function in $\mathscr{S}$. Then $(x_j + iD_j)\phi = 0$ so $(-D_j + i\xi_j)\hat{\phi}(\xi) = 0$, $j = 1, \ldots, n$. Hence $\hat{\phi} = c_1 \phi$ where $c_1 = \hat{\phi}(0) = (2\pi)^{n/2}$ by (3.4.1)'. Thus $F^2\phi = c_1^2\phi$, so $c = c_1^2 = (2\pi)^n$ which completes the proof.

Another interesting determination of the constant c is as follows. Assume $n = 1$, which is sufficient since we can otherwise take for f a product $f_1(x_1)\dots f_n(x_n)$. Splitting the integral (7.1.3) at 0 and integrating by parts we obtain $\hat{f}(\xi)=F_+(\xi)+F_-(\xi)$ where

$$F_+(\xi)=\int_0^\infty f(x)e^{-ix\xi}dx=f(0)/i\xi+f'(0)/(i\xi)^2+\int_0^\infty f''(x)e^{-ix\xi}dx/(i\xi)^2$$
$$=f(0)/i\xi+O(|\xi|^{-2}) \qquad \text{as } \xi\to\infty,\ \operatorname{Im}\xi\leqq 0,$$
$$F_-(\xi)=\int_{-\infty}^0 f(x)e^{-ix\xi}dx=-f(0)/i\xi+O(|\xi|^{-2}) \qquad \text{as } \xi\to\infty,\ \operatorname{Im}\xi\geqq 0.$$

Assume $f\in C_0^\infty$ which makes F_- and F_+ entire analytic functions. If γ_R is the circle $|\zeta|=R$ oriented counterclockwise and $\gamma_R^\pm$ are the half circles in the upper and lower half planes respectively, then Cauchy's integral formula gives

$$\int\hat{f}(\xi)d\xi=\lim_{R\to\infty}\int_{\gamma_R^-}F_+(\zeta)d\zeta-\int_{\gamma_R^+}F_-(\zeta)d\zeta$$
$$=\lim_{R\to\infty}\int_{\gamma_R}f(0)/i\zeta\, d\zeta=2\pi f(0).$$

The constants 2π in Cauchy's integral formula and in the inversion formula are therefore "the same", and one is often free to choose between using the Fourier inversion formula or Cauchy's integral formula.

Instead of relying on Lemma 7.1.4 we could also have verified directly that Fourier's inversion formula must be valid for some constant c in place of $(2\pi)^n$. What is involved is computing the double integral

$$\int e^{i\langle x,\xi\rangle}d\xi\int\phi(y)e^{-i\langle y,\xi\rangle}dy, \qquad \phi\in\mathscr{S}.$$

Since the double integral does not converge absolutely, the order of integration cannot be inverted so we must introduce a factor which is a function of ξ to produce convergence. Thus choose $\psi\in\mathscr{S}$ with $\psi(0)=1$ and note that by dominated convergence

$$\int\hat{\phi}(\xi)e^{i\langle x,\xi\rangle}d\xi=\lim_{\varepsilon\to 0}\int\psi(\varepsilon\xi)\hat{\phi}(\xi)e^{i\langle x,\xi\rangle}d\xi$$
$$=\lim_{\varepsilon\to 0}\iint\psi(\varepsilon\xi)\phi(y)e^{i\langle x-y,\xi\rangle}d\xi\, dy.$$

In this absolutely convergent double integral we integrate first with respect to ξ and obtain

$$\int\hat{\phi}(\xi)e^{i\langle x,\xi\rangle}d\xi=\lim_{\varepsilon\to 0}\int\phi(y)\hat{\psi}((y-x)/\varepsilon)dy/\varepsilon^n$$
$$=\lim_{\varepsilon\to 0}\int\phi(x+\varepsilon z)\hat{\psi}(z)dz=\phi(x)\int\hat{\psi}(z)dz.$$

We shall now prove some fundamental properties of the Fourier transformation on $\mathscr{S}$.

Theorem 7.1.6. *If ϕ and ψ are in $\mathscr{S}$, then*

$$\int \hat{\phi}\psi\, dx = \int \phi\hat{\psi}\, dx, \tag{7.1.7}$$

$$\int \phi\bar{\psi}\, dx = (2\pi)^{-n} \int \hat{\phi}\bar{\hat{\psi}}\, dx \qquad \text{(Parseval's formula)}, \tag{7.1.8}$$

$$\widehat{\phi * \psi} = \hat{\phi}\hat{\psi}, \tag{7.1.9}$$

$$\widehat{\phi\psi} = (2\pi)^{-n}\hat{\phi} * \hat{\psi}. \tag{7.1.10}$$

Proof. Both sides of (7.1.7) are equal to the double integral

$$\int\int \phi(x)\psi(\xi)e^{-i\langle x,\xi\rangle}\, dx\, d\xi.$$

To prove (7.1.8) we set $\chi = (2\pi)^{-n}\bar{\hat{\psi}}$ and obtain using the Fourier inversion formula

$$\overline{\hat{\chi}(\xi)} = (2\pi)^{-n} \int \hat{\psi}(x) e^{i\langle x,\xi\rangle}\, dx = \psi(\xi).$$

Hence (7.1.8) follows if we apply (7.1.7) with ψ replaced by χ. The proof of (7.1.9) is as elementary as that of (7.1.7) and is left for the reader. To obtain (7.1.10) finally we note that the Fourier transform of $\widehat{\phi\psi}$ is $(2\pi)^n\phi(-x)\psi(-x)$ and that the Fourier transform of $\hat{\phi} * \hat{\psi}$ is $(2\pi)^n\phi(-x)(2\pi)^n\psi(-x)$ in view of (7.1.9) and the Fourier inversion formula. The proof is complete.

Definition 7.1.7. A continuous linear form u on $\mathscr{S}$ is called a temperate distribution. The set of all temperate distributions is denoted by $\mathscr{S}'$.

The restriction of a temperate distribution to $C_0^\infty(\mathbb{R}^n)$ is obviously a distribution in $\mathscr{D}'(\mathbb{R}^n)$. We can in fact *identify* $\mathscr{S}'$ with a subspace of $\mathscr{D}'(\mathbb{R}^n)$ since the following lemma shows that a distribution $u \in \mathscr{S}'$ which vanishes on $C_0^\infty(\mathbb{R}^n)$ must also vanish on $\mathscr{S}$.

Lemma 7.1.8. *C_0^∞ is dense in $\mathscr{S}$.*

Proof. Let $\phi \in \mathscr{S}$ and take $\psi \in C_0^\infty$ such that $\psi(x) = 1$ when $|x| \leqq 1$. Put $\phi_\varepsilon(x) = \phi(x)\psi(\varepsilon x)$. Then it is clear that $\phi_\varepsilon \in C_0^\infty$, and since

$$\phi_\varepsilon(x) - \phi(x) = \phi(x)(\psi(\varepsilon x) - 1) = 0 \qquad \text{if } |x| < 1/\varepsilon,$$

we conclude that $\phi_\varepsilon \to \phi$ in $\mathscr{S}$ when $\varepsilon \to 0$.

It is obvious that $\mathscr{E}' \subset \mathscr{S}'$. Other examples of elements in $\mathscr{S}'$ are measures $d\mu$ such that for some m

$$\int (1 + |x|)^{-m} |d\mu(x)| < \infty.$$

In particular, this implies that $L^p(\mathbb{R}^n)\subset\mathscr{S}'$ for every p. It is also clear that $\mathscr{S}'$ is closed under differentiation and under multiplication by polynomials or functions in $\mathscr{S}$.

Definition 7.1.9. If $u\in\mathscr{S}'$, the Fourier transform $\hat{u}$ is defined by

$$\hat{u}(\phi)=u(\hat{\phi}),\qquad \phi\in\mathscr{S}. \tag{7.1.11}$$

It follows from Lemma 7.1.3 that $\hat{u}\in\mathscr{S}'$, and since the proof of (7.1.7) is valid for all $\phi,\psi\in L^1$ the preceding definition agrees with (7.1.3) if $f\in L^1$.

Fourier's inversion formula as proved in Theorem 7.1.5 states that

$$\hat{\hat{\phi}}=(2\pi)^n\check{\phi}\quad \text{if } \phi\in\mathscr{S},\quad \check{\phi}(x)=\phi(-x).$$

If u is in $\mathscr{S}'$ we obtain

$$\hat{\hat{u}}(\phi)=u(\hat{\hat{\phi}})=(2\pi)^n u(\check{\phi})=(2\pi)^n\check{u}(\phi).$$

Here $\check{u}$ is of course the composition of u and $x\to -x$. Thus we have

Theorem 7.1.10. *The Fourier transformation is an isomorphism of $\mathscr{S}'$ (with the weak topology), and Fourier's inversion formula $\hat{\hat{u}}=(2\pi)^n\check{u}$ is valid for every $u\in\mathscr{S}'$.*

In particular, $f\in L^1$ and $\hat{f}\in L^1$ then the inversion formula (7.1.4) is valid for almost every x.

Theorem 7.1.11. *If $u\in L^2(\mathbb{R}^n)$ then the Fourier transform $\hat{u}$ is also in $L^2(\mathbb{R}^n)$ and Parseval's formula (7.1.8) is valid for all $\phi,\psi\in L^2$.*

Proof. Choose a sequence $u_j\in C_0^\infty$ such that $u_j\to u$ in L^2 norm. Then

$$\|\hat{u}_j-\hat{u}_k\|_{L^2}^2=(2\pi)^n\|u_j-u_k\|_{L^2}^2\to 0$$

by (7.1.8) which is already proved in $\mathscr{S}$. In view of the Riesz-Fischer theorem it follows that there is a function $U\in L^2$ with $\hat{u}_j\to U$ in L^2, and $U=\hat{u}$ by the continuity of the Fourier transformation in $\mathscr{S}'$. Now both sides of (7.1.8) are continuous functions of ϕ and ψ in the L^2 norms, so (7.1.8) is valid for arbitrary L^2 functions.

If $u\in L^p$ and $1\leqq p\leqq 2$, we can write u as the sum of a function in L^2 and one in L^1, so the Fourier transform is in L^2_{loc}. A better result follows from the Riesz-Thorin convexity theorem:

Theorem 7.1.12. *If T is a linear map from $L^{p_1}\cap L^{p_2}$ to $L^{q_1}\cap L^{q_2}$ such that*

(7.1.12) $$\|Tf\|_{q_j} \leqq M_j \|f\|_{p_j}, \quad j=1,2,$$

and if $1/p = t/p_1 + (1-t)/p_2$, $1/q = t/q_1 + (1-t)/q_2$, *for some* $t \in (0,1)$, *then*

(7.1.12)′ $$\|Tf\|_q \leqq M_1^t M_2^{1-t} \|f\|_p, \quad f \in L^{p_1} \cap L^{p_2}.$$

Proof. We may assume $p < \infty$ for otherwise $p_1 = p_2 = \infty$ and (7.1.12)′ follows then from Hölder's inequality. The method of proof is similar to that of Theorem 4.5.1. First we write (7.1.12) in the form

$$|\langle Tf, g\rangle| \leqq M_j \|f\|_{p_j} \|g\|_{q_j'}, \quad 1/q_j + 1/q_j' = 1.$$

If $0 \leqq F$, $G \in L^1$ and $\|F\|_{L^1} = \|G\|_{L^1} = 1$ then the absolute value of

$$\Phi(z) = \langle T(f_0 F^{z/p_1 + (1-z)/p_2}), g_0 G^{z/q_1' + (1-z)/q_2'}\rangle M_1^{-z} M_2^{z-1}$$

is $\leqq 1$ when $\operatorname{Re} z = 0$ or $\operatorname{Re} z = 1$ provided that $|f_0| \leqq 1$, $|g_0| \leqq 1$. If we take for F and G functions which take only finitely many values, it is clear that Φ is analytic and bounded when $0 \leqq \operatorname{Re} z \leqq 1$. By the Phragmén-Lindelöf theorem the bound 1 is also valid in the interior, thus

$$|\langle T(f_0 F^{1/p}), g_0 G^{1/q'}\rangle| \leqq M_1^t M_2^{1-t}, \quad 1/q + 1/q' = 1.$$

This proves (7.1.12)′ for a dense subset of $L^{p_1} \cap L^{p_2}$. By hypothesis both sides are continuous in $L^{p_1} \cap L^{p_2}$ which completes the proof.

Theorem 7.1.13 (Hausdorff-Young). *If* $f \in L^p$ *and* $1 \leqq p \leqq 2$ *then* $\hat{f} \in L^{p'}$, $1/p + 1/p' = 1$, *and*

(7.1.13) $$\|\hat{f}\|_{L^{p'}} \leqq (2\pi)^{n/p'} \|f\|_{L^p}.$$

Proof. (7.1.13) follows from Parseval's formula when $p = 2$ and is obvious when $p = 1$, so it follows in general from Theorem 7.1.12.

In Section 7.6 we shall see that the Fourier transform of a function in L^p may have positive order if $p > 2$.

Already Lemma 7.1.3 indicates that the Fourier transformation exchanges local smoothness properties and growth properties at ∞. Another case of this is the following

Theorem 7.1.14. *The Fourier transform of a distribution* $u \in \mathscr{E}'(\mathbb{R}^n)$ *is the function*

(7.1.14) $$\hat{u}(\xi) = u_x(e^{-i\langle x, \xi\rangle}).$$

The right-hand side is also defined for every complex vector $\xi \in \mathbb{C}^n$ *and is an entire analytic function of* ξ, *called the Fourier-Laplace transform of* u.

Proof. If $\phi\in C_0^\infty(\mathbb{R}^n)$ we have

$$\hat{u}(\phi)=u(\hat{\phi})=(u_x\otimes\phi_\xi)(e^{-i\langle x,\xi\rangle})=\int\phi(\xi)\,u(e^{-i\langle .,\xi\rangle})\,d\xi$$

by Theorem 5.1.1 or rather its analogue for distributions of compact support and test functions in C^∞. It follows that $\hat{u}$ is equal to the function (7.1.14) which is a C^∞ function in $\mathbb{C}^n$ by Theorem 2.1.3. It satisfies the Cauchy-Riemann equations since we may differentiate with respect to ξ directly on the exponential, and this proves the theorem.

Note that $u*e^{i\langle .,\zeta\rangle}=\hat{u}(\zeta)\,e^{i\langle .,\zeta\rangle}$. The properties of the entire analytic function $\hat{u}$ will be discussed further in Section 7.3.

Theorem 7.1.15. *If $u_1\in\mathscr{S}'$ and $u_2\in\mathscr{E}'$, it follows that $u_1*u_2\in\mathscr{S}'$ and that the Fourier transform of u_1*u_2 is $\hat{u}_1\hat{u}_2$.*

The product is defined since Theorem 7.1.14 shows that $\hat{u}_2\in C^\infty$.

Proof. If $\phi\in C_0^\infty$ we have by definition of the convolution

$$(u_1*u_2)(\phi)=u_1*u_2*\check{\phi}(0)=u_1(\check{u}_2*\phi).$$

The right-hand side is a continuous linear form on $\mathscr{S}$, for if $\phi\in\mathscr{S}$ and u_2 is of order k then $\check{u}_2*\phi\in\mathscr{S}$ and

$$\sum_{|\alpha+\beta|\leqq j}\sup|x^\alpha\partial^\beta\check{u}_2*\phi|\leqq C_j\sum_{|\alpha+\beta|\leqq j+k}\sup|x^\alpha\partial^\beta\phi|.$$

To compute the Fourier transform of u_1*u_2 we note that

$$(u_1*u_2)(\hat{\phi})=u_1(\check{u}_2*\hat{\phi}),\qquad \phi\in\mathscr{S}.$$

If $\phi\in C_0^\infty$ then $\check{u}_2*\hat{\phi}$ is the Fourier transform of $\hat{u}_2\phi$, for this is

$$\int e^{-i\langle x,\xi\rangle}u_2(e^{-i\langle .,\xi\rangle})\,\phi(\xi)\,d\xi=u_2(\hat{\phi}(x+.))$$

by Theorem 5.1.1. Hence

$$(u_1*u_2)(\hat{\phi})=\hat{u}_1(\hat{u}_2\phi)=(\hat{u}_1\hat{u}_2)(\phi),\qquad \phi\in C_0^\infty,$$

which proves the statement.

Theorem 7.1.15 contains the extension to $\mathscr{S}'$ of the basic properties of the Fourier transformation observed in Lemma 7.1.3,

$$\widehat{D_j u}=\xi_j\hat{u},\quad \widehat{x_j u}=-D_j\hat{u},\quad u\in\mathscr{S}',\tag{7.1.15}$$

for differentiation can be written as convolution with a derivative of the Dirac measure at 0. We can also obtain (7.1.15) by differentiation of

(7.1.16) $$\widehat{\delta_h * u} = e^{-i\langle h,.\rangle}\hat{u}, \quad (e^{i\langle .,h\rangle}u)\hat{} = \delta_h * \hat{u}.$$

Alternatively we could also argue that (7.1.15) must be valid in $\mathscr{S}'$ since it is valid in the dense subset $\mathscr{S}$ and both sides are continuous in $\mathscr{S}'$. This argument shows also that if T is a linear bijection $\mathbb{R}^n \to \mathbb{R}^n$ then

(7.1.17) $$\widehat{T^* u} = ({}^tT^{-1})^* \hat{u} |\det T|^{-1}.$$

Theorem 7.1.16. *If $u \in \mathscr{S}'(\mathbb{R}^n)$ is a homogeneous distribution of degree a, then $\hat{u}$ is homogeneous of degree $-a-n$.*

Proof. If M_t denotes multiplication by t then $M_t^* u = t^a u$ so (7.1.17) gives for $t>0$

$$t^a \hat{u} = M_{1/t}^* \hat{u} t^{-n}, \quad \text{that is,} \quad M_s^* \hat{u} = s^{-n-a}\hat{u}, \quad s>0.$$

Remark. We shall see below (Theorem 7.1.18) that every homogeneous $u \in \mathscr{D}'(\mathbb{R}^n)$ is automatically in $\mathscr{S}'(\mathbb{R}^n)$.

Example 7.1.17. With the notation in Section 3.2 the Fourier transform of $\chi_\pm^a$ is $e^{\mp i\pi(a+1)/2}(\xi \mp i0)^{-a-1}$ and that of $(x \pm i0)^a$ is $2\pi e^{\pm i\pi a/2}\chi_\mp^{-a-1}(\xi)$ for every $a \in \mathbb{C}$. If k is an integer then the Fourier transform of $\underline{x}^{-1-k}$ is $\pi i^{-1-k}(\operatorname{sgn}\xi)\,\xi^k/k!$ if $k \geqq 0$ and $2\pi i^{-1-k}\delta^{(-k-1)}$ if $k<0$.

The second statement follows from the first by the Fourier inversion formula, and the third follows from the second and (3.2.10)′, (3.2.17)′. To prove the first statement we observe that when $\varepsilon>0$ and $\operatorname{Re} a > -1$ the Fourier transform of $e^{-\varepsilon x}\chi_+^a(x)$ is

$$\xi \to \int_0^\infty x^a e^{-x(\varepsilon+i\xi)}\,dx/\Gamma(a+1) = (\varepsilon+i\xi)^{-a-1}\int_0^\infty z^a e^{-z}\,dz/\Gamma(a+1)$$

where the last integral is taken on the ray with direction $\varepsilon+i\xi$ and z^a is defined in $\mathbb{C}$ slit along $\mathbb{R}_-$ so that $1^a=1$. In view of the Cauchy integral formula the integral can be taken along $\mathbb{R}_+$ so it is equal to $\Gamma(a+1)$. Hence the Fourier transform is

$$\xi \to (\varepsilon+i\xi)^{-a-1} = e^{-i\pi(a+1)/2}(\xi - i\varepsilon)^{-a-1}.$$

(Note that this explains (3.4.10).) When $\varepsilon \to 0$ it follows that the Fourier transform of χ_+^a has the stated form, that is,

$$\langle \chi_+^a, \hat{\phi}\rangle = e^{-i\pi(a+1)/2}\langle(\xi - i0)^{-a-1}, \phi\rangle, \quad \phi \in \mathscr{S},$$

if $\operatorname{Re} a > -1$. Both sides are entire analytic functions of a so the equality must hold for all $a \in \mathbb{C}$. The remaining part of the first statement follows when x and ξ are replaced by $-x$ and $-\xi$.

We shall now consider homogeneous distributions in $\mathbb{R}^n$ or in $\mathbb{R}^n \setminus 0$.

Theorem 7.1.18. *If $u \in \mathscr{D}'(\mathbb{R}^n)$ and the restriction to $\mathbb{R}^n \setminus 0$ is homogeneous of degree a, then $u \in \mathscr{S}'$. If in addition $u \in C^\infty(\mathbb{R}^n \setminus 0)$ then $\hat{u} \in C^\infty(\mathbb{R}^n \setminus 0)$.*

Proof. Choose $\psi \in C_0^\infty(\mathbb{R}^n \setminus 0)$ satisfying (3.2.22). Since

$$\psi_0(x) = 1 - \int_1^\infty \psi(x/t)\, dt/t$$

is in C_0^∞ we can write

$$\begin{aligned} u(\phi) &= u(\psi_0 \phi) + \int_1^\infty u(\psi(./t)\, \phi)\, dt/t \\ &= u(\psi_0 \phi) + \int_1^\infty u(\psi\, \phi(t.))\, t^{a+n-1}\, dt, \qquad \phi \in C_0^\infty. \end{aligned}$$

If $K_0 = \operatorname{supp} \psi_0$, $K = \operatorname{supp} \psi$, and k is the order of u in a neighborhood, then

$$\begin{aligned} |u(\phi)| &\leqq C\Big(\sum_{|\alpha| \leqq k} \sup_{K_0} |D^\alpha \phi| + \int_1^\infty \sum_{|\alpha| \leqq k} t^{|\alpha| + \operatorname{Re} a + n - 1} \sup_{tK} |D^\alpha \phi|\, dt\Big) \\ &\leqq C' \sum_{|\alpha| \leqq k} \sum_{|\beta| \leqq |\alpha| + M} \sup |x^\beta D^\alpha \phi|, \qquad \phi \in C_0^\infty(\mathbb{R}^n), \end{aligned}$$

if M is an integer $> \operatorname{Re} a + n$. This proves that $u \in \mathscr{S}'$. Assume now that $u \in C^\infty$ in $\mathbb{R}^n \setminus 0$. If $\operatorname{Re} a < -n$ then $\hat{u}$ is continuous, for $u = \psi_0 u + (1 - \psi_0) u$ where the first term has compact support and the second is integrable. In general we conclude that $D^\beta \xi^\alpha \hat{u}$, which is the Fourier transform of $(-x)^\beta D^\alpha u$, is a continuous function if $\operatorname{Re} a - |\alpha| + |\beta| < -n$. Since this is true for any $|\beta|$ if $|\alpha| > \operatorname{Re} a + |\beta| + n$, we obtain $\hat{u} \in C^\infty(\mathbb{R}^n \setminus 0)$.

Theorems 7.1.16 and 7.1.18 show that the Fourier transformation is a bijection of homogeneous distributions of degree a in $\mathbb{R}^n$ which are C^∞ in $\mathbb{R}^n \setminus 0$ on such distributions of degree $-n-a$. Let us now consider the Fourier transform U of a distribution $\dot{u}$ satisfying (3.2.24)′, that is,

$$\dot{u} = t^{-k-n} M_{1/t}^* \dot{u} + \log t \sum_{|\alpha| = k} S(x^\alpha u)(-1)^k \partial^\alpha \delta_0 / \alpha!.$$

By (7.1.17) we have in $\mathbb{R}^n$

$$U = t^{-k} M_t^* U + Q \log t$$

where Q is the homogeneous polynomial of degree k

$$Q(\xi) = S_x(\langle -ix, \xi\rangle^k u/k!). \tag{7.1.18}$$

(Recall that S_x denotes integration over the unit sphere.) Thus

$$U_0 = U + Q \log|.|$$

is homogeneous of degree k, for

$$t^{-k} M_t^* U_0 = t^{-k} M_t^* U + Q(\log t + \log|.|) = U + Q \log|.| = U_0.$$

It follows that U_0 is a bounded function in a neighborhood of 0, and homogeneous of degree k and C^∞ in $\mathbb{R}^n \setminus 0$; we have

$$U(\xi) = U_0(\xi) - Q(\xi) \log|\xi|. \tag{7.1.19}$$

We shall apply this to construct fundamental solutions, but first we must introduce some terminology.

Definition 7.1.19. A homogeneous polynomial $P(\xi)$ in n variables, with complex coefficients, is called elliptic if $P(\xi) \neq 0$, $0 \neq \xi \in \mathbb{R}^n$. An inhomogeneous polynomial is called elliptic if the homogeneous part of highest degree is elliptic.

Theorem 7.1.20. *If P is a homogeneous elliptic polynomial of degree m in $\mathbb{R}^n$, then $P(D)$ has a fundamental solution E such that*

$$E = E_0 - Q(x) \log|x|.$$

Here E_0 is homogeneous of degree $m-n$ and C^∞ in $\mathbb{R}^n \setminus 0$, and Q is a polynomial which is identically 0 when $n > m$ and is defined when $n \leqq m$ by

$$Q(x) = S_\xi(\langle ix, \xi\rangle^{m-n}/P(\xi))(2\pi)^{-n}/(m-n)!. \tag{7.1.20}$$

Proof. We define E so that $\hat{E} = (1/P)^{\cdot}$ with the notation in Theorems 3.2.3 and 3.2.4. Then $P\hat{E} = 1^{\cdot} = \hat{\delta}$, hence $P(D)E = \delta$, and $(2\pi)^n \check{E} = \hat{\hat{E}}$ has the properties just described.

Remark. Using the analyticity of P it is easy to show that E_0 is analytic in $\mathbb{R}^n \setminus 0$. We shall prove this in a general context in Section 8.6. (See also a remark after Theorem 7.1.22.)

Theorem 7.1.20 shows that every homogeneous elliptic operator with constant coefficients satisfies the hypothesis in Theorem 4.4.1.

However, the main difficulty in the proof of Theorem 7.1.20, the definition of $(1/P)^{\cdot}$, can be avoided in this kind of application. Indeed, the proofs of Theorems 4.4.1 and 4.4.2 work equally well if one just has a parametrix:

Definition 7.1.21. If $P(D)$ is a differential operator with constant coefficients, then $E \in \mathscr{D}'(\mathbb{R}^n)$ is called a parametrix of $P(D)$ if

$$P(D)\,E = \delta + \omega, \qquad \omega \in C^\infty(\mathbb{R}^n).$$

When constructing a parametrix one does not have to pay attention to the definition of $1/P$ at 0 since the Fourier transform of a distribution of compact support is always in C^∞. More generally, we have

Theorem 7.1.22. *Every elliptic operator $P(D)$ with constant coefficients has a parametrix E which is a C^∞ function in $\mathbb{R}^n \setminus 0$.*

Proof. We can write

$$P(\xi) = P_m(\xi) + P_{m-1}(\xi) + \ldots + P_0$$

where P_j is homogeneous of degree ξ and $P_m(\xi) \neq 0$ when $\xi \neq 0$. Then $|P_m(\xi)| \geqq c > 0$ when $|\xi| = 1$, so the homogeneity gives

$$|P_m(\xi)| = |\xi|^m \, |P_m(\xi/|\xi|)| \geqq c\,|\xi|^m, \qquad \xi \in \mathbb{R}^n.$$

It follows that for some constants C and R

$$|P(\xi)| \geqq |P_m(\xi)| - |P_{m-1}(\xi)| - \ldots \geqq c\,|\xi|^m - C(|\xi|^{m-1} + \ldots + 1) \geqq c\,|\xi|^m/2$$

when $\xi \in \mathbb{R}^n$ and $|\xi| \geqq R$. Since a derivative of order k of $1/P(\xi)$ is of the form $Q(\xi)/P(\xi)^{k+1}$ with Q of degree $\leqq (m-1)\,k$, as is immediately verified by induction, we conclude that when $|\xi| > R$

$$|\xi^\beta D^\alpha(1/P(\xi))| \leqq C_{\alpha\beta} |\xi|^{|\beta| - |\alpha| - m}. \tag{7.1.21}$$

Choose $\chi \in C_0^\infty(\mathbb{R}^n)$ equal to 1 in $\{\xi; |\xi| < R\}$. Then $(1-\chi(\xi))/P(\xi)$ is a bounded C^∞ function, hence the Fourier transform of a distribution $E \in \mathscr{S}'$. We have $P(D)\,E = \delta + \omega$ where $\hat{\omega} = -\chi$. Hence $\omega \in \mathscr{S}$, and as in the proof of Theorem 7.1.18 it follows from (7.1.21) that $D^\beta x^\alpha E$ is continuous when $|\beta| - |\alpha| - m < -n$. The proof is complete.

Remark. The error term ω obtained in the proof has an analytic extension to $\mathbb{C}^n$ by Theorem 7.1.14. It is also easy to show by deforming the integration contour into the complex domain that E has an analytic extension to a conic neighborhood of $\mathbb{R}^n \setminus 0$. We leave the proof as an exercise for the reader since more general results will be given in Section 8.6. - The proof of Theorem 7.1.22 remains valid for all P such that $P^{(\alpha)}(\xi)/P(\xi)\to 0$ for all $\alpha\neq 0$ when $\xi\to\infty$ in $\mathbb{R}^n$. Here $P^{(\alpha)}(\xi)=\partial^\alpha P(\xi)$. One just has to show that $P^{(\alpha)}(\xi)/P(\xi)=O(|\xi|^{-|\alpha|c})$ for some $c>0$ as $\xi\to\infty$ and then verify that $E(\xi)=1/P(\xi)$ inherits the same property for all α. (Thus c is independent of α.) The converse is also true. We refer to Chapter XI for these and many related results.

We shall now determine the Fourier transforms of the homogeneous distributions studied in Section 6.2.

Theorem 7.1.23. *Under the assumptions in Theorems 6.2.1 the Fourier transform of* $(A\pm i0)^{(2-n)/2}$ *is* $(n-2)\,c_n|\det A|^{-\frac{1}{2}}e^{\mp\pi i n_-/2}(B\mp i0)^{-1}$, *and that of* $A^*\chi_\pm^{(2-n)/2}$ *is*

$$2\pi^{(n-2)/2}\,|\det A|^{-\frac{1}{2}}(i\,e^{\pm\pi i n_\pm/2}(B-i0)^{-1}-i\,e^{\mp\pi i n_\pm/2}(B+i0)^{-1}).$$

Proof. If A had positive definite real part instead, then (6.2.1)″ would be valid. Taking Fourier transforms we would then find that the Fourier transform of $A^{(2-n)/2}$ is equal to

$$(n-2)\,c_n(\det A)^{-\frac{1}{2}}/B(\xi)$$

outside 0. By Theorem 3.2.3 and Theorem 7.1.16 this must then be true in the whole space. Now assume that A is just real and non-degenerate, and apply this result to $A_\varepsilon=-iA+\varepsilon|x|^2$, $\varepsilon>0$. As in the proof of Theorem 6.2.1 we obtain when $\varepsilon\to 0$ that the Fourier transform of

$$i^{-1}e^{\pi i n/4}(A+i0)^{(2-n)/2}$$

is equal to $|\det A|^{-\frac{1}{2}}e^{\pi i(\operatorname{sgn}A)/4}(n-2)\,c_n$ times the limit of $1/iF_\varepsilon(\xi)$ where $F_\varepsilon(\xi)$ is the quadratic form with matrix

$$(a_{jk}+\varepsilon i\delta_{jk})^{-1}=(b_{jk})-\varepsilon i(b_{jk})^2+O(\varepsilon^2).$$

The limit is $i^{-1}(B(\xi)-i0)^{-1}$. In fact, since we have functions homogeneous of degree $-2>-n$ it is sufficient to verify this in $\mathbb{R}^n\setminus 0$ (by Theorem 3.2.3), and then it follows from Lemma 6.2.2. We have therefore proved the first statement in Theorem 7.1.23. The others are then derived as at the beginning of the proof of Theorem 6.2.1.

We shall now compute the Fourier transform of a distribution $u\in\mathscr{D}'(\mathbb{R}^n)$ such that for some integer k

$$\langle u,\phi\rangle=\operatorname{sgn}t\,t^{-k}\langle u,\phi(t.)\rangle,\quad \phi\in C_0^\infty(\mathbb{R}^n),\quad t\neq 0. \tag{7.1.22}$$

This means that u is homogeneous of degree $-n-k$ and has parity opposite to k (cf. (3.2.18)′). By Theorem 3.2.4 restricting to $\mathbb{R}^n\setminus 0$ gives a bijection of such distributions in $\mathbb{R}^n$ on the distributions in $\mathbb{R}^n \setminus 0$ having the same property when $\phi \in C_0^\infty(\mathbb{R}^n \setminus 0)$. Moreover, we have by (3.2.23)″

$$\langle u, \phi \rangle = S(u \langle \underline{t}^{-k-1}, \phi(t.)\rangle)/2, \qquad \phi \in C_0^\infty(\mathbb{R}^n).$$

The formula extends by continuity to all $\phi \in \mathscr{S}$. To compute $\langle \hat{u}, \phi \rangle = \langle u, \hat{\phi} \rangle$ we first calculate $\langle \underline{t}^{-k-1}, \hat{\phi}(tx)\rangle$ in terms of ϕ when $x=(1, 0, \ldots, 0)$. Then $\hat{\phi}(t, 0, \ldots, 0)$ is the Fourier transform of $\int \phi(\xi_1, \xi')\, d\xi'$ where $\xi' = (\xi_2, \ldots, \xi_n)$. The Fourier transform of $\underline{t}^{-k-1}$ is $2\pi i^{-1-k} \sigma_k$ where

$$\sigma_k(\tau) = 2^{-1}(\operatorname{sgn} \tau)\, \tau^k/k!, \qquad k = 0, 1, \ldots; \tag{7.1.23}$$
$$\sigma_k(\tau) = \delta^{(-k-1)}, \qquad k = -1, -2, \ldots$$

by example 7.1.17. Hence

$$\begin{aligned}\langle \underline{t}^{-k-1}, \hat{\phi}(tx)\rangle &= 2\pi i^{-1-k} \int \langle \sigma_k(\tau), \phi(\tau, \xi')\rangle\, d\xi' \\ &= 2\pi i^{-1-k} \langle \sigma_k(\langle x, \xi \rangle), \phi(\xi)\rangle.\end{aligned}$$

Note that σ_k is homogeneous of degree k so the last expression is homogeneous in x of degree k. The orthogonal invariance shows that it is equal to $\langle \underline{t}^{-k-1}, \hat{\phi}(tx)\rangle$ for every $x \neq 0$, so we obtain for all $\phi \in \mathscr{S}$

$$\langle \hat{u}, \phi \rangle = \pi i^{-1-k} S_x(u(x) \langle \sigma_k(\langle x, \xi \rangle), \phi(\xi)\rangle) \tag{7.1.24}$$

or formally

$$\hat{u}(\xi) = \pi i^{-1-k} S_x(u(x)\, \sigma_k(\langle x, \xi \rangle)). \tag{7.1.24$'$}$$

In Sections 8.2 and 12.6 we shall give a precise meaning to such formulas. Note that when $k<0$ they show that $\hat{u}$ is determined at ξ by the restriction of u to a neighborhood of the orthogonal plane; for $k \geqq 0$ this is true for the singularities only. In Chapter XII we shall return to (7.1.24) which is the essence of the Herglotz-Petrovsky formula for the fundamental solution of hyperbolic equations. To have a reference then we sum up the preceding results:

Theorem 7.1.24. *If $u \in \mathscr{S}'(\mathbb{R}^n)$ satisfies* (7.1.22), *that is, u is homogeneous of degree $-n-k$ and of parity opposed to the integer k, then the Fourier transform is given by* (7.1.24) *with σ_k defined in* (7.1.23).

We shall close this section by studying Fourier transforms of densities (simple layers) on submanifolds. First recall that the Fourier transform of δ_0 is the Lebesgue measure dx, so by Fourier's inversion formula the Fourier transform of dx is $(2\pi)^n \delta_0$. More generally:

Theorem 7.1.25. *If $V \subset \mathbb{R}^n$ is a linear subspace and $V^\perp$ is the orthogonal space, then the Fourier transform of the Euclidean surface measure in V is $(2\pi)^{\dim V}$ times the Euclidean surface measure in $V^\perp$.*

Proof. By an orthogonal transformation we can make V defined by $x'' = (x_{k+1}, \dots, x_n) = 0$. Set $x' = (x_1, \dots, x_k)$. Then

$$\int \hat{\phi}(x', 0)\, dx' = (2\pi)^k \int \phi(0, x'')\, dx'', \qquad \phi \in \mathscr{S}, \tag{7.1.25}$$

for if $\Phi(x') = \int \phi(x', x'')\, dx''$, then $\Phi \in \mathscr{S}(\mathbb{R}^k)$, $\hat{\Phi}(\xi') = \hat{\phi}(\xi', 0)$, so (7.1.25) is just Fourier's inversion formula for Φ. The theorem is proved.

Let dS_V and $dS_{V^\perp}$ be the surface measures in V and in $V^\perp$, and let $u_0 \in L^2(dS_V)$. Then $u = u_0\, dS_V \in \mathscr{S}'(\mathbb{R}^n)$ and the Fourier transform is $\hat{u}_0(\xi')$ with the coordinates just used. Hence Parseval's formula gives if $\phi \in C_0(\mathbb{R}^n)$

$$\begin{aligned} R^{k-n} \int |\hat{u}(\xi)|^2\, \phi(\xi/R)\, d\xi &= \int |\hat{u}_0(\xi')|^2\, \phi(\xi'/R, \xi'')\, d\xi \\ &\to \int |\hat{u}_0(\xi')|^2\, d\xi' \int \phi(0, \xi'')\, d\xi'' = (2\pi)^k \int |u_0|^2\, dS_V \int_{V^\perp} \phi\, dS_{V^\perp}, \qquad R \to \infty. \end{aligned}$$

We shall now extend these facts to general surfaces. As indicated by the preceding formula, it is convenient to change notation and let k be the codimension of V instead.

Theorem 7.1.26. *Let K be a compact subset of a C^1 manifold $M \subset \mathbb{R}^n$ of codimension k, with Euclidean surface area denoted by dS. If $u = u_0\, dS$ where u_0 is supported by K and square integrable with respect to dS, then*

$$\int_{|\xi| < R} |\hat{u}(\xi)|^2\, d\xi \leq C R^k \int |u_0|^2\, dS, \qquad R > 0, \tag{7.1.26}$$

where C is independent of u and R.

Proof. After using a partition of unity we may assume that in a neighborhood of K the manifold M is of the form

$$x'' = h(x'); \qquad x' = (x_1, \dots, x_{n-k}) \in \mathbb{R}^{n-k}, \quad x'' = (x_{n-k+1}, \dots, x_n) \in \mathbb{R}^k.$$

Here $h \in C^1$. Then $dS = a(x')\, dx'$ where a is a positive continuous function; u_0 is a function of x' and

$$\hat{u}(\xi) = \int e^{-i(\langle x', \xi' \rangle + \langle h(x'), \xi'' \rangle)}\, u_0(x')\, a(x')\, dx'.$$

For a fixed ξ'' Parseval's formula gives

$$\int |\hat{u}(\xi)|^2\, d\xi' = (2\pi)^{n-k} \int |u_0|^2 a^2\, dx' = (2\pi)^{n-k} \int |u_0|^2\, a\, dS.$$

Integration with respect to ξ'' for $|\xi''| < R$ leads to (7.1.26).

In spite of the simplicity of the proof, the estimate (7.1.26) is optimal:

Theorem 7.1.27. *Let* $u \in \mathscr{S}'$, $\hat{u} \in L^2_{\mathrm{loc}}$ *and assume that*

$$\limsup_{R\to\infty} \int_{|\xi|<R} |\hat{u}(\xi)|^2\, d\xi / R^k < \infty. \tag{7.1.27}$$

If the restriction of u *to an open subset* X *of* $\mathbb{R}^n$ *is supported by a* C^1 *submanifold* M *of codimension* k, *then it is an* L^2 *density* $u_0 dS$ *on* M *and*

$$\int_M |u_0|^2\, dS \leqq C \limsup_{R\to\infty} \int_{|\xi|<R} |\hat{u}(\xi)|^2\, d\xi/R^k \tag{7.1.28}$$

where C *only depends on* n.

Proof. Choose an even function $\chi \in C_0^\infty(\mathbb{R}^n)$ with support in the unit ball and $\int \chi\, dx = 1$. The Fourier transform of $u_\varepsilon = u * \chi_\varepsilon$, where $\chi_\varepsilon(x) = \varepsilon^{-n}\chi(x/\varepsilon)$, is $\hat{u}(\xi)\hat{\chi}(\varepsilon\xi)$. Hence, with $\|\ \|$ denoting L^2 norm,

$$\|u_\varepsilon\|^2 = (2\pi)^{-n} \int |\hat{u}(\xi)|^2 |\hat{\chi}(\varepsilon\xi)|^2\, d\xi \leqq C' \varepsilon^{-k} K(\varepsilon)$$

where

$$K(\varepsilon) = \sup_{\varepsilon R > 1} (2\pi)^{-n} \int_{|\xi|<R} |\hat{u}(\xi)|^2\, d\xi / R^k,$$

$$C' = \sup_{|\xi|<1} |\hat{\chi}(\xi)|^2 + \sum_1^\infty \sup_{|\xi|>2^{j-1}} |\hat{\chi}(\xi)|^2 2^{kj}.$$

The intersection of $\operatorname{supp} u_\varepsilon$ with a compact subset of X belongs to the set M_ε of points at distance $<\varepsilon$ from $\operatorname{supp} u \subset M$ when ε is small. If $\psi \in C_0^\infty(X)$ it is geometrically evident and easily proved by a change of variables that

$$\varepsilon^{-k} \int_{M_\varepsilon} |\psi(x)|^2\, dx \to C_k \int_{\operatorname{supp} u} |\psi|^2\, dS, \qquad \varepsilon \to 0,$$

where C_k is the volume of the unit ball in $\mathbb{R}^k$. Hence

$$|\langle u, \psi\rangle|^2 = \lim_{\varepsilon\to 0} |\langle u_\varepsilon, \psi\rangle|^2 \leqq C'' \int_{\operatorname{supp} u} |\psi|^2\, dS \lim_{\varepsilon\to 0} K(\varepsilon).$$

This proves that there is an L^2 density $u_0 dS$ on M such that for $\psi \in C_0^\infty(X)$

$$\langle u, \psi\rangle = \int u_0 \psi\, dS, \qquad \int |u_0|^2\, dS \leqq C'' \limsup_{R\to\infty} \int_{|\xi|<R} |\hat{u}(\xi)|^2\, d\xi/R^k.$$

The proof is complete.

The following is a substitute for Parseval's formula:

Theorem 7.1.28. *Let $\phi \in C_0(\mathbb{R}^n)$. If $u = u_0\, dS$ is an L^2 density with compact support on a C^1 manifold $M \subset \mathbb{R}^n$ of codimension k, then*

$$\lim_{R\to\infty} \int |\hat{u}(\xi)|^2 \phi(\xi/R)\, d\xi/R^k = (2\pi)^{n-k} \int_M |u_0(x)|^2 \Big(\int_{N_x} \phi(\xi)\, d\sigma(\xi)\Big) dS(x) \tag{7.1.29}$$

where dS is the Euclidean surface element on M and $d\sigma$ is the Euclidean integration element in the normal plane N_x of M at x, passing through 0.

Proof. In view of Theorem 7.1.26 it suffices to prove (7.1.29) when u_0 is continuous and $\phi \in C_0^\infty(\mathbb{R}^n)$. Let $\phi(\xi) = \hat{\Phi}(\xi)$, thus $\Phi \in \mathscr{S}$, and set

$$\Phi_R(x) = R^{n-k} \Phi(Rx).$$

Then the Fourier transform of Φ_R is $R^{-k}\phi(\xi/R)$ so the Fourier transform of $u * \Phi_R$ is $\hat{u}(\xi)\phi(\xi/R)R^{-k}$. Hence by Parseval's formula,

$$\int |\hat{u}(\xi)|^2 \phi(\xi/R)\, d\xi/R^k = (2\pi)^n (u * \Phi_R, u).$$

It remains to compute the limit of $u * \Phi_R$ on M. In the formula

$$u * \Phi_R(x) = \int u_0(y) \Phi(R(x-y)) R^{n-k}\, dS(y), \qquad x \in M,$$

we may assume without restriction that M is given by $x'' = h(x')$ as in the proof of Theorem 7.1.26, for the integral tends rapidly to zero outside the support of u_0. With the notation used there we can write

$$\begin{aligned} u * \Phi_R(x', h(x')) &= \int u_0(y') \Phi(R(x'-y'), R(h(x') - h(y'))) R^{n-k} a(y')\, dy' \\ &= \int u_0(x' - y'/R) \Phi(y', R(h(x') - h(x' - y'/R))) a(x' - y'/R)\, dy' \\ &\to u_0(x') \int \Phi(y', h'(x')y') a(x')\, dy' \end{aligned}$$

by dominated convergence. Clearly the convergence is locally uniform. But the integral is the integral of Φ over the tangent plane T of M at $(x', h(x'))$, taken with respect to the Euclidean area, so by Theorem 7.1.25 it is equal to

$$(2\pi)^{-k} \int_{T^\perp} \phi(\xi)\, d\sigma(\xi).$$

The formula (7.1.29) follows at once.

Theorem 7.1.28 also occurs in a different guise which will be useful in Chapter XIV.

Theorem 7.1.29. *Let $F \in C^1(\mathbb{R}^n)$ be real valued, let $v \in L^2$ and denote by V_t the Fourier transform of $e^{itF}v$, $t \in \mathbb{R}$. If $\phi \in C^0(\mathbb{R}^n) \cap L^\infty(\mathbb{R}^n)$ then*

$$(2\pi)^{-n} \int |V_t(\xi)|^2 \phi(\xi/t)\, d\xi \to \int |v(x)|^2 \phi(F'(x))\, dx, \qquad t \to \infty. \tag{7.1.30}$$

Proof. First we assume that $v \in C_0^\infty$ and that $\phi \in \mathscr{S}$. Then $\phi(\xi/t) V_t(\xi)$ is the Fourier transform of the convolution of $t^n \psi(t.)$ and $v e^{itF}$, if $\hat{\psi} = \phi$. The product by the complex conjugate of $v e^{itF}$ is

$$\begin{aligned}
&\overline{v(x)}\, e^{-itF(x)} \int v(x-y) e^{itF(x-y)} t^n \psi(ty)\, dy \\
&\quad = \overline{v(x)} \int v(x-y/t) e^{it(F(x-y/t)-F(x))} \psi(y)\, dy \\
&\quad \to |v(x)|^2 \int e^{-i\langle y, F'(x)\rangle} \psi(y)\, dy = |v(x)|^2 \phi(F'(x)), \qquad t \to \infty.
\end{aligned}$$

We also have a majorant $C|v(x)|$ so (7.1.30) follows. If $\phi = 1$ we obtain (7.1.30) from Parseval's formula which also extends (7.1.30) to all ϕ in the closure of $\mathscr{S}$ in the maximum norm, that is, all $\phi \in C^0(\mathbb{R})$ converging to 0 at ∞. If $0 \leqq \phi \leqq 1$, $\phi \in C_0^\infty$ and $\phi = 1$ in $\{F'(x);\ x \in \operatorname{supp} v\}$, it follows that

$$\int |V_t(\xi)|^2 (1 - \phi(\xi/t))\, dx \to 0, \qquad t \to \infty.$$

Hence (7.1.30) is valid for every bounded ϕ vanishing in a sufficiently large compact set. Thus we have proved (7.1.30) for all $\phi \in C^0(\mathbb{R}^n) \cap L^\infty(\mathbb{R}^n)$ if $v \in C_0^\infty$. Since C_0^∞ is dense in L^2 it follows that (7.1.30) is valid for all $v \in L^2$.

The following corollary is essentially identical to Theorem 7.1.28 with $k=1$. The case of a higher codimension k can be discussed in the same way.

Corollary 7.1.30. *Let V_t be defined as in Theorem 7.1.29 and let $\Phi \in C_0(\mathbb{R}^{n+1})$, Then*

$$(2\pi)^{-n} \iint_{t \gtrless 0} |V_t(\xi)|^2 \Phi(\xi/R, t/R)\, d\xi\, dt/R \to \iint_{t \gtrless 0} |v(x)|^2 \Phi(tF'(x), t)\, dt\, dx. \tag{7.1.31}$$

Proof. With $s = t/R$ as new variable the left-hand side can be written

$$(2\pi)^{-n} \int ds \int |V_{Rs}(\xi)|^2 \Phi(\xi/R, s)\, d\xi.$$

If we replace t by R, F by sF and $\phi(\xi)$ by $\Phi(\xi, s)$ in Theorem 7.1.29 it follows that the inner integral converges boundedly to

$$(2\pi)^n \int |v(x)|^2 \Phi(sF'(x), s)\, dx.$$

This implies (7.1.31) since the integration with respect to s is taken over a finite interval.

7.2. Poisson's Summation Formula and Periodic Distributions

In Section 7.1 we determined the Fourier transform of the Lebesgue measure on any linear subspace of $\mathbb{R}^n$. Our first purpose here is to determine the Fourier transform of the sum of the Dirac measures at the points in a discrete subgroup of $\mathbb{R}^n$.

Theorem 7.2.1. *If* u_a *is the sum of Dirac measures*

$$u_a = \sum_{g \in \mathbb{Z}^n} \delta_{ag}, \qquad 0 \neq a \in \mathbb{R},$$

then $\hat{u}_a = (2\pi/a)^n u_{2\pi/a}$.

Proof. Since $\delta_{ag} * u_a = u_a$ if $g \in \mathbb{Z}^n$, we have

$$(e^{i\langle ag, .\rangle} - 1)\hat{u}_a = 0, \qquad g \in \mathbb{Z}^n.$$

All the factors do not vanish except at points in $(2\pi/a)\mathbb{Z}^n$. At the origin for example $\hat{u}_a$ must just be a multiple of the Dirac measure, for $(\sin a x_j/2)\hat{u}_a = 0$, $j = 1, \ldots, n$, so this follows from Theorem 3.1.16 if we take $\sin a x_j/2$ as new variables. Now $\hat{u}_a$ is invariant under translations in $(2\pi/a)\mathbb{Z}^n$ because $e^{2\pi i\langle ., g\rangle/a} u_a = u_a$. This implies that u_a is a measure with the same mass at every point in $(2\pi/a)\mathbb{Z}^n$, thus

$$\hat{u}_a = c_a u_{2\pi/a}.$$

Explicitly this means that

$$\sum \hat{\phi}(ag) = c_a \sum \phi(2\pi g/a), \qquad \phi \in \mathscr{S}, \tag{7.2.1}$$

or if we replace ϕ by a translation of ϕ

$$\sum \hat{\phi}(ag)\, e^{i\langle ag, x\rangle} = c_a \sum \phi(2\pi g/a + x), \qquad \phi \in \mathscr{S}.$$

Now integrate both sides for $0 < x_j < 2\pi/a$, $j = 1, \ldots, n$. All terms with $g \neq 0$ drop out on the left-hand side then and we obtain

$$\hat{\phi}(0)(2\pi/a)^n = c_a \int \phi(x)\, dx = c_a \hat{\phi}(0),$$

which completes the proof.

Note that the preceding argument did not use Fourier's inversion formula. Since Theorem 7.2.1 implies that

$$\hat{\hat{u}}_a = (2\pi/a)^n a^n u_a = (2\pi)^n u_a$$

we obtain another determination of the constant in Fourier's inversion formula which has the advantage that the constant is directly related to the volume of the period of the exponentials $e^{i\langle x,g\rangle}$, $g \in \mathbb{Z}^n$. (7.2.1) can now be written

$$\sum \hat{\phi}(ag) = (2\pi/a)^n \sum \phi(2\pi g/a), \qquad \phi \in \mathscr{S}, \tag{7.2.1$'$}$$

and is called *Poisson's summation formula.* Note that as $a \to 0$ or ∞ we obtain as special cases the Fourier transform of 1 and δ_0.

Let us now consider a distribution u which is periodic with period 1 in each variable, that is,

$$u(x-g) = u(x), \qquad g \in \mathbb{Z}^n.$$

Such a distribution is automatically temperate. For let ϕ be a function in $C_0^\infty(\mathbb{R}^n)$ with

$$\sum \phi(x-g) = 1 \tag{7.2.2}$$

(see Theorem 1.4.6). If $\psi \in C_0^\infty$ then

$$\langle u, \psi\rangle = \sum \langle u, \psi\phi(.-g)\rangle = \sum \langle u, \psi(.+g)\phi\rangle$$

by the periodicity. Since

$$\mathscr{S} \ni \psi \to \phi \sum \psi(.+g) \in C_0^\infty$$

is continuous, this proves that u is temperate. To determine $\langle \hat{u}, \psi\rangle = \langle u, \hat{\psi}\rangle$ we apply Poisson's summation formula to $\psi(.)e^{-i\langle x,.\rangle}$ which gives

$$\phi(x) \sum \hat{\psi}(x+g) = (2\pi)^n \sum \psi(2\pi g)\, e^{-2\pi i\langle x,g\rangle} \phi(x).$$

Hence

$$\langle \hat{u}, \psi\rangle = (2\pi)^n \sum c_g \psi(2\pi g)$$

where

$$c_g = \langle u, \phi e^{-2\pi i\langle .,g\rangle}\rangle.$$

Note that if u is a continuous function then

$$\int u\phi e^{-2\pi i\langle .,g\rangle}\, dx = \int_I u e^{-2\pi i\langle .,g\rangle}\, dx$$

where $I = \{x; 0 \leqq x_j < 1\}$ is the unit cube, for we can just integrate over the integer translations of I and sum using the periodicity of u and (7.2.2). A general periodic u can be regarded as a distribution on the torus $T^n = \mathbb{R}^n/\mathbb{Z}^n$ and as a limiting case we just have the integral over the torus then. Thus

$$\hat{u} = (2\pi)^n \sum c_g \delta_{2\pi g}, \tag{7.2.3}$$

$$c_g = \langle u, e^{-2\pi i\langle .,g\rangle}\rangle_{T^n}, \tag{7.2.4}$$

$$c_g = \widehat{(u\phi)}(2\pi g) = O(|g|^k) \quad \text{if } u \in \mathscr{D}'^k.$$

By Fourier's inversion formula

$$u=\sum c_g e^{2\pi i\langle x,g\rangle} \tag{7.2.5}$$

with convergence in $\mathscr{S}'$. Thus we have recovered the basic facts on Fourier series. If $u\in C^k$ then $c_g=\widehat{u\phi}(2\pi g)=O(|g|^{-k})$ so (7.2.5) is uniformly convergent if $k>n$.

Theorem 7.2.2. *If $u\in C^k(\mathbb{R}^n)$, $k>n$, and u is periodic with period 1 in each variable, then* (7.2.5) *is valid with uniform convergence and with the coefficients given by* (7.2.4).

That c_g must be given by (7.2.4) follows at once by integration of (7.2.5) over the torus, so the contents of the theorem are just the existence of such a series expansion. This obvious determination of the Fourier coefficients is of course the reason behind the determination of the constant in Fourier's inversion formula after Theorem 7.2.1.

Theorem 7.2.3. *If $u\in L^2(T^n)$ then* (7.2.5) *is valid with the coefficients* (7.2.4) *and convergence in $L^2(T^n)$. Parseval's formula is valid,*

$$\int_{T^n} |u|^2\,dx=\sum |c_g|^2. \tag{7.2.6}$$

Conversely, if $\sum |c_g|^2<\infty$ then (7.2.5) *converges in $L^2(T^n)$, and the Fourier coefficients of the sum u are equal to c_g.*

Proof. If $u\in C^{n+1}(T^n)$ we have (7.2.5) with absolute convergence. Hence

$$\int_{T^n}|u|^2\,dx=\int_{T^n}\sum c_g\bar{c}_{g'}\,e^{2\pi i\langle x,g-g'\rangle}\,dx=\sum |c_g|^2.$$

Hence the map $L^2(T^n)\ni u\to\{c_g\}\in l^2(\mathbb{Z}^n)$ is isometric. The range contains the dense subset $l^1(\mathbb{Z}^n)$ so the map is unitary.

Closely related to Poisson's summation formula is the formula

$$\sum_{0\leqq g_j<k} f(ag/k)=(k/a)^n\sum_g\int_{0<x_j<a} f(x)\,e^{-2\pi i\langle g,x\rangle k/a}\,dx \tag{7.2.7}$$

which is valid if $f\in C^{n+1}(\mathbb{R}^n)$ is periodic with period a in each variable and k is a positive integer. It follows if we observe that

$$F(x)=\sum_{0\leqq g_j<k} f(x+ag/k)$$

is periodic with period a/k and has the Fourier coefficients in the right-hand side of (7.2.7). When $a=k\to\infty$ we obtain formally Poisson's formula again, and one can justify this limiting procedure.

We shall now indicate a slightly different path to Poisson's summation formula when $n=1$, which is parallel to our proof of Fourier's inversion formula by Cauchy's integral formula. Let $\psi \in \mathscr{S}$ and form

$$\hat{\psi}(0)/2+\sum_1^\infty \hat{\psi}(2\pi k)=\langle\hat{\psi}, u\rangle=\langle\psi, \hat{u}\rangle$$

where u is the measure with mass $1/2$ at 0 and 1 at $2\pi k$ when k is a positive integer. Since u is the limit of $u e^{-\varepsilon x}$ as $\varepsilon \to +0$, we have with $\mathscr{S}'$ convergence

$$\begin{aligned}\hat{u}(\xi)&=\lim_{\varepsilon\to 0}\left(1/2+\sum_1^\infty e^{-2\pi i k\xi-2\pi\varepsilon k}\right)\\ &=\lim_{\varepsilon\to 0}(1/2+e^{-2\pi i\xi-2\pi\varepsilon}/(1-e^{-2\pi i\xi-2\pi\varepsilon}))\\ &=\lim_{\varepsilon\to 0}\frac{1}{2i}\cot\pi(\xi-\varepsilon i)=\frac{1}{2i}\cot\pi(\xi-i0).\end{aligned}$$

Hence

$$\hat{\psi}(0)/2+\sum_1^\infty \hat{\psi}(2\pi k)=\langle\cot\pi(.-i0), \psi/2i\rangle \tag{7.2.8}$$

and replacing ψ by $\check{\psi}$ we obtain

$$\hat{\psi}(0)/2+\sum_{-\infty}^{-1} \hat{\psi}(2\pi k)=-\langle\cot\pi(.+i0), \psi/2i\rangle. \tag{7.2.8$'$}$$

Now we have

$$\cot\pi(\xi-i0)-\cot\pi(\xi+i0)=2i\sum_{-\infty}^{\infty}\delta_k(\xi)$$

by Example 3.1.13 since $\pi\cot\pi z$ has a simple pole with residue 1 at every integer. Adding (7.2.8) and (7.2.8)′ we obtain Poisson's summation formula.

The fact that $\pi\cot\pi z$ has a pole with residue 1 at every integer can often be used to compute sums of the form $\sum\psi(k)$ directly as the sum of residues of $\psi(z)\pi\cot\pi z$. Take for example $\psi(z)=1/(z-w)$ where w is not an integer, and integrate $(2\pi i)^{-1}\psi(z)\pi\cot\pi z$ around a rectangle with sides given by $|\mathrm{Re}\, z|=N+1/2$ or $|\mathrm{Im}\, z|=N$ where N is an integer. By Cauchy's integral formula the integral is equal to

$$-\sum_{-N}^{N}(w-k)^{-1}+\pi\cot\pi w$$

when N is large. The integral of $z^{-1}\cot\pi z$ around the rectangle is 0 since the integrand is an even function so that contributions from opposite points cancel. Now $\cot\pi z$ is uniformly bounded and we have $z^{-1}-(z-w)^{-1}=O(N^{-2})$ on the rectangle so we conclude that the integral $\to 0$ as $N \to \infty$. Hence we obtain the familiar formula

$$\pi \cot \pi w = \lim_{N \to \infty} \sum_{-N}^{N} (w-k)^{-1},$$

which is closely related to (3.4.11).

7.3. The Fourier-Laplace Transformation in $\mathscr{E}'$

We have seen in Theorem 7.1.14 that the Fourier transform $\hat{u}$ of any distribution $u \in \mathscr{E}'(\mathbb{R}^n)$ can be extended to an entire analytic function in $\mathbb{C}^n$ called the Fourier-Laplace transform of u. We shall now discuss its properties in greater detail. If $u \in L^1$ and H is the supporting function of supp u, defined in (4.3.1), then it is clear that

$$|\hat{u}(\zeta)| \leqq \|u\|_{L^1} \exp H(\operatorname{Im} \zeta). \tag{7.3.1}$$

Theorem 7.3.1 (Paley-Wiener-Schwartz). *Let K be a convex compact subset of $\mathbb{R}^n$ with supporting function H. If u is a distribution of order N with support contained in K, then*

$$|\hat{u}(\zeta)| \leqq C(1+|\zeta|)^N e^{H(\operatorname{Im} \zeta)}, \quad \zeta \in \mathbb{C}^n. \tag{7.3.2}$$

Conversely, every entire analytic function in $\mathbb{C}^n$ satisfying an estimate of the form (7.3.2) *is the Fourier-Laplace transform of a distribution with support contained in K. If $u \in C_0^\infty(K)$ there is for every N a constant C_N such that*

$$|\hat{u}(\zeta)| \leqq C_N(1+|\zeta|)^{-N} e^{H(\operatorname{Im} \zeta)}, \quad \zeta \in \mathbb{C}^n. \tag{7.3.3}$$

Conversely, every entire analytic function in $\mathbb{C}^n$ satisfying (7.3.3) *for every N is the Fourier-Laplace transform of a function in $C_0^\infty(K)$.*

Proof. To prove the *necessity* of (7.3.2) we choose $\chi_\delta \in C_0^\infty(K_\delta)$ where $K_\delta = \{x+y;\ x \in K,\ |y| \leqq \delta\}$, so that $\chi_\delta = 1$ in $K_{\delta/2}$ and $|D^\alpha \chi_\delta| \leqq C_\alpha \delta^{-|\alpha|}$ (see (1.4.2)). Then we have if $u \in \mathscr{E}'^N(K)$

$$\begin{aligned} |\hat{u}(\zeta)| = |u(\chi_\delta e^{-i\langle \cdot, \zeta \rangle})| &\leqq C \sum_{|\alpha| \leqq N} \sup |D^\alpha(\chi_\delta e^{-i\langle \cdot, \zeta \rangle})| \\ &\leqq C' \exp(H(\operatorname{Im} \zeta) + \delta |\operatorname{Im} \zeta|) \sum_{|\alpha| \leqq N} \delta^{-|\alpha|}(1+|\zeta|)^{N-|\alpha|}. \end{aligned}$$

The estimate (7.3.2) follows when we take $\delta = 1/(1+|\zeta|)$. To prove (7.3.3) when $u \in C_0^\infty(K)$ it suffices to note that by (7.3.1)

$$|\zeta^\alpha \hat{u}(\zeta)| \leqq \|D^\alpha u\|_{L^1} \exp H(\operatorname{Im} \zeta).$$

Next we prove the *sufficiency* of (7.3.3). Thus let U be an entire analytic function satisfying (7.3.3) for every N. Then the restriction of U to $\mathbb{R}^n$ is the Fourier transform of the C^∞ function

$$u(x)=(2\pi)^{-n}\int e^{i\langle x,\xi\rangle}U(\xi)\,d\xi$$

so all we have to prove is that $\operatorname{supp} u\subset K$. To do so we note that for any choice of $\eta\in\mathbb{R}^n$

$$u(x)=(2\pi)^{-n}\int e^{i\langle x,\xi+i\eta\rangle}U(\xi+i\eta)\,d\xi \tag{7.3.4}$$

for the rapid decrease of U at infinity allows us to shift integration in the direction η in $\mathbb{R}^n$ to a parallel in the complex plane. (Alternatively, differentiation with respect to η_j under the integral sign is equivalent to i times differentiation with respect to ξ_j, and the integral of a derivative is 0.) Estimation of (7.3.4) by means of (7.3.3) with $N=n+1$ gives

$$|u(x)|\leqq e^{-\langle x,\eta\rangle+H(\eta)}\,C_{n+1}\int(1+|\xi|)^{-n-1}\,d\xi.$$

If we replace η by $t\eta$ and let $t\to+\infty$ it follows that $u(x)=0$ unless $H(\eta)\geqq\langle x,\eta\rangle$. But if this is true for every η we have $x\in K$ by Theorem 4.3.2, which proves that $\operatorname{supp} u\subset K$.

To prove the sufficiency of (7.3.2) we first note that the restriction to $\mathbb{R}^n$ of a function U satisfying (7.3.2) is the Fourier transform of a distribution $u\in\mathscr{S}'$. Choose $\phi\in C_0^\infty$ with support in the unit ball and $\int\phi\,dx=1$, and set $\phi_\delta(x)=\delta^{-n}\phi(x/\delta)$. Then the Fourier transform of $u*\phi_\delta$ is $\hat{u}\hat{\phi}_\delta$ which has an entire analytic extension $U\hat{\phi}_\delta$ such that

$$|U(\zeta)\hat{\phi}_\delta(\zeta)|\leqq C_{N,\delta}(1+|\zeta|)^{-N}\exp(H(\operatorname{Im}\zeta)+\delta|\operatorname{Im}\zeta|),\qquad N=1,2,\ldots.$$

In fact, (7.3.3) is valid for $\hat{\phi}_\delta$ with H replaced by the supporting function of the ball with radius δ and center at 0. From the results already proved it follows therefore that $\operatorname{supp} u*\phi_\delta\subset K_\delta$. When $\delta\to 0$ we have $u*\phi_\delta\to u$ which proves that $\operatorname{supp} u\subset K$. The proof is complete.

Note that the special case of Theorem 4.3.3 where $u_1=u_2$ in (4.3.5) is an immediate consequence of Theorem 7.3.1. We shall give some further applications to differential operators with constant coefficients. Let P be a polynomial with complex coefficients in $(\zeta_1,\ldots,\zeta_n)$ and let $P(D)$ be the differential operator obtained when ζ_j is replaced by $-i\partial/\partial x_j$.

Theorem 7.3.2. *If* $f\in\mathscr{E}'(\mathbb{R}^n)$ *then the equation*

$$P(D)u=f \tag{7.3.5}$$

has a solution $u\in\mathscr{E}'(\mathbb{R}^n)$ *if and only if* $\hat{f}(\zeta)/P(\zeta)$ *is an entire function. The solution is then uniquely determined and*

$$ch\operatorname{supp} u = ch\operatorname{supp} f. \tag{7.3.6}$$

Proof. If (7.3.5) has a solution $u\in\mathscr{E}'$ we obtain

$$P(\zeta)\,\hat{u}(\zeta)=\hat{f}(\zeta)$$

by taking the Fourier-Laplace transform on both sides, so $\hat{f}(\zeta)/P(\zeta)$ is the entire function $\hat{u}(\zeta)$. The other half of the proof requires a lemma.

Lemma 7.3.3. *If* $h(z)$ *is an analytic function of* $z\in\mathbb{C}$ *and* $p(z)$ *is a polynomial with leading coefficient a, then*

$$|a h(0)|\leq \max_{|z|=1}|h(z)\,p(z)|.$$

Proof. Set $q(z)=z^m\,\bar{p}(1/z)$ where m is the degree of p and $\bar{p}$ is obtained from p by conjugation of the coefficients. Then $q(0)=\bar{a}$ and by the maximum principle

$$|a h(0)|=|q(0)\,h(0)|\leq \max_{|z|=1}|q(z)\,h(z)|=\max_{|z|=1}|p(z)\,h(z)|$$

which proves the lemma.

End of proof of Theorem 7.3.2. We can choose the coordinates so that the coefficient a of ζ_1^m in $P(\zeta)$ is not 0, when m is the degree of P. Then $P(\zeta_1+z,\zeta_2,\ldots)$ has leading coefficient a so the lemma gives if $g=\hat{f}/P$ is entire

$$\begin{aligned}|a|\,|g(\zeta)|&\leq \sup_{|z|=1}|P(\zeta_1+z,\zeta_2,\ldots)\,g(\zeta_1+z,\zeta_2,\ldots)|\\ &=\sup_{|z|=1}|\hat{f}(\zeta_1+z,\zeta_2,\ldots)|.\end{aligned}$$

When $\hat{f}$ satisfies (7.3.2) we obtain the same estimate for g but with another constant, so $g=\hat{u}$ where $u\in\mathscr{E}'$ and $ch\operatorname{supp} u\subset ch\operatorname{supp} f=ch\operatorname{supp} Pu$. The opposite inclusion is trivial so we obtain (7.3.6), which is also a consequence of Theorem 4.3.3.

As an application we shall now prove a refinement of the Asgeirsson mean value theorem.

Theorem 7.3.4. *Let* $u(x,y)$ *be a continuous solution of the equation*

$$(\Delta_x-\Delta_y)\,u=0$$

in a neighborhood of $K=\{(x,y);\ x,y\in\mathbb{R}^n;\ |x|+|y|\leq R\}$. *Then we have*

$$\int_{|x|=R} u(x,0)\,dS(x) = \int_{|y|=R} u(0,y)\,dS(y).$$

If n is odd and $\neq 1$ it suffices to assume that u is a solution near ∂K.

Proof. By regularization the proof is reduced to the case where $u \in C^\infty$. Let f be the measure

$$f(\phi) = \int_{|x|=R} \phi(x,0)\,dS(x) - \int_{|y|=R} \phi(0,y)\,dS(y).$$

Then

$$\hat{f}(\xi,\eta) = G(\langle \xi,\xi\rangle) - G(\langle \eta,\eta\rangle)$$

where G is an entire function of one complex variable, for an even entire function of one complex variable z is an entire function of z^2. Now

$$G(z) - G(w) = (z-w)\int_0^1 G'(w+t(z-w))\,dt$$

so $(G(z)-G(w))/(z-w)$ is an entire function in $\mathbb{C}^2$. Hence

$$\hat{f}(\xi,\eta)/(\langle \xi,\xi\rangle - \langle \eta,\eta\rangle)$$

is an entire function and therefore the Fourier transform of a distribution μ with support in K such that $(\Delta_x - \Delta_y)\mu = -f$. Hence

$$-\langle f,u\rangle = \langle \mu, (\Delta_x - \Delta_y)u\rangle = 0$$

if u is a solution in a neighborhood of K. The stronger statement for odd n requires a more detailed study of μ. Since $\mu = E * f$ by (4.4.2) where E is one of the fundamental solutions in Theorem 6.2.1 (with $n_+ = n_- = n$) we have by Theorem 4.2.5

$$\operatorname{sing\,supp}\mu \subset \{(x+x', y+y');\ |x| = |y|,\ (x',y') \in \operatorname{supp} f\}.$$

Thus $|x'| + |y'| = R$ and $|x'| = 0$ or $|y'| = 0$. By the triangle inequality, this implies $|x+x'| + |y+y'| \geqq R$ and since $\operatorname{supp}\mu \subset K$ it follows that

$$\operatorname{sing\,supp}\mu \subset \{(x,y);\ |x| + |y| = R\} = \partial K.$$

If n is odd, $n \neq 1$, then (6.2.1)′ gives a fundamental solution E such that $|x| = |y|$ if $(x,y) \in \operatorname{supp} E$, so the same inclusion is valid for the support then. This proves the theorem. Note that for even n we have

$$E(x,y)(-1)^{1+n/2} \gtrless 0 \quad \text{when } |x| \gtrless |y|$$

for the fundamental solution given by (6.2.1). Since $\mu(x,y) = E * f(x,y)$ has the sign $(-1)^{1+n/2}$ when $|x| + |y| < R$, the two mean values having opposite signs the Asgeirsson theorem fails for a regularization of E.

The special case of Theorem 7.3.4 when u is independent of y is just the mean value property of harmonic functions (cf. Theorem

4.1.8). Let us next consider a solution of the wave equation

$$(\Delta_x - \partial^2/\partial t^2)u = 0$$

in $\mathbb{R}^{3+1}$, assuming that $u \in C^1$ near the double cone defined by $|x| + |t| \leq R$. We can apply Asgeirsson's theorem to

$$U(x_1, x_2, x_3, y_1, y_2, y_3) = u(x_1, x_2, x_3, y_1)$$

and obtain

$$2\pi \int_{-R}^{R} u(0,t)\,dt = R \int_{|\omega|=1} u(R\omega, 0)\,dS(\omega). \tag{7.3.7}$$

Differentiation with respect to R gives

$$2\pi(u(0,R) + u(0,-R)) = \int_{|\omega|=1} (u(R\omega,0) + R\langle u'_x(R\omega,0), \omega\rangle)\,dS(\omega)$$

and if we apply (7.3.7) to $\partial u/\partial t$ instead we have

$$2\pi(u(0,R) - u(0,-R)) = R \int_{|\omega|=1} u'_t(R\omega,0)\,dS(\omega).$$

Elimination of $u(0,-R)$ gives after change of notation Kirchoff's formula

$$u(0,t) = \frac{1}{4\pi} \int_{|\omega|=1} (u(t\omega,0) + t(\langle u'_x(t\omega,0), \omega\rangle + u'_t(t\omega,0)))\,dS(\omega). \tag{7.3.8}$$

This is of course a special case of (6.2.9).

After this digression to show that a general result like Theorem 7.3.2 may contain some quite specific information, we prove an approximation theorem related to Theorem 4.4.5.

Definition 7.3.5. A solution u of the differential equation $P(D)u = 0$ in $\mathbb{R}^n$ is called an exponential solution if it can be written in the form

$$u(x) = f(x)\,e^{i\langle x, \zeta\rangle}$$

where $\zeta \in \mathbb{C}^n$ and f is a polynomial.

Theorem 7.3.6. *If $X \subset \mathbb{R}^n$ is convex, then the closed linear hull in $C^\infty(X)$ of the exponential solutions of the equation $P(D)u = 0$ consists of all its solutions in $C^\infty(X)$.*

Proof. By the Hahn-Banach theorem we must show that if $v \in \mathscr{E}'(X)$ is orthogonal to the exponential solutions then v is orthogonal to all solutions in $C^\infty(X)$. To do so it is sufficient to show that $\hat{v}(\zeta)/P(-\zeta)$ is an entire function, for then Theorem 7.3.2 shows that $v = P(-D)\mu$ for some $\mu \in \mathscr{E}'(X)$, hence

$$\langle v, u\rangle = \langle P(-D)\mu, u\rangle = \langle \mu, P(D)u\rangle = 0$$

if $u \in C^\infty(X)$ and $P(D)u=0$. The proof is therefore completed by the following

Lemma 7.3.7. *If* $v \in \mathscr{E}'(\mathbb{R}^n)$ *is orthogonal to all exponential solution of* $P(D)u=0$, *then* $\hat{v}(\zeta)/P(-\zeta)$ *is an entire analytic function.*

Proof. Choose a fixed vector $\theta \neq 0$ such that $P(-t\theta-\zeta)$ is not independent of t for any ζ. This is true in particular if $P_m(\theta)\neq 0$ where P_m is the principal part of P, that is, the homogeneous part of highest degree. From the hypothesis it follows then that $\hat{v}(t\theta+\zeta)/P(-t\theta-\zeta)$ is an analytic function of t for fixed ζ. In fact, if $P(-t\theta-\zeta)$ considered as a polynomial in t has a zero of order k for $t=t_0$, we obtain by differentiating the identity

$$P(D)\,e^{-i\langle x,t\theta+\zeta\rangle}=P(-t\theta-\zeta)\,e^{-i\langle x,t\theta+\zeta\rangle}$$

with respect to t that

$$P(D)(\langle x,\theta\rangle^j e^{-i\langle x,t_0\theta+\zeta\rangle})=0, \qquad j<k.$$

Hence $v(\langle x,\theta\rangle^j e^{-i\langle x,t_0\theta+\zeta\rangle})=0$, $j<k$, which means that $\hat{v}(t\theta+\zeta)$ has a zero of order k at least at t_0. For every ζ we can now define

$$F(\zeta)=\lim_{t\to 0}\hat{v}(t\theta+\zeta)/P(-t\theta-\zeta).$$

That F is entire follows from Weierstrass' preparation theorem (Theorem 7.5.1) but we can also give a direct proof. For a fixed ζ_0 choose r so that $P(-t\theta-\zeta_0)\neq 0$ when $|t|=r$. Then $P(-t\theta-\zeta)\neq 0$ when $|t|=r$ for all ζ in a neighborhood of ζ_0. Hence

$$F(\zeta)=(2\pi i)^{-1}\int_{|t|=r}\hat{v}(t\theta+\zeta)/P(-t\theta-\zeta)\,dt/t$$

is an analytic function of ζ in a neighborhood of ζ_0. The proof is complete.

There is a result parallel to Theorem 7.3.1 which describes the convex hull of the singular support.

Theorem 7.3.8. *Let* $u \in \mathscr{E}'(\mathbb{R}^n)$ *and let* K *be a convex non-empty compact subset of* $\mathbb{R}^n$ *with supporting function* H. *In order that* $\operatorname{sing\,supp} u \subset K$ *it is necessary and sufficient that there exists a constant* N *and a sequence of constants* C_m *such that for* $m=1,2,\ldots$

$$|\hat{u}(\zeta)| \leq C_m(1+|\zeta|)^N e^{H(\operatorname{Im}\zeta)} \quad \text{if } |\operatorname{Im}\zeta| \leq m\log(|\zeta|+1), \quad \zeta \in \mathbb{C}^n. \tag{7.3.9}$$

Proof. To show that (7.3.9) is *necessary* we split u into a sum $u=u_1+u_2$ where $\operatorname{supp} u_1 \subset K_{1/m}$ and $u_2 \in C_0^\infty(\mathbb{R}^n)$. (Here K_δ is defined as in the proof of Theorem 7.3.1.) If u is of order $N-1$ it follows from Theorem 7.3.1 that

$$|\hat{u}_1(\zeta)| \leqq C_m(1+|\zeta|)^{N-1} e^{H(\operatorname{Im}\zeta)+|\operatorname{Im}\zeta|/m}.$$

Hence

$$|\hat{u}_1(\zeta)| \leqq C_m(1+|\zeta|)^N e^{H(\operatorname{Im}\zeta)} \quad \text{if } |\operatorname{Im}\zeta| \leqq m \log(|\zeta|+1),$$

and by (7.3.3) we have $\hat{u}_2(\zeta) e^{-H(\operatorname{Im}\zeta)} \to 0$ when $\zeta \to \infty$ in this set. The estimate (7.3.9) is therefore valid with some larger C_m.

To prove the sufficiency of (7.3.9) we shall make a change of integration contour as in the proof of Theorem 7.3.1 but it has to be chosen differently in order to fit the set where (7.3.9) is applicable. Thus define Γ_η to be the cycle

$$\mathbb{R}^n \ni \xi \to \zeta(\xi) = \xi + i\eta \log(1+|\xi|^2).$$

An immediate computation gives that $d\zeta_1 \wedge \ldots \wedge d\zeta_n = F(\xi)\, d\xi_1 \wedge \ldots \wedge d\xi_n$ where $F \to 1$ as $\xi \to \infty$. We have $|\operatorname{Im}\zeta| < 2|\eta| \log(1+|\zeta|)$ on Γ_η. Choose ϕ_δ as in the proof of Theorem 7.3.1. Since $\hat{u}(\zeta)\hat{\phi}_\delta(\zeta)$ is rapidly decreasing we obtain by application of Stokes' formula to the homotopic cycles Γ_0 and Γ_η

$$u * \phi_\delta(x) = (2\pi)^{-n} \int_{\Gamma_\eta} e^{i\langle x, \zeta\rangle} \hat{u}(\zeta) \hat{\phi}_\delta(\zeta)\, d\zeta_1 \wedge \ldots \wedge d\zeta_n. \tag{7.3.10}$$

(Actually we are only changing the integration contour in one direction so this could be done by means of Cauchy's integral formula.) On Γ_η we have

$$\begin{aligned} |e^{i\langle x,\zeta\rangle} \hat{u}(\zeta)| &\leqq C_\eta \exp(H(\operatorname{Im}\zeta) - \langle x, \operatorname{Im}\zeta\rangle)(1+|\zeta|)^N \\ &= C_\eta (1+|\xi|^2)^{H(\eta)-\langle x,\eta\rangle}(1+|\zeta|)^N. \end{aligned} \tag{7.3.11}$$

If $x_0 \notin K$ we can choose η so that

$$H(\eta) - \langle x, \eta\rangle < -1$$

for all x in a neighborhood X_0 of x_0. If we replace η by $t\eta$ and $2t > n + N$ it follows from (7.3.11) that the integral in (7.3.10) is absolutely convergent for $x \in X_0$ even if the decreasing factor $\hat{\phi}_\delta$ is omitted. Since $\hat{\phi}_\delta(\zeta) = \hat{\phi}(\delta\zeta) \to 1$ boundedly as $\delta \to 0$ we conclude that the restriction of u to X_0 is the function

$$u(x) = (2\pi)^{-n} \int_{\Gamma_{t\eta}} e^{i\langle x,\zeta\rangle} \hat{u}(\zeta)\, d\zeta_1 \wedge \ldots \wedge d\zeta_n, \quad x \in X_0, \tag{7.3.12}$$

if $2t > n + N$. If $2t > n + N + j$ the integral (7.3.12) remains absolutely and uniformly convergent when $x \in X_0$ after at most j differentiations under the integral sign. Hence $u \in C^j(X_0)$ for every j, so $u \in C^\infty(\complement K)$ as was to be proved.

The following application to differential operators is parallel to (7.3.6).

Theorem 7.3.9. *If* $u \in \mathscr{E}'(\mathbb{R}^n)$ *then*

$$ch\operatorname{sing\,supp} u = ch\operatorname{sing\,supp} P(D)u. \tag{7.3.13}$$

Proof. Set $f=P(D)u$ and let H be the supporting function of $\operatorname{sing\,supp} f$. Then (7.3.9) is valid for $\hat{f}$ and follows for $\hat{u}$ by means of Lemma 7.3.3, just as in the proof of Theorem 7.3.2. Hence the left-hand side of (7.3.13) is contained in the right-hand side, and the opposite inclusion is obvious.

In particular Theorem 7.3.9 states that $u \in C^\infty$ if $u \in \mathscr{E}'$ and $P(D)u \in C^\infty$. This is not true for arbitrary differential operators with variable C^∞ coefficients. For example, $x\,du/dx \in C^\infty$ if u is a function which is 0 for $x<0$, equal to 1 for $0<x<1$ and goes smoothly to 0 for $1<x<2$. There is no analogue of the theorem of supports (Theorem 4.3.3) with supports replaced by singular supports which is valid for arbitrary distributions. In fact, one can find $u_1, u_2 \in \mathscr{E}'(\mathbb{R}^n)$ so that $u_1 * u_2 \in C^\infty$ but neither factor is in C^∞. An example is $u_1(x)=\chi(x)/(x+i0)$, $u_2(x)=\chi(x)/(x-i0)$ if $\chi \in C_0^\infty(\mathbb{R})$. However, (7.3.13) gives such a result when the support of one factor is a point; it is also valid when the support of one factor is finite. These matters will be studied further in Section 16.3.

It follows from Theorem 7.3.1 that fairly general Laplace transforms define distributions in $\mathbb{R}^n$. To motivate the definition we recall that

$$\langle u, \phi\rangle = (2\pi)^{-n}\langle \hat{u}, \hat{\phi}(-\,.)\rangle, \qquad \text{if } u \in \mathscr{S}' \text{ and } \phi \in \mathscr{S}.$$

Given a measure $d\mu$ in $\mathbb{C}^n$ we now try to define $u \in \mathscr{D}'(\mathbb{R}^n)$ by the similar formula

$$u(\phi) = (2\pi)^{-n}\int \hat{\phi}(-\zeta)\,d\mu(\zeta), \qquad \phi \in C_0^\infty(\mathbb{R}^n). \tag{7.3.14}$$

This is a valid definition if for some $m \geqq 0$ and C, N

$$|\operatorname{Im}\zeta| \leqq m\log(1+|\zeta|)+C \qquad \text{when } \zeta \in \operatorname{supp} d\mu, \tag{7.3.15}$$

$$\int (1+|\zeta|)^{-N}\,|d\mu(\zeta)| < \infty. \tag{7.3.16}$$

In fact, if $\phi \in C_0^\infty(\{x; |x| \leqq R\})$ we have by (7.3.3) or rather by the derivation of (7.3.3) from (7.3.1)

$$(1+|\zeta|)^{N+k}\,|\hat{\phi}(\zeta)| \leqq C_{k,R}\, e^{R|\operatorname{Im}\zeta|} \sum_{|\alpha| \leqq N+k} \sup|D^\alpha \phi|.$$

If $k \geqq Rm$ the exponential can be estimated by $(1+|\zeta|)^k$ in $\operatorname{supp} d\mu$. Hence the integral (7.3.14) converges and

$$|u(\phi)| \leqq C_R \sum_{|\alpha| \leqq N+k} \sup |D^\alpha \phi|, \quad \phi \in C_0^\infty(\{x; |x| \leqq R\}).$$

Thus (7.3.14) defines a distribution which is of order $\leqq N+Rm+1$ when $|x|<R$. It is of order N if $|\operatorname{Im} \zeta|$ is bounded in $\operatorname{supp} d\mu$.

If P is a polynomial, then

$$\langle P(D)u, \phi \rangle = \langle u, P(-D)\phi \rangle$$

and the Fourier-Laplace transform of $P(-D)\phi$ is $P(-\zeta)\hat{\phi}(\zeta)$. Hence

$$\langle P(D)u, \phi \rangle = (2\pi)^{-n} \int \hat{\phi}(-\zeta) P(\zeta)\, d\mu(\zeta). \tag{7.3.17}$$

As an application we shall now show that every differential operator with constant coefficients has a fundamental solution. We must then choose the measure $d\mu$ so that the integral on the right-hand side of (7.3.17) is equal to the integral of $\hat{\phi}$ over $\mathbb{R}^n$.

Theorem 7.3.10. *For every polynomial $P \not\equiv 0$ in n variables one can find a distribution $E \in \mathscr{D}'_F(\mathbb{R}^n)$ such that $P(D)E = \delta$.*

We prepare the proof with two lemmas used to construct the measure $d\mu$.

Lemma 7.3.11. *If $\Phi \in C_0^\infty(\mathbb{C}^n)$ and*

$$\Phi(e^{i\theta}\zeta) = \Phi(\zeta), \quad \theta \in \mathbb{R}; \quad \int \Phi(\zeta)\, d\lambda(\zeta) = 1, \tag{7.3.18}$$

where $d\lambda$ is the Lebesgue measure in $\mathbb{C}^n$, then

$$\int F(\zeta)\Phi(\zeta)\, d\lambda(\zeta) = F(0) \tag{7.3.19}$$

for any entire analytic function F.

Proof. By Cauchy's integral formula

$$\int F(\zeta e^{i\theta})\, d\theta = 2\pi F(0).$$

If we multiply by $\Phi(\zeta)$ and integrate, (7.3.19) follows from (7.3.18).

With a fixed positive integer m we denote by $\operatorname{Pol}(m)$ the complex vector space of polynomials of degree $\leqq m$ in n variables and by $\operatorname{Pol}^\circ(m)$ the vector space with the origin removed. A norm in $\operatorname{Pol}(m)$ is given by $Q \to \tilde{Q}(0)$ where

$$\tilde{Q}(\xi) = \Big(\sum_\alpha |Q^{(\alpha)}(\xi)|^2\Big)^{\frac{1}{2}}. \tag{7.3.20}$$

Note that this function of ξ is bounded from below if $Q \neq 0$, for some derivative of Q is a constant $\neq 0$.

Lemma 7.3.12. *For every ball $Z \subset \mathbb{C}^n$ with center at 0 one can find a non-negative function $\Phi \in C^\infty(\mathrm{Pol}^\circ(m) \times \mathbb{C}^n)$ such that*

(i) *$\Phi(Q,\zeta)$ is absolutely homogeneous of degree 0 with respect to Q.*

(ii) *$\Phi(Q,\zeta)$ satisfies (7.3.18) for fixed Q and vanishes when $\zeta \notin Z$.*

(iii) *there is a constant C such that*

$$\tilde{Q}(0) \leqq C\,|Q(\zeta)| \quad \text{if } \Phi(Q,\zeta) \neq 0. \tag{7.3.21}$$

If F is analytic in Z and Q is a polynomial of degree $\leqq m$ it follows that

$$\tilde{Q}(0)\,|F(0)| \leqq C_Z \int_Z |F(\zeta)\,Q(\zeta)|\,d\lambda(\zeta).$$

Proof. For a fixed Q_0 the existence of such a Φ is quite obvious, for we can find $w \in \mathbb{C}^n$ such that $Q_0(w) \neq 0$. We can then choose $r>0$ so that $rw \in Z$ and $Q_0(zw) \neq 0$ when $|z|=r$, which excludes at most m values of r. If $\Psi \geqq 0$ and Ψ has support near rw and integral 1, then

$$\Phi(\zeta) = \int \Psi(e^{i\theta}\zeta)\,d\theta/2\pi$$

has the required properties. The same Φ can be used for all $Q \in \mathrm{Pol}^\circ(m)$ such that aQ is close to Q_0 for some $a \in \mathbb{C}$. Piecing such local constructions together by a partition of unity in the projective space of $\mathrm{Pol}^\circ(m)$, that is, using a partition of unity where the terms are absolutely homogeneous functions of degree 0, we obtain Φ with the stated properties. The last statement follows from (7.3.19) and (7.3.21) for

$$\tilde{Q}(0)\,|F(0)| = \left|\int F(\zeta)\,\tilde{Q}(0)\,\Phi(Q,\zeta)\,d\lambda(\zeta)\right| \leqq C_Z \int_Z |F(\zeta)\,Q(\zeta)|\,d\lambda(\zeta).$$

Proof of Theorem 7.3.10. Let P_ξ be the polynomial $\zeta \to P(\xi+\zeta)$ obtained by translation of P and set

$$E(\phi) = (2\pi)^{-n} \int d\xi \int \hat{\phi}(-\xi-\zeta)/P(\xi+\zeta)\,\Phi(P_\xi,\zeta)\,d\lambda(\zeta). \tag{7.3.22}$$

This is of the form (7.3.14) and

$$\begin{aligned}\int (1+|\zeta|)^{-N}\,|d\mu(\zeta)| &\leqq \int d\xi \int (1+|\xi+\zeta|)^{-N}\,|\Phi(P_\xi,\zeta)|/|P_\xi(\zeta)|\,d\lambda(\zeta)\\ &\leqq C \int d\xi \int_Z (1+|\xi+\zeta|)^{-N}\,\tilde{P}(\xi)^{-1}\,d\lambda(\zeta) < \infty\end{aligned}$$

if $N>n$. (Recall that $\tilde{P}(\xi)$ is bounded from below.) Since $\operatorname{Im}\zeta$ is bounded in $\operatorname{supp} d\mu$ the distribution E is of order at most $n+1$. From (7.3.17) and (7.3.19) we obtain

$$\langle P(D)E, \phi\rangle = (2\pi)^{-n}\int d\xi \int \hat{\phi}(-\xi-\zeta)\,\Phi(P_\xi, \zeta)\,d\lambda(\zeta)$$
$$= (2\pi)^{-n}\int \hat{\phi}(-\xi)\,d\xi = \phi(0),$$

which completes the proof.

From the preceding construction one can extract quite precise regularity properties of E both for fixed P and as a function of P. This will be done in Section 10.2.

7.4. More General Fourier-Laplace Transforms

For distributions u with compact support we have defined the Fourier-Laplace transform by

(7.4.1) $$\hat{u}(\zeta) = \langle u, e^{-i\langle\cdot,\zeta\rangle}\rangle, \quad \zeta \in \mathbb{C}^n.$$

Now it may be possible to define $\hat{u}(\zeta)$ at least in some subsets of $\mathbb{C}^n$ for more general distributions u. For fixed $\eta = \operatorname{Im}\zeta$ we can at least define (7.4.1) as a distribution in $\xi = \operatorname{Re}\zeta$ if $e^{\langle\cdot,\eta\rangle}u \in \mathscr{S}'$. We shall therefore start by studying for given $u \in \mathscr{D}'(\mathbb{R}^n)$ the set

(7.4.2) $$\Gamma_u = \{\eta \in \mathbb{R}^n; e^{\langle\cdot,\eta\rangle}u \in \mathscr{S}'\}.$$

Lemma 7.4.1. *If $v \in \mathscr{S}'$ and $\phi \in C^\infty$ has bounded derivatives of all orders, then $\phi v \in \mathscr{S}'$. If $\psi \in \mathscr{S}$ then the Fourier transform of ψv is the C^∞ function*

(7.4.3) $$\xi \to \langle v, e^{-i\langle\cdot,\xi\rangle}\psi\rangle.$$

Proof. The first statement follows since multiplication by ϕ is a continuous map in $\mathscr{S}$. The second follows from Theorem 7.1.14 if $\psi \in C_0^\infty$. In general we just choose using Lemma 7.1.8 a sequence $\psi_k \in C_0^\infty$ with $\psi_k \to \psi$ in $\mathscr{S}$ and obtain $\psi_k v \to \psi v$ in $\mathscr{S}'$, hence $\widehat{\psi_k v} \to \widehat{\psi v}$ in $\mathscr{S}'$, and $e^{-i\langle\cdot,\xi\rangle}\psi_k \to e^{-i\langle\cdot,\xi\rangle}\psi$ in $\mathscr{S}$, locally uniformly in ξ. Thus (7.4.3) is the distribution limit of the corresponding function with ψ replaced by ψ_k, which proves the statement since (7.4.3) defines a C^∞ function by the first part of the proof.

It follows at once from the lemma that the set Γ_u defined by (7.4.2) is convex. In fact, if $\eta_1, \eta_2 \in \Gamma_u$ and we set $u_\eta = e^{\langle\cdot,\eta\rangle}u$, then we have for $\eta = t\eta_1 + (1-t)\eta_2$, $0<t<1$,

$$u_\eta = \phi(u_{\eta_1} + u_{\eta_2})$$

where

$$\phi(x)=e^{\langle x,\eta\rangle}/(e^{\langle x,\eta_1\rangle}+e^{\langle x,\eta_2\rangle})$$

is bounded by 1 and has bounded derivatives of all orders. Now assume that Γ_u has a non-empty interior Γ_u°. Choose $\eta_0,\dots,\eta_n\in\Gamma_u^\circ$ affinely independent. Then

$$u_\eta=\psi_\eta\sum_0^n u_{\eta_j}$$

where

$$\psi_\eta(x)=e^{\langle x,\eta\rangle}/\sum e^{\langle x,\eta_j\rangle}$$

is in $\mathscr{S}$ if η is in the interior of the simplex spanned by $\eta_0,\dots,\eta_n$. In fact,

$$\psi_\eta(x)=(\sum e^{\langle x,\eta_j-\eta\rangle})^{-1}$$

and if 0 is in the interior of the convex hull of $\eta_j-\eta$, then

$$|x|\leqq C\max\langle x,\eta_j-\eta\rangle,$$

so ψ_η and its derivatives are exponentially decreasing. This is true uniformly when η is in a compact subset of the interior of the simplex. It follows that $\hat{u}_\eta$ is then a C^∞ function

$$\hat{u}_\eta(\xi)=\sum_j\langle u_{\eta_j}, e^{-i\langle .,\xi+i\eta\rangle}/\sum e^{\langle .,\eta_j\rangle}\rangle$$

of (ξ,η). It is of course analytic since the Cauchy-Riemann equations $(\partial/\partial\xi_\nu+i\partial/\partial\eta_\nu)\hat{u}_\eta(\xi)=0$ follow immediately by differentiation.

Theorem 7.4.2. *If* $u\in\mathscr{D}'(\mathbb{R}^n)$ *then* (7.4.2) *defines a convex set* Γ_u. *If the interior* Γ_u° *is not empty then there is an analytic function* $\hat{u}$ *in* $\mathbb{R}^n+i\Gamma_u^\circ$ *such that the Fourier transform of* $e^{\langle .,\eta\rangle}u$ *is* $\hat{u}(.+i\eta)$ *for all* $\eta\in\Gamma_u^\circ$. *For every compact set* $M\subset\Gamma_u^\circ$ *there is an estimate*

$$|\hat{u}(\zeta)|\leqq C(1+|\zeta|)^N,\quad \operatorname{Im}\zeta\in M. \tag{7.4.4}$$

Conversely, if Γ *is an open convex set in* $\mathbb{R}^n$ *and* U *is an analytic function in* $\mathbb{R}^n+i\Gamma$ *with bounds of the form* (7.4.4) *for every* $M\Subset\Gamma$, *then there is a distribution* u *such that* $e^{\langle .,\eta\rangle}u\in\mathscr{S}'$ *and has Fourier transform* $U(.+i\eta)$ *for every* $\eta\in\Gamma$.

Proof. Only the last statement remains to be proved. Let u_η be the inverse Fourier transform of $U(.+i\eta)$. Then $\partial u_\eta/\partial\eta_\nu$ is the inverse Fourier transform of

$$\partial U(\xi+i\eta)/\partial\eta_\nu=i\partial U(\xi+i\eta)/\partial\xi_\nu,$$

so $\partial u_\eta/\partial\eta_\nu=x_\nu u_\eta$ and $e^{-\langle x,\eta\rangle}u_\eta=u$ is independent of η. The proof is complete.

Now assume that u has support in a convex closed set K which is no longer assumed to be compact. We can still define the supporting function by

$$H_K(\xi)=\sup_{x\in K}\langle x,\xi\rangle,$$

and it is a convex, positively homogeneous function with values in $(-\infty,\infty]$. However, H_K may not be continuous, but as the supremum of a family of continuous functions it is lower semi-continuous. Conversely, if H is a function with these properties, then there is precisely one convex closed set K such that $H=H_K$, and

$$K=\{x;\langle x,\xi\rangle\leqq H(\xi),\ \xi\in\mathbb{R}^n\}.$$

The proof of Theorem 4.3.2 gives this with no essential change, for the lower semi-continuity of H means precisely that $\{(\tau,\xi)\in\mathbb{R}^{n+1};\ \tau\geqq H(\xi)\}$ is closed.

Now it is clear that when $\operatorname{supp} u\subset K$ we have

$$\eta\in\Gamma_u\Rightarrow\eta+\theta\in\Gamma_u\qquad\text{if } H_K(\theta)<\infty,$$

and $\{\theta;H_K(\theta)<\infty\}$ is a convex cone. In fact, if $\chi\in C^\infty(\mathbb{R})$ is 1 on $(-\infty,1/2)$ and 0 on $(1,\infty)$, then

$$e^{\langle x,\theta\rangle-H_K(\theta)}=e^{\langle x,\theta\rangle-H_K(\theta)}\chi(\langle x,\theta\rangle-H_K(\theta))\qquad\text{near }\operatorname{supp} u \tag{7.4.5}$$

and on the right-hand side we have a function of x with bounded derivatives.

Theorem 7.4.3. *If u satisfies the hypotheses in Theorem 7.4.2 and in addition* $\operatorname{supp} u\subset K$, *then*

$$|\hat u(\zeta)|\leqq C(1+|\zeta|)^N e^{H_K(\operatorname{Im}\zeta-\eta)},\quad\text{if } \eta\in M,\quad H_K(\operatorname{Im}\zeta-\eta)<\infty, \tag{7.4.6}$$

where M is a compact subset of Γ_u°. Conversely, if there is some η for which (7.4.6) *is valid, then* $\operatorname{supp} u\subset K$ *if K is closed and convex.*

Proof. To prove (7.4.6) we set $\zeta=\xi+i(\theta+\eta)$. Then we have $H_K(\theta)<\infty$ and

$$\hat u(\zeta)=\langle u_\eta, e^{-i\langle x,\xi\rangle}e^{\langle x,\theta\rangle}\chi(\langle x,\theta\rangle-H_K(\theta))\rangle$$

by (7.4.5). For the function on which u_η acts the derivatives of order k are bounded by $C_k(1+|\zeta|)^k e^{H_K(\theta)}$. Hence (7.4.6) follows from the estimates for u_η in the proof of Theorem 7.4.2. Assume now that (7.4.6) is valid for some η. If $\phi\in C_0^\infty$ and h is the supporting function of $\operatorname{supp}\phi$, then

$$\begin{aligned}(2\pi)^n|\langle u_\eta,\phi\rangle|&=|\textstyle\int\hat u(\xi+i(\theta+\eta))\,\hat\phi(-\xi-i\theta)\,d\xi|\\&\leqq C\textstyle\int(1+|\xi|+|\theta|)^N e^{H_K(\theta)+h(-\theta)}(1+|\xi|+|\theta|)^{-N-n-1}\,d\xi\\&\leqq C' e^{H_K(\theta)+h(-\theta)}.\end{aligned}$$

Replacing θ by $t\theta$ we obtain $\langle u_\eta, \phi\rangle = 0$ if $H_K(\theta)+h(-\theta)<0$. Hence $u=0$ in a neighborhood of x if $H_K(\theta)-\langle x,\theta\rangle<0$ for some θ. If $x\in\operatorname{supp} u$ it follows that $\langle x,\theta\rangle \leqq H_K(\theta)$ for every θ, that is, $\operatorname{supp} u \subset K$. This completes the proof which is of course just a slight variation of the second part of that of Theorem 7.3.1.

Remark. In Theorem 7.4.2 we discussed $\hat{u}$ in the interior of $\mathbb{R}^n + i\Gamma_u$ only. However, the continuity of the Fourier transformation in $\mathscr{S}'$ shows that if $\eta\in\Gamma_u \setminus \Gamma_u^\circ$ then the Fourier transform of $e^{\langle\cdot,\eta\rangle}u$ is the limit in $\mathscr{S}'$ of $\xi\to\hat{u}(\xi+i(1-t)\eta+it\theta)$ as $t\to 0$, if θ belongs to a compact subset M of Γ_u°.

As an application we shall now compute the Fourier transform of the advanced fundamental solution E of the wave operator

$$\square = c^{-2}\,\partial^2/\partial t^2 - \Delta$$

in $\mathbb{R}^{n+1}$, constructed in Section 6.2. The support is in

$$K=\{(t,x);\, ct\geqq |x|\},$$

$$H_K(\tau,\xi)=\sup_K (t\tau+\langle x,\xi\rangle)=\sup_{t>0}(t\tau+ct|\xi|),$$

so $H_K(\tau,\xi)=0$ if $\tau+c|\xi|\leqq 0$ and $H_K(\tau,\xi)=\infty$ if $\tau+c|\xi|>0$. It follows that the Fourier-Laplace transform $\hat{E}(\tau,\zeta)$ must be analytic when $\operatorname{Im}\tau < -c|\operatorname{Im}\zeta|$. Since

$$(\zeta^2-\tau^2/c^2)\hat{E}=1, \quad \text{if } \zeta^2=\langle\zeta,\zeta\rangle,$$

it follows that

$$\hat{E}(\tau,\zeta)=(\zeta^2-\tau^2/c^2)^{-1} \tag{7.4.7}$$

there. The Fourier transform of E is therefore the limit (Lemma 6.2.2)

$$\begin{aligned}\lim_{\varepsilon\to+0}(\xi^2-(\tau-i\varepsilon)^2/c^2)^{-1} &= (\xi^2-\tau^2/c^2+i0)^{-1} \quad \text{if } \tau>0\\ &= (\xi^2-\tau^2/c^2-i0)^{-1} \quad \text{if } \tau<0;\end{aligned}$$

at the origin it is determined by the homogeneity.

It is also easy to construct E starting from (7.4.7). We must then show that $\langle\zeta,\zeta\rangle-\tau^2/c^2 \neq 0$ when $\operatorname{Im}\tau<-c|\operatorname{Im}\zeta|$. To do so we observe that the equation

$$q(s)=(\operatorname{Re}\zeta+s\operatorname{Im}\zeta)^2-(\operatorname{Re}\tau+s\operatorname{Im}\tau)^2/c^2=0$$

has real roots then for the quadratic form $\xi^2-\tau^2/c^2$ would otherwise be negative in a real two dimensional plane. Hence $|q(i)|$ is at least as large as the absolute value of the leading coefficient in q, that is,

$$|\zeta^2-\tau^2/c^2|\geqq|\operatorname{Im}\tau|^2/c^2-|\operatorname{Im}\zeta|^2. \tag{7.4.8}$$

From (7.4.8) and Theorem 7.4.3 it follows at once that the right-hand side of (7.4.7) is the Fourier-Laplace transform of a distribution E_+ with support in the forward cone $\{(t,x); ct \geqq |x|\}$ so we obtain another construction of the forward fundamental solution. It has a much wider scope than the earlier one. (See Section 12.5.)

7.5. The Malgrange Preparation Theorem

Decomposition of the Fourier transform of a function $\phi \in \mathscr{S}$ by a partition of unity gives a representation of ϕ as a sum of entire analytic functions. This simple observation will be used in this section to derive the Malgrange preparation theorem from the classical analytical analogue which we first recall.

Theorem 7.5.1 (The Weierstrass preparation theorem). *Let $f(t,z)$ be an analytic function of $(t,z) \in \mathbb{C}^{1+n}$ in a neighborhood of $(0,0)$ such that*

(7.5.1) $f = \partial f/\partial t = \ldots = \partial^{k-1} f/\partial t^{k-1} = 0, \quad \partial^k f/\partial t^k \neq 0$ at $(0,0)$.

Then there is a unique factorization

$$f(t,z) = c(t,z)(t^k + a_{k-1}(z)\, t^{k-1} + \ldots + a_0(z)) \tag{7.5.2}$$

where a_j and c are analytic in a neighborhood of 0 and $(0,0)$ respectively, $c(0,0) \neq 0$ and $a_j(0) = 0$.

Proof. Choose $r > 0$ so that f is analytic at $(t,0)$ when $|t| < 2r$ and $f(t,0) \neq 0$ when $0 < |t| < 2r$. Then choose $\delta > 0$ so that $f(t,z)$ is analytic when $|t| < 3r/2$, $|z| < \delta$ and $f(t,z) \neq 0$ when $|t| = r$, $|z| < \delta$. For every z with $|z| < \delta$ the equation $f(t,z) = 0$ has then precisely k roots t_j with $|t_j| < r$ (counted with multiplicity). If there is a factorization (7.5.2) we must therefore have

$$\begin{aligned} t^k + a_{k-1}(z)\, t^{k-1} + \ldots + a_0(z) &= \prod_1^k (t - t_j) \\ &= \exp((2\pi i)^{-1} \int_{|s|=r} ((\partial f(s,z)/\partial s)/f(s,z)) \log(t-s)\, ds). \end{aligned}$$

Here the last equality assumes that $|t| > r$ so that the logarithm has an analytic branch when $|s| < r$. The exponential is an analytic function of t and z and a polynomial in t so using for example the Lagrange interpolation formula we conclude that $a_j(z)$ are analytic functions of z. If

$$p(t,z) = t^k + a_{k-1}(z)\, t^{k-1} + \ldots + a_0(z)$$

the quotient $f(t,z)/p(t,z)$ is analytic when $|t|\leqq r$ for fixed z, $|z|<\delta$. Hence

$$f(t,z)/p(t,z)=(2\pi i)^{-1}\int_{|s|=r} f(s,z)\,p(s,z)^{-1}(s-t)^{-1}\,ds$$

if $|t|<r$ and $|z|<\delta$. The right-hand side is then an analytic function of (t,z), which completes the proof.

By the Weierstrass preparation theorem one often means the following more general result, also known as the division theorem or the Weierstrass formula.

Theorem 7.5.2. *If f satisfies (7.5.1) and g is analytic in a neighborhood of $(0,0)$ in $\mathbb{C}^{1+n}$ then*

$$g(t,z)=q(t,z)f(t,z)+\sum_0^{k-1} t^j r_j(z) \tag{7.5.3}$$

where q and r_j are analytic near $(0,0)$ and 0. Both the quotient q and the remainder are uniquely determined.

Proof. By Theorem 7.5.1 we may assume that f is a polynomial in t of degree k. If g is an analytic function when $|t|<2r$ and $|z|<\delta$, and if $f(t,z)\neq 0$ when $|t|\geqq r$, $|z|<\delta$, then (7.5.3) implies

$$q(t,z)=(2\pi i)^{-1}\int_{|s|=r} g(s,z)f(s,z)^{-1}(s-t)^{-1}\,ds, \qquad |t|<r,\quad |z|<\delta, \tag{7.5.4}$$

for the integral

$$\int_{|s|=r}\Big(\sum_0^{k-1} s^j r_j(z)\Big) f(s,z)^{-1}(s-t)^{-1}\,ds$$

is equal to 0 since the integration may be made over an arbitrarily large circle and the integrand is $O(|s|^{-2})$ as $s\to\infty$. Hence (7.5.3) implies

$$\sum_0^{k-1} t^j r_j(z)=(2\pi i)^{-1}\int_{|s|=r} g(s,z)(f(s,z)-f(t,z))/(s-t)\,ds/f(s,z). \tag{7.5.5}$$

The quotient $(f(s,z)-f(t,z))/(s-t)$ is a polynomial of degree $k-1$ in s and t, so the right hand side does define a polynomial in t for all z. The coefficients are analytic functions of z for $|z|<\delta$. Since (7.5.4) also defines an analytic function when $|z|<\delta$ and $|t|<r$, the existence and uniqueness of the decomposition (7.5.3) is proved.

We shall now discuss the corresponding results for C^∞ functions. The difficulty is then that zeros may be lost. For example, t^2+x has two real zeros when $x<0$ but none when $x>0$. To be able to keep track of the zeros we shall therefore start by examining the division theorem for functions analytic in thin strips around the real axis when f is a polynomial. We omit the parameters x but insist on uniform estimates instead. Thus let

$$p(t)=t^k+a_{k-1}t^{k-1}+\ldots+a_0 \tag{7.5.6}$$

be a polynomial in $t\in\mathbb{C}$ of fixed degree k. Assume that $\sum|a_j|<1$, and let g be a bounded analytic function in the strip $|\operatorname{Im} t|<\varepsilon$. We want to make a division

$$g(t)=q(t)\,p(t)+\sum_0^{k-1} t^j r_j \tag{7.5.7}$$

so that q and r_j have bounds in terms of $M=\sup\{|g(t)|;|\operatorname{Im} t|<\varepsilon\}$ in a smaller strip $|\operatorname{Im} t|<c\varepsilon$. In addition the decomposition must depend analytically on p if we do not change p very much. To achieve this we first choose j with $1\leqq j\leqq k+1$ so that

$$p(t)\neq 0 \quad \text{when } ||\operatorname{Im} t|-\varepsilon j/(k+2)|<\varepsilon/(2(k+2)). \tag{7.5.8}$$

This is possible since p has only k zeros. If ω is a bounded open set and g is analytic in $\bar\omega$, p is a polynomial with no zero on $\partial\omega$, then the proof of Theorem 7.5.2 gives if $\partial\omega\in C^1$

$$g(t)=q(t)\,p(t)+r(t), \qquad t\in\omega,$$

where

$$q(t)=(2\pi i)^{-1}\int_{\partial\omega} g(s)\,p(s)^{-1}(s-t)^{-1}\,ds, \qquad t\in\omega,$$

$$r(t)=\sum_0^{k-1} t^j r_j=(2\pi i)^{-1}\int_{\partial\omega} g(s)\,((p(s)-p(t))/(s-t))\,p(s)^{-1}\,ds.$$

Note that r is determined by the additional fact that $r(t)=0$ if $p(t)=0$ and $t\notin\omega$. When g is analytic for $|\operatorname{Im} t|<\varepsilon$ and (7.5.8) is valid, we obtain

$$q(t)=(2\pi i)^{-1}\int_{\gamma_t} g(s)\,p(s)^{-1}(s-t)^{-1}\,ds, \tag{7.5.9}$$

$$\sum_0^{k-1} t^\nu r_\nu=(2\pi i)^{-1}\int_\gamma g(s)\,((p(s)-p(t))/(s-t))\,p(s)^{-1}\,ds. \tag{7.5.10}$$

Here $|\operatorname{Im} t|<\varepsilon j/(k+2)$ in (7.5.9), γ is the boundary of the rectangle $\{s;|\operatorname{Im} s|<\varepsilon j/(k+2), |\operatorname{Re} s|<2\}$ and γ_t is the boundary of the union of this rectangle and a congruent one with center at $\operatorname{Re} t$. The bounds

$$|q(t)|\leqq CM\varepsilon^{-k-1} \quad \text{when } |\operatorname{Im} t|<\varepsilon/(2(k+2)); \quad |r_\nu|\leqq CM\varepsilon^{-k}; \tag{7.5.11}$$

are obvious since $|s|<1$ when $p(s)=0$, hence $|p(s)|\geqq c\varepsilon^k$ on $\partial\gamma$ and $\partial\gamma_t$.

We shall now examine what happens when the coefficients of p are changed. Let

$$p(t,b)=t^k+b_{k-1}t^{k-1}+\ldots+b_0. \tag{7.5.12}$$

Since (7.5.8) implies

$$|p(t,a)|>c\,\varepsilon^k \quad \text{if } |\operatorname{Im} t|=\varepsilon j/(k+2)$$

we still have

$$|p(t,b)|>c\,\varepsilon^k/2 \quad \text{if } |\operatorname{Im} t|=\varepsilon j/(k+2), \quad |a-b|<c_1\varepsilon^k.$$

We can therefore use the formulas (7.5.9), (7.5.10) with the same j to divide $g(t)$ by $p(t,b)$ for all b with $|a-b|<c_1\varepsilon^k$. Hence

$$g(t)=q(t,b)\,p(t,b)+\sum_0^{k-1} t^j r_j(b), \quad |b-a|<c_1\varepsilon^k \tag{7.5.7$'$}$$

where $q(t,b)$ and $r_j(b)$ are analytic in b also. When $|b-a|<c_2\varepsilon^k$, $c_2<c_1$, it follows from Cauchy's inequalities that

$$\begin{aligned}&|\partial_t^\alpha \partial_b^\beta q(t,b)|\leqq C_{\alpha\beta} M\,\varepsilon^{-\alpha-1-k(|\beta|+1)}, \quad |\operatorname{Im} t|<\varepsilon/(2(k+2));\\ &|\partial_b^\beta r_j(b)|\leqq C_\beta M\,\varepsilon^{-k(|\beta|+1)}.\end{aligned} \tag{7.5.11$'$}$$

To piece the preceding local solutions of (7.5.7)$'$ together we shall use the partition of unity in Theorem 1.4.6 for $\mathbb{C}^k=\mathbb{R}^{2k}$. If K is the diameter of the support of the function ϕ in that lemma, then

$$1=\sum \phi_G(a), \quad \phi_G(a)=\phi(Ka/c_2\varepsilon^k-G), \quad a\in\mathbb{C}^k.$$

The diameter of $\operatorname{supp}\phi_G$ is at most $c_2\varepsilon^k$. For any lattice point G in $\mathbb{R}^{2k}$ with

$$\operatorname{supp}\phi_G\cap\{a;\textstyle\sum|a_j|<1\}\neq\emptyset \tag{7.5.13}$$

we choose a point a_G in the intersection and apply the preceding construction with $a=a_G$. This gives $q_G(t,b)$ and $r_{j,G}(b)$ satisfying (7.5.7)$'$ and (7.5.11)$'$ when $b\in\operatorname{supp}\phi_G$. It follows that

$$q^\varepsilon(t,b,g)=\sum\phi_G(b)\,q_G(t,b), \quad r_j^\varepsilon(b,g)=\sum\phi_G(b)\,r_{j,G}(b)$$

with summation over all G satisfying (7.5.13) has the properties in the following lemma where b is regarded as a variable in $\mathbb{R}^{2k}$.

Lemma 7.5.3. *For every bounded analytic function g in $\{t\in\mathbb{C};|\operatorname{Im} t|<\varepsilon\}$ one can find $q^\varepsilon(t,b,g)\in C^\infty(\mathbb{R}\times\mathbb{R}^{2k})$ and $r_j^\varepsilon(b,g)\in C^\infty(\mathbb{R}^{2k})$ depending linearly on g such that*

$$g(t)=q^\varepsilon(t,b,g)\,p(t,b)+\sum_0^{k-1} t^j r_j^\varepsilon(b,g), \quad \text{if } \sum|b_j|<1, \tag{7.5.7$''$}$$

(7.5.11)''
$$|\partial_t^\alpha \partial_b^\beta q^\varepsilon(t,b,g)| \leqq C_{\alpha\beta} M \varepsilon^{-\alpha-1-k(|\beta|+1)},$$
$$|\partial_b^\beta r_j^\varepsilon(b,g)| \leqq C_\beta M \varepsilon^{-k(|\beta|+1)}.$$

Here $M = \sup\{|g(s)|; |\operatorname{Im} s| < \varepsilon\}$ *and the constants are independent of* ε.

We shall now eliminate the hypothesis that g is analytic by using the decomposition of C^∞ functions into sums of analytic functions mentioned at the beginning of this section.

Theorem 7.5.4. *For every* $g \in \mathscr{S}(\mathbb{R})$ *one can find* $q(t,b,g) \in C^\infty(\mathbb{R} \times \mathbb{R}^{2k})$ *and* $r_j(b,g) \in C^\infty(\mathbb{R}^{2k})$ *depending linearly on* g *such that* (7.5.7)'' *is valid with* q^ε *replaced by* q *and* r_j^ε *replaced by* r_j, *and*

(7.5.14)
$$|\partial_t^\alpha \partial_b^\beta q(t,b,g)| \leqq C_{\alpha\beta} \int (|g| + |g^{(\nu)}|)\, dt, \quad \nu = 3 + \alpha + k(|\beta|+1),$$
$$|\partial_b^\beta r_j(b,g)| \leqq C_\beta \int (|g| + |g^{(\nu)}|)\, dt, \quad \nu = 2 + k(|\beta|+1).$$

Proof. Choose a function $\psi \in C_0^\infty(\mathbb{R})$ such that $0 \leqq \psi \leqq 1$ and $\psi(\tau) = 1$ for $|\tau| < 1$, $\psi(\tau) = 0$ for $|\tau| > 2$. Set

$$g_0(t) = (2\pi)^{-1} \int \hat{g}(\tau)\, \psi(\tau)\, e^{it\tau}\, d\tau,$$
$$g_j(t) = (2\pi)^{-1} \int \hat{g}(\tau)(\psi(2^{-j}\tau) - \psi(2^{1-j}\tau))\, e^{it\tau}\, d\tau, \quad j = 1, 2, \dots.$$

It is then clear that g_j is analytic and that $g = \sum g_j$ in $\mathscr{S}$. Since

$$|\tau^\nu \hat{g}(\tau)| \leqq \int |g^{(\nu)}(t)|\, dt$$

we have

$$|g_j(t)| \leqq C 2^{j(1-\nu)} \int (|g| + |g^{(\nu)}|)\, dt \quad \text{if } |\operatorname{Im} t| < \varepsilon_j = 2^{-j}.$$

Hence an application of Lemma 7.5.3 gives

$$g_j(t) = q^{\varepsilon_j}(t,b,g_j)\, p(t,b) + \sum_0^{k-1} t^i r_i^{\varepsilon_j}(b,g_j), \quad \sum |b_j| < 1,$$
$$|\partial_t^\alpha \partial_b^\beta q^{\varepsilon_j}(t,b,g_j)| \leqq C_{\alpha\beta} 2^{-j(\nu-1-\alpha-1-k(|\beta|+1))} \int (|g| + |g^{(\nu)}|)\, dt,$$
$$|\partial_b^\beta r_i^{\varepsilon_j}(b,g)| \leqq C_\beta 2^{-j(\nu-1-k(|\beta|+1))} \int (|g| + |g^{(\nu)}|)\, dt.$$

With ν chosen as in the theorem we obtain a convergent geometric series on the right-hand side and conclude that

$$q(t,b,g) = \sum_{j=0}^\infty q^{\varepsilon_j}(t,b,g_j), \quad r_i(b,g) = \sum_{j=0}^\infty r_i^{\varepsilon_j}(b,g_j)$$

are C^∞ functions with the stated properties.

Remark. If $g(t,x) \in \mathscr{S}(\mathbb{R}^{1+n})$ then

$$q(t,b,g(.,x)) \in C^\infty(\mathbb{R}^{1+n+2k})$$

and

$$r_j(b,g(.,x)) \in C^\infty(\mathbb{R}^{n+2k}).$$

In fact, differentiation with respect to x may be performed directly on g by the continuity and linearity with respect to g.

We are now ready to prove the part of the Malgrange preparation theorem which corresponds to Theorem 7.5.1.

Theorem 7.5.5. *Let $f(t,x)$ be a C^∞ function of $(t,x)\in\mathbb{R}^{1+n}$ near $(0,0)$ which satisfies (7.5.1). Then there exists a factorization*

$$f(t,x)=c(t,x)(t^k+a_{k-1}(x)\,t^{k-1}+\ldots+a_0(x)) \tag{7.5.2$'$}$$

where a_j and c are C^∞ functions near 0 and $(0,0)$ respectively, $c(0,0)\neq 0$ and $a_j(0)=0$. When f is real the factorization can be chosen real.

Proof. We may assume that $f\in C_0^\infty$ since the statement is local. By Theorem 7.5.4 and the remark after its proof we have in a neighborhood of $(0,0,0)$

$$f(t,x)=Q(t,x,b)\,p(t,b)+\sum_0^{k-1} t^j R_j(x,b)$$

where Q and R_j are in C^∞. Taking $x=b=0$ we obtain

$$f(t,0)=Q(t,0,0)\,t^k+\sum_0^{k-1} t^j R_j(0,0).$$

Hence $R_j(0,0)=0$ and $Q(0,0,0)\neq 0$ by (7.5.1). If we differentiate with respect to b and put $x=0$, $b=0$ afterwards, we obtain when $b=0$

$$0=d_b Q(t,0,b)\,t^k+Q(t,0,0)\sum_0^{k-1} t^j\,db_j+\sum_0^{k-1} t^j d_b R_j(0,b).$$

Since $Q(0,0,0)\neq 0$ we have $a_0=\ldots=a_{k-1}=0$ if $Q(t,0,0)\sum_0^{k-1} t^j a_j=O(t^k)$. Hence the differential of the map $(b_0,\ldots,b_{k-1})\to(R_0,\ldots,R_{k-1})$ is bijective at $(0,0)$. By the implicit function theorem it follows that the equations

$$R_j(x,b)=0,\qquad j=0,\ldots,k-1,$$

define C^∞ functions $b_j(x)$, $j=0,\ldots,k-1$, in a neighborhood of 0. Since

$$f(t,x)=Q(t,x,b(x))\,p(t,b(x))$$

and $b(0)=0$ we have obtained (7.5.2)$'$. If f is real we can take Q and R real when $b\in\mathbb{R}^k$ and apply the implicit function theorem with $b\in\mathbb{R}^k$ instead of $b\in\mathbb{C}^k=\mathbb{R}^{2k}$, which completes the proof.

The following division theorem is analogous to Theorem 7.5.2. Note that no uniqueness is valid in Theorems 7.5.5 and 7.5.6.

Theorem 7.5.6 (The Malgrange preparation theorem). *If f satisfies the hypothesis in Theorem 7.5.5 and $g(t,x)$ is a C^∞ function in a neighborhood of $(0,0)$ then*

$$g(t,x)=q(t,x)f(t,x)+\sum_0^{k-1} t^j r_j(x) \tag{7.5.3'}$$

where q and r_j are C^∞ functions in a neighborhood of $(0,0)$ and 0.

Proof. By Theorem 7.5.5 we may assume that

$$f(t,x)=t^k+\sum_0^{k-1} t^j a_j(x)=p(t,a(x)).$$

Assuming as we may that $g\in C_0^\infty$ we can take

$$q(t,x)=q(t,a(x),g(.,x)),\quad r_j(x)=r_j(a(x),g(.,x))$$

with the notation in Theorem 7.5.4.

Note that Theorem 7.5.5 is actually the special case of Theorem 7.5.6 with $g(t,x)=t^k$. In the proof of Theorem 7.5.6 it was only the application of Theorem 7.5.5 which required shrinking the neighborhood of $(0,0)$. More precisely, if $g\in C^\infty(\mathbb{R}^{1+n})$ then (7.5.3)′ follows in a neighborhood of $(0,0)$ independent of g.

The Malgrange preparation theorem is highly non-trivial even when $k=1$. This is in fact the case which we shall use most frequently. In Section 7.7 we shall need the extension of it to several t variables.

Theorem 7.5.7. *Let $f_j(t,x)$, $j=1,\dots,m$, be complex valued C^∞ functions in a neighborhood of 0 in $\mathbb{R}^{m+n}$ with $f_j(0,0)=0$, $j=1,\dots,m$, and $\det \partial f_j(0,0)/\partial t_k \neq 0$. If $g\in C^\infty$ in a neighborhood of $(0,0)$ we can then find $q_j(t,x)\in C^\infty$ at $(0,0)$ and $r(x)\in C^\infty$ at 0 so that*

$$g(t,x)=\sum_1^m q_j(t,x)f_j(t,x)+r(x).$$

Proof. When $m=1$ this is a special case of Theorem 7.5.6. We can always label the functions f_j so that $\partial f_1(0,0)/\partial t_1 \neq 0$. Writing $t'=(t_2,\dots,t_n)$ we then obtain from Theorem 7.5.6 with $k=1$

$$g(t,x)=q_1(t,x)f_1(t,x)+h(t',x),$$
$$f_j(t,x)=q_j(t,x)f_1(t,x)+F_j(t',x),\quad j=2,\dots,m.$$

Since $df_j=q_j\,df_1+dF_j$ at 0, the differentials of $F_2,\dots,F_m$ with respect to t' are linearly independent at 0. If the lemma is already proved for $m-1$ variables we obtain

$$h(t',x)=\sum_2^m p_j(t',x)\,F_j(t',x)+r(x),$$

$$g(t,x)=\left(q_1(t,x)-\sum_2^m p_j(t',x)\,q_j(t,x)\right)f_1(t,x)+\sum_2^m p_j(t',x)\,f_j(t,x)+r(x)$$

which proves the theorem.

To state an analogue of Theorem 7.5.5 in this situation we introduce the ideal $I=I(f_1,\dots,f_m)$ generated by $f_1,\dots,f_m$. This is the set of all C^∞ functions g in a neighborhood of 0 such that

$$g(t,x)=\sum_1^m q_j(t,x)\,f_j(t,x)$$

in a neighborhood of 0 for some $q_j\in C^\infty$.

Lemma 7.5.8. *If $F_1,\dots,F_m\in I(f_1,\dots,f_m)$ and $dF_1,\dots,dF_m$ are linearly independent at 0, $f_1=\dots=f_m=0$ at 0, then*

$$I(F_1,\dots,F_m)=I(f_1,\dots,f_m).$$

Proof. By hypothesis

$$F_i=\sum_1^m g_{ij}f_j,\qquad i=1,\dots,m.$$

Since $f_j(0,0)=0$ we have $dF_i=\sum g_{ij}\,df_j$ at 0. Hence $\det g_{ij}\neq 0$ so

$$f_j=\sum (g^{-1})_{ji}\,F_i\in I(F_1,\dots,F_m).$$

Here is now an extension of Theorem 7.5.5 with $k=1$.

Theorem 7.5.9. *If $f_1,\dots,f_m$ satisfy the hypotheses in Theorem 7.5.7 then*

$$I(f_1,\dots,f_m)=I(t_1-T_1(x),\dots,t_m-T_m(x))$$

for some $T_j\in C^\infty$ vanishing at 0.

Proof. Theorem 7.5.7 applied to the coordinates t_i gives

$$t_i=\sum q_{ij}(t,x)\,f_j(t,x)+T_i(x),$$

and it follows from Lemma 7.5.8 that $t_i-T_i(x)$ can be used as generators.

We shall now study the indetermination of the functions $T_i(x)$. It is clear that one can add to them any function of x which is in I, so we need to characterize such functions.

Lemma 7.5.10. *If* $R(x) \in I(t_1 - T_1(x), \dots, t_m - T_m(x))$ *then*

$$|R(x)| \leqq C_N |\operatorname{Im} T(x)|^N, \quad N = 1, 2, \dots \tag{7.5.15}$$

in a neighborhood of 0.

Proof. There exist C^∞ functions $R_\alpha(x)$ and $Q_\alpha(t, x)$ such that

$$R(x) = \sum_{0<|\alpha|<N} (t - T(x))^\alpha R_\alpha(x)/\alpha! + \sum_{|\alpha|=N} (t - T(x))^\alpha Q_\alpha(t, x)/\alpha! \tag{7.5.16}$$

in a fixed neighborhood of $(0, 0)$ for $N = 1, 2, \dots$. For $N = 1$ this follows from the hypothesis that $R \in I$. If (7.5.16) is known for one value of N it follows with N replaced by $N+1$ if we divide Q_α by $t_1 - T_1(x), \dots, t_m - T_m(x)$ using Theorem 7.5.7. (As observed after the proof of Theorem 7.5.6 the neighborhood can be chosen independent of N.) With $|x|$ small we set

$$t = \operatorname{Re} T(x) + s \operatorname{Im} T(x), \quad 0 \leqq s \leqq 1.$$

Then it follows from (7.5.16) that for $0 \leqq s \leqq 1$

$$|R(x) - \sum_{0<|\alpha|<N} (s - i)^{|\alpha|} (\operatorname{Im} T(x))^\alpha R_\alpha(x)/\alpha!| \leqq C_N |\operatorname{Im} T(x)|^N.$$

Any interpolation formula shows now that the coefficients of the polynomial in $s - i$ on the left-hand side have similar bounds, which gives (7.5.15) with another constant.

Lemma 7.5.11. *Let* $F \geqq 0$ *be Lipschitz continuous in a neighborhood of* 0 *in* $\mathbb{R}^n$. *If* $g \in C^\infty$ *and* $|g| \leqq C_N F^N$ *for every* N, *it follows that*

$$|D^\alpha g| \leqq C_{N,\alpha} F^N, \quad \textit{for all } \alpha, N.$$

Proof. Since $F(x + x') \leqq C F(x)$ if $|x'| \leqq F(x)$, we obtain

$$|\sum_{|\alpha|<N} g^{(\alpha)}(x) x'^\alpha F(x)^{|\alpha|}/\alpha!| \leqq C'_N F(x)^N, \quad |x'| \leqq 1,$$

if $g(x + F(x)x')$ is expanded by means of Taylor's formula. Hence we have an upper bound

$$|g^{(\alpha)}(x) F(x)^{|\alpha|}| \leqq C_{N,\alpha} F(x)^N, \quad |\alpha| < N,$$

for the coefficients. Since N is arbitrary the lemma is proved.

Theorem 7.5.12. *If* $I = I(t_1 - T_1(x), \dots, t_m - T_m(x))$ *where* $T_1(0) = \dots = T_m(0) = 0$ *and if* $R(x) \in C^\infty$ *then the following conditions are equivalent:*

(i) $R \in I$ *at* $(0,0)$.
(ii) (7.5.15) *is valid in a neighborhood of* 0.
(iii) $R \in I^\infty = \bigcap I^N$ *at* $(0,0)$. *More precisely, there is a neighborhood* V *of* $(0,0)$ *and functions* $q_\alpha \in C^\infty(\mathbb{R}^{m+n})$ *such that for every* N

$$R(x) = \sum_{|\alpha|=N} q_\alpha(t,x)(t-T(x))^\alpha/\alpha!, \qquad (t,x)\in V.$$

Proof. (i) $\Rightarrow$ (ii) by Lemma 7.5.10. From (ii) it follows by Lemma 7.5.11 that for all β and ν we have in a neighborhood of 0

$$|R^{(\alpha)}(x)| \leqq C_{N,\beta} F^\nu, \qquad F=\sum |f_j|^2, \qquad f_j = t_j - T_j(x).$$

Hence

$$R = \left(\sum f_j \bar f_j/F\right)^N R = \sum_{|\alpha|=N} f^\alpha q_\alpha$$

where $q_\alpha = c_\alpha \bar f^\alpha F^{-|\alpha|} R$ when $F \neq 0$. Now q_α and all of its derivatives converge to 0 when $F \to 0$, so repeated application of Corollary 1.1.2 shows that $q_\alpha \in C^\infty$ if q_α is defined as 0 when $F=0$. Since (iii) obviously implies (i) the proof of the theorem is complete.

In Section 7.7 we shall also need the following application of Theorem 7.5.6 which shows that if f is real valued, then the normal form in Theorem 7.5.5 can also be reached by a change of variables without multiplying by a function.

Theorem 7.5.13. *Let* $f(t,x)$ *be a real valued* C^∞ *function near* $(0,0)$ *in* $\mathbb{R}^{1+n}$ *such that*

$$f = \partial f/\partial t = \ldots = \partial^{k-1} f/\partial t^{k-1} = 0, \qquad \partial^k f/\partial t^k > 0 \quad \text{at } (0,0). \tag{7.5.17}$$

Then one can find a real valued C^∞ *function* $T(t,x)$ *with*

$$T=0, \qquad \partial T/\partial t > 0 \quad \text{at } (0,0)$$

and C^∞ *functions* $a_j(x)$ *vanishing at* 0 *such that*

$$f(t,x) = T^k/k + \sum_0^{k-2} a_j(x)\, T^j \tag{7.5.18}$$

in a neighborhood of $(0,0)$.

Proof. Since $f(t,0) = t^k c(t)/k$ where $c \in C^\infty$ and $c(0)>0$, we can introduce $t c(t)^{1/k}$ as a new variable. Thus we may assume that $f(t,0) = t^k/k$. If $n>1$ and the theorem is already proved for smaller values of n, we may if $x' = (x_1, \ldots, x_{n-1})$ assume that

$$f(t,x',0) = t^k/k + \sum_0^{k-2} b_j(x')\, t^j.$$

Set

$$F(t, x, a)=f(t, x)+\sum_0^{k-2} a_j t^j.$$

Then $F(t, x, 0)=f(t, x)$ and we want to let t and a vary with x_n so that $F(t, x, a)$ remains constant. This means that we want to have

$$\left(\partial f/\partial t+\sum_1^{k-2} j a_j t^{j-1}\right) dt/dx_n+\partial f/\partial x_n+\sum_0^{k-2} da_j/dx_n t^j=0. \tag{7.5.19}$$

Now Theorem 7.5.6 gives the decomposition

$$\partial f/\partial x_n=q(t, x, a)\left(\partial f/\partial t+\sum_1^{k-2} j a_j t^{j-1}\right)+\sum_0^{k-2} r_j(x, a)\, t^j$$

in a neighborhood of 0, and (7.5.19) is fulfilled if

$$dt/dx_n=-q(t, x, a), \quad da_j/dx_n=-r_j(x, a). \tag{7.5.20}$$

We integrate these equations with initial data $t=T$ and $a=A$ when $x_n=0$ and obtain C^∞ functions $t(T, x, A)$ and $a(x, A)$ defined near 0 such that $t(T, x, A)=T$ and $a(x, A)=A$ when $x_n=0$. In a neighborhood of 0 we have

$$\begin{aligned} &f(t(T, x, A), x)+\sum_0^{k-2} a_j(x, A)\, t(T, x, A)^j \\ &=f(T, x', 0)+\sum_0^{k-2} A_j T^j=T^k/k+\sum_0^{k-2} (A_j+b_j(x'))\, T^j. \end{aligned}$$

Since the Jacobian matrices $\partial a/\partial A$ and $\partial(t, x, a)/\partial(T, x, A)$ are equal to the identity when $x_n=0$ we can in a neighborhood of 0 express A as a function of x and $a(x, A)$, and T as a function of $t(T, x, A)$, x and $a(x, A)$. Hence

$$f(t, x)+\sum_0^{k-2} a_j t^j=T(t, x, a)^k/k+\sum_0^{k-2} (A_j(x, a)+b_j(x'))\, T(t, x, a)^j.$$

When we put $a=0$ the theorem is proved.

7.6. Fourier Transforms of Gaussian Functions

If A is a symmetric $n\times n$ matrix with complex elements, it is clear that the function

$$\mathbb{R}^n \ni x \to \exp(-\langle Ax, x\rangle/2) \tag{7.6.1}$$

is in $\mathscr{S}'$ if and only if $\langle \operatorname{Re} Ax, x\rangle \geqq 0$, $x\in\mathbb{R}^n$.

Theorem 7.6.1. *If A is a non-singular symmetric matrix and* $\operatorname{Re} A \geqq 0$ *then the Fourier transform of* (7.6.1) *is another Gaussian function*

$$\mathbb{R}^n \ni \xi \to (2\pi)^{n/2} (\det B)^{\frac{1}{2}} \exp(-\langle B\xi, \xi\rangle/2) \tag{7.6.2}$$

where $B = A^{-1}$ and the square root is defined as explained in Section 3.4. If $A = -iA_0$ where A_0 is real, symmetric and non-singular then (7.6.2) *becomes*

$$\mathbb{R}^n \ni \xi \to (2\pi)^{n/2} |\det A_0|^{-\frac{1}{2}} \exp(\pi i(\operatorname{sgn} A_0)/4 - i\langle A_0^{-1}\xi, \xi\rangle/2). \tag{7.6.2$'$}$$

Proof. If $u(x) = \exp(-\langle Ax, x\rangle/2)$, then

$$D_j u = i(Ax)_j u; \quad j = 1, \dots, n. \tag{7.6.3}$$

Conversely, if u is a distribution satisfying these equations then the derivatives of $u \exp(\langle Ax, x\rangle/2)$ are 0, so u is a constant multiple of (7.6.1). If we pass to Fourier transforms in (7.6.3) we obtain

$$\xi_j \hat{u} = -i(AD)_j \hat{u}, \quad j = 1, \dots, n,$$

or if we multiply by B

$$i(B\xi)_j \hat{u} = D_j \hat{u}, \quad j = 1, \dots, n.$$

It follows that

$$\hat{u} = C \exp(-\langle B\xi, \xi\rangle/2).$$

When $\operatorname{Re} A$ is positive definite we have $C = (2\pi)^{n/2} (\det A)^{-\frac{1}{2}} = (2\pi)^{n/2} (\det B)^{\frac{1}{2}}$ by (3.4.1)$''$, and when $\operatorname{Re} A$ is just non-negative the same result follows if we replace A by $A + \varepsilon I$ and let $\varepsilon \to 0$. The special case (7.6.2)$'$ is obtained from (3.4.6).

Remark. The first part of the proof could of course be replaced by a shift of the integration contour. – The quadratic forms $\langle Ax, x\rangle$ and $\langle B\xi, \xi\rangle$ are said to be dual. Since $\langle B\xi, \xi\rangle = \langle x, \xi\rangle^2 / \langle Ax, x\rangle$ when $\partial_x \langle x, \xi\rangle^2 / \langle Ax, x\rangle = 0$, this is an invariant correspondence between quadratic forms in dual spaces.

Theorem 3.3.5 is now perfectly obvious, for $E(x, t)$ is for $t > 0$ just the inverse Fourier transform of $\exp -t\langle A\xi, \xi\rangle$ so it tends to δ in $\mathscr{S}'$ as $t \to 0$. The solution of the Cauchy problem

$$\partial u/\partial t + \langle AD, D\rangle u = 0, \quad t > 0; \quad u(.,0) = v \in \mathscr{S}(\mathbb{R}^n) \tag{7.6.4}$$

which is given by

$$u(x, t) = \int E(x - y, t) v(y) \, dy$$

has the Fourier transform

$$\hat{u}(\xi, t) = \hat{v}(\xi) \exp(-t\langle A\xi, \xi\rangle)$$

with respect to x. This solution is also immediately obtained by taking Fourier transforms with respect to x in (7.6.4).

We shall now study operators of this form in some detail to prepare for some applications in Chapter XVIII. If A is any quadratic form in $\mathbb{R}^n$ with $\operatorname{Re} A \geqq 0$ we write

$$e^{-A(D)}v = (2\pi)^{-n} \int e^{i\langle x,\xi\rangle - A(\xi)} \hat{v}(\xi)\, d\xi, \qquad v \in \mathscr{S},$$

for the inverse Fourier transform of $e^{-A(\xi)}\hat{v}(\xi)$. It is clear that $e^{-A(D)}$ maps $\mathscr{S}$ into $\mathscr{S}$, $\mathscr{S}'$ into $\mathscr{S}'$ and L^2 into L^2 continuously. Note that

$$e^{-A(D)} e^{i\langle\cdot,\xi\rangle} = e^{-A(\xi)} e^{i\langle\cdot,\xi\rangle}.$$

If we make a linear change of coordinates it is therefore clear that $e^{-A(D)}$ is transformed to $e^{-A'(D)}$ if $A'(D)$ is the differential operator $A(D)$ expressed in the new coordinates.

Theorem 7.6.2. *If* $\operatorname{Re} A \geqq 0$ *we have for every integer* $k \geqq 0$

$$(7.6.5) \qquad \|e^{-A(D)}u - \sum_{j<k} (-A(D))^j u/j!\|_{L^2} \leqq \|A(D)^k u/k!\|_{L^2}, \qquad u \in \mathscr{S}.$$

Proof. If we take Fourier transforms on both sides, the inequality reduces by Parseval's formula to the inequality

$$|e^w - \sum_{j<k} w^j/j!| \leqq |w|^k/k!, \qquad \operatorname{Re} w \leqq 0$$

which follows from Taylor's formula.

To pass to maximum norms in (7.6.5) we need an elementary case of Sobolev's inequalities (Section 4.5). We shall give a direct proof and a statement which will be useful also for later reference.

Lemma 7.6.3. *If* s *is an integer* $> n/2$ *then*

$$(7.6.6) \qquad |u(x)|^2 \leqq C \int_{|x-y|<1} \Big(\sum_{|\alpha|=s} |D^\alpha u(y)|^2 + |u(y)|^2\Big) dy, \qquad u \in \mathscr{S}.$$

Proof. By Theorem 7.1.22 the operator $(-\Delta)^s$ has a parametrix E with support in the open unit ball. Thus $(-\Delta)^s E = \delta + \omega$, $\omega \in C_0^\infty$, so $|\xi|^{2s}\hat{E}(\xi)$ is a bounded function. Hence $|\xi^\alpha \hat{E}(\xi)| \leqq C(1+|\xi|)^{-s} \in L^2$ if $|\alpha| = s$. Since

$$u = u*(-\Delta)^s E - u*\omega = \sum_{|\alpha|=s} \frac{s!}{\alpha!} D^\alpha u * D^\alpha E - u*\omega$$

the estimate (7.6.6) follows by Cauchy-Schwarz' inequality.

If we apply (7.6.5) to $D^\alpha u$ when $|\alpha|=s$ also, we obtain in view of (7.6.6) since $e^{-A(D)}$ commutes with D^α

$$\sup|e^{-A(D)}u(x)-\sum_{j<k}(-A(D))^j u(x)/j!|^2 \le C/k! \sum_{|\alpha|\le s}\|A(D)^k D^\alpha u\|^2_{L^2}, \quad u\in\mathscr{S}. \tag{7.6.7}$$

Hence $e^{-A(D)}u(x)-u(x)\to 0$ uniformly when $\|A\|\to 0$, and the difference is $O(\|A\|^k)$ for every k outside supp u. We shall now prove a more precise estimate there.

Lemma 7.6.4. *If* $\operatorname{Re} A\ge 0$, $\|A\|\le 1$, *and the Euclidean distance* $d(x)$ *from* x *to* supp u *is at least* 1, *then we have for every integer* $k\ge 0$

$$|e^{-A(D)}u(x)|\le C_k\|A\|^{k+s}d(x)^{-k}\sum_{|\alpha|\le k+2s}\sup|D^\alpha u|, \quad u\in\mathscr{S}(\mathbb{R}^n). \tag{7.6.8}$$

Proof. We may assume that $x=0$, thus $u(y)=0$ when $|y|<d=d(0)$. Since $|y_j|/|y|\ge 1/\sqrt{n}$ for some j if $y\neq 0$, we can choose a partition of unity $1=\sum\chi_j(y)$ in $\mathbb{R}^n\setminus 0$ so that $\chi_j\in C^\infty$ is homogeneous of degree 0 and $\langle t_j, y\rangle\ge|y|/2\sqrt{n}$ when $y\in\operatorname{supp}\chi_j$ with $\pm t_j$ equal to one of the basis vectors. Since

$$\sum_j\sum_{|\alpha|\le k+2s}|D^\alpha\chi_j u|\le C_k\sum_{|\alpha|\le k+2s}|D^\alpha u|,$$

the estimate (7.6.8) will follow for u if it is known for each $\chi_j u$. In proving (7.6.8) we may therefore assume that

$$\langle t, y\rangle\ge|y|/2\sqrt{n} \quad \text{if } y\in\operatorname{supp} u, \tag{7.6.9}$$

where t is a unit vector. Now we have if B is the dual form of A

$$e^{-A(D)}u(0)=c\int e^{-B(y)/4}u(y)\,dy,$$
$$\langle y, t\rangle e^{-B(y)/4}=2A(t,-\partial/\partial y)\,e^{-B(y)/4}.$$

Hence repeated integrations by parts give for $j=0,1,\dots$

$$\begin{aligned}e^{-A(D)}u(0)&=2^j c\int e^{-B(y)/4}(A(t,\partial/\partial y)\langle y,t\rangle^{-1})^j u(y)\,dy\\&=2^j e^{-A(D)}(A(t,\partial/\partial y)\langle y,t\rangle^{-1})^j u(y)|_{y=0}.\end{aligned}$$

If we now apply (7.6.7) with $k=0$ it follows that

$$|e^{-A(D)}u(0)|^2\le 4^j C\sum_{|\alpha|\le s}\|D^\alpha(A(t,\partial/\partial y)\langle y,t\rangle^{-1})^j u(y)\|^2_{L^2}.$$

If we recall (7.6.9) and that $|y|\ge d\ge 1$ in supp u, we obtain if $2j>n$

$$|e^{-A(D)}u(0)|^2\le C_j d^{n-2j}\|A\|^{2j}(\sum_{|\alpha|\le s+j}\sup|D^\alpha u|)^2.$$

If we take $j=k+s$, the estimate (7.6.8) is proved.

Theorem 7.6.5. *Let* $\operatorname{Re} A \geqq 0$, $\|A\| \leqq 1$, *and let* s *be an integer* $> n/2$. *Then we have for every integer* $k \geqq s$ *that* $e^{-A(D)}u$ *is continuous and*

$$(7.6.10) \qquad |e^{-A(D)}u(x) - \sum_{j<k} (-A(D))^j u(x)/j!| \leqq C_k \|A\|^k \sum_{|\alpha| \leqq s+2k} \sup |D^\alpha u|$$

if $u \in C^{s+2k}$ *and the right-hand side is finite. At Euclidean distance* $d(x)$ *from* $\operatorname{supp} u$ *greater than* 1, *the stronger bound* (7.6.8) *is valid.*

Proof. Assume first that $u \in C_0^\infty$. Let $x=0$ and choose $\chi \in C_0^\infty$ equal to 1 in the unit ball. If we apply (7.6.7) to χu and (7.6.8) to $(1-\chi)u$ it follows that (7.6.10) is valid for $x=0$, hence for every x. When $u \in C_0^{s+2k}$ we can apply this result to the regularizations of u or rather their differences to conclude that $e^{-A(D)}u$ is continuous and that (7.6.8), (7.6.10) remain valid for u. If only $u \in C^{s+2k}$ and the right hand side of (7.6.10) is finite we apply this result to $u_R = \chi(./R)u$ and conclude when $R \to \infty$ that $e^{-A(D)}u \in L^\infty$ and that (7.6.8), (7.6.10) hold almost everywhere. (Recall that $e^{-A(D)}$ is continuous in $\mathscr{S}'$, hence $e^{-A(D)}u_R \to e^{-A(D)}u$ in $\mathscr{S}'$.) But $e^{-A(D)}u_R \in C$ and $e^{-A(D)}(u-u_R) \to 0$ in L^∞ on every compact set by (7.6.8) so it follows that $e^{-A(D)}u$ is in fact continuous.

Note that when $\operatorname{Re} A$ is not positive definite it is in no way obvious a priori that $e^{-A(D)}u$ must be a continuous function when the derivatives of u are bounded but do not tend to 0 at ∞. Theorem 7.6.5 can be regarded as a statement on the Cauchy problem (7.6.4). As already mentioned the preceding results will be important in Chapter XVIII. The local properties of $e^{-A(D)}$ will be developed further in Section 18.4. Another aspect of the preceding results is the method of stationary phase which will be studied in Section 7.7.

The following application of Theorem 7.6.1 was mentioned after Theorem 7.1.13.

Theorem 7.6.6. *If* $\hat{u}$ *is a distribution of order* $\leqq k$ *for every* $u \in L^p(\mathbb{R}^n)$ *then*

$$(7.6.11) \qquad k \geqq n(1/2 - 1/p).$$

Proof. The assumption implies if K is a compact set in $\mathbb{R}^n$ that

$$(7.6.12) \qquad |\langle \hat{u}, \phi \rangle| \leqq C \|u\|_{L^p} \sum_{|\alpha| \leqq k} \sup |D^\alpha \phi|; \quad u \in L^p, \quad \phi \in C_0^\infty(K).$$

In fact, $\langle \hat{u}, \phi \rangle$ is continuous with respect to $u \in L^p$ for fixed $\phi \in C_0^\infty(K)$ since $\langle \hat{u}, \phi \rangle = \langle u, \hat{\phi} \rangle$ and $\hat{\phi} \in \mathscr{S}$. By hypothesis $\langle \hat{u}, \phi \rangle$ is also continuous with respect to ϕ in the C^k norm for fixed $u \in L^p$. A separately continuous bilinear form in the product of a Fréchet space and a metrizable space is always continuous, so (7.6.12) follows. Choose now $u \neq 0$ in $\mathscr{S}$ so that $\hat{u}$ has compact support K, and apply (7.6.12) to u_t and ϕ_t where

$$\hat{u}_t(\xi) = \hat{u}(\xi)\, e^{it|\xi|^2}, \qquad \phi_t(\xi) = \overline{\hat{u}_t(\xi)}$$

and t is a large positive number. Then u_t is the convolution of u and $ct^{-n/2} e^{-i|x|^2/4t}$, with a constant c which can be obtained from (7.6.2), so we have if $p > 2$

$$\|u_t\|_{L^2} = \|u\|_{L^2}, \qquad \|u_t\|_{L^\infty} \leqq ct^{-n/2} \|u\|_{L^1}.$$

Hence

$$\|u_t\|_{L^p}^p = \int |u_t|^{p-2} |u_t|^2 \, dx \leqq C_u t^{-n(p-2)/2}.$$

The left-hand side of (7.6.12) is a constant when $u = u_t$, $\phi = \phi_t$, so we have

$$1 \leqq C' t^{n(1/p - \frac{1}{2})} t^k$$

for large t, which proves (7.6.11). (Note that (7.6.11) is void if $p \leqq 2$.)

In Section 7.9 we shall prove that $\hat{u} \in \mathscr{D}'^k$ for every $u \in L^p$ if there is strict inequality in (7.6.11).

We shall now discuss two examples which show that Theorem 7.6.1 can sometimes be used to solve differential equations with variable coefficients also. The first example is the construction of a fundamental solution for the Kolmogorov equation

$$\partial^2 u/\partial x^2 + x\, \partial u/\partial y - \partial u/\partial t = 0, \quad t > 0; \qquad u(t,.) \to \delta_{(x_0, y_0)} \quad \text{as } t \to 0. \tag{7.6.13}$$

Assume that a Fourier transform U with respect to (x, y) exists and behaves well as $t \to 0$. Then we obtain by taking Fourier transforms in (7.6.13)

$$-\xi^2 U - \eta\, \partial U/\partial \xi - \partial U/\partial t = 0, \qquad t > 0.$$

Thus $dU = -\xi^2 U\, dt$ if $d\eta = 0$ and $d\xi = \eta\, dt$, so

$$\begin{aligned} U(t, \xi + \eta t, \eta) &= U(0, \xi, \eta) \exp \int_0^t -(\xi + \eta s)^2 \, ds \\ &= U(0, \xi, \eta) \exp(-\xi^2 t - \xi \eta t^2 - \eta^2 t^3/3). \end{aligned}$$

With $B = \begin{pmatrix} 2t & t^2 \\ t^2 & 2t^3/3 \end{pmatrix}$ we have $\det B = t^4/3$ and

$$A = B^{-1} = \begin{pmatrix} 2/t & -3/t^2 \\ -3/t^2 & 6/t^3 \end{pmatrix}$$

so the exponential is the Fourier transform of

$$(x, y) \to 3^{\frac{1}{2}}/(2\pi t^2)\exp(-x^2/t+3xy/t^2-3y^2/t^3).$$

By inverting the Fourier transformation we obtain when $U(0, \xi, \eta) = e^{-ix_0\xi - iy_0\eta}$

$$u(t, x, y-tx) = 3^{\frac{1}{2}}/(2\pi t^2)\exp(-(x-x_0)^2/t \\ +3(x-x_0)(y-y_0)/t^2 - 3(y-y_0)^2/t^3).$$

Writing $u(t, x, y) = E(t, x, y; x_0, y_0)$ we have

$$E(t, x, y; x_0, y_0) = 3^{\frac{1}{2}}/(2\pi t^2)\exp(-(x-x_0)^2/t \\ +3(x-x_0)(y+tx-y_0)/t^2 - 3(y+tx-y_0)^2/t^3).$$

It is now easy to reverse the argument to show that E is a C^∞ function when $t>0$ and satisfies (7.6.13). The map

$$C_0^\infty(\mathbb{R}^3) \ni f \to Ef \in C^\infty(\mathbb{R}^3),$$
$$Ef(t, x, y) = \int_{s<t} E(t-s, x, y; x_0, y_0) f(s, x_0, y_0)\, dx_0\, dy_0\, ds$$

is a twosided fundamental solution of the Kolmogorov operator in (7.6.13), that is, a left and right inverse of the operator.

Our second example is the construction of a fundamental solution for the operator $-D_0^2 + 2D_0D_1 + x_1^2 D_2^2$. Because of translation invariance in x_0 and x_2 it suffices to find a solution u of the equation

$$(-D_0^2 + 2D_0D_1 + x_1^2 D_2^2)\, u = \delta_{(0,a,0)}. \tag{7.6.14}$$

We want u to have support where $x_0>0$ which will lead to solution of the Cauchy problem. Let U be the Fourier transform with respect to x_2, and the Fourier-Laplace transform with respect to x_0 which we expect to exist when $\operatorname{Im}\xi_0<0$. Then we obtain the equation

$$(-\xi_0^2 + 2\xi_0 D_1 + x_1^2\xi_2^2)\, U = \delta(x_1-a). \tag{7.6.15}$$

For the Eq. (7.6.14) without the $x_1^2 D_2^2$ term the fundamental solution has support where $x_1<a$, $2x_0+x_1-a>0$, so it is natural to expect that u shall vanish when $x_1>a$. Integration of (7.6.15) gives when $x_1<a$

$$U = C(\xi_0, \xi_2)\exp(i(\xi_0^2 x_1 - x_1^3\xi_2^2/3)/2\xi_0)$$

and if this together with $U=0$ for $x_1>a$ shall satisfy (7.6.15) we must have $2\xi_0 C i \exp(i(\xi_0^2 a - a^3\xi_2^2/3)/2\xi_0) = 1$, hence

$$U(\xi_0, x_1, \xi_2) = (-i/2\xi_0)\exp(i(\xi_0^2(x_1-a) - \xi_2^2(x_1^3-a^3)/3)/2\xi_0), \qquad x_1<a.$$

Note that the real part of the exponent is

$$\operatorname{Im}\xi_0((a-x_1)/2+\xi_2^2(a^3-x_1^3)/6|\xi_0|^2)\leqq 0 \quad \text{when } \operatorname{Im}\xi_0<0.$$

If we now use Theorem 7.6.1 to invert the Fourier transform with respect to x_2 we obtain the function

$$V(\xi_0,x_1,x_2)=(-i/2\,\xi_0)(3\,i\,\xi_0/(a^3-x_1^3))^{\frac{1}{2}}(2\,\pi)^{-\frac{1}{2}}\exp iE,$$
$$E=\xi_0((x_1-a)+3x_2^2/(x_1^3-a^3))/2.$$

Here $(i\xi_0)^{\frac{1}{2}}$ is considered as lying in the right half plane, so $(i\xi_0)^{-\frac{1}{2}}=e^{-\pi i/4}\xi_0^{-\frac{1}{2}}$ when $-\pi<\arg\xi_0<0$. By Example 7.1.17 this is the Fourier-Laplace transform with respect to x_0 of $x_{0+}^{-\frac{1}{2}}/\Gamma(1/2)=x_{0+}^{-\frac{1}{2}}/\pi^{\frac{1}{2}}$. The factor e^{iE} gives a translation so we obtain

$$u(x)=\sqrt{3}(2\pi)^{-1}((2x_0+x_1-a)(a^3-x_1^3)-3x_2^2)^{-\frac{1}{2}}$$

when $x_0>0$, $x_1<a$, and the expression under the square root is positive. Otherwise $u=0$. The integral with respect to x_2 is independent of x_0 and x_1 so this is a locally integrable function. It is an easy exercise to go back through the calculations now to show that u does satisfy (7.6.14).

In a discussion of Gaussian functions it would not be natural to omit the central limit theorem although it is outside our main topic:

Theorem 7.6.7. *Let μ be a positive measure in $\mathbb{R}^n$ such that*

$$\int d\mu=1, \quad \int|x|^2\,d\mu<\infty, \quad \int x\,d\mu=0.$$

Suppose that no hyperplane through 0 contains the support of μ and set

$$A_{jk}=\int x_j x_k\,d\mu(x).$$

A is then a positive definite symmetric matrix. If μ_ν is $\nu^{n/2}$ times the composition of the ν-fold convolution $\mu*\ldots*\mu$ with the map $x\to x\nu^{\frac{1}{2}}$, it follows that in the weak topology of measures

$$\mu_\nu\to|\det 2\pi A|^{-\frac{1}{2}}\exp(-\langle A^{-1}x,x\rangle/2).$$

Proof. Since

$$\sum A_{jk}y_j y_k=\int\langle x,y\rangle^2\,d\mu(x)$$

is positive unless $\langle x,y\rangle=0$ when $x\in\operatorname{supp}\mu$, it is clear that A is positive definite. The convolution of two positive measures μ' and μ'' with total mass 1 is defined by

$$(\mu'*\mu'')(\phi)=\iint\phi(x+y)\,d\mu'(x)\,d\mu''(y)$$

if ϕ is a bounded continuous function. As observed in Section 5.1 this agrees with our earlier definition when μ'' and ϕ have compact support. It is clear that $\mu' * \mu''$ has total mass 1 and that the Fourier transform is $\hat{\mu}'\hat{\mu}''$. Hence the Fourier transform of μ_ν is (see (7.1.17))

$$\hat{\mu}_\nu(\xi) = \hat{\mu}(\xi/\sqrt{\nu})^\nu.$$

Since $\int d\mu_\nu(x) = 1$ for every ν it follows that $|\hat{\mu}_\nu(\xi)| \leqq 1$. We may differentiate twice under the integral sign in

$$\hat{\mu}(\xi) = \int e^{-i\langle x, \xi\rangle} d\mu(x)$$

and conclude that $\hat{\mu} \in C^2$, $D_j D_k \hat{\mu}(0) = A_{jk}$, $D_j \hat{\mu}(0) = 0$. Hence

$$\hat{\mu}(\xi) = 1 - \langle A\xi, \xi\rangle/2 + o(|\xi|^2)$$

so

$$\hat{\mu}(\xi/\sqrt{\nu}) = 1 - \langle A\xi, \xi\rangle/2\nu + o(1/\nu) = \exp(-\langle A\xi, \xi\rangle/2\nu + o(1/\nu))$$

uniformly on any compact set. Hence it follows by dominated convergence that

$$\hat{\mu}_\nu \to \exp(-\langle A\xi, \xi\rangle/2) \quad \text{in } \mathscr{S}'.$$

Now the continuity in $\mathscr{S}'$ of the inverse Fourier transformation gives

$$\mu_\nu \to |\det 2\pi A|^{-\frac{1}{2}} \exp(-\langle A^{-1}x, x\rangle/2)$$

in $\mathscr{S}'$. But all μ_ν are positive measures so we have convergence in the weak topology of measures by Theorem 2.1.9.

So far we have only discussed the Fourier transform of $\exp Q$ when Q is a quadratic polynomial. We shall end this section by a brief study of the simplest cubic polynomial Q in one variable. To conform with standard notation we consider an inverse Fourier transform:

Definition 7.6.8. The Airy function $Ai(x)$ on $\mathbb{R}$ is the inverse Fourier transform of $\xi \to \exp(i\xi^3/3)$.

A priori Ai is just in $\mathscr{S}'$, but we shall prove that Ai is a C^∞ function defined by

$$Ai(x) = (2\pi)^{-1} \int_{\operatorname{Im}\zeta = \eta > 0} \exp(i\zeta^3/3 + i\zeta x)\, d\zeta. \tag{7.6.16}$$

To do so we observe that for real x we have if $\zeta = \xi + i\eta$

$$\operatorname{Re}(i\zeta^3/3 + i\zeta x) = -\xi^2\eta + \eta^3/3 - \eta x.$$

This proves that the integral converges and is independent of η. Moreover,

$$\exp(i(\xi+i\eta)^3/3)\to\exp(i\xi^3/3) \quad \text{in } \mathscr{S}' \text{ as } \eta\to+0,$$

so (7.6.16) follows when $\eta\to+0$. Since $\xi^2\eta$ grows faster than any linear function of ξ when $\xi\to\infty$ the integral in (7.6.16) converges for all $x\in\mathbb{C}$. Hence $Ai(x)$ is an entire analytic function of $x\in\mathbb{C}$. It satisfies the Airy differential equation

(7.6.17) $$Ai''(x)-x\,Ai(x)=0$$

because

$$\int(\zeta^2+x)\exp(i\zeta^3/3+i\zeta x)\,d\zeta=0.$$

From the equation it follows that $Ai''(0)=0$. A change of integration contour gives

$$\begin{aligned}Ai(0)&=(2\pi)^{-1}\int_{\mathbb{R}+i}\exp(i\zeta^3/3)\,d\zeta=\operatorname{Re}\pi^{-1}\int_0^\infty e^{\pi i/6}e^{-t^3/3}\,dt\\&=3^{-\frac{1}{6}}\Gamma(1/3)/2\pi,\\ Ai'(0)&=(2\pi)^{-1}\int i\zeta e^{i\zeta^3/3}\,d\zeta=-3^{\frac{1}{6}}\Gamma(2/3)/2\pi.\end{aligned}$$

If ω is a cubic root of unity, that is, $\omega^3=1$, then it is clear that $x\to Ai(\omega x)$ is another solution of (7.6.17) with the same value at 0 but with the derivative $\omega Ai'(0)$. Hence any two of these solutions form a basis for the solutions of Airy's differential equation. The linear relation between the three of them is

(7.6.18) $$\sum\omega\,Ai(\omega x)=0$$

which follows since the value of this solution at 0 is $Ai(0)\sum\omega=0$ and the derivative is $Ai'(0)\sum\omega^2=0$.

When $x\to+\infty$ we obtain an asymptotic expansion for the Airy function by choosing η in (7.6.16) so that the derivative of $\zeta^3/3+\zeta x$ vanishes at $i\eta$, that is, $\eta^2=x$. By Taylor's formula we have then

$$(\xi+i\eta)^3/3+x(\xi+i\eta)=\xi^3/3+i\sqrt{x}\,\xi^2+2ix^{\frac{3}{2}}/3,$$

so we obtain

(7.6.19) $$Ai(x)=\exp(-2x^{\frac{3}{2}}/3)(2\pi)^{-1}\int_{-\infty}^{\infty}e^{-\xi^2\sqrt{x}+i\xi^3/3}\,d\xi.$$

As in the proof of Theorem 7.6.2 we expand $e^{i\xi^3/3}$ in a finite Taylor series and obtain an asymptotic series with the terms

$$\begin{aligned}&\int_{-\infty}^{\infty}e^{-\xi^2\sqrt{x}}(-1)^k\xi^{6k}3^{-2k}/(2k)!\,d\xi\\&\qquad=(-1)^k3^{-2k}\Gamma(3k+\tfrac{1}{2})/(2k)!\,x^{-(6k+1)/4}.\end{aligned}$$

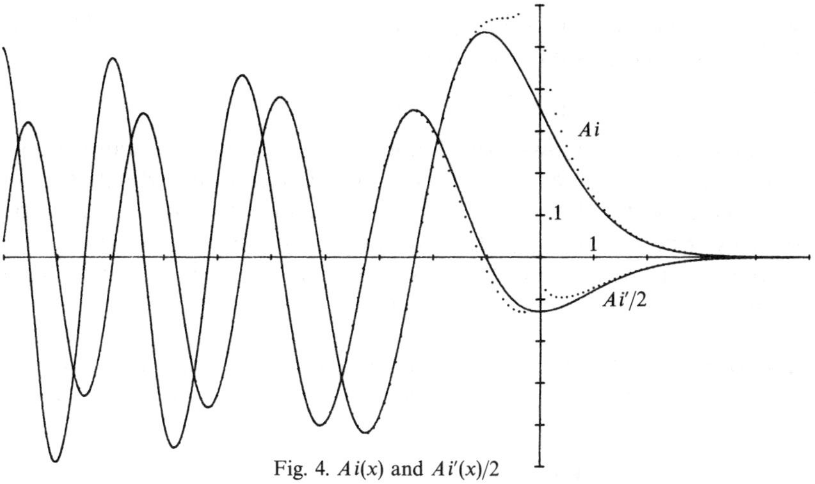

Fig. 4. $Ai(x)$ and $Ai'(x)/2$

Hence

(7.6.20) $$Ai(x) \sim (2\pi)^{-1} \exp(-2x^{\frac{3}{2}}/3)\, x^{-\frac{1}{4}} \sum_{0}^{\infty} (-9)^{-k} \Gamma(3k+\tfrac{1}{2})/(2k)!\, x^{-3k/2}$$

which means that the difference between $Ai(x)$ and a partial sum is smaller in absolute value than the first term left out. The preceding result is not only valid for positive x. If $\varepsilon > 0$ we still have (7.6.19) when $|\arg x| < \pi - \varepsilon$, and (7.6.20) remains valid although the error must be estimated by the next term with $\sqrt{x}$ replaced by $\operatorname{Re}\sqrt{x}$. Note that $Ai(x)$ is exponentially decreasing when $|\arg x| < (\pi - \varepsilon)/3$, oscillatory when $\arg x = \pm\pi/3$, and exponentially increasing when $(\pi+\varepsilon)/3 < |\arg x| < \pi - \varepsilon$. To see what happens when $\arg x$ is close to $-\pi$ we can use (7.6.18) which gives in particular when $r > 0$

(7.6.21) $$Ai(-r) = \pi^{-\frac{1}{2}} r^{-\frac{1}{4}} (\sin(\tfrac{2}{3} r^{\frac{3}{2}} + \pi/4) + O(r^{-\frac{3}{2}})).$$

This proves that there is a zero of $Ai(-r)$ with $2r^{\frac{3}{2}}/3 + \pi/4$ close to $n\pi$ when n is a large positive integer. In fact there is one for $n = 1, 2, \ldots$ and these are all the zeros of the Airy function. (See Fig. 4 where the dotted curves give the leading term in the asymptotic expansion.)

7.7. The Method of Stationary Phase

In this section we shall make a systematic study of the asymptotic behavior of integrals of the form

$$\int u e^{i\omega f}\, dx$$

where f and u are smooth, $\operatorname{Im} f \geqq 0$ and $\omega \to +\infty$. If $u \in C_0^\infty$, $f \in C^\infty$ is real valued and $f' \neq 0$ in $\operatorname{supp} u$, then

$$\int u e^{i\omega f} dx = O(\omega^{-k}), \qquad \omega \to \infty,$$

for every k. This follows from Lemma 7.1.3 if u has so small support that f can be taken as a coordinate in a new local coordinate system in a neighborhood of $\operatorname{supp} u$. We can reduce to this situation by decomposing u with a partition of unity. In the following theorem we elaborate the integration by parts in Lemma 7.1.3 to obtain uniform bounds and to cover the case of complex valued f also.

Theorem 7.7.1. *Let $K \subset \mathbb{R}^n$ be a compact set, X an open neighborhood of K and j, k non-negative integers. If $u \in C_0^k(K)$, $f \in C^{k+1}(X)$ and $\operatorname{Im} f \geqq 0$ in X, then*

$$\omega^{j+k} \left| \int u(x) (\operatorname{Im} f(x))^j e^{i\omega f(x)} dx \right| \leqq C \sum_{|\alpha| \leqq k} \sup |D^\alpha u| (|f'|^2 + \operatorname{Im} f)^{|\alpha|/2 - k}, \qquad \omega > 0. \tag{7.7.1}$$

Here C is bounded when f stays in a bounded set in $C^{k+1}(X)$. When f is real valued, the estimate (7.7.1) *reduces to*

$$\omega^k \left| \int u(x) e^{i\omega f(x)} dx \right| \leqq C \sum_{|\alpha| \leqq k} \sup |D^\alpha u| \, |f'|^{|\alpha| - 2k}, \qquad \omega > 0. \tag{7.7.1$'$}$$

Proof. When $k=0$ the assertion is obvious since $t^j e^{-t}$ is bounded for $t>0$. In the proof we may therefore assume that $k>0$ and that (7.7.1) is already proved for smaller values of k as well as for smaller values of j and the same k if $j>0$. It may also be assumed that

$$N = |f'|^2 + \operatorname{Im} f$$

is positive in K, for if (7.7.1) is proved then we can always replace f by $f + \varepsilon i$ in (7.7.1) and let $\varepsilon \to +0$.

To be able to raise j or integrate by parts we observe that

$$N u = u \sum_1^n |\partial f / \partial x_\nu|^2 + u \operatorname{Im} f,$$

hence

$$u = \sum_1^n u_\nu \partial f / \partial x_\nu + u_0 \operatorname{Im} f \tag{7.7.2}$$

if $N u_\nu = u \partial \bar{f} / \partial x_\nu$ when $\nu \neq 0$ and $N u_0 = u$. Since

$$i\omega \partial f / \partial x_\nu e^{i\omega f} = \partial_\nu e^{i\omega f}$$

we obtain after an integration by parts

$$\int u(\operatorname{Im} f)^j e^{i\omega f}\,dx = \int u_0(\operatorname{Im} f)^{j+1} e^{i\omega f}\,dx$$
$$+ i/\omega \sum_\nu \Big(\int (\partial_\nu u_\nu)(\operatorname{Im} f)^j e^{i\omega f}\,dx + \int j u_\nu(\partial_\nu \operatorname{Im} f)(\operatorname{Im} f)^{j-1} e^{i\omega f}\,dx\Big).$$

With the notation

$$|u|_\mu = \sum_{|\alpha|=\mu} |\partial^\alpha u|$$

we have by the inductive hypothesis

(7.7.3) $$\omega^{j+k}\Big|\int u(\operatorname{Im} f)^j e^{i\omega f}\,dx\Big|$$
$$\leqq C \sup\Big(\sum_{\mu=0}^{k-1}\Big(|u_0|_\mu + \sum_{\nu=1}^{n} |u_\nu|_{\mu+1}\Big) N^{\mu/2-k+1}$$
$$+ \sum_{\mu=0}^{k}\sum_\nu |u_\nu \partial_\nu \operatorname{Im} f|_\mu N^{\mu/2-k}\Big).$$

To estimate the right-hand side we need a lemma.

Lemma 7.7.2. *If* $g \in C^2(-\delta, \delta)$ *is non-negative, then*

$$\delta^2 |g'(0)|^2 \leqq g(0)\big(g(0) + 2 \sup_{|x|<\delta} \delta^2 |g''(x)|\big).$$

Proof. The proof is reduced to the case $\delta = 1$ if x/δ is introduced as a new variable. We may also assume $g(0)=1$. By Taylor's formula

$$1 + g'(0)\,x + Mx^2/2 \geqq g(x) \geqq 0, \quad |x|<1; \qquad M = \sup|g''|.$$

If $M \leqq 2$ we take $x^2=1$ and obtain $|g'(0)| \leqq 1 + M/2 \leqq (1+2M)^{\frac12}$ since $M^2/4 \leqq M$, and if $M>2$ we take $x^2 = 2/M$ and obtain $|g'(0)| \leqq 2/|x| = (2M)^{\frac12}$. This proves the lemma.

End of proof of Theorem 7.7.1. Lemma 7.7.2 applied to $\operatorname{Im} f$ gives

(7.7.4) $$|\partial_\nu \operatorname{Im} f|^2 \leqq C \operatorname{Im} f, \quad |\partial_\nu N|^2 \leqq CN \quad \text{in } K.$$

We shall now prove that when $\mu \leqq k$ we have

(7.7.5) $$N^{\frac12} \sum_{\nu=1}^{n} |u_\nu|_\mu + N|u_0|_\mu \leqq C \sum_{r=0}^{\mu} |u|_r N^{(r-\mu)/2}.$$

This follows from the definition of N and u_ν if $\mu=0$ so we may assume that $0<\mu\leqq k$ and that the estimate is proved for smaller values of μ. Let $|\alpha|=\mu$ and apply ∂^α to the equations

$$N u_\nu = u\,\partial \bar f/\partial x_\nu \quad \text{when } \nu \neq 0, \qquad N u_0 = u.$$

Estimating all terms except $N\partial^\alpha u_\nu$ in the left-hand side by means of (7.7.5) we obtain when $\nu \neq 0$ in view of (7.7.4)

$$\begin{aligned} N|u_\nu|_\mu &\leqq C(|N|_1 |u_\nu|_{\mu-1} + |u_\nu|_{\mu-2} + \ldots + |u_\nu|_0 + |u|_\mu |f|_1 + |u|_{\mu-1} + \ldots + |u|_0) \\ &\leqq C'(N^{\frac{1}{2}}|u_\nu|_{\mu-1} + |u_\nu|_{\mu-2} + \ldots + |u|_\mu N^{\frac{1}{2}} + |u|_{\mu-1} + \ldots) \\ &\leqq C'' \sum_0^\mu |u|_r N^{(r-\mu+1)/2} \end{aligned}$$

which gives (7.7.5) then. Similarly

$$\begin{aligned} N|u_0|_\mu &\leqq C(|N|_1 |u_0|_{\mu-1} + |u_0|_{\mu-2} + \ldots + |u_0|_0 + |u|_\mu) \\ &\leqq C \sum_0^\mu |u|_r N^{(r-\mu)/2} \end{aligned}$$

which completes the proof of (7.7.5). From (7.7.5) and (7.7.4) we obtain

$$(7.7.6) \qquad |u_\nu \partial_\nu \operatorname{Im} f|_\mu \leqq C \sum_0^\mu |u|_r N^{(r-\mu)/2}, \qquad \nu \neq 0, \quad \mu \leqq k.$$

If the estimates (7.7.5) and (7.7.6) are used in the right-hand side of (7.7.3) we obtain (7.7.1).

Theorem 7.7.1 shows that the integral

$$\int u e^{i\omega f}\, dx, \qquad \omega > 0,$$

is a rapidly decreasing function of ω if $u \in C_0^\infty$, $f \in C^\infty$, $\operatorname{Im} f \geqq 0$, and there is no point in $\operatorname{supp} u$ with $\operatorname{Im} f = 0$ and $f' = 0$. Thus the essential contributions must always come from points where the phase f is *real and stationary*. This is the basis for the stationary phase method which in addition describes the contributions from the simplest types of such critical points. The special case where f is a quadratic form was discussed in Section 7.6 but we restate the result in a form which is more convenient here.

Lemma 7.7.3. *Let A be a symmetric non-degenerate matrix with $\operatorname{Im} A \geqq 0$. Then we have for every integer $k > 0$ and integer $s > n/2$*

$$(7.7.7) \qquad \left|\int u(x) e^{i\omega\langle Ax,x\rangle/2} dx - (\det(\omega A/2\pi i))^{-\frac{1}{2}} T_k(\omega)\right| \leqq C_k (\|A^{-1}\|/\omega)^{n/2+k} \sum_{|\alpha| \leqq 2k+s} \|D^\alpha u\|_{L^2}, \qquad u \in \mathscr{S},$$

$$(7.7.8) \qquad T_k(\omega) = \sum_0^{k-1} (2i\omega)^{-j} \langle A^{-1} D, D\rangle^j u(0)/j!.$$

Proof. By Theorem 7.6.1 the Fourier transform of $x \to \exp i\omega\langle Ax, x\rangle/2$ is

$$\xi \to \exp(-i\langle A^{-1}\xi, \xi\rangle/2\omega)(2\pi)^{n/2}(\det(i\omega^{-1}A^{-1}))^{\frac{1}{2}}$$
$$= \exp(-i\langle A^{-1}\xi, \xi\rangle/2\omega)(\det(\omega A/2\pi i))^{-\frac{1}{2}}.$$

Hence

$$\int e^{i\omega\langle Ax, x\rangle/2} u(x)\,dx = (\det(\omega A/2\pi i))^{-\frac{1}{2}} e^{-i\langle A^{-1}D, D\rangle/2\omega} u(0)$$

so (7.7.7) follows from (7.6.7).

Remark. If B is another symmetric matrix with $\operatorname{Im} B \geqq 0$ we obtain by replacing A with $A+tB$ in (7.7.7) and expanding the t derivative of the integral that

(7.7.9) $$(2i\omega)^{-j}\left[\frac{d}{dt}\langle (A+tB)^{-1}D, D\rangle^j u(0)/j! - \tfrac{1}{2}\operatorname{Tr} B(A+tB)^{-1}\right]$$
$$= (2i\omega)^{-j-1}\langle (A+tB)^{-1}D, D\rangle^{j+1}(i\omega\langle Bx, x\rangle u(x)/2)_{x=0}/(j+1)!$$

when t is small. This is of course an identity which remains valid for all symmetric matrices with $A+tB$ non-singular.

The calculation of sums like (7.7.8) is occasionally simplified by rewriting them as if part of the exponential in (7.7.7) were a factor of u. The algebraic contents of this are described in the following lemma.

Lemma 7.7.4. *Let A be a symmetric non-singular matrix, and let B be a symmetric matrix such that $\det(A+tB)$ is independent of t. This means precisely that $A^{-1}B$ is nilpotent, that is, $(A^{-1}B)^k = 0$ for some k. With the same k we have when $\omega \to \infty$*

(7.7.10) $$\sum_{j<N} (2i\omega)^{-j}\langle (A+B)^{-1}D, D\rangle^j u(0)/j!$$
$$- \sum_{j<N} (2i\omega)^{-j}\langle A^{-1}D, D\rangle^j (e^{i\omega\langle Bx, x\rangle/2}u)(0)/j! = O(\omega^{-N/k}).$$

Proof. If λ_i are the eigenvalues of $A^{-1}B$ then

$$\det(A+tB) = \det A \det(I+tA^{-1}B) = \det A \prod(1+t\lambda_i).$$

This is independent of t if and only if all λ_i are 0, that is, $A^{-1}B$ is nilpotent. If $(A^{-1}B)^k = 0$ we have

$$(A+tB)^{-1} = (I+tA^{-1}B)^{-1}A^{-1} = \sum_{\nu=0}^{k-1}(-tA^{-1}B)^\nu A^{-1}.$$

If we expand the polynomial $\langle (A+tB)^{-1}D, D\rangle^j u(0)$ in t by Taylor's formula and use (7.7.9) we can write the first sum in (7.7.10) in the form

$$(7.7.11)\quad \sum_{j=0}^{N-1}\sum_{\nu=0}^{j(k-1)} (2i\omega)^{-j-\nu}\langle A^{-1}D, D\rangle^{j+\nu}((i\omega\langle Bx,x\rangle/2)^\nu u)(0)/(j+\nu)!\,\nu!.$$

By Taylor's formula the second sum in (7.7.10) is equal to

$$\sum_{j<N}(2i\omega)^{-j}\langle A^{-1}D, D\rangle^{j}((i\omega\langle Bx,x\rangle/2)^\nu u)(0)/j!\,\nu!.$$

The terms with $\nu>j$ vanish so we can replace j by $j+\nu$ where $j\geqq 0$, $\nu\geqq 0$ and $j+\nu<N$. This means that we have the terms in (7.7.11) with $j+\nu<N$. For the missing terms $j+\nu\geqq N$ and $\nu\leqq j(k-1)$, hence $N\leqq jk$. This means that they are $O(\omega^{-N/k})$ which completes the proof.

Using Taylor's formula it is easy to extend (7.7.7) to more general phase functions:

Theorem 7.7.5. *Let $K\subset\mathbb{R}^n$ be a compact set, X an open neighborhood of K and k a positive integer. If $u\in C_0^{2k}(K)$, $f\in C^{3k+1}(X)$ and $\operatorname{Im} f\geqq 0$ in X, $\operatorname{Im} f(x_0)=0$, $f'(x_0)=0$, $\det f''(x_0)\neq 0$, $f'\neq 0$ in $K\smallsetminus\{x_0\}$ then*

$$(7.7.12)\quad \Big|\int u(x)\,e^{i\omega f(x)}dx - e^{i\omega f(x_0)}(\det(\omega f''(x_0)/2\pi i))^{-\frac12}\sum_{j<k}\omega^{-j}L_j u\Big| \leqq C\omega^{-k}\sum_{|\alpha|\leqq 2k}\sup|D^\alpha u|,\qquad \omega>0.$$

Here C is bounded when f stays in a bounded set in $C^{3k+1}(X)$ and $|x-x_0|/|f'(x)|$ has a uniform bound. With

$$g_{x_0}(x)=f(x)-f(x_0)-\langle f''(x_0)(x-x_0), x-x_0\rangle/2$$

which vanishes of third order at x_0 we have

$$L_j u=\sum_{\nu-\mu=j}\ \sum_{2\nu\geqq 3\mu} i^{-j}2^{-\nu}\langle f''(x_0)^{-1}D, D\rangle^\nu (g_{x_0}^\mu u)(x_0)/\mu!\,\nu!.$$

This is a differential operator of order $2j$ acting on u at x_0. The coefficients are rational homogeneous functions of degree $-j$ in $f''(x_0),\dots,f^{(2j+2)}(x_0)$ with denominator $(\det f''(x_0))^{3j}$. In every term the total number of derivatives of u and of f'' is at most $2j$.

Proof. First note that since

$$f'(x)=f'(x)-f'(x_0)=f''(x_0)(x-x_0)+O(|x-x_0|^2)$$

we have

$$|x-x_0|\leqq\|f''(x_0)^{-1}\|(|f'(x)|+C|x-x_0|^2).$$

Hence $|x-x_0| \leqq 2\|f''(x_0)^{-1}\| \, |f'(x)|$ if $|x-x_0|$ is small enough, so the hypothesis made on f in (7.7.12) implies that $|x-x_0|/|f'(x)|$ is bounded near x_0. To have C bounded we have just added a condition which excludes that f' is close to 0 at some other point in K.

If $D^\alpha u(x_0)=0$, $|\alpha|<2k$, then Taylor's formula gives

$$|D^\alpha u(x)| \leqq C|x-x_0|^{2k-|\alpha|} M, \qquad M = \sum_{|\beta| \leqq 2k} \sup |D^\beta u|.$$

Since $|x-x_0|/|f'(x)|$ is bounded it follows from Theorem 7.7.1 that

$$|\int u e^{i\omega f} dx| \leqq CM\omega^{-k}.$$

If $\rho \in C_0^\infty$ is equal to 1 in a neighborhood of x_0 and u_1 is the product of ρ and the Taylor expansion of order $2k$ at x_0 of any $u \in C_0^{2k}(K)$ we can apply this result to $u_0 = u - u_1$. Hence it remains only to estimate $\int u_1 e^{i\omega f} dx$. All derivatives of u_1 can be estimated by M, and the support is close to x_0.

Assuming as we may that $f(x_0)=0$ we set

$$f_s(x) = \langle f''(x_0)(x-x_0), x-x_0 \rangle / 2 + s g_{x_0}(x).$$

Thus $f_1(x)=f(x)$ and $f_0(x)$ is a quadratic form in $x-x_0$ with $\operatorname{Im} f_0 \geqq 0$. If $\operatorname{supp} \rho$ is sufficiently close to x_0 we have a uniform bound for $|x-x_0|/|f_s'(x)|$ when $0 \leqq s \leqq 1$ and $x \in \operatorname{supp} \rho$. Differentiating

$$I(s) = \int u_1(x) \, e^{i\omega f_s(x)} dx$$

$2k$ times we obtain

$$I^{(2k)}(s) = (i\omega)^{2k} \int u_1(x) \, g_{x_0}(x)^{2k} e^{i\omega f_s(x)} dx.$$

When $|\alpha| \leqq 3k$ we have

$$|D^\alpha(u_1 g_{x_0}^{2k})(x)| \, |x-x_0|^{|\alpha|-6k} \leqq CM$$

so it follows from Theorem 7.7.1 with k replaced by $3k$ that

$$|I^{(2k)}(s)| \leqq CM\omega^{-k}, \qquad 0 \leqq s \leqq 1.$$

By Taylor's formula

$$|I(1) - \sum_{\mu<2k} I^{(\mu)}(0)/\mu!| \leqq \sup_{0<s<1} |I^{(2k)}(s)|/(2k)!$$

which has the desired bound. If G_{x_0} is the Taylor expansion of g_{x_0} of order $3k$ then

$$g_{x_0}^\mu - G_{x_0}^\mu = O(|x-x_0|^{3k+3\mu-2}) = O(|x-x_0|^{2k+2\mu})$$

and this is a function in $C^{3k} \subset C^{k+\mu}$ so another application of Theorem 7.7.1 gives

$$|I^{(\mu)}(0) - \int (i\omega G_{x_0})^\mu u_1 e^{i\omega f_0} dx| \leqq CM\omega^{-k}.$$

If we apply Lemma 7.7.3 to the integral it follows that

$$|\int u e^{i\omega f} dx - (\det(\omega f''(x_0)/2\pi i))^{-\frac{1}{2}} \sum_{\mu<2k} \sum_{\nu \leqq \mu+k} (2i\omega)^{-\nu}$$
$$\times \langle f''(x_0)^{-1} D, D\rangle^\nu (i\omega G_{x_0})^\mu u(x_0)/\nu!\mu!| \leqq CM\omega^{-k}.$$

To get a non-zero term in the sum we must have $2\nu \geqq 3\mu$. When $\nu-\mu = j$ this implies $3j = 3\nu - 3\mu \geqq \nu$, $2j = 2\nu - 2\mu \geqq \mu$ so there is just a finite number of terms for each j. Altogether 3μ derivatives are required to remove the zero of G_{x_0} at x_0, and $2\nu - 3\mu = 2j - \mu$ derivatives then remain. If $\mu = 0$ we obtain $2j$ derivatives acting on u. If $\mu > 0$ we have altogether $2j - \mu$ derivatives at most acting on u and on $(G'''_{x_0})^\mu$, that is, at most $2j$ derivatives acting on u and on f'', which completes the proof.

Remark. When f is real one can also prove Theorem 7.7.5 by using the Morse lemma to change variables so that $f(x) - f(x_0)$ becomes a quadratic form. This gives a result of the same form as Lemma 7.7.3 when $f \in C^{2k+2+s}$, and a similar improvement can be made when f is complex. However, we are primarily interested in the C^∞ case and have chosen a proof which assumes higher regularity of f than necessary but gives an effective method for computing the quite complicated expressions L_j.

In practice the functions u and f usually depend on parameters. As we shall now see this causes no problem when f is real valued. For the sake of simplicity the result is stated in the C^∞ case only.

Theorem 7.7.6. *Let $f(x, y)$ be a real valued C^∞ function in a neighborhood of $(0,0)$ in $\mathbb{R}^{n+m}$. Assume that $f'_x(0,0) = 0$ and that $f''_{xx}(0,0)$ is non-singular, with signature σ. Denote by $x(y)$ the solution of the equation $f'_x(x, y) = 0$ with $x(0) = 0$ given by the implicit function theorem. Then there exist differential operators $L_{f,j,y}$ in x of order $2j$ with the properties in Theorem 7.7.5 such that when $u \in C_0^\infty(K)$, K close to $(0,0)$,*

$$|\int u(x, y) e^{i\omega f(x,y)} dx - e^{i\omega f(x(y),y)} |\det(\omega f''_{xx}(x(y), y)/2\pi)|^{-\frac{1}{2}} \tag{7.7.13}$$
$$\times e^{\pi i\sigma/4} \sum_0^{k-1} L_{f,j,y} u(x(y), y)\,\omega^{-j}| \leqq C\omega^{-k} \sum_{|\alpha| \leqq 2k} \sup_x |D_x^\alpha u(x, y)|.$$

The theorem is an immediate consequence of Theorem 7.7.5. It is of course sufficient to sum for $j + n/2 < k$. Sometimes it is advantageous to sum the terms in a different order as in Lemma 7.7.4. We give an example which will be useful in Section 18.1.

Theorem 7.7.7. *Let f be a real valued function in $C^\infty(\mathbb{R}^{n+m})$. If K is a compact subset of $\mathbb{R}^{2n+m}$ and $u \in C_0^\infty(K)$, $k \geqq n$, then*

$$(7.7.14)\quad \Big|\int u(x,\xi,y)\, e^{i\omega(f(x,y)-\langle x,\xi\rangle)}\,dx\,d\xi - e^{i\omega f(0,y)}(2\pi/\omega)^n \sum_0^{k-n} \langle iD_x/\omega, D_\xi\rangle^\nu (e^{i\omega r_y(x)} u(x,\xi,y))_{x=0,\,\xi=f'_x(0,y)}\Big| \leqq C\omega^{-(k+n)/2} \sum_{|\alpha|\leqq 2k} \sup_{x,\xi} |D^\alpha_{x,\xi} u(x,\xi,y)|.$$

Here $r_y(x) = f(x,y) - f(0,y) - \langle f'_x(0,y), x\rangle$ vanishes of second order when $x=0$.

In the sum in (7.7.14) the ξ derivatives must act on u, and x derivatives acting on $e^{i\omega r_y(x)}$ bring out with ω a derivative of r_y vanishing at $x=0$. Another x derivative must act on it to give a non-zero contribution. This shows that the terms in the sum are $O(\omega^{-\nu/2})$, for at most $\nu/2$ derivations bring out a factor ω.

Proof. The differential of the phase $f(x,y) - \langle x,\xi\rangle$ with respect to x and ξ is equal to 0 precisely when $x=0$ and $\xi = f'_x(0,y)$. The Hessian there is the sum of the Hessian $B = \begin{pmatrix} f''_{xx} & 0 \\ 0 & 0 \end{pmatrix}$ of $f(x,y)$ and the Hessian $A = \begin{pmatrix} 0 & -I \\ -I & 0 \end{pmatrix}$ of $-\langle x,\xi\rangle$. Note that $|\det A| = 1$, $\operatorname{sgn} A = 0$, $A = A^{-1}$ and $(A^{-1}B)^2 = 0$. The asymptotic expansion of the integral in (7.7.14) is according to Theorem 7.7.6 given by

$$e^{i\omega f(0,y)}(2\pi/\omega)^n \sum_{\nu<k-n} (2i\omega)^{-\nu} \langle (A+B)^{-1}D, D\rangle^\nu (e^{i\omega R_y(x)} u)/\nu!_{x=0}$$

where $R_y(x) = r_y(x) - \langle f''_{xx}(0,y)\,x, x\rangle/2$ and the form $\langle (A+B)^{-1}D, D\rangle = -2\langle D_x, D_\xi\rangle - \langle f''_{xx}(0,y)\,D_\xi, D_\xi\rangle$ is of order 1 in x. As observed above the terms left out in the sum are $O(\omega^{-(k-n)/2})$. By Lemma 7.7.4 and the same argument the difference between the sum and

$$\sum_{\nu<k-n} (2i\omega)^{-\nu} \langle A^{-1}D, D\rangle^\nu (e^{i\omega r_y(x)} u)/\nu!$$

can be estimated by $C\omega^{-(k-n)/2} \sum_{|\alpha|\leqq 2k} |D^\alpha_{x,\xi} u|$ which proves (7.7.14).

Remark. It is easy to give a direct proof similar to that of Theorem 7.7.5.

When the phase f is allowed to be complex valued there is a difficulty already in stating an analogue of Theorem 7.7.6, for the equation $f'_x(x,y) = 0$ no longer defines a critical point $x \in \mathbb{R}^n$ when

$y \neq 0$. To cope with this problem we can use the Malgrange preparation theorem proved in Section 7.5. Assume that $f(x,y) \in C^\infty$ in a neighborhood of $(0,0)$ in $\mathbb{R}^{n+m}$ and that

$$\operatorname{Im} f \geqq 0; \quad \operatorname{Im} f(0,0)=0, \quad f'_x(0,0)=0, \quad \det f''_{xx}(0,0) \neq 0. \tag{7.7.15}$$

Using Theorem 7.5.9, where the variables (t,x) correspond to (x,y) now, we can find $X_j(y) \in C^\infty$ at 0 in $\mathbb{R}^m$ so that

$$I = I(\partial f/\partial x_1, \dots, \partial f/\partial x_n) = I(x_1 - X_1(y), \dots, x_n - X_n(y)).$$

If $X(y)$ is real then $x \to f(x,y)$ has a critical point at $X(y)$. Otherwise there is no critical point near 0, but it is suggestive to think of $X(y)$ as a critical point which has become complex although this is only a figure of speech. Repeated use of Theorem 7.5.7 gives (see the proof of Lemma 7.5.10)

$$f(x,y) = \sum_{|\alpha|<N} f^\alpha(y)(x - X(y))^\alpha/\alpha! \mod I^N \tag{7.7.16}$$

for some $f^\alpha \in C^\infty$. Since $\partial_x^\alpha f(x,y) - f^\alpha(y) \in I$ we can think of $f^\alpha(y)$ as the value of $\partial_x^\alpha f(x,y)$ at the critical point. When $|\alpha|=1$ we have in particular $f^\alpha(y) \in I$, hence $f^\alpha(y) \in I^\infty$ by Theorem 7.5.12. We can therefore choose $f^\alpha = 0$ in (7.7.16) when $|\alpha|=1$, and $f - f^0 \in I^2$.

Lemma 7.7.8. *The hypothesis* (7.7.15) *implies that near* 0, *for some* $C>0$,

$$\operatorname{Im} f^0(y) \geqq C|\operatorname{Im} X(y)|^2. \tag{7.7.17}$$

Neither f^0 nor X are uniquely determined. However, Theorem 7.5.12 shows that another choice would only change the two sides by terms which are $O(|\operatorname{Im} X|^N)$ for every N, and this does not affect the validity of (7.7.17) near 0 since $X(0)=0$.

Proof of Lemma 7.7.8. By hypothesis and (7.7.16) we have

$$0 \leqq \operatorname{Im} f(x,y) = \operatorname{Im}(f^0(y) + \sum_{|\alpha|=2} f^\alpha(y)(x - X(y))^\alpha/\alpha!) + O(|x - X(y)|^3).$$

Choose $x = \operatorname{Re} X(y) - t|\operatorname{Im} X(y)|$ with $t \in \mathbb{R}^n$, $|t|<1$. If $X(y)$ is real we obtain (7.7.17) at once. Otherwise we have with $\eta = \operatorname{Im} X(y)/|\operatorname{Im} X(y)|$

$$0 \leqq \operatorname{Im}(f^0(y) + |\operatorname{Im} X(y)|^2 \sum_{|\alpha|=2} f^\alpha(y)(t + i\eta)^\alpha/\alpha!) + O(|\operatorname{Im} X(y)|^3).$$

Hence

$$\operatorname{Im} f^0(y) \geqq |\operatorname{Im} X(y)|^2 (\sup_{|t|<1} - \operatorname{Im} \sum_{|\alpha|=2} f^\alpha(y)(t + i\eta)^\alpha/\alpha! - C|\operatorname{Im} X(y)|).$$

Now $f^\alpha(y) \to f^\alpha(0) = \partial_x^\alpha f(0,0)$ when $y \to 0$. If $A_{jk} = \partial^2 f(0)/\partial x_j \partial x_k$ then $\det A \neq 0$ and $\operatorname{Im} A \geqq 0$ by (7.7.15). The proof is therefore completed by the following

Lemma 7.7.9. *If A is an invertible symmetric $n \times n$ matrix with $\operatorname{Im} A \geqq 0$ then there is a positive constant c such that, with t, $\eta \in \mathbb{R}^n$,*

$$\sup_{|t|<1} -\operatorname{Im}\langle A(t+i\eta), t+i\eta\rangle \geqq c, \quad |\eta|=1. \tag{7.7.18}$$

Proof. For compactness reasons it suffices to prove this for a fixed η. Write $A = A_1 + iA_2$. Then

$$-\operatorname{Im}\langle A(t+i\eta), t+i\eta\rangle = \langle A_2\eta, \eta\rangle - \langle A_2 t, t\rangle - 2\langle A_1 t, \eta\rangle.$$

If $\langle A_2\eta, \eta\rangle \neq 0$ we just take $t=0$. If $\langle A_2\eta, \eta\rangle = 0$ then $A_2\eta = 0$ since A_2 is semi-definite, so $A_1\eta \neq 0$ since $A\eta \neq 0$. Then we can take $t = -\varepsilon A_1 \eta$ with a small positive ε.

By analytic continuation from the real case we shall now extend the stationary phase formula in Theorem 7.7.6 to the complex case. Set as above

$$f''_{xx}(0,0) = A = A_1 + iA_2.$$

Then $\det A \neq 0$ and we have $\det(A_1 + zA_2) \neq 0$ when $\operatorname{Im} z \neq 0$. In fact, if $x \in \mathbb{C}^n$ and $(A_1 + zA_2)x = 0$ then

$$0 = \operatorname{Im}\langle (A_1 + zA_2)x, \bar{x}\rangle = \operatorname{Im} z \langle A_2 x, \bar{x}\rangle.$$

This implies $A_2 x = 0$ if $\operatorname{Im} z \neq 0$, for A_2 is semi-definite. Hence $A_1 x = 0$ so $Ax = 0$ which implies $x = 0$. The equation $\det(A_1 + zA_2) = 0$ can have at most n real roots z. Thus we can choose $\lambda \in [0,1]$ so that $\det(A_1 + \lambda A_2) \neq 0$. Then

$$\det(A_1 + zA_2) \neq 0, \quad z \in Z,$$

if Z is a sufficiently small neighborhood of the line segment $[\lambda, i]$ between λ and i. Now introduce

$$F(x,y,z) = \operatorname{Re} f(x,y) + z \operatorname{Im} f(x,y), \quad z \in Z.$$

If x and y are sufficiently small then $\det F''_{xx}(x,y,z) \neq 0$, $z \in Z$, and we have

$$\operatorname{Im} F(x,y,z) \geqq 0 \quad \text{if } z \in Z_+ = \{z \in Z;\ \operatorname{Im} z \geqq 0\}.$$

Let K be a small compact neighborhood of $(0,0)$ in $\mathbb{R}^{n+m}$ and let $u \in C_0^\infty(K)$. To obtain an asymptotic expansion of

$$s(\omega, y) = \int u(x,y)\, e^{i\omega f(x,y)}\, dx$$

when $\omega \to +\infty$ we shall more generally consider

$$S(\omega, y, z) = \int u(x, y) e^{i\omega F(x,y,z)} dx, \qquad z \in Z_+.$$

An expansion of S is given by Theorem 7.7.6 if $z \in Z_0 = Z \cap \mathbb{R}$. Since $\operatorname{Im} F(x, y, z) \geqq 0$ we also have a uniform bound for S when $z \in Z_+$. This will lead to an asymptotic expansion of $s(\omega, y) = S(\omega, y, i)$ after we have examined the analyticity of the expansion (7.7.13). When $z \in Z_0$ the terms in the asymptotic expansion are of the form

$$(\det(\omega F''_{xx}/2\pi i))^{-\frac{1}{2}} e^{i\omega F} \omega^{-j} L_{F,j} u \tag{7.7.19}$$

evaluated at the critical point defined by $F'_x = 0$. Here $L_{F,j}$ is a differential operator in x of order $2j$ such that the coefficients multiplied by $(\det F''_{xx})^{3j}$ are polynomials of degree $3nj - j$ in $D^\alpha_x F$, $2 \leqq |\alpha| \leqq 2j+2$. The square root of the determinant is of course defined as explained in Section 3.4.

To extend the definition of (7.7.19) to Z_+ we shall consider the ideal I generated by $\partial F(x, y, z)/\partial x_j$, $j = 1, \dots, n$, in a neighborhood of

$$M = \{(0, 0, t\lambda + (1-t)i);\ 0 \leqq t \leqq 1\}$$

in $\mathbb{R}^{n+m} \times \mathbb{C}$. Since $F'_x = 0$ on M, it follows from Theorem 7.5.9 by a partition of unity in z that there exist generators of the form $x_j - X_j(y, z)$. Here X_j satisfy the Cauchy-Riemann equation approximately; more generally we have

Lemma 7.7.10. *If $\Psi(x, y, z)$ is a C^∞ function in a neighborhood of M and $\partial \Psi/\partial \bar{z} = 0$, then*

$$\Psi(x, y, z) = \sum \psi_j(x, y, z) \partial F(x, y, z)/\partial x_j + \Psi^0(y, z)$$

where $\partial \Psi^0/\partial \bar{z} \in I^\infty$, hence

$$|\partial \Psi^0/\partial \bar{z}| \leqq C_N |\operatorname{Im} X(y, z)|^N, \qquad N = 1, 2, \dots$$

for small $y \in \mathbb{R}^n$ and z in a neighborhood of $[\lambda, i]$.

Proof. Since $\partial \Psi/\partial \bar{z} = \partial F/\partial \bar{z} = 0$ we obtain by differentiation

$$\partial \Psi^0/\partial \bar{z} = -\sum \partial \psi_j/\partial \bar{z}\, \partial F/\partial x_j \in I.$$

Hence the lemma follows from Theorem 7.5.12.

We can take Z so small that the lemma is always applicable when $z \in \bar{Z}$. Now denote by $S_j(\omega, y, z)$ the function obtained from (7.7.19) when every derivative $\partial^\alpha_x F$ and $\partial^\beta_x u$ is replaced by a function $F^\alpha(y, z)$ or $u^\beta(y, z)$ in the same residue class mod I. When $z \in Z_0$ these are precisely the terms in the asymptotic expansion given by Theorem 7.7.6,

hence

(7.7.20) $$|S(\omega, y, z) - \sum_{j<N} S_j(\omega, y, z)| \leqq C_N \omega^{-N-n/2}, \quad z \in Z_0.$$

It is clear that $S(\omega, y, z)$ is an analytic function of z, and

(7.7.21) $$|\partial S_j(\omega, y, z)/\partial \bar{z}| \leqq C_{j,\nu}\, \omega^{-\nu}, \quad \nu = 1, 2, \ldots, z \in Z_+,$$

by Lemma 7.7.8 and Lemma 7.7.10, for

$$|\operatorname{Im} X|^{2\nu} e^{-c\omega |\operatorname{Im} X|^2} \leqq C_\nu \omega^{-\nu}.$$

Choose Z so that the boundary of Z_+ is piecewise in C^1 and set

$$S_j^a(\omega, y, z) = (2\pi i)^{-1} \int_{\partial Z_+} S_j(\omega, y, \zeta)\, d\zeta/(\zeta - z), \quad z \in Z_+$$

S_j^a is then analytic in Z_+, and Cauchy's integral formula (3.1.11) shows that

$$|S_j - S_j^a| \leqq C_{j,\nu}\, \omega^{-\nu}, \quad \nu = 1, 2, \ldots.$$

Thus (7.7.20) remains valid with S_j replaced by S_j^a. It is obvious that we have a bound

$$|S(\omega, y, z) - \sum_{j<N} S_j^a(\omega, y, z)| \leqq C, \quad z \in Z_+,$$

for $\operatorname{Im} F^0(y, z) \geqq 0$, $z \in Z_+$. It follows that there is a constant $\gamma > 0$ such that

(7.7.22) $$|s(\omega, y) - \sum_{j<N} S_j^a(\omega, y, i)| \leqq C \omega^{-\gamma(N+n/2)},$$

for we have the following elementary lemma:

Lemma 7.7.11. *Let Z_+ be simply connected. Then there exists a positive harmonic function $\gamma(z)$ in Z_+ such that if g is analytic and $|g| \leqq 1$ in Z_+ while $|g| \leqq \varepsilon < 1$ on Z_0 it follows that $|g(z)| \leqq \varepsilon^{\gamma(z)}$, $z \in Z_+$.*

Proof. We can choose a conformal mapping $z \to w(z)$ of Z_+ on the upper half plane in $\mathbb{C}$ mapping Z_0 to $(-1, 1)$. Set $g(z) = G(w(z))$. Then it follows from the maximum principle that

$$\log |G(w)| - (\log \varepsilon)(\arg(w - a) - \arg(w + a))/\pi \leqq 0, \quad \operatorname{Im} w > 0,$$

if $0 < a < 1$, for this is true at the boundary. Letting $a \to 1$ we obtain $|g(z)| \leqq \varepsilon^{\gamma(z)}$ where $\gamma(z) = (\arg(w(z) - 1) - \arg(w(z) + 1))/\pi$.

It follows from (7.7.22) that

(7.7.22)′ $$|s(\omega, y) - \sum_{j<N} S_j(\omega, y, i)| \leqq C \omega^{-(N+n/2)}.$$

In fact, this follows from (7.7.22) if we sum for all $j<\nu$ with ν so large that $\gamma(\nu+n/2)>N+n/2$, for it is clear that the terms with $j\geqq N$ can be estimated by the right hand side of (7.7.22)'. Hence we have proved

Theorem 7.7.12. *Let $f(x,y)$ be a complex valued C^∞ function in a neighborhood of $(0,0)$ in $\mathbb{R}^{n+m}$, satisfying (7.7.15), and let $u\in C_0^\infty(K)$ where K is a small neighborhood of $(0,0)$ in $\mathbb{R}^{n+m}$. Then*

$$\Big|\int u(x,y)e^{i\omega f(x,y)}dx-((\det(\omega f''_{xx}/2\pi i))^0)^{-\frac{1}{2}}e^{i\omega f^0}\sum_0^{N-1}(L_{f,j}u)^0\,\omega^{-j}\Big| \leqq C_N\omega^{-N-n/2}, \tag{7.7.23}$$

where for functions $G(x,y)$ the notation $G^0(y)$ stands for a function of y only which is in the same residue class modulo the ideal generated by $\partial f/\partial x_j$, $j=1,\dots,n$. When $y=0$ the definition of the square root is as in Section 3.4 and for small $y\neq 0$ it is determined by continuity.

So far we have not really proved that (7.7.23) is valid for every choice of the representatives for the residue classes but only those which were obtained in the proof. However, this is a consequence of the following result.

Proposition 7.7.13. *Let the hypotheses of Theorem 7.7.12 be fulfilled and let $\tilde f^0$, $\tilde G^0$ be in the same residue classes as f^0, G^0. Then we have*

$$|G^0e^{i\omega f^0}-\tilde G^0 e^{i\omega\tilde f^0}|\leqq C_N\omega^{-N},\qquad N=1,2,\dots. \tag{7.7.24}$$

Proof. Integration of e^z from z to w shows that

$$|e^z-e^w|\leqq|z-w|\exp\max(\operatorname{Re} z, \operatorname{Re} w).$$

Hence it follows from Lemmas 7.5.10 and 7.7.8 that for every N

$$|G^0e^{i\omega f^0}-\tilde G^0 e^{i\omega\tilde f^0}|\leqq C_N\omega\,|\operatorname{Im} X|^N e^{-c\omega|\operatorname{Im} X|^2}\leqq C'_N\omega^{1-N/2},$$

which proves the proposition.

The importance of the preceding results will be manifest in Chapter XVIII and particularly in Chapter XXV which rests entirely on them. (Theorem 7.7.12 will only be required in Section 25.4.) However, we shall interrupt the flow of technical results to give a simple application already here.

Theorem 7.7.14. *Let $u=a\,dS$ be a C_0^∞ density on a C^∞ hypersurface Σ in $\mathbb{R}^n$ of total curvature $K\neq 0$, dS denoting Euclidean surface measure. Then*

(7.7.25) $|\tau^{(n-1)/2}\hat{u}(\tau\xi)-\sum a(x)|K(x)|^{-\frac{1}{2}}(2\pi)^{(n-1)/2}e^{-i\langle x,\tau\xi\rangle-\pi i\sigma/4}|\leqq C/\tau;$

$$|\xi|=1,\ \tau>1.$$

Here the sum is taken over all $x\in\operatorname{supp}u$ where ξ is normal to Σ, and σ denotes the number of centers of curvature at x in the direction ξ minus the number of centers of curvature in the direction $-\xi$.

Proof. If $n(x)$ is normal to Σ at x, then the Gauss map $\Sigma\ni x\to n(x)\in S^{n-1}$ is a local diffeomorphism since the total curvature is not 0. We can assume that u has so small support that Σ can be parametrized in a neighborhood by parameters $t\in\mathbb{R}^{n-1}$ and the Gauss map is a diffeomorphism. Writing $a\,dS=b(t)dt$ we have

$$\hat{u}(\tau\xi)=\int e^{-i\tau\langle x(t),\xi\rangle}b(t)\,dt,\qquad |\xi|=1.$$

The phase function $f(t,\xi)=-\langle x(t),\xi\rangle$ has a critical point as a function of t if ξ is a normal to Σ at $x(t)$, and then the Hessian $f''_{tt}=-\langle x''(t),\xi\rangle$ is the second fundamental form with respect to the normal direction $-\xi$. Thus the eigenvalues with respect to the first fundamental form

$$\sum g_{jk}\,dt_j\,dt_k=|dx(t)|^2$$

are the principal curvatures $K_1,\ldots,K_{n-1}$, so the signature is the number of centers of curvature in the direction $-\xi$ minus the number in the direction ξ, and

$$\det(f''_{tt})/\det(g_{jk})=\prod K_i=K.$$

Hence

$$|\det(\tau f''/2\pi)|^{-\frac{1}{2}}=(2\pi/\tau)^{(n-1)/2}|K|^{-\frac{1}{2}}|\det(g_{jk})|^{-\frac{1}{2}}.$$

Since $b=a|\det(g_{jk})|^{\frac{1}{2}}$ the estimate (7.7.25) follows from (7.7.13) when $k>1+n/2$.

Theorem 7.7.14 is much more precise than Theorems 7.1.28 and 7.1.29 when it is applicable, but those results make no hypothesis on the curvature.

Corollary 7.7.15. *Let $X\subset\mathbb{R}^n$ be a bounded open convex set with C^∞ boundary of strictly positive curvature. If χ is the characteristic function of X then*

(7.7.26) $||\xi|^{(n+1)/2}\hat{\chi}(\xi)-(2\pi)^{(n-1)/2}(K(x_+)^{-\frac{1}{2}}e^{-i\langle x_+,\xi\rangle+\pi i(n+1)/4}$

$$+K(x_-)^{-\frac{1}{2}}e^{-i\langle x_-,\xi\rangle-\pi i(n+1)/4})|<C/|\xi|,\qquad |\xi|>1,$$

where x_+, x_- are the points on ∂X where the exterior normal is ξ, $-\xi$.

Proof. By (3.1.5) we have $\partial_j\chi=n_j dS$ where n is the interior normal on X. Thus $i\xi_j\hat{\chi}(\xi)$ is the Fourier transform of the density $n_j dS$ for $j=1$,

..., n. If we write down (7.7.25) for these densities, the estimate (7.7.26) follows.

The following estimate for the number of lattice points in convex sets in $\mathbb{R}^n$ is a classical application of Corollary 7.7.15.

Theorem 7.7.16. *Let X have the properties in Corollary 7.7.15 and assume that $0 \in X$. If $N(t)$ is the number of points in $\mathbb{Z}^n \cap tX$, $t>0$, then*

$$|N(t)-t^n m(X)| \leqq C t^{n-2n/(n+1)}, \quad t>1. \tag{7.7.27}$$

Proof. We may assume $n>1$. The characteristic function of tX is $\chi(x/t)$, so

$$N(t)=\sum_{g\in\mathbb{Z}^n} \chi(g/t).$$

It is tempting to apply Poisson's summation formula here, but χ is not smooth enough. Let us therefore form a regularization

$$\chi_\varepsilon(x)=\int \chi(x-\varepsilon y)\phi(y)\,dy$$

with $0 \leqq \phi \in C_0^\infty$, $\int \phi\, dy=1$. If $\operatorname{supp}\phi \subset (X\cap(-X))/2$ then

$$\chi(x(1+\varepsilon)) \leqq \chi_\varepsilon(x) \leqq \chi(x(1-\varepsilon)), \quad 0<\varepsilon<1.$$

This implies that

$$N(t/(1+\varepsilon)) \leqq \sum \chi_\varepsilon(g/t) \leqq N(t/(1-\varepsilon)). \tag{7.7.28}$$

Poisson's summation formula can now be applied and it gives

$$\sum \chi_\varepsilon(g/t)=t^n \sum \hat{\chi}_\varepsilon(2\pi g t)=t^n \sum \hat{\chi}(2\pi g t)\hat{\phi}(2\pi g \varepsilon t).$$

The term with $g=0$ is $t^n m(X)$, so it is the main term in (7.7.27). By (7.7.26) we can estimate the sum for $g \neq 0$ by

$$C t^n \sum_{g\neq 0} |g t|^{-(n+1)/2}(1+|g\varepsilon t|)^{-N}$$

where N may be taken arbitrary. We want ε to be so small that εt is small too. Then the second factor plays no role when we sum for $|g|<1/\varepsilon t$, and when $|g|>1/\varepsilon t$ we replace it by $|g\varepsilon t|^{-N}$. Comparing the two sums obtained with the corresponding integrals we can estimate the sum by

$$C' t^n (\varepsilon t)^{-n} \varepsilon^{(n+1)/2} = C' \varepsilon^{-(n-1)/2}.$$

This allows us to rewrite (7.7.28) in the form

$$t^n(1-\varepsilon)^n m(X) - C\varepsilon^{-(n-1)/2} < N(t) < t^n(1+\varepsilon)^n m(X) + C\varepsilon^{-(n-1)/2}.$$

To minimize the difference between the estimates we choose ε now so that $t^n\varepsilon=\varepsilon^{-(n-1)/2}$, that is, $\varepsilon=t^{-2n/(n+1)}$, which gives (7.7.27).

So far we have only studied the asymptotic behavior of the integral

$$\int u e^{i\omega f}\,dx$$

extended over the whole space. We shall now discuss what happens if we only integrate over an open set X with C^∞ boundary ∂X. The results will not be used in this book so we shall be quite brief. Since they are only local we can make a local change of variables such that X becomes the half space in $\mathbb{R}^n$ defined by $x_n<0$. We write $x'=(x_1,\dots,x_{n-1})$.

Theorem 7.7.17. *Let $f\in C^\infty(\mathbb{R}^n)$ and $\operatorname{Im} f\geqq 0$ when $x_n<0$, and let $u\in C_0^\infty(\mathbb{R}^n)$ have support close to 0. Then*

(i) *if $\partial f(0)/\partial x'\neq 0$ we have*

$$\int_{x_n<0} u e^{i\omega f}\,dx=O(\omega^{-N}),\qquad \omega\to+\infty,\quad N=1,2,\dots.$$

(ii) *if $\partial f(0)/\partial x'=0$, $\det\partial^2 f(0)/\partial x'\partial x'\neq 0$ and $\partial f(0)/\partial x\neq 0$, we have*

$$\int_{x_n<0} u(x)e^{i\omega f(x)}\,dx\sim e^{i\omega f(0)}\omega^{-(n+1)/2}\sum_0^\infty a_j\omega^{-j}$$

where $a_0=(\det(f''_{x'x'}(0)/2\pi i))^{-\frac12}(i\partial f(0)/\partial x_n)^{-1}u(0)$.

(iii) *if $\partial f(0)/\partial x=0$ but $\det\partial^2 f(0)/\partial x'\partial x'\neq 0$ and $\det\partial^2 f(0)/\partial x\,\partial x\neq 0$ we have*

$$\int_{x_n<0} u(x)e^{i\omega f(x)}\,dx\sim e^{i\omega f(0)}\omega^{-n/2}\sum_0^\infty b_j\omega^{-j/2}$$

where $b_0=(\det(f''_{xx}(0)/2\pi i))^{-\frac12}u(0)/2$.

Proof. (i) follows at once from Theorem 7.7.1 for the integral with respect to x' is rapidly decreasing uniformly with respect to x_n. In cases (ii) and (iii) we have by Theorem 7.7.12 an asymptotic expansion

$$I(x_n,\omega)=\int u(x)e^{i\omega f(x)}\,dx'\sim\sum\omega^{-(n-1)/2-j}u_j(x_n)e^{i\omega f^0(x_n)}.$$

Here $\operatorname{Im} f^0(x_n)\geqq 0$ for small $x_n\leqq 0$, and $u_j\in C^\infty(\{x_n;\,x_n\leqq 0\})$ has support close to 0 and $u_0(0)=u(0)\,(\det(f''_{x'x'}(0)/2\pi i))^{-\frac12}$. Since

$$f(x)-f^0(x_n)\in I(\partial f/\partial x_1,\dots,\partial f/\partial x_{n-1})^2,$$

we have $\partial f/\partial x_n-df^0/dx_n\in I$, so $f^{0\prime}(0)=\partial f(0)/\partial x_n\neq 0$ in case (ii). Hence

$$\begin{aligned}J(\omega)&=\int_{-\infty}^0 u_0(x_n)e^{i\omega f^0(x_n)}\,dx_n\\&=u_0(0)e^{i\omega f^0(0)}/i\omega f^{0\prime}(0)+\int_{-\infty}^0 i/\omega(u_0/f^{0\prime})'e^{i\omega f^0}\,dx_n.\end{aligned}$$

The partial integration can be continued and the other terms in the expansion of $I(x_n, \omega)$ can be discussed in the same way which proves the statement (ii). In case (iii) we note that if $L_j = d\partial f/\partial x_j$ at 0 then

$$\langle f''(0)x, x\rangle = f^{0\prime\prime}(0)x_n^2 + \sum_{j,k=1}^{n-1} a_{jk} L_j(x) L_k(x).$$

Hence $f^{0\prime\prime}(0) \neq 0$, $L_j(x', 0)$ are linearly independent and

$$\begin{aligned}\det f''_{xx}(0) &= (\det(\partial L_j/\partial x_k)_{j,k=1}^{n-1})^2 f^{0\prime\prime}(0) \det a_{jk}\\ &= f^{0\prime\prime}(0) \det f''_{x'x'}(0).\end{aligned}$$

It is clear that if $c = f^{0\prime\prime}(0)$ then $\operatorname{Im} c \geqq 0$.

If $v \in C_0^\infty(\mathbb{R})$ an asymptotic expansion of

$$\int_{-\infty}^{0} e^{i\omega c t^2/2} v(t)\,dt$$

is obtained by writing $v = v_0 + v_1$ where $v_0(t) = (v(t) + v(-t))/2$ is even and $v_1(t) = (v(t) - v(-t))/2$ is odd. From the Taylor expansion it is easy to see that $v_1(t) = t w(t^2)$ where $w \in C^\infty$, $w^{(\nu)}(0)/\nu! = v^{(2\nu+1)}(0)/(2\nu+1)!$. Hence

$$\begin{aligned}\int_{-\infty}^{0} e^{i\omega c t^2/2} v(t)\,dt &= \int_{-\infty}^{\infty} e^{i\omega c t^2/2} v_0(t)\,dt/2 - \int_0^\infty e^{i\omega c s} w(2s)\,ds\\ &\sim \tfrac{1}{2}(c\omega/2\pi i)^{-\frac{1}{2}} \sum_0^\infty (2 i\omega c)^{-\nu} D^{2\nu} v(0)/\nu!\\ &\quad + (\omega c)^{-1} \sum_0^\infty (i\omega c/2)^{-\nu} D^{2\nu+1} v(0)\,\nu!/(2\nu+1)!\end{aligned}$$

The proof of Theorem 7.7.5 shows that we have a similar expansion if the phase function is replaced by f^0, and this completes the proof in case (iii).

When f is real one can study the effect of parameter dependence in the preceding situation by means of Theorem 7.5.13 with $k=2$. This leads to Fresnel integrals. However, we leave this for the reader since we shall discuss a similar but more complicated problem below.

Finally we shall discuss briefly the asymptotic behavior of

$$\int u(x) e^{i\omega f(x)}\,dx$$

when f is real valued but has a degenerate critical point x_0. From now on integration will always be taken over the whole space so there will be no boundary problems to investigate. The simplest case is when the rank of f'' is $n-1$ at x_0. We can choose coordinates so that

$f''_{x'x'}$ is non-degenerate, if $x'=(x_1,\dots,x_{n-1})$. An asymptotic expansion of the integral with respect to x' is obtained from Theorem 7.7.6. We are then left with an integral with respect to x_n. After changing notation our problem is therefore to study the asymptotic behavior of

$$\int_{-\infty}^{\infty} u(t)e^{i\omega f(t)}\,dt$$

when $u\in C_0^\infty$ has support close to 0 and $f'(0)=f''(0)=0$.

If f' has a zero of finite order $k-1$ at 0 we can write

$$f(t)=f(0)+t^k a(t), \qquad a(0)\neq 0.$$

Introducing $t|a(t)|^{1/k}$ as a new variable and changing notation again, we are led to studying the integral

$$\int u(t)e^{i\omega t^k}\,dt$$

or its complex conjugate. First assume that $k=3$. Since $e^{i\omega t^3}$ is the Fourier transform of $Ai(\tau(3\omega)^{-\frac{1}{3}})(3\omega)^{-\frac{1}{3}}$ (see Section 7.6) we have

$$\int u(t)e^{i\omega t^3}\,dt=\int \hat u(\tau)Ai(\tau(3\omega)^{-\frac{1}{3}})(3\omega)^{-\frac{1}{3}}\,d\tau.$$

If we introduce a finite part of the Taylor series expansion of Ai, it follows when $u\in C_0^\infty$ that

(7.7.29)
$$\int u(t)e^{i\omega t^3}\,dt = \sum_{j<N} Ai^{(j)}(0)/j!\,(3\omega)^{-(j+1)/3}2\pi D^j u(0)+O(\omega^{-(N+1)/3}).$$

The derivatives $Ai^{(j)}(0)$ can be expressed in terms of the Γ function. Also for arbitrary k the Fourier transform of e^{it^k} is easily seen to be an entire function if we write it as an integral over a contour where it^k is negative. With $u\in C_0^\infty$ we obtain for *odd* k

(7.7.30)
$$\int u(t)e^{i\omega t^k}\,dt = \sum_{j<N} 2k^{-1}\Gamma((j+1)/k)\left(\sin\frac{(k-1)(j+1)\pi}{2k}\right)\omega^{-(j+1)/k} \times(-D)^j u(0)/j!+O(\omega^{-(N+1)/k}), \qquad \omega\to\infty,$$

and for *even* k

(7.7.31)
$$\int u(t)e^{i\omega t^k}\,dt = \sum_{j<N} 2k^{-1}\Gamma((2j+1)/k)e^{\pi i(2j+1)/2k}\omega^{-(2j+1)/k} \times(iD)^{2j}u(0)/(2j)!+O(\omega^{-(2N+1)/k}), \qquad \omega\to\infty.$$

The simple details of the proof are left as an exercise for the reader. Note that in (7.7.30) the terms for which k divides $j+1$ are missing.

All this is quite elementary, but the situation becomes rather complicated if there are parameters y present, for the multiplicity of the zeros of $f_t'(t,y)$ as a function of t may vary with y. We shall only discuss the case $k=3$ where the Airy function turns up and leave the extension to general k for the reader.

Theorem 7.7.18. *Let f be a real valued C^∞ function near 0 in $\mathbb{R}^{1+n}$ such that $\partial f/\partial t = \partial^2 f/\partial t^2 = 0$ but $\partial^3 f/\partial t^3 \neq 0$ at 0. Then there exist C^∞ real valued functions $a(y)$, $b(y)$ near 0 such that $a(0)=0$, $b(0)=f(0)$ and*

$$\int u(t,y)e^{i\omega f(t,y)}dt \sim e^{i\omega b(y)}\Big(Ai(a(y)\omega^{\frac{2}{3}})\omega^{-\frac{1}{3}}\sum_0^\infty u_{0\nu}(y)\omega^{-\nu} + Ai'(a(y)\omega^{\frac{2}{3}})\omega^{-\frac{2}{3}}\sum_0^\infty u_{1\nu}(y)\omega^{-\nu}\Big), \tag{7.7.32}$$

provided that $u\in C_0^\infty$ and $\operatorname{supp} u$ is sufficiently close to 0. Here $u_{j\nu}\in C_0^\infty$.

Proof. Replacing t by $-t$ if necessary we may assume that $\partial^3 f(0)/\partial t^3 > 0$. Application of Theorem 7.5.13 with $k=3$ gives a function $T(t,y)$ with $T(0,0)=0$, $\partial T(t,0)/\partial t > 0$ such that

$$f(t,y) = T^3/3 + a(y)T + b(y),$$

where $a, b\in C^\infty$. Introducing T as a new integration variable instead of t we obtain when $\operatorname{supp} u$ is sufficiently close to 0

$$I(\omega,y) = \int u(t,y)e^{i\omega f(t,y)}dt = \int v(T,y)e^{i\omega(T^3/3+a(y)T+b(y))}dT$$

where $v\in C_0^\infty$ and $\operatorname{supp} v$ is close to 0. Choose $\chi\in C_0^\infty(\mathbb{R})$ equal to 1 in a neighborhood of 0 and with small support. Then we have

$$I(\omega,y) = \int \chi(T)v(T,y)\, e^{i\omega(T^3/3+a(y)T+b(y))}dT.$$

Here we divide v by $T^2+a(y)$ using Theorem 7.5.6,

$$v(T,y) = (T^2+a(y))q(T,y) + r_1(y)T + r_0(y).$$

After an integration by parts we then obtain

$$I(\omega,y) = i/\omega \int \frac{d}{dT}(\chi(T)q(T,y))e^{i\omega(T^3/3+a(y)T+b(y))}dT + \int \chi(T)(r_1(y)T+r_0(y))e^{i\omega(T^3/3+a(y)T+b(y))}dT.$$

The first integral on the right hand side is of the same form as the one we started with apart from the factor $1/\omega$ in front. We have

$$\int e^{i\omega(T^3/3+a(y)T)}\chi(T)dT = \int e^{i(T^3/3+a(y)\omega^{\frac{2}{3}}T)}\chi(T\omega^{-\frac{1}{3}})dT/\omega^{\frac{1}{3}}$$
$$= 2\pi Ai(a(y)\omega^{\frac{2}{3}})\omega^{-\frac{1}{3}} - \int e^{i(T^3/3+a(y)\omega^{\frac{2}{3}}T)}(1-\chi(T\omega^{-\frac{1}{3}}))dT/\omega^{\frac{1}{3}}.$$

In the last integral the integration outside the support of $\chi(T\omega^{-\frac{1}{3}})$ should be moved into the upper half plane, and it is easily seen by means of Theorem 7.7.1 that it is rapidly decreasing as $\omega\to+\infty$.

Similarly

$$\int e^{i\omega(T^3/3+a(y)T)}\chi(T)\,T\,dT-(2\pi/i)\,A\,i'(a(y)\,\omega^{\frac{2}{3}})\,\omega^{-\frac{2}{3}}$$

is rapidly decreasing. Iteration of this argument proves the theorem.

It is easy to extend Theorem 7.7.18 to several integration variables:

Theorem 7.7.19. *Let $f(x,y)$ be a real valued C^∞ function near 0 in $\mathbb{R}^{n+m}$ such that*

$$f_x'(0,0)=0, \quad \operatorname{rank} f_{xx}''(0,0)=n-1, \tag{7.7.33}$$
$$\langle X,\partial/\partial x\rangle^3 f(0,0)\neq 0 \quad \text{if } 0\neq X\in\operatorname{Ker} f_{xx}''(0,0).$$

Then there exist real valued C^∞ functions $a(y)$, $b(y)$ near 0 such that $a(0)=0$, $b(0)=f(0)$ and

$$\int u(x,y)e^{i\omega f(x,y)}dx \sim e^{i\omega b(y)}\omega^{-(n-1)/2}\Big(A\,i(a(y)\omega^{\frac{2}{3}})\omega^{-\frac{1}{3}} \tag{7.7.32$'$}$$
$$\times\sum_0^\infty u_{0\nu}(y)\omega^{-\nu}+A\,i'(a(y)\omega^{\frac{2}{3}})\omega^{-\frac{2}{3}}\sum_0^\infty u_{1\nu}(y)\omega^{-\nu}\Big),$$

provided that $u\in C_0^\infty$ and supp u *is sufficiently close to 0. Here $u_{j\nu}\in C_0^\infty$.*

Proof. Let us first verify the invariance of the last condition in (7.7.33). To do so we let $x\to x(s)$ be any smooth curve with $x(0)=0$ and $x'(0)=X$. When $s=0$ we have

$$d^3 f(x(s),0)/ds^3=\langle X,\partial/\partial x\rangle^3 f(0,0)+3\langle f_{xx}''(0,0)X,x''(0)\rangle.$$

The second term vanishes since $X\in\operatorname{Ker} f_{xx}''(0,0)$. Thus the last part of (7.7.33) means that

$$d^3 f(x(s),0)/ds^3\neq 0, \quad s=0, \tag{7.7.33$'$}$$

and this condition is independent of the choice of local coordinates. Now we may assume the coordinates labelled so that

$$\det(\partial^2 f/\partial x_i\partial x_j)_{i,j=2}^n\neq 0 \quad \text{at } (0,0).$$

Then the equations $\partial f(x,y)/\partial x_j=0, j=2,\dots,n$, determine x_j as C^∞ functions $X_j(x_1,y), j=2,\dots,n$. If we introduce x_j-X_j as new variables instead of x_j, $j=2,\dots,n$, we have reduced the proof to the case where $\partial f(x,y)/\partial x_j=0$, $j=2,\dots,n$, when $x_2=\dots=x_n=0$. By Theorem 7.7.6 the integral (7.7.32)′ with respect to the variables $x_2,\dots,x_n$ only is of the form

$$\omega^{-(n-1)/2}e^{i\omega f(x_1,0)}U(x_1,y,\omega)$$

where U has an asymptotic expansion in powers of $1/\omega$ as $\omega \to \infty$. If we apply Theorem 7.7.18 to each term in the expansion, we obtain (7.7.32)'. (Note that we can take $X=(1,0,\dots,0)$ because $\partial^2 f(0,0)/\partial x_1 \partial x_j = 0$ for all j.)

In view of (7.6.20) the right-hand side of (7.7.32)' is rapidly decreasing when $a(y)>0$. On the other hand, it follows from (7.6.21) that (7.7.32)' agrees with the expansion given by Theorem 7.7.6 (with two critical points) when $a(y)<0$. When $a(y)=0$ we have an asymptotic expansion in powers of $\omega^{-\frac{1}{3}}$, and this remains true in any zone where $a(y)\omega^{\frac{2}{3}}$ is bounded. However, the amplitude quickly becomes very small when this quantity grows positive. In Section 12.2 we shall see that these expansions represent in optics the behavior of a wave in respectively the shadow region, the illuminated region and the penumbra at a caustic. The role of the Airy function is to describe the transition between the different types of asymptotic expansions.

7.8. Oscillatory Integrals

Theorem 7.7.1 allows one to define the useful notion of oscillatory integrals. Let $X \subset \mathbb{R}^n$ be open and let Γ be an open cone in $X \times (\mathbb{R}^N \setminus \{0\})$ for some N. This means that Γ is invariant under multiplication by positive scalars of the component in $\mathbb{R}^N$. We shall say that a function $\phi \in C^\infty(\Gamma)$ is a *phase function* in Γ if

(i) $\phi(x, t\theta) = t\phi(x,\theta)$ if $(x,\theta) \in \Gamma$, $t>0$.
(ii) $\operatorname{Im}\phi \geqq 0$ in Γ,
(iii) $d\phi \neq 0$ in Γ.

We wish to show that an integral of the form

$$\int e^{i\phi(x,\theta)} a(x,\theta)\, d\theta \tag{7.8.1}$$

defines a distribution in X even if a is large, provided that a oscillates more slowly than the factor $e^{i\phi}$.

Definition 7.8.1. Let m, ρ, δ be real numbers with $0<\rho\leqq 1$ and $0 \leqq \delta < 1$. Then we denote by $S^m_{\rho,\delta}(X \times \mathbb{R}^N)$ the set of all $a \in C^\infty(X\times\mathbb{R}^N)$ such that for every compact set $K \subset X$ and all α, β the estimate

$$|D_x^\beta D_\theta^\alpha a(x,\theta)| \leqq C_{\alpha,\beta,K}(1+|\theta|)^{m-\rho|\alpha|+\delta|\beta|}, \quad x\in K, \quad \theta \in \mathbb{R}^N, \tag{7.8.2}$$

is valid for some constant $C_{\alpha,\beta,K}$.

The elements of $S^m_{\rho,\delta}$ are called *symbols* of order m and type ρ, δ. The best possible constants in (7.8.2) are semi-norms in $S^m_{\rho,\delta}$ which make it a Fréchet space.

Theorem 7.8.2. *Let ϕ be a phase function in the open cone $\Gamma \subset X \times \mathbb{R}^N$ and let F be a closed cone $\subset \Gamma \cup (X \times \{0\})$. Then the functional I_ϕ defined by*

$$I_\phi(au) = \int e^{i\phi(x,\theta)} a(x,\theta)\, u(x)\, dx\, d\theta \tag{7.8.3}$$

when the integral is absolutely convergent can be extended in a unique way to all $a \in \bigcup_{m,\rho,\delta} S^m_{\rho,\delta}(X \times \mathbb{R}^N)$ with support in F and all $u \in C_0^\infty(X)$, so that $I_\phi(au)$ is a continuous linear function of $a \in S^m_{\rho,\delta}$ for every fixed $u \in C_0^\infty(X)$, $m \in \mathbb{R}$, $\rho \in (0,1]$ and $\delta \in [0,1)$. The linear form

$$I_{\phi,a}\colon\ u \to I_\phi(au)$$

is a distribution of order $\leqq k$ if $a \in S^m_{\rho,\delta}$ and

$$m - k\rho < -N,\ m - k(1-\delta) < -N.$$

Proof. Choose $\chi \in C_0^\infty(\mathbb{R}^N)$ so that $\chi(\theta) = 1$ when $|\theta| < 1$ and $\chi(\theta) = 0$ when $|\theta| > 2$, and set (see the proof of Theorem 7.5.4)

$$\chi_\nu(\theta) = \chi(2^{-\nu}\theta) - \chi(2^{1-\nu}\theta), \quad \nu > 0; \qquad \chi_0(\theta) = \chi(\theta).$$

Then we have

$$\sum_0^\infty \chi_\nu = 1 \quad \text{and} \quad 2^{\nu-1} \leqq |\theta| \leqq 2^{\nu+1} \quad \text{when } \theta \in \operatorname{supp} \chi_\nu, \qquad \nu \neq 0.$$

If $a \in S^m_{\rho,\delta}$ and $x \in K \Subset X$ we obtain

$$|D_x^\beta D_\theta^\alpha \chi_\nu(\theta)\, a(x,\theta)| \leqq C_{\alpha,\beta,K} (1 + |\theta|)^{m - \rho|\alpha| + \delta|\beta|}$$

since $|D^\alpha \chi_\nu(\theta)| \leqq C_\alpha (1 + |\theta|)^{-|\alpha|}$ with a constant independent of ν. Hence the series $\sum \chi_\nu\, a$ converges to a in $S^{m'}_{\rho,\delta}$ if $m' > m$, for at most two terms in the series have overlapping supports. If there is an extension of I_ϕ with the required properties, it follows that it must be given by

$$I_\phi(au) = \sum I_\phi(\chi_\nu a u). \tag{7.8.4}$$

The theorem will therefore be proved if we show that the series on the right-hand side converges and that the sum has the properties listed in the theorem.

To do so we write for $\nu \geqq 0$

$$\begin{aligned} I_\phi(\chi_{\nu+1} a u) &= \int e^{i\phi(x,\theta)} \chi_1(2^{-\nu}\theta)\, a(x,\theta)\, u(x)\, dx\, d\theta \\ &= 2^{N\nu} \int e^{i\omega\phi(x,\theta)} \chi_1(\theta)\, a(x,\omega\theta)\, u(x)\, dx\, d\theta \end{aligned}$$

where $\omega=2^\nu$. By hypothesis $\gamma=\max(\delta, 1-\rho)<1$, and if

$$|D_x^\beta D_\theta^\alpha a(x,\theta)|\leqq M(1+|\theta|)^{m-\rho|\alpha|+\delta|\beta|}, \quad |\alpha+\beta|\leqq k, \quad x\in K,$$

we obtain

$$|D_{x\theta}^\alpha a(x, 2^\nu\theta)|\leqq CM\,2^{\nu(m+\gamma|\alpha|)} \quad \text{if } x\in K, \quad 1/2<|\theta|<2, \quad |\alpha|\leqq k.$$

Hence Theorem 7.7.1 gives the estimate

$$|I_\phi(\chi_{\nu+1}au)|\leqq CM\,2^{\nu(N+m+\gamma k-k)}\sum_{|\alpha|\leqq k}\sup|D^\alpha u|, \qquad u\in C_0^\infty(K).$$

When $(1-\gamma)k>N+m$ we conclude that (7.8.4) converges and that $u\to I_\phi(au)$ is a distribution of order $\leqq k$ as stated.

It is convenient to use the notation (7.8.3) for $I_\phi(au)$ even when the integral is not convergent. The extended definition of (7.8.3) will be called an *oscillatory integral*. For the distribution $u\to I_\phi(au)$ the notation

$$\int e^{i\phi(x,\theta)}a(x,\theta)\,d\theta$$

will often be used.

An important example of an oscillatory integral is the formula

$$\int_{\mathbb{R}^n} e^{i\langle x,\theta\rangle}d\theta=\delta_0(x)(2\pi)^n. \tag{7.8.5}$$

By definition we must for the proof consider the oscillatory integral

$$\iint e^{i\langle x,\theta\rangle}u(x)\,dx\,d\theta, \qquad u\in C_0^\infty(\mathbb{R}^n).$$

If $\chi\in C_0^\infty$ is equal to 1 in a neighborhood of 0 we have $\chi(./t)\to 1$ in $S_{1,0}^m$ for any $m>0$ as $t\to\infty$. Thus the oscillatory integral is the limit of the convergent double integral

$$\begin{aligned}\iint e^{i\langle x,\theta\rangle}\chi(\theta/t)\,u(x)\,dx\,d\theta&=\int t^n\hat\chi(-tx)\,u(x)\,dx\\ &=\int\hat\chi(-x)\,u(x/t)\,dx\to u(0)\int\hat\chi(-x)\,dx\\ &=u(0)(2\pi)^n\chi(0)=u(0)(2\pi)^n.\end{aligned}$$

This proves (7.8.5) which is thus another way of expressing Fourier's inversion formula. That the left-hand side of (7.8.5) is a distribution which is singular only at 0 is also a consequence of the following theorem.

Theorem 7.8.3. *For the distribution $I_{\phi,a}$ defined by* (7.8.3) *we have*

$$\operatorname{sing\,supp} I_{\phi,a}\subset\{x\in X;\ \phi_\theta'(x,\theta)=0 \text{ for some } (x,\theta)\in F\}=S.$$

The restriction of $I_{\phi,a}$ to $X\setminus S$ is the C^∞ function

$$x\to\int e^{i\phi(x,\theta)}a(x,\theta)\,d\theta \tag{7.8.6}$$

which is defined for fixed x as an oscillatory integral.

Proof. The definition of S means that $\phi(x,\theta)$ is a phase function of θ when x is fixed in $X\setminus S$, so the oscillatory integral in (7.8.6) is defined then. It is a continuous function of x, for the proof of the existence of the oscillatory integral shows that it is the limit, locally uniformly with respect to x, of the C^∞ function

$$\int e^{i\phi(x,\theta)} a(x,\theta)\,\chi(\theta/t)\,d\theta.$$

The derivative with respect to x is

$$\int e^{i\phi(x,\theta)}(i\phi'_x(x,\theta)\,a(x,\theta)+a'_x(x,\theta))\,\chi(\theta/t)\,d\theta$$

which converges to the oscillatory integral obtained by differentiating (7.8.6) under the integral sign. Thus the function (7.8.6) is in $C^1(X\setminus S)$ and the derivative may be computed by formal differentiation under the integral sign. Since $\phi'_x(x,\theta)\,a(x,\theta)(1-\chi(\theta))\in S^{m+1}_{\rho,\delta}$ we may repeat the procedure and conclude that (7.8.6) defines a function in $C^\infty(X\setminus S)$. This function is equal to the distribution $u\to I_\phi(au)$ there, for if $u\in C_0^\infty(X\setminus S)$ we have

$$\begin{aligned} I_\phi(au) &= \lim_{t\to\infty}\int u(x)\,dx\left(\int e^{i\phi(x,\theta)}a(x,\theta)\,\chi(\theta/t)\,d\theta\right)\\ &= \int u(x)\,dx\left(\int e^{i\phi(x,\theta)}a(x,\theta)\,d\theta\right). \end{aligned}$$

The arguments used in the preceding proof show quite generally that one may operate on oscillatory integrals as with standard integrals; differentiation can be performed under the integral sign, orders of integration can be interchanged and so on. We leave for the reader to contemplate this extension of integral calculus and give instead an example.

Example 7.8.4. The Cauchy problem

$$c^{-2}\partial^2 E/\partial t^2-\Delta E=0 \quad \text{in } \mathbb{R}^{1+n}, \tag{7.8.7}$$
$$E=0,\quad \partial E/\partial t=\delta_0 \quad \text{when } t=0,$$

has the solution

$$E(t,x)=(2\pi)^{-n}\int\left(e^{i(ct|\xi|+\langle x,\xi\rangle)}-e^{i(-ct|\xi|+\langle x,\xi\rangle)}\right)d\xi/2i|\xi|c.$$

The integral is absolutely convergent when $|\xi|<1$, and when $|\xi|>1$ it is defined as the difference between two oscillatory integrals with phase functions $\langle x,\xi\rangle\pm ct|\xi|$. The differential with respect to x is never 0, so $E(t,x)$ is a C^∞ function of t with values in $\mathscr{D}'(\mathbb{R}^n)$. We obtain (7.8.7) by differentiating under the integral sign if we recall (7.8.5). By Theorem 7.8.3 we have

$$\begin{aligned} \operatorname{sing\,supp} E &\subset \{(t,x);\, x\pm ct\xi/|\xi|=0 \text{ for some } \xi\neq 0\}\\ &= \{(t,x);\, |x|=c|t|\}. \end{aligned}$$

This is the double light cone. We leave as an exercise for the reader to verify that $2c^2E = E_+ - E_-$ where E_+ and E_- are the advanced (retarded) fundamental solutions in Theorem 6.2.3.

7.9. $H_{(s)}$, L^p and Hölder Spaces

In Theorem 7.1.11 we proved that the Fourier transformation maps L^2 onto itself. By Theorem 7.1.13 we also know that L^p is mapped into $L^{p'}$ if $1 \leqq p < 2$. However, the range is much smaller than $L^{p'}$ then for Theorem 7.6.6 shows that the Fourier transform of $L^{p'}$ contains distributions of positive order when $p' > 2$. We shall now round off these results by studying some spaces of distributions which are closely related to L^2 and therefore possible to keep track of when using the Fourier transformation.

Definition 7.9.1. If s is a real number, then $H_{(s)}(\mathbb{R}^n)$ denotes the space of all $u \in \mathscr{S}'$ such that $\hat{u} \in L^2_s$, the L^2 space with respect to the measure $(1+|\xi|^2)^s d\xi/(2\pi)^n$. In $H_{(s)}$ we define the norm

$$\|u\|_{(s)} = ((2\pi)^{-n} \int |\hat{u}(\xi)|^2 (1+|\xi|^2)^s d\xi)^{\frac{1}{2}}. \tag{7.9.1}$$

The Fourier transformation is an isomorphism $H_{(s)} \to L^2_s$ since $L^2_s \subset \mathscr{S}'$. We shall now give a description of $H_{(s)}$ which does not refer to $\hat{u}$. By Theorem 7.1.11 we have $H_{(0)} = L^2$, and it is obvious that $H_{(s)}$ decreases when s increases. Let s be a positive integer for a moment. Expanding $(1+|\xi|^2)^s = (1+\xi_1^2+\ldots+\xi_n^2)^s$ we then obtain

$$\|u\|^2_{(s)} = \sum_{|\alpha| \leqq s} c_\alpha \|D^\alpha u\|^2_{L^2} \tag{7.9.2}$$

where c_α are polynomial coefficients. Thus $H_{(s)}$ is the space of all $u \in L^2$ such that $D^\alpha u \in L^2$ when $|\alpha| \leqq s$. (If L^2 is replaced by L^p here we obtain the general *Sobolev spaces* usually denoted by W^p_s.) Next let $0 < s < 1$. Then

$$\|u\|^2_{(s)} \leqq (2\pi)^{-n} \int |\hat{u}(\xi)|^2 (1+|\xi|^{2s})\, d\xi \leqq 2\|u\|^2_{(s)}, \tag{7.9.3}$$

and we shall also prove that

$$(2\pi)^{-n} \int |\hat{u}(\xi)|^2 (1+|\xi|^{2s})\, d\xi = \int |u|^2 dx + A_s \iint |u(x)-u(y)|^2 |x-y|^{-n-2s} dx\, dy. \tag{7.9.4}$$

This implies that $H_{(s)}$ consists of all L^2 functions for which the right-hand side of (7.9.4) is finite. In view of (7.9.1) the estimates (7.9.3) are equivalent to the elementary inequalities

$$(1+|\xi|^2)^s \leqq 1+|\xi|^{2s} \leqq 2(1+|\xi|^2)^s, \qquad 0<s<1.$$

(The second one is trivial and the first follows if we divide by $(1+|\xi|^2)^s$ and note that $a^s \geqq a$ if $0 \leqq a \leqq 1$ and $0<s<1$.) To prove (7.9.4) we use Parseval's formula and obtain

$$\begin{aligned}\iint |u(x)-u(y)|^2 |x-y|^{-n-2s} dx\,dy &= \iint |u(x+y)-u(y)|^2 |x|^{-n-2s} dx\,dy \\ &= (2\pi)^{-n} \iint |e^{i\langle x,\xi\rangle}-1|^2 |x|^{-n-2s} |\hat{u}(\xi)|^2 dx\,d\xi\end{aligned}$$

for the Fourier transform of $y \to u(x+y)-u(y)$ is $\hat{u}(\xi)(e^{i\langle x,\xi\rangle}-1)$. Now

$$A_s^{-1} = |\xi|^{-2s} \int |e^{i\langle x,\xi\rangle}-1|^2 |x|^{-n-2s} dx$$

is independent of ξ, for it is obviously a function of $|\xi|$ only and if we replace x by tx we find that the value at ξ is equal to the value at $t\xi$. This proves (7.9.3) and (7.9.4). (It is easy to see that $A_s/s(1-s)$ has finite limits as $s \to 0$ or $s \to 1$.) If $0<s<1$ and k is a positive integer it follows now as in our discussion of $H_{(k)}$ above that $H_{(s+k)}$ consists of all $u \in L^2$ such that $D^\alpha u \in L^2$ when $|\alpha| \leqq k$ and

$$\iint |D^\alpha u(x)-D^\alpha u(y)|^2 |x-y|^{-n-2s} dx\,dy < \infty, \qquad |\alpha|=k. \tag{7.9.5}$$

The norm $\|u\|_{(s+k)}$ in $H_{(s+k)}$ is equivalent to

$$\sum_{|\alpha| \leqq k} \|D^\alpha u\|_{L^2} + \sum_{|\alpha|=k} \left(\iint |D^\alpha u(x)-D^\alpha u(y)|\,|x-y|^{-n-2s} dx\,dy\right)^{\frac{1}{2}}.$$

Having described the $H_{(s)}$ spaces with $s \geqq 0$ we observe that $\mathscr{S}$ is dense in L^2_s, hence in $H_{(s)}$, for every s. If $u \in \mathscr{S}'$ it follows that

$$\sup_{\phi \in \mathscr{S}} |(u,\phi)|/\|\phi\|_{(-s)} = \sup_{\phi \in \mathscr{S}} |(2\pi)^{-n}(\hat{u},\hat{\phi})|/((2\pi)^{-n} \int |\hat{\phi}(\xi)|^2 (1+|\xi|^2)^{-s} d\xi)^{\frac{1}{2}}$$

is finite if and only if $u \in H_{(s)}$ and then it is equal to $\|u\|_{(s)}$. Thus $H_{(s)}$ is the dual of $H_{(-s)}$ which gives a description also when $s<0$.

The space C^γ of Hölder continuous functions in $\mathbb{R}^n$ of order $\gamma \in (0,1)$ was defined in Theorem 4.5.12 to be the space of continuous functions such that for every compact set K

$$\sup_{x,y \in K} |u(x)-u(y)|/|x-y|^\gamma < \infty.$$

We shall say that $u \in C^{k+\gamma}$, where k is a positive integer and $0<\gamma<1$, if $u \in C^k$ and for every compact set K

$$\sup_{x,y \in K} |D^\alpha u(x)-D^\alpha u(y)|/|x-y|^\gamma < \infty, \qquad |\alpha|=k.$$

It is clear that the subspace $C_0^{k+\gamma}$ of elements in $C^{k+\gamma}$ with compact support satisfies (7.9.5) if $s<\gamma$. Thus

(7.9.6) $$C_0^{k+\gamma}\subset H_{(s)} \quad \text{if } 0<s<k+\gamma.$$

If $u\in H_{(-s)}$ and $0<s<k+\gamma$ we obtain

$$|(u,\phi)|\leqq \|u\|_{(-s)}\|\phi\|_{(s)}\leqq C\sum_{|\alpha|=k}\sup|D^\alpha\phi(x)-D^\alpha\phi(y)|/|x-y|^\gamma,$$

when $\phi\in C_0^\infty(K)$ where K is a compact set in $\mathbb{R}^n$. From the obvious inclusion $C_0^k\subset H_{(k)}$ we also conclude that

(7.9.7) $$H_{(-s)}\subset\mathscr{D}'^k \quad \text{if } s\leqq k \text{ and } k \text{ is an integer.}$$

Next we shall relate the Fourier transform L_s^2 of $H_{(s)}$ to L^p spaces.

Lemma 7.9.2. *$L_s^2\subset L^q$ if and only if $q=2$ and $s\geqq 0$ or $1\leqq q<2$ and $s>n(1/q-1/2)$.*

Proof. The case $q=2$ is obvious. It is also clear that $L_s^2\subset L^q$ implies $q\leqq 2$, for all functions in L_s^2 are not in L_{loc}^q when $q>2$. When $1\leqq q<2$ we obtain by Hölder's inequality

$$\begin{aligned}\int|v|^q d\xi=\int(|v|^2)^{q/2}d\xi &\leqq \left(\int|v|^2(1+|\xi|^2)^s d\xi\right)^{q/2}\left(\int(1+|\xi|^2)^{-qs/(2-q)}d\xi\right)^{1-q/2}\\ &\leqq C\left(\int|v|^2(1+|\xi|^2)^s d\xi\right)^{q/2}\end{aligned}$$

if $2qs/(2-q)>n$, that is, $s>n(1/q-1/2)$. If $s=n(1/q-1/2)$ we can take $v=(1+|\xi|^2)^{-n/2q}(\log(2+|\xi|))^{-a}$ and obtain $v\in L^q$ if and only if $qa>1$, and $v\in L_s^2$ if and only if $2a>1$. When we take $a\in(1/2,1/q)$ we find that L_s^2 is not contained in L^q.

Theorem 7.9.3. *The Fourier transform of $H_{(s)}$ is contained in L^q if $1\leqq q<2$ and $s>n(1/q-1/2)$. The Fourier transform of L^p is contained in $H_{(-s)}$ if $2<p\leqq\infty$ and $s>n(1/2-1/p)$.*

Proof. The first statement follows immediately from Lemma 7.9.2. To prove the second one let $u\in L^p$ and note that when $\phi\in\mathscr{S}$

$$|\langle\hat{u},\phi\rangle|=|\langle u,\hat{\phi}\rangle|\leqq\|u\|_{L^p}\|\hat{\phi}\|_{L^{p'}}\leqq C\|\phi\|_{(s)}$$

by the first part of the proof, so $\hat{u}\in H_{(-s)}$.

If we combine the second part of Theorem 7.9.3 with (7.9.7) we conclude that for $p>2$ the Fourier transform of L^p is in $\mathscr{D}'^j$ if $j>n(1/2-1/p)$. (Recall that by Theorem 7.6.6 this would be false if $j<n(1/2-1/p)$.) We have actually proved a great deal more for there is a considerable margin in the inclusion (7.9.7).

A particularly important special case of Theorem 7.9.3 is the following Bernstein theorem:

Corollary 7.9.4. *The Fourier transform of $H_{(s)}$ is contained in L^1 if $s>n/2$, and $H_{(s)}$ is then contained in the space of continuous functions on $\mathbb{R}^n$ tending to 0 at ∞.*

Corollary 7.9.4 is of course a slightly stronger version of Lemma 7.6.3. In estimates such as (7.6.10) we could therefore have used $H_{(s)}$ norms for any $s>n/2$ instead of the smallest integer $s>n/2$ as we actually did.

Our discussion so far shows that one cannot express the L^p norm of u very well in terms of the Fourier transform $\hat{u}$. To prove continuity of maps in L^p spaces one can therefore seldom use Fourier transforms except in L^2. However, we shall now prove some rather precise estimates supplementing those in Section 4.5 by combining the methods used there with the Fourier transformation in L^2.

Theorem 7.9.5. *Let $k\in\mathscr{S}'(\mathbb{R}^n)$ and assume that $\hat{k}\in L^1_{\mathrm{loc}}$,*

$$\sum_{|\alpha|\leqq s}\int_{R/2<|\xi|<2R} |R^{|\alpha|}D^\alpha \hat{k}(\xi)|^2\,d\xi/R^n \leqq C<\infty, \qquad R>0, \tag{7.9.8}$$

where s is an integer $>n/2$. Then it follows that for $1<p<\infty$

$$\|k*u\|_{L^p}\leqq C_p\|u\|_{L^p}, \qquad u\in L^p\cap\mathscr{E}'. \tag{7.9.9}$$

In addition

$$\tau m\{x;\,|k*u(x)|>\tau\}\leqq C\|u\|_{L^1}, \qquad u\in L^2\cap\mathscr{E}'. \tag{7.9.10}$$

Proof. Choose a function $\psi\in C_0^\infty(\{\xi;\,|\xi|\leqq 2\})$ which is equal to 1 when $|\xi|\leqq 1$. Then we have for $\xi\neq 0$

$$1=\sum_{-\infty}^{\infty}(\psi(2^{-j}\xi)-\psi(2^{1-j}\xi)), \tag{7.9.11}$$

which we shall use to decompose $\hat{k}$. If we set

$$\hat{k}_R(\xi)=(\psi(\xi)-\psi(2\xi))\,\hat{k}(R\xi)$$

it follows from (7.9.8) that (with another C)

$$\sum_{|\alpha|\leqq s}\int |D^\alpha \hat{k}_R(\xi)|^2\,d\xi\leqq C. \tag{7.9.12}$$

Hence $\sup|\hat{k}_R|\leqq C'$ by Lemma 7.6.3, so $|\hat{k}(R\xi)|\leqq C'$ when $|\xi|=1$, which means that

$$|\hat{k}(\xi)|\leqq C' \tag{7.9.13}$$

when $\xi \neq 0$. Since we have assumed that $\hat{k} \in L^1_{\text{loc}}$ it follows that $\hat{k} \in L^\infty$. Parseval's formula now gives (7.9.9) when $p=2$ with $C_2 = C'$. Moreover, Corollary 7.9.4 shows that $\hat{k}_R$ is the Fourier transform of a function $k_R \in L^1$ with $\|k_R\|_{L^1} \leqq C''$ and $k_R \in C^\infty$. More precisely we have

$$\int |k_R(x)|^2 (1+|x|^2)^s \, dx \leqq C_3$$

so Cauchy-Schwarz' inequality gives

$$\int_{|x|>t} |k_R(x)| \, dx \leqq C_4 t^{n/2-s}. \tag{7.9.14}$$

Bounds of the same form are valid for $\xi_j \hat{k}_R$, hence for $D_j k_R$, so we have

$$\int |k'_R(x)| \, dx \leqq C_5 \tag{7.9.15}$$

which implies

$$\int |k_R(x+y) - k_R(x)| \, dx \leqq C_5 |y|. \tag{7.9.16}$$

We are now ready to prove the analogue of (4.5.16),

$$\int_{\complement I^*} |k * w| \, dx \leqq C \int |w| \, dx \quad \text{if } w \in C_0^\infty(I) \text{ and } \int w \, dx = 0. \tag{7.9.17}$$

Here I is a cube and I^* the "doubled cube" as in Lemma 4.5.6. We may assume in the proof that the center is at 0 and that the norm in $\mathbb{R}^n$ is the maximum norm so that I is defined by $|x|<t$ and I^* by $|x|<2t$. Since the Fourier transform of $k_R(Rx) R^n$ is $\hat{k}_R(\xi/R)$ it follows from (7.9.11) that

$$k = \sum_{-\infty}^{\infty} k_{2^j}(2^j \, .) \, 2^{nj}$$

with $\mathscr{S}'$ convergence. Now (7.9.14) and (7.9.16) give since $\operatorname{supp} w \subset I$

$$\int_{x \notin I^*} R^n |k_R(.R) * w| \, dx \leqq \int |w| \, dx \int_{|x|>t} |k_R(Rx)| \, d(Rx) \leqq C \int |w| \, dx \, (tR)^{n/2-s},$$

$$\int_{x \notin I^*} R^n |k_R(.R) * w| \, dx \leqq \iint |(k_R((x-y)R) - k_R(xR)) \, w(y)| R^n \, dx \, dy$$
$$\leqq C \int |w| \, dx \, tR.$$

Hence the triangle inequality gives

$$\int_{x \notin I^*} |k * w| \, dx \leqq C \int |w| \, dx \Big(\sum_{2^j t \geqq 1} (2^j t)^{n/2-s} + \sum_{2^j t < 1} 2^j t \Big) \leqq C'' \int |w| \, dx$$

which proves (7.9.17).

We shall now prove (7.9.10). In doing so we decompose u according to Lemma 4.5.5 with s replaced by τ (since s has a different meaning now). All terms are in L^2. Since (7.9.9) is already proved when $p=2$ we have

$$\tau^2 m\{x;\, |k*v(x)|>\tau/2\} \leqq 4\|k*v\|_{L^2}^2 \leqq C\|v\|_{L^2}^2 \leqq C'\tau\|v\|_{L^1}.$$

If $O=\bigcup I_k^*$ then $\tau m(O)\leqq 2^n \|u\|_{L^1}$ by (4.5.13), and (7.9.17) gives

$$m\{x;\, x\notin O, \sum |k*w_j(x)|>\tau/2\}\tau/2 \leqq \int_{\complement O} \sum |k*w_j|\,dx \leqq C\int \sum |w_j|\,dx \leqq 3C\|u\|_{L^1}.$$

Since $|k*u(x)|\leqq\tau$ unless $|k*v(x)|>\tau/2$ or $x\in O$ or $x\notin O$ and $\sum|k*w_j(x)| >\tau/2$, we have proved the weak type estimate (7.9.10).

It suffices to prove (7.9.9) when $u\in C_0^\infty$. If (7.9.9) is known for some p then it follows for the conjugate exponent p', $1/p+1/p'=1$. In fact

$$|k*u*v(0)|=|k*v*u(0)|\leqq\|k*v\|_{L^p}\|u\|_{L^{p'}}\leqq C\|v\|_{L^p}\|u\|_{L^{p'}}$$

when u, $v\in C_0^\infty$. This implies $k*u\in L^{p'}$ and that (7.9.9) is valid with p replaced by p'. Thus we may assume $1<p<2$. The proof is then a case of the Marcinkiewicz interpolation theorem like that of Theorem 4.5.3. For $\tau>0$ we write $u=u_\tau+U_\tau$ where $u_\tau=u$ when $|u|<\tau$ and $U_\tau=u$ when $|u|\geqq\tau$. Both terms are then in L^2 and

$$\begin{aligned} m\{x;\, |k*u(x)|>\tau\} &\leqq m\{x;\, |k*u_\tau(x)|>\tau/2\}+m\{x;\, |k*U_\tau(x)|>\tau/2\}\\ &\leqq C(\tau^{-2}\|u_\tau\|_{L^2}^2+\tau^{-1}\|U_\tau\|_{L^1}), \end{aligned}$$

by (7.9.9) with $p=2$ and by (7.9.10). Hence

$$\begin{aligned} \|k*u\|_{L^p}^p &= p\int_0^\infty \tau^{p-1} m\{x;\, |k*u(x)|>\tau\}\,d\tau\\ &\leqq C\Big(\iint_{|u(x)|<\tau} |u(x)|^2\tau^{p-3}\,dx\,d\tau+\iint_{|u(x)|\geqq\tau} |u(x)|\,\tau^{p-2}\,dx\,d\tau\Big)\\ &= C((2-p)^{-1}+(p-1)^{-1})\int |u(x)|^p\,dx \end{aligned}$$

which completes the proof.

A partly parallel but much more elementary argument gives estimates in Hölder spaces also. Set for $0<\gamma<1$

$$|u|_\gamma=\sup_{x\neq y}|u(x)-u(y)|/|x-y|^\gamma.$$

Theorem 7.9.6. *If k satisfies the hypotheses of Theorem 7.9.5 and $0<\gamma<1$, then*

$$|k*u|_\gamma\leqq C_\gamma|u|_\gamma,\qquad u\in C_0^\gamma(\mathbb{R}^n). \tag{7.9.18}$$

Proof. By using the partition of unity (7.9.11) we can decompose u,

$$u=\sum u_j,\qquad \hat u_j(\xi)=\hat\phi(\xi/2^j)\,\hat u(\xi).$$

Here we have written $\hat{\phi}(\xi)=\psi(\xi)-\psi(2\xi)$, so $\phi\in\mathscr{S}$. Explicitly

$$u_j(x)=\int u(x-2^{-j}y)\,\phi(y)\,dy=\int (u(x-2^{-j}y)-u(x))\,\phi(y)\,dy$$

since

$$\int \phi(y)\,dy=\hat{\phi}(0)=0.$$

Hence

$$|u_j|\leqq C2^{-\gamma j}|u|_\gamma, \tag{7.9.19}$$

and since

$$\partial_\nu u_j(x)=2^j\int u(x-2^{-j}y)\,\partial_\nu\phi(y)\,dy$$

we also have

$$|u_j'|\leqq C2^{(1-\gamma)j}|u|_\gamma. \tag{7.9.20}$$

Choose $\chi\in C_0^\infty(\mathbb{R}^n\setminus 0)$ equal to 1 in $\operatorname{supp}\hat{\phi}$. Then

$$\hat{k}(\xi)\,\hat{u}_j(\xi)=\hat{k}(\xi)\,\chi(\xi/2^j)\,\hat{u}_j(\xi).$$

The estimate (7.9.12) is valid for $\hat{k}(R\xi)\,\chi(\xi)$, uniformly in R, so this is the Fourier transform of a function k_R with uniformly bounded L^1 norm. With $R=2^j$ we have

$$k*u_j=R^n k_R(R.)*u_j.$$

Hence it follows from (7.9.19), (7.9.20) that

$$|k*u_j|\leqq C'2^{-\gamma j}|u|_\gamma, \tag{7.9.19$'$}$$

$$|k*u_j'|\leqq C'2^{(1-\gamma)j}|u|_\gamma. \tag{7.9.20$'$}$$

Combining (7.9.20)′ with the mean value theorem we obtain

$$\begin{aligned}|k*u(x)-k*u(y)|&\leqq\sum|k*u_j(x)-k*u_j(y)|\\&\leqq C'|u|_\gamma\Big(2\sum_{2^{-j}<|x-y|}2^{-\gamma j}+|x-y|\sum_{2^{-j}\geqq|x-y|}2^{(1-\gamma)j}\Big)\\&\leqq C''|u|_\gamma|x-y|^\gamma/(\gamma(1-\gamma)).\end{aligned}$$

The proof is complete.

The following is a typical application of Theorems 7.9.5 and 7.9.6 in the theory of partial differential equations.

Theorem 7.9.7. *Let $P(D)$ be an elliptic differential operator in $\mathbb{R}^n$ of order m. If $X\subset\mathbb{R}^n$ is an open set and $u\in\mathscr{D}'(X)$, then $P(D)u\in L^p_{\text{loc}}(X)$ implies $D^\alpha u\in L^p_{\text{loc}}(X)$ when $|\alpha|=m$, if $1<p<\infty$, and $P(D)u\in C^\gamma(X)$ implies $D^\alpha u\in C^\gamma(X)$ when $|\alpha|=m$ if $0<\gamma<1$.*

Proof. Let E be the parametrix constructed in Theorem 7.1.22. Since $\hat{E}\in C^\infty$ and (7.1.21) gives

$$|\xi^\beta D^\alpha \hat{E}(\xi)| \leqq C_{\alpha\beta}|\xi|^{|\beta|-m-|\alpha|}$$

the hypotheses of Theorems 7.9.5 and 7.9.6 are fulfilled by $D^\beta E$ when $|\beta|=m$. Let Y be a relatively compact open subset of X and choose $\chi\in C_0^\infty(X)$ equal to 1 in Y. Then

$$P(D)(\chi u)=f_1+f_2$$

where $f_1=\chi P(D)u\in L^p$ (resp. C^γ) and $f_2=0$ in Y. Thus

$$\chi u+\omega*(\chi u)=E*f_1+E*f_2$$

so $D^\alpha u-(D^\alpha E)*f_1$ is a C^∞ function in Y. Since $D^\alpha E*f_1\in L^p$ (resp. C^γ) when $|\alpha|=m$ by Theorems 7.9.5 and 7.9.6, the proof is complete.

The conclusion in Theorem 7.9.7 is of course also valid when $|\alpha|<m$ but combination of Theorem 7.9.7 and Theorem 4.5.13 gives a better result then. The following theorem shows that it is not possible to extend Theorem 7.9.7 to the excluded limiting cases. The result justifies a remark made in the introduction concerning the flaws in the classical notion of solution of a differential equation.

Theorem 7.9.8. *Let X be an open set in $\mathbb{R}^n$, $n>1$, and let $P(D)$ be a differential operator with constant coefficients of order $m>0$. Then one can find $u\in C_0^{m-1}(X)\setminus C_0^m(X)$ so that $P(D)u\in C_0^0(X)$.*

Proof. We may assume that $0\in X$. Let $K=\{x;\ |x|\leqq R\}\subset X$ be a compact ball with center at 0. The set of all $u\in C_0^{m-1}(K)$ with $P(D)u\in C_0^0(K)$ is a Banach space with the norm

$$\sup|P(D)u|+\sum_{|\alpha|<m}\sup|D^\alpha u|.$$

If all such functions are in C_0^m it follows from Banach's theorem that

$$\sum_{|\alpha|=m}\sup|D^\alpha u|\leqq C(\sup|P(D)u|+\sum_{|\alpha|<m}\sup|D^\alpha u|),\qquad u\in C_0^m(K).$$

If P_m is the principal symbol of P we conclude that

$$\sum_{|\alpha|=m}\sup|D^\alpha u|\leqq C'(\sup|P_m(D)u|+\sum_{|\alpha|<m}\sup|D^\alpha u|),\qquad u\in C_0^m(K). \tag{7.9.21}$$

Let $U\in C^\infty(\mathbb{R}^n)$ be a solution of the equation $P_m(D)U=0$, choose $\chi\in C_0^\infty(K)$ so that $\chi(x)=1$ when $|x|<R/2$ and set

$$u_j(x)=2^{-mj}(\chi U)(2^j x).$$

We have

$$\sup|D^\alpha u_j| \leqq C 2^{(|\alpha|-m)j}, \qquad |\alpha| \leqq m,$$

and $R/2 \leqq 2^j |x| \leqq R$ in $\operatorname{supp} P_m(D) u_j$ so these supports are disjoint. If we apply (7.9.21) to $u = \sum_0^N u_j$ it follows that

$$N \sum_{|\alpha|=m} |D^\alpha U(0)| \leqq C.$$

When $N \to \infty$ we conclude that $D^\alpha U(0) = 0$, $|\alpha| = m$, for every $U \in C^\infty$ satisfying the equation $P_m(D) U = 0$. Taking $U(x) = e^{i\langle x, \zeta\rangle}$ we find that $P_m(\zeta) = 0$ implies $\zeta = 0$. This is not possible when $n > 1$, which proves the theorem.

Remark. If Lemma 7.3.7 is used at the end of the proof one can conclude that for a given α with $|\alpha| = m$ there is a function $u \in C_0^{m-1}$ with $P(D) u \in C_0^0$ and $D^\alpha u \notin C_0^0$ unless $P_m(D)$ is a multiple of D^α. By a simple category argument it follows that u can be chosen so that $D^\alpha u \notin C_0^0$ for every α with $|\alpha| = m$ such that $P_m(D)$ is not a multiple of D^α.

Notes

The basic facts on the Fourier transformation in Section 7.1 go back to Fourier in a more or less precise form. However, the idea of Schwartz to start from the dense function space $\mathscr{S}$ meant a great simplification also of the classical foundations of the subject. The dual definition of the Fourier transformation in $\mathscr{S}'$ absorbed a number of earlier generalizations, in particular that of Bochner [1], while preserving the classical ease of calculation. Apart from this and the precision derived from the Lebesgue integral the first result in Section 7.1 dating from this century is Theorem 7.1.13 for which the Fourier series analogue is due to Young for even integer p' and to Hausdorff [1] in general. Theorem 7.1.12 and its application to the Hausdorff-Young theorem for Fourier series as well as Fourier integrals is due to M. Riesz [2], but our proof of Theorem 7.1.12 is due to Thorin [1]. The best possible constant in Theorem 7.1.13 has been found by Beckner [1]. Fourier transforms of homogeneous distributions were discussed in much greater detail by Gelfand and Shilov [2] and by Gårding [4]. Theorem 7.1.24 is very close to the Herglotz-Petrovsky formula as given in Atiyah, Bott and Gårding [1] (see Section 12.6). Fourier transforms of densities on manifolds in $\mathbb{R}^n$ occur frequently in scattering theory and will be useful in such contexts in Chapter XIV. Theorem 7.1.26 to Corollary 7.1.30 can essentially be found in Hörmander [32] and Agmon and Hörmander [1] with references to earlier literature.

The term Paley-Wiener-Schwartz theorem for Theorem 7.3.1 is well established although perhaps not quite appropriate; Paley and Wiener [1] actually proved a case of Theorem 7.4.2 dealing with Fourier-Laplace transforms of L^2 functions on a half line and Schwartz [5] proved Theorems 7.4.2 and 7.4.3 in full. Theorems 7.3.2 and 7.3.6 are from Malgrange [1]. Asgeirsson [1] proved Theorem 7.3.4 in a quite different way, the refinement is due to Lewy [2]. Theorem 7.3.8 was first stated and proved in the predecessor of this book, but the main idea comes from Ehrenpreis [2] (see also Malgrange [1] and Hörmander [14]). Theorem 7.3.10 is due to Ehrenpreis [1] and Malgrange [1] who proved it by duality. A constructive proof was already attempted by Cauchy [1]; unfortunately it involves Fourier transforms of exponentially increasing functions but this flaw is not hard to correct. For large classes of differential operators fundamental solutions were constructed long ago (see e.g. Fredholm [1], Herglotz [1], Zeilon [1]). A general construction yielding good regularity properties was given in the predecessor of this book. (See Agranovich [1] for an earlier one.) The improved version here was published by Hörmander [29].

For a history of the Weierstrass preparation theorem and criticism of current terminology and general trends in mathematics we refer to Siegel [1]. For earlier proofs of the Malgrange preparation theorem the reader should consult Malgrange [6], Mather [1] and Nirenberg [3]. The proof given here is close to that of Mather [1]. Theorem 7.5.13 is due to Chester, Friedmann and Ursell [1] and Levinson [1] in the analytic case. The theory of unfolding of singularities has given a great simplification and an extension to C^∞ by means of the Malgrange preparation theorem. (See Guillemin and Schaefer [1], Duistermaat [1] and the references there.) We have proved the earlier results here in this spirit.

The Fourier transform of Gaussians is of course classical. As examples we have given the fundamental solution of the Kolmogorov equation (see Kolmogorov [1]) which is important in the theory of Brownian motion. The operator (7.6.14) is one of the simplest normal forms of hyperbolic operators with double characteristics (cf. Hörmander [36]). The Airy function was introduced by Airy [1] to study light near a caustic (see Section 12.2). The proof by Stokes [1] of the asymptotic expansion of Ai might be considered as a forerunner of the method of stationary phase in Section 7.7. This has been very popular with the physicists in this century under the name of the (J)WKB method for (Jeffreys), Wentzel, Kramers, Brillouin, or sometimes, from a somewhat different analytical point of view, the name of the saddle point method. We refer to Fröman [1] for a systematic discussion. For several variables the method seems to have appeared first in Hlawka [1] who proved Theorems 7.7.14, 7.7.16 and Corollary 7.7.15. The presentation here is close to Hörmander [26, 34] up to and including Theorem 7.7.6 although we have avoided using the Morse lemma here. It can be found in a suitable form in Hörmander [26]. The extension to complex valued phase functions is due to Melin and Sjöstrand [1]. They used the notion of almost analytic continuation, which is a systematic development of the arguments used here in the proof of Theorem 3.1.15. The Malgrange preparation theorem was proved with such techniques by Nirenberg [3]. It is therefore not surprising that we have been able to replace the almost analytic machinery by the Malgrange preparation theorem. However, the reader should be aware that the methods of Melin and Sjöstrand are more precise from the point of view of the number of derivatives required. Theorem 7.7.18 goes back to Airy [1]. A complete proof in the analytic case was given by Chester, Friedmann and Ursell [1]; see also Ludwig [2]. Here we have followed the simpler and more general modern approach of Guillemin and Schaeffer [1], Duistermaat [1] who built on the progress in singularity theory by Thom, Arnold and others.

The notion of oscillatory integral in Section 7.8 was introduced as here in Hörmander [26]. It will be convenient particularly in Chapter XXV to be able to use this suggestive notation which is of course common in applied mathematics without precise definitions.

As already mentioned in the text the spaces $H_{(s)}$ are special cases of the spaces of Sobolev [2] when s is a non-negative integer. For negative integers s they arose in the theory of partial differential equations in connection with duality methods (see e.g. Lax [2]) and for half integers in connection with boundary problems (see e.g. Aronszajn [1]). They have been very much generalized and studied during the last decades. Examples of these more general Besov spaces will be encountered in Chapters XIV and XXX but we refer to Peetre [4] for a general discussion. Theorem 7.9.5 is essentially due to Mihlin [1]. In the form given here it was proved in Hörmander [13]. However, the prototype is the theorem of M. Riesz [3] on conjugate functions and its n dimensional generalization by Calderón and Zygmund [1] whose proof is followed here. We refer the reader who wants to study these matters further to Stein [1]. In the theory of linear partial differential equations to which this book is devoted we shall have very few occasions to refer to the L^p or Hölder theory at all.

Chapter VIII. Spectral Analysis of Singularities

Summary

In Chapter VII we have seen that a distribution u of compact support is smooth if and only if the Fourier transform $\hat{u}$ is rapidly decreasing. If u is not smooth we can use the set of directions where $\hat{u}$ is not rapidly decreasing to describe which are the high frequency components of u causing the singularities. This analysis turns out to have an invariant and local character. For a distribution $u \in \mathscr{D}'(X)$ on a C^∞ manifold X we are therefore led to define a set

$$WF(u) \subset T^*(X) \setminus 0$$

with projection in X equal to sing supp u, which is conic with respect to multiplication by positive scalars in the fibers of $T^*(X)$. We call it the *wave front set* of u by analogy with the classical Huyghens construction of a propagating wave. In this construction one assumes that the location and oriented tangent plane of a wave is known at one instant of time and concludes that at a later time it has been translated in the normal direction with the speed of light. The data are thus precisely rays in the cotangent bundle.

The advantages of the notion of wave front set are manifold. First of all it allows one to extend a number of operations on distributions. For example, the restriction of $u \in \mathscr{D}'(X)$ to a submanifold Y of X can always be defined when the normal bundle of Y does not meet $WF(u)$, that is, high frequency components of u remain of high frequency after restriction to Y. Secondly, differential operators and to some extent their fundamental solutions are local even with respect to the wave front set. This leads to important simplifications in their study known as microlocal analysis.

Section 8.1 gives the basic definitions of the wave front set and some important examples. In Section 8.2 we then reconsider the operations defined in Chapters III–VI from our new point of view. Thus we obtain extended definitions of composition and multiplication as well as more precise information on the singularities of the results of these operations. In Section 8.3 we prove the simplest facts on the wave front set of solutions of linear partial differential equations, in particular that the wave front set is included in the union of the characteristic set and the wave front set of the right-hand side. Note that since the characteristic set $\subset T^*(X)\setminus 0$ usually projects onto X it is not possible to give a satisfactory statement of this result without the notion of wave front set. When the principal part is real and has constant coefficients we also show that the wave front set is invariant under the bicharacteristic flow, which in the case of the wave equation reduces to the Huyghens construction above and so justifies our terminology.

One can also consider a stricter classification of singularities, such as the set $\operatorname{sing\,supp}_A u$ of points where u is not a real analytic function. This set too admits a spectral decomposition to a set $WF_A(u)\subset T^*(X)\setminus 0$, which is defined in Section 8.4 and studied in Sections 8.5 and 8.6. In particular this notion allows one to state a more precise form of the uniqueness of analytic continuation: If a distribution vanishes on one side of a C^1 hypersurface Y and the normal of Y at y is not in $WF_A(u)$, then u vanishes in a neighborhood of y. In other words, the normals of the boundary of $\operatorname{supp} u$ must be in $WF_A(u)$ where the boundary is in C^1. In Section 8.5 we also give a notion of normal to the boundary of a general closed set making this statement valid in general. This concept is discussed geometrically at some length in preparation for some later applications. The first comes already in Section 8.6 where various generalizations of the theorem of Holmgren on unique continuation of solutions to partial differential equations with analytic coefficients are given.

In Section 8.7 finally we discuss the analytic wave front set for distributions obtained as limits of $F(x+iy)^{-1}$ where F is analytic and $y\to 0$ in a cone such that the zeros of F are only encountered in the limit. The results are useful in the study of the Cauchy problem in Chapter XII.

8.1. The Wave Front Set

If $v\in\mathscr{E}'(\mathbb{R}^n)$ we can decide whether v is in C_0^∞ by examining the behavior of the Fourier transform $\hat v$ at ∞. In fact, if $v\in C_0^\infty(\mathbb{R}^n)$ then

$$|\hat v(\xi)|\leqq C_N(1+|\xi|)^{-N},\qquad N=1,2,\ldots,\ \xi\in\mathbb{R}^n, \tag{8.1.1}$$

by Lemma 7.1.3. Conversely, if (8.1.1) is fulfilled then $v\in C_0^\infty$ by Fourier's inversion formula (7.1.4). (See also Theorem 7.3.1.) For a general $v \in \mathscr{E}'$ we have defined $\operatorname{sing\,supp} v$ as the set of points having no neighborhood where v is in C^∞ (Definition 2.2.3). Similarly we can introduce the cone $\Sigma(v)$ of all $\eta \in \mathbb{R}^n \setminus 0$ having no conic neighborhood V such that (8.1.1) is valid when $\xi \in V$. It is clear that $\Sigma(v)$ is then a closed cone in $\mathbb{R}^n \setminus 0$, and we have $\Sigma(v) = \emptyset$ if and only if $v \in C_0^\infty$.

While $\operatorname{sing\,supp} v$ only describes the location of the singularities, the cone $\Sigma(v)$ describes only the directions of the high frequencies causing them. We can combine the two types of information by using the following lemma.

Lemma 8.1.1. *If* $\phi \in C_0^\infty(\mathbb{R}^n)$ *and* $v \in \mathscr{E}'(\mathbb{R}^n)$ *then*

$$\Sigma(\phi v) \subset \Sigma(v). \tag{8.1.2}$$

Proof. The Fourier transform of $u = \phi v$ is the convolution

$$\hat{u}(\xi) = (2\pi)^{-n} \int \hat{\phi}(\eta)\, \hat{v}(\xi - \eta)\, d\eta$$

where $\hat{\phi} \in \mathscr{S}$. For some $M \geqq 0$ we have

$$|\hat{v}(\xi)| \leqq C(1 + |\xi|)^M.$$

Let $0 < c < 1$ and split the integral into the parts where $|\eta| < c|\xi|$ and $|\eta| \geqq c|\xi|$. In the second case $|\xi - \eta| \leqq (1 + c^{-1})|\eta|$. Hence

$$\begin{aligned}(2\pi)^n |\hat{u}(\xi)| \leqq &\sup_{|\eta - \xi| < c|\xi|} |\hat{v}(\eta)|\, \|\hat{\phi}\|_{L^1} \\ &+ C \int_{|\eta| > c|\xi|} |\hat{\phi}(\eta)|(1 + c^{-1})^M (1 + |\eta|)^M\, d\eta.\end{aligned} \tag{8.1.3}$$

If Γ is an open cone where (8.1.1) is valid and Γ_1 is a closed cone $\subset \Gamma \cup \{0\}$ we can choose c so that $\eta \in \Gamma$ if $\xi \in \Gamma_1$ and $|\xi - \eta| < c|\xi|$, for this is obviously possible when $|\xi| = 1$. Since $|\eta| \geqq (1 - c)|\xi|$ it follows from (8.1.3) and (8.1.1) that $\hat{u}$ is rapidly decreasing in Γ_1. In fact, we have for $N \geqq 0$

$$\begin{aligned}\sup_{\Gamma_1} (1 + |\xi|)^N |\hat{u}(\xi)| \leqq &(1 - c)^{-N} \sup_{\Gamma} |\hat{v}(\eta)|(1 + |\eta|)^N \|\hat{\phi}\|_{L^1} \\ &+ C(1 + c^{-1})^{N+M} \int |\hat{\phi}(\eta)|(1 + |\eta|)^{N+M}\, d\eta.\end{aligned} \tag{8.1.3)'$$

The lemma is proved.

If X is an open set in $\mathbb{R}^n$ and $u \in \mathscr{D}'(X)$, we set for $x \in X$

$$\Sigma_x(u) = \bigcap_\phi \Sigma(\phi u); \quad \phi \in C_0^\infty(X), \quad \phi(x) \neq 0. \tag{8.1.4}$$

From Lemma 8.1.1 it follows that

$$\Sigma(\phi u) \to \Sigma_x(u) \quad \text{if } \phi \in C_0^\infty(X),\ \phi(x) \neq 0 \text{ and } \operatorname{supp} \phi \to \{x\}. \tag{8.1.5}$$

In fact, if V is an open cone $\supset \Sigma_x(u)$, the compactness of the unit sphere shows that we can find $\phi_1, \dots, \phi_j \in C_0^\infty(X)$ with

$$\phi_1(x) \dots \phi_j(x) \neq 0, \quad \bigcap_1^j \Sigma(\phi_i u) \subset V.$$

When $\phi \in C_0^\infty(X)$ and $\operatorname{supp} \phi$ is so close to x that $\phi_1 \dots \phi_j \neq 0$ there, we can write $\phi = \psi \phi_1 \dots \phi_j$ with $\psi \in C_0^\infty(X)$ and obtain from (8.1.2)

$$\Sigma(\phi u) \subset \bigcap_1^j \Sigma(\phi_i u) \subset V.$$

This proves (8.1.5) since by definition $\Sigma(\phi u) \supset \Sigma_x(u)$ when $\phi(x) \neq 0$.

In particular (8.1.5) implies that $\Sigma_x(u) = \emptyset$ if and only if $\phi u \in C^\infty$ for some $\phi \in C_0^\infty(X)$ with $\phi(x) \neq 0$, that is, $x \notin \operatorname{sing\,supp} u$.

Definition 8.1.2. If $u \in \mathscr{D}'(X)$, then the closed subset of $X \times (\mathbb{R}^n \setminus 0)$ defined by

$$WF(u) = \{(x, \xi) \in X \times (\mathbb{R}^n \setminus 0);\ \xi \in \Sigma_x(u)\}$$

is called the wave front set of u. The projection in X is $\operatorname{sing\,supp} u$.

The set $WF(u)$ is conic in the sense that it is invariant under multiplication of the second variable by positive scalars. It could therefore be considered as a subset of $X \times S^{n-1}$ where S^{n-1} is the unit sphere.

Proposition 8.1.3. *If $u \in \mathscr{E}'(\mathbb{R}^n)$ then the projection of $WF(u)$ on the second variable is $\Sigma(u)$.*

Proof. The projection W is contained in $\Sigma(u)$ by the definition of $WF(u)$. It is closed since the intersection with the unit sphere is the projection of a compact set in $\mathbb{R}^n \times S^{n-1}$. If V is a conic neighborhood of W then every $x \in \mathbb{R}^n$ has a neighborhood U_x such that

$$\Sigma(\phi u) \subset V \quad \text{if } \phi \in C_0^\infty(U_x).$$

We can cover $\operatorname{supp} u$ by a finite number of such neighborhoods U_{x_j} and choose $\phi_j \in C_0^\infty(U_{x_j})$ with $\Sigma \phi_j = 1$ near $\operatorname{supp} u$. But then it follows that

$$\Sigma(u) = \Sigma(\sum \phi_j u) \subset \bigcup \Sigma(\phi_j u) \subset V$$

which proves the proposition.

Proposition 8.1.3 shows that $WF(u)$ contains all information in $\operatorname{sing\,supp} u$ and in $\Sigma(u)$. However, the projection in Proposition 8.1.3 is of limited interest since it is not invariant under a change of variables.

Theorem 8.1.4. *If X is an open set in $\mathbb{R}^n$ and S a closed conic subset of $X \times (\mathbb{R}^n \setminus 0)$ then one can find $u \in \mathscr{D}'(X)$ with $WF(u) = S$.*

Proof. It is sufficient to prove the statement when $X = \mathbb{R}^n$ for otherwise we can apply this case to the closure of S in $\mathbb{R}^n \times (\mathbb{R}^n \setminus 0)$. Choose a sequence $(x_k, \theta_k) \in S$ with $|\theta_k| = 1$ so that every $(x, \theta) \in S$ with $|\theta| = 1$ is the limit of a subsequence. Let $\phi \in C_0^\infty$ and $\hat{\phi}(0) = 1$. Then

$$u(x) = \sum_1^\infty k^{-2} \phi(k(x - x_k)) e^{ik^3 \langle x, \theta_k \rangle} \tag{8.1.6}$$

is a continuous function in $\mathbb{R}^n$, and we shall prove that $WF(u) = S$. First we prove that $WF(u) \subset S$. If $(x_0, \xi_0) \notin S$ we can choose an open neighborhood U of x_0 and an open conic neighborhood V of ξ_0 such that

$$(U \times V) \cap S = \emptyset. \tag{8.1.7}$$

Write $u = u_1 + u_2$ where u_1 is the sum of the terms in (8.1.6) with $x_k \notin U$ and u_2 the sum of terms with $x_k \in U$. Then $u_1 \in C^\infty$ in a neighborhood U_1 of x_0 because all but a finite number of terms vanish in U_1 if $\bar{U}_1 \subset U$. Now

$$\hat{u}_2(\xi) = \sum_{x_k \in U} k^{-2-n} \hat{\phi}((\xi - k^3 \theta_k)/k) e^{i \langle x_k, k^3 \theta_k - \xi \rangle}. \tag{8.1.8}$$

Here $\theta_k \notin V$ because of (8.1.7). If V_1 is another conic neighborhood of ξ_0 and $\bar{V}_1 \subset V \cup \{0\}$ then $|\xi - \eta| \geqq c(|\xi| + |\eta|)$ when $\xi \in V_1$ and $\eta \notin V$, for some $c > 0$, since this is true when $|\xi| + |\eta| = 1$. Thus

$$|\xi - k^3 \theta_k| \geqq c(|\xi| + k^3) \geqq c|\xi|^{\frac{2}{3}} k, \qquad \xi \in V_1,$$

and since $\hat{\phi} \in \mathscr{S}$ it follows that $\hat{u}_2$ is rapidly decreasing in V_1. Thus (x_0, ξ_0) is not in $WF(u)$.

Now let $(x_0, \xi_0) \in S$. Choose $\chi \in C_0^\infty$ equal to 1 near x_0. To prove that $(x_0, \xi_0) \in WF(u)$ we must show that $\widehat{\chi u}$ cannot decrease rapidly in a conic neighborhood of ξ_0. To do so we first observe that

$$\chi(x) \phi(k(x - x_k)) = \phi_k(k(x - x_k))$$

where $\phi_k(x) = \chi(x/k + x_k) \phi(x)$ belongs to a bounded set in $\mathscr{S}$. The Fourier transform of χu is a sum of the form (8.1.8) with ϕ replaced by ϕ_k. If x_k is close to x_0 and k is large then $\phi_k = \phi$, and we obtain for any N

$$|\widehat{\chi u}(k^3 \theta_k)| \geqq k^{-n-2} - C_N \sum_{j \neq k} j^{-n-2} (|k^3 \theta_k - j^3 \theta_j|/j)^{-N}.$$

Here

$$|k^3 \theta_k - j^3 \theta_j| \geqq |k^3 - j^3| \geqq k^2 + kj + j^2 \geqq kj \quad \text{ if } k \neq j$$

so the sum is $O(k^{-N})$. If we choose $N>n+2$ we obtain for large k that

$$|\widehat{\chi u}(k^3\theta_k)| \geqq k^{-n-2}/2$$

if x_k is close to x_0. Since $(x_0, \xi_0/|\xi_0|)$ is a limit point of the sequence (x_k, θ_k) it follows that $\widehat{\chi u}$ cannot decrease rapidly in a conic neighborhood of ξ_0 and the theorem is proved.

We shall now determine the wave front set for some classes of distributions which occur very frequently.

Theorem 8.1.5. *Let V be a linear subspace of $\mathbb{R}^n$ and $u=u_0\,dS$, where $u_0\in C^\infty(V)$ and dS is the Euclidean surface measure. Then*

$$WF(u)=\operatorname{supp} u \times (V^\perp \setminus 0).$$

Proof. If $\chi\in C_0^\infty$ then

$$(\widehat{\chi u})(\xi)=\int_V e^{-i\langle x,\xi\rangle}\chi(x)u_0(x)\,dS(x).$$

If we write $\xi=\xi'+\xi''$ where $\xi'\in V$ and $\xi''\in V^\perp$, then this is a rapidly decreasing function of ξ' which does not vanish on any open set unless $\chi u=0$. Hence $\widehat{\chi u}$ does not decrease rapidly in any open cone meeting $V^\perp$ unless $\chi u=0$, but there is rapid decrease in every cone where $|\xi|\leqq C|\xi'|$. This proves the assertion.

It would have been sufficient to prove Theorem 8.1.5 for dS itself, for we have always

$$WF(a\,u)\subset WF(u) \quad \text{if } a\in C^\infty. \tag{8.1.9}$$

This follows at once from the definition. Another important general fact is that for all α

$$WF(D^\alpha u)\subset WF(u). \tag{8.1.10}$$

To prove this we take $\chi\in C_0^\infty$ equal to 1 near x and $\chi_1\in C_0^\infty$ equal to 1 in a neighborhood of $\operatorname{supp}\chi$. Then we have

$$\Sigma_x(D^\alpha u)\subset\Sigma(\chi D^\alpha u)=\Sigma(\chi D^\alpha \chi_1 u)\subset\Sigma(D^\alpha\chi_1 u)\subset\Sigma(\chi_1 u).$$

When $\operatorname{supp}\chi_1\to\{x\}$ it follows that (8.1.10) is valid. Summing up, we have

$$WF(Pu)\subset WF(u) \tag{8.1.11}$$

if P is any linear differential operator with C^∞ coefficients.

Next we shall examine the boundary values of analytic functions as defined in Theorem 3.1.15. There Γ is an open convex cone. Let

$$\Gamma^\circ = \{\xi \in \mathbb{R}^n;\ \langle y, \xi\rangle \geqq 0 \text{ for all } y \in \Gamma\} \tag{8.1.12}$$

be the *dual cone*. It is closed, convex and *proper*, that is, it contains no straight line, for Γ would otherwise be contained in a hyperplane and lack interior points. Conversely, every closed convex proper cone Γ° is the dual cone of precisely one open convex cone Γ. It is defined by

$$\Gamma = \{y \in \mathbb{R}^n;\ \langle y, \xi\rangle > 0 \text{ for every } \xi \in \Gamma^\circ \setminus 0\}. \tag{8.1.13}$$

The proof by the Hahn-Banach theorem is very close to that of Theorem 4.3.2 and is left for the reader. Instead we shall prove

Theorem 8.1.6. *If the hypotheses of Theorem 3.1.15 are fulfilled, then*

$$WF(f_0) \subset X \times (\Gamma^\circ \setminus 0) \tag{8.1.14}$$

where Γ° is the dual cone of Γ.

Proof. If $\phi \in C_0^\infty(X)$ the representation (3.1.20) of $\langle f_0, \phi\rangle$ is valid with N replaced by any integer $\nu \geqq N$ provided that N is also replaced by ν in the definition (3.1.18) of Φ. Hence

$$\begin{aligned}(\widehat{\phi f_0})(\xi) &= \langle f_0 e^{-i\langle \cdot, \xi\rangle}, \phi\rangle = \int \Phi(x, Y) f(x+iY) e^{-i\langle x+iY, \xi\rangle} dx \\ &+ (\nu+1) \iint_{0<t<1} f(x+itY) e^{-i\langle x+itY, \xi\rangle} \sum_{|\alpha|=\nu+1} \partial^\alpha \phi(x)(iY)^\alpha/\alpha!\, t^\nu\, dx\, dt.\end{aligned} \tag{8.1.15}$$

When $\langle Y, \xi\rangle < 0$ it follows that

$$\begin{aligned}|\widehat{\phi f_0}(\xi)| &\leqq C_{\phi,\nu}\Big(e^{\langle Y, \xi\rangle} + \int_0^\infty e^{t\langle Y, \xi\rangle} t^{\nu-N} dt\Big) \\ &= C_{\phi,\nu}(e^{\langle Y, \xi\rangle} + (\nu-N)!\,\langle -Y, \xi\rangle^{N-\nu-1}).\end{aligned} \tag{8.1.16}$$

The right-hand side in $O(|\xi|^{N-\nu-1})$ in a conic neighborhood of any point in the half space $\langle Y, \xi\rangle < 0$. Hence $\Sigma(\phi f_0) \subset \{\xi;\ \langle Y, \xi\rangle \geqq 0\}$ for every $Y \in \Gamma$ with $|Y| < \gamma$, so $\Sigma(\phi f_0) \subset \Gamma^\circ$ which proves the theorem.

The hypotheses in the theorem can be weakened in various ways. In particular it is sufficient to assume f analytic for $z \in X_1 + i\Gamma_1$ and $|\mathrm{Im}\, z|$ small when $X_1 \Subset X$ and $\bar{\Gamma}_1 \subset \Gamma \cup \{0\}$. We could also have added to f_0 a C^∞ term since this does not affect $WF(f_0)$. A converse result is then valid, and it will be proved in Section 8.4. We shall also prove then that Theorem 8.1.6 remains valid when singularities are defined as points of non-analyticity.

To prepare for a discussion of the wave front set for homogeneous distributions we shall now prove a modification of Lemma 8.1.1.

Lemma 8.1.7. *If $v \in \mathscr{S}'$ then $WF(v) \subset \mathbb{R}^n \times F$ where F is the limit cone of* $\operatorname{supp} \hat{v}$ *at* ∞, *consisting of all limits of sequences $t_j x_j$ with $x_j \in \operatorname{supp} \hat{v}$ and $0 < t_j \to 0$.*

Proof. F is obviously closed. For every closed cone Γ with $\Gamma \cap F = \{0\}$ we can choose $\varepsilon > 0$ and C so that

$$|\xi - \eta| \geqq \varepsilon |\xi| \quad \text{if } \xi \in \Gamma, \quad \eta \in \operatorname{supp} \hat{v} \text{ and } |\xi| > C.$$

In fact, we could otherwise choose $\xi_j \in \Gamma$ and $\eta_j \in \operatorname{supp} \hat{v}$ so that $|\xi_j - \eta_j| < |\xi_j|/j$ and $|\xi_j| > j$. The sequence $\eta_j / |\xi_j|$ will then have a limit point $\theta \in \Gamma \cap F$ with $|\theta| = 1$ which is a contradiction. If $\phi \in C_0^\infty(\mathbb{R}^n)$ then the Fourier transform of $u = \phi v$ is $(2\pi)^{-n} \hat{\phi} * \hat{v}$. Choose $\psi \in C^\infty(\mathbb{R}^n)$ so that $\psi(\xi) = 1$ when $|\xi| > 1$ and $\psi(\xi) = 0$ when $|\xi| < 1/2$. Then $\Phi_R(\xi) = \hat{\phi}(\xi) \psi(\xi/R)$ is equal to $\hat{\phi}(\xi)$ when $|\xi| \geqq R$, hence

$$(2\pi)^n \hat{u}(\xi) = \hat{v}_\eta(\Phi_R(\xi - \eta)) \quad \text{if } \xi \in \Gamma \text{ and } R \leqq \varepsilon |\xi|, \quad |\xi| > C.$$

Since $\hat{v} \in \mathscr{S}'$ it follows that for some N, C', C'' we have when $\xi \in \Gamma$, $|\xi| > C, R \leqq \varepsilon |\xi|$

$$\begin{aligned} |\hat{u}(\xi)| &\leqq C' \sum_{|\alpha + \beta| \leqq N} \sup |\eta^\alpha D_\eta^\beta \Phi_R(\xi - \eta)| \\ &\leqq C''(1 + |\xi|)^N \sum_{|\alpha + \beta| \leqq N} \sup_{|\eta| > R/2} |\eta^\alpha D^\beta \hat{\phi}(\eta)|. \end{aligned}$$

If we choose $R = \varepsilon |\xi|$ the right-hand side is rapidly decreasing since $\hat{\phi} \in \mathscr{S}$. This proves the lemma.

Theorem 8.1.8. *If $u \in \mathscr{D}'(\mathbb{R}^n)$ is homogeneous in $\mathbb{R}^n \setminus 0$ then*

$$(x, \xi) \in WF(u) \Leftrightarrow (\xi, -x) \in WF(\hat{u}) \quad \text{if } \xi \neq 0 \text{ and } x \neq 0, \tag{8.1.17}$$

$$x \in \operatorname{supp} u \Leftrightarrow (0, -x) \in WF(\hat{u}) \quad \text{if } x \neq 0, \tag{8.1.18}$$

$$\xi \in \operatorname{supp} \hat{u} \Leftrightarrow (0, \xi) \in WF(u) \quad \text{if } \xi \neq 0. \tag{8.1.19}$$

Proof. Assume first that u is homogeneous in $\mathbb{R}^n$. To prove (8.1.17) it is sufficient to show that if $x_0 \neq 0$, $\xi_0 \neq 0$ then

$$(x_0, \xi_0) \notin WF(u) \Rightarrow (\xi_0, -x_0) \notin WF(\hat{u}), \tag{8.1.17$'$}$$

for $\hat{u}$ is also homogeneous and (8.1.17)$'$ applied to $\hat{u}$ gives the reversed implication since $\hat{\hat{u}} = (2\pi)^n \check{u}$. Choose $\chi \in C_0^\infty(\mathbb{R}^n)$ equal to 1 in a neighborhood of ξ_0 and $\psi \in C_0^\infty(\mathbb{R}^n)$ equal to 1 in a neighborhood of x_0 so small that

$$(\operatorname{supp} \psi \times \operatorname{supp} \chi) \cap WF(u) = \emptyset. \tag{8.1.20}$$

We have to estimate the Fourier transform of $v = \chi \hat{u}$ in a conic neighborhood of $-x_0$. Let $\psi(x) = 1$ when $|x - x_0| < 2r$ and consider

$\hat{v}(-tx)$ when $|x-x_0|<r$ and t is large. If u is homogeneous of degree a in $\mathbb{R}^n$ then

$$\hat{v}(-tx)=\hat{\chi}*\check{u}(-tx)=\langle u,\hat{\chi}(-tx+.)\rangle=t^{a+n}\langle u,\hat{\chi}(t(.-x))\rangle.$$

Set $\psi u=u_0$ and $(1-\psi)u=u_1$. Then $\Sigma(u_0)\cap\operatorname{supp}\chi=\emptyset$ by Proposition 8.1.3 and (8.1.20). Hence

$$\langle u_0,\hat{\chi}(t(.-x))\rangle=\int\hat{u}_0(\xi)\chi(\xi/t)e^{i\langle x,\xi\rangle}d\xi/t^n$$

is rapidly decreasing as $t\to\infty$, for $t^N\hat{u}_0(t\xi)\chi(\xi)$ is bounded for every N. Moreover,

$$\langle u_1,\hat{\chi}(t(.-x))\rangle=\langle u,(1-\psi)\hat{\chi}(t(.-x))\rangle$$

is also rapidly decreasing, for

$$y\to t^N(1-\psi(y))\hat{\chi}(t(y-x))$$

is bounded in $\mathscr{S}$ for any N. In fact, $|x-x_0|<r$ by hypothesis, and $|y-x_0|>2r$ in $\operatorname{supp}(1-\psi(y))$, hence $t\leqq t|y-x|/r$ and $|y|\leqq|y-x|+|x_0|+r$. Since $\hat{\chi}\in\mathscr{S}$ this completes the proof of (8.1.17)$'$.

In general it follows from (7.1.19) and (7.1.18) that we can write $u=w+w_0+Q(D)w_1$, $\hat{u}=\hat{w}+\hat{w}_0+Q(\xi)\hat{w}_1$ where w is homogeneous, $\operatorname{supp}w_0\subset\{0\}$, $w_1(x)=|x|^{-n}/c_n$ when $x\neq 0$, $\hat{w}_1(\xi)=-\log|\xi|$, and Q is a polynomial. Since $u-w$ and $\hat{u}-\hat{w}$ are in C^∞ except at the origin, we obtain (8.1.17) in general.

To prove (8.1.18) we first observe that since $\hat{\hat{u}}=(2\pi)^n\check{u}$ it follows from Lemma 8.1.7 with $v=\hat{u}$ that $x\notin\operatorname{supp}u\Rightarrow(0,-x)\notin WF(\hat{u})$. Assume now that $(0,-x_0)\notin WF(\hat{u})$. Choose $\chi\in C_0^\infty$ equal to 1 at 0 so that the Fourier transform of $\chi\hat{u}$ is rapidly decreasing in a conic neighborhood Γ of $-x_0$. Adding to u a term with support at 0 does not affect (8.1.18) so we may assume that u is homogeneous of degree a in $\mathbb{R}^n$ unless $a=-n-k$ and (3.2.24) is valid for an integer $k\geqq 0$. Hence the Fourier transform of $\chi\hat{u}$ at tx is

$$\hat{\chi}*\check{u}(tx)=\langle u,\hat{\chi}(.+tx)\rangle=t^a\langle u,\phi_t(.+x)\rangle+\log t\sum_{|\alpha|=k}c_\alpha(\partial^\alpha\hat{\chi})(tx)/\alpha!$$

where $\phi_t(x)=t^n\hat{\chi}(tx)$ and the sum should be omitted unless $k=-n-a$ is an integer $\geqq 0$. When $x\in\Gamma$ the left-hand side tends rapidly to 0 as $t\to\infty$, and so does the sum. Thus

$$\langle u,\phi_t(.+x)\rangle=\check{u}*\phi_t(x)\to 0\quad\text{in }\Gamma\text{ as }t\to\infty.$$

The convolution converges to $(2\pi)^n\check{u}$ in $\mathscr{S}'(\mathbb{R}^n)$. Hence $\check{u}=0$ in Γ so $x_0\notin\operatorname{supp}u$ and (8.1.18) is proved.

If u and therefore $\hat{u}$ is homogeneous in $\mathbb{R}^n$ then (8.1.19) follows if (8.1.18) is applied to $\hat{u}$. If u is not homogeneous then $u(t.)-t^a u$ is a distribution $\neq 0$ supported by 0 for some $t>0$. Hence $(0, \xi)$ is in $WF(u)$ for every $\xi \neq 0$, and $\xi \in \operatorname{supp} \hat{u}$ since $\hat{u}=U+V$ where U is of the form (7.1.19) with U_0 and Q homogeneous, $Q \not\equiv 0$ and V is a polynomial. The proof is complete.

Our final example concerns the distributions defined by oscillatory integrals in Section 7.8.

Theorem 8.1.9. *For the distribution*

$$A=\int e^{i\phi(.,\theta)} a(., \theta) d\theta$$

defined in Theorem 7.8.2 we have

$$WF(A) \subset \{(x, \phi_x'(x, \theta));\ (x, \theta) \in F \text{ and } \phi_\theta'(x, \theta)=0\}. \tag{8.1.21}$$

Before the proof we observe that $\phi_\theta'(x, \theta)=0$ implies $\phi(x, \theta)=0$ since ϕ is homogeneous of degree 1 with respect to θ. By hypothesis $\operatorname{Im} \phi \geqq 0$ so it follows that $\operatorname{Im} \phi_x'(x, \theta)=0$. Thus $\phi_x'(x, \theta)$ is real in (8.1.21).

Proof. Let $\psi \in C_0^\infty(X)$. Then the definition of A means that

$$\widehat{\psi A}(\xi)=\iint e^{i(\phi(x, \theta)-\langle x, \xi\rangle)} \psi(x) a(x, \theta) d x d \theta$$

as an oscillatory integral. We want to show that this is rapidly decreasing in any closed cone $V \subset \mathbb{R}^n$ which does not intersect

$$\{\phi_x'(x, \theta);\ (x, \theta) \in F,\ x \in \operatorname{supp} \psi,\ \phi_\theta'(x, \theta)=0\}.$$

Then we have for some $c>0$

$$|\xi-\phi_x'(x, \theta)|+|\theta||\phi_\theta'(x, \theta)| \geqq c(|\xi|+|\theta|) \quad \text{if } (x, \theta) \in F, \quad x \in \operatorname{supp} \psi, \quad \xi \in V. \tag{8.1.22}$$

To prove (8.1.22) we first observe that $\phi_x'(x, \theta)$ and $|\theta| \phi_\theta'(x, \theta)$ are continuous in F with the value 0 when $\theta=0$. By the homogeneity it suffices to prove (8.1.22) when $|\xi|+|\theta|=1$. By the compactness we only have to show then that the left-hand side is never 0 when $(x, \theta) \in F$, $x \in \operatorname{supp} \psi$, $\xi \in V$. If $\theta=0$ we have $|\xi-\phi_x'(x, \theta)|=1$, and when $\theta \neq 0$, $\phi_\theta'(x, \theta)=0$ we have $\xi \neq \phi_x'(x, \theta)$ since $\xi \in V$, which proves (8.1.22).

Expressing the oscillatory integral by means of the partition of unity in θ used in the proof of Theorem 7.8.2 we have

$$\widehat{\psi A}(\xi)=\sum_0^\infty \iint e^{i(\phi(x, \theta)-\langle x, \xi\rangle)} \psi(x) \chi_\nu(\theta) a(x, \theta) d x d \theta.$$

Each term is in $\mathscr{S}$. With $R=2^{\nu-1}$ the terms with $\nu \neq 0$ can be written

$$R^N \iint e^{i(R \phi(x, \theta)-\langle x, \xi\rangle)} \psi(x)\, \chi_1(\theta) a(x, R \theta) d x d \theta. \tag{8.1.23}$$

If $\Phi(x,\theta)=(R\phi(x,\theta)-\langle x,\xi\rangle)/(R+|\xi|)$ and $\xi\in V$ we have by (8.1.22)

$$|\Phi'_x|+|\Phi'_\theta|\geqq c(R|\theta|+|\xi|)/(R+|\xi|)\geqq c$$

in the support of $\psi(x)\chi_1(\theta)a(x,R\theta)$. With $\gamma=\max(1-\rho,\delta)<1$ we have

$$|D^\alpha_\theta D^\beta_x\psi(x)\chi_1(\theta)a(x,R\theta)|\leqq C_{\alpha\beta}R^{m+\gamma(|\alpha|+|\beta|)}.$$

By Theorem 7.7.1 it follows that (8.1.23) is estimated for large k by

$$C_k R^{m+N+k\gamma}(R+|\xi|)^{-k}\leqq C_k R^{-1}|\xi|^{m+N+1+(\gamma-1)k} \quad \text{if } \xi\in V.$$

Since $\sum_1^\infty 2^{1-\nu}=2$ we conclude that $\widehat{\chi A}(\xi)$ is rapidly decreasing in V, which completes the proof of the theorem.

Theorem 8.1.5 is a very special case of Theorem 8.1.9. In fact, let M be a C^∞ manifold in $\mathbb{R}^n$ defined near a point $x_0\in M$ by

$$\phi_1(x)=\ldots=\phi_k(x)=0$$

where $d\phi_1,\ldots,d\phi_k$ are linearly independent at x_0. If $a\in C_0^\infty(\mathbb{R}^n)$ has support near x_0 then

$$A=\int a(.)e^{i\phi(.,\theta)}d\theta, \qquad \phi(x,\theta)=\sum_1^k \phi_j(x)\theta_j$$

is by (7.8.5) equal to

$$(2\pi)^k a(x)\delta(\phi_1,\ldots,\phi_k)$$

where δ is the δ "function" at 0 in $\mathbb{R}^k$. This is an arbitrary smooth density on M with support near x_0. Theorem 8.1.9 gives now

$$WF(A)\subset\{(x,\phi'_x(x,\theta));\ \phi_j(x)=0, j=1,\ldots,k,\ x\in\operatorname{supp} a\}.$$

In the right-hand side we have the conormal bundle of M at $\operatorname{supp} a$. (We shall see in Section 8.2 that equality is valid.) We leave as an exercise for the reader to apply Theorem 8.1.9 to the distributions in Example 7.8.4. Theorem 8.1.9 is in fact one of the basic results leading to Lagrangian distributions (see Chap. XXV).

8.2. A Review of Operations with Distributions

Singularities made it impossible to give a general definition of multiplication of distributions and composition with maps. We shall now show that the definition of both operations can be extended when one takes into account the more refined description of the singularities given by the wave front set. As in Section 6.1 we shall always define such operations by continuous extension from the smooth case, so the first point to discuss is the topology in the space of distributions with a given bound for the wave front set.

Let X be an open set in $\mathbb{R}^n$, let Γ be a closed cone in $X \times (\mathbb{R}^n \setminus 0)$ and set

$$\mathscr{D}'_\Gamma(X) = \{u \in \mathscr{D}'(X), WF(u) \subset \Gamma\}.$$

Lemma 8.2.1. *A distribution $u \in \mathscr{D}'(X)$ is in $\mathscr{D}'_\Gamma(X)$ if and only if for every $\phi \in C_0^\infty(X)$ and every closed cone $V \subset \mathbb{R}^n$ with*

$$\Gamma \cap (\operatorname{supp} \phi \times V) = \emptyset \tag{8.2.1}$$

we have

$$\sup_V |\xi|^N |\widehat{\phi u}(\xi)| < \infty, \qquad N = 1, 2, \ldots. \tag{8.2.2}$$

Proof. (8.2.2) implies that $(x, \xi) \notin WF(u)$ if $\phi(x) \neq 0$ and ξ is in the interior of V, so the condition is sufficient. On the other hand, if $u \in \mathscr{D}'_\Gamma(X)$ and $\xi \in \Sigma(\phi u)$ then $(x, \xi) \in \Gamma$ for some $x \in \operatorname{supp} \phi$, by Proposition 8.1.3 and (8.1.9), so $\xi \notin V$ by (8.2.1). This proves (8.2.2).

Definition 8.2.2. For a sequence $u_j \in \mathscr{D}'_\Gamma(X)$ and $u \in \mathscr{D}'_\Gamma(X)$ we shall say that $u_j \to u$ in $\mathscr{D}'_\Gamma(X)$ if

$$u_j \to u \quad \text{in } \mathscr{D}'(X) \text{ (weakly)} \tag{i}$$

$$\sup_V |\xi|^N |\widehat{\phi u}(\xi) - \widehat{\phi u_j}(\xi)| \to 0, \quad j \to \infty, \tag{ii}$$

for $N = 1, 2, \ldots$ if $\phi \in C_0^\infty(X)$ and V is a closed cone in $\mathbb{R}^n$ such that (8.2.1) is valid.

Since (i) implies that $\widehat{\phi u_j} \to \widehat{\phi u}$ uniformly on every compact set and N is arbitrary in (ii), we can replace (ii) by

$$\sup_j \sup_{\xi \in V} |\xi|^N |\widehat{\phi u_j}(\xi)| < \infty, \qquad N = 1, 2, \ldots. \tag{ii)'}$$

The following is an extension of Theorem 4.1.5.

Theorem 8.2.3. *For every $u \in \mathscr{D}'_\Gamma(X)$ there is a sequence $u_j \in C_0^\infty(X)$ such that $u_j \to u$ in $\mathscr{D}'_\Gamma(X)$.*

Proof. As in the proof of Theorem 4.1.5 we take $u_j = (\chi_j u) * \phi_j$ where

a) $\chi_j \in C_0^\infty(X)$ and $\chi_j = 1$ on any compact set in X for large j,

b) $0 \leq \phi_j \in C_0^\infty(\mathbb{R}^n)$, $\int \phi_j dx = 1$, and $\operatorname{supp} \phi_j$ is so small that (4.1.4) holds.

Then we know already that $u_j \in C_0^\infty(X)$ and that $u_j \to u$ in $\mathscr{D}'(X)$. If ϕ and V satisfy (8.2.1) we can find $\psi \in C_0^\infty(X)$ equal to 1 in a neighborhood of $\operatorname{supp}\phi$ and a closed cone W with interior containing $V \setminus 0$ so that

$$\Gamma \cap (\operatorname{supp}\psi \times W) = \emptyset.$$

For large j we have $\phi u_j = \phi w_j$ where $w_j = \phi_j * (\psi u)$, hence

$$|\hat{w}_j| = |\hat{\phi}_j|\,|\widehat{\psi u}| \leqq |\widehat{\psi u}|.$$

Since $|\widehat{\psi u}|$ is rapidly decreasing in W, the proof of Lemma 8.1.1 gives (ii)′ (see (8.13)′), and the theorem is proved.

Theorem 8.2.4. *Let X and Y be open subsets of $\mathbb{R}^m$ and $\mathbb{R}^n$ respectively and let $f: X \to Y$ be a C^∞ map. Denote the set of normals of the map by*

$$N_f = \{(f(x), \eta) \in Y \times \mathbb{R}^n;\ {}^t f'(x)\,\eta = 0\}.$$

Then the pullback f^*u can be defined in one and only one way for all $u \in \mathscr{D}'(Y)$ with

$$N_f \cap WF(u) = \emptyset \tag{8.2.3}$$

so that $f^*u = u \circ f$ when $u \in C^\infty$ and for any closed conic subset Γ of $Y \times (\mathbb{R}^n \setminus 0)$ with $\Gamma \cap N_f = \emptyset$ we have a continuous map $f^*: \mathscr{D}'_\Gamma(Y) \to \mathscr{D}'_{f^*\Gamma}(X)$,

$$f^*\Gamma = \{(x, {}^t f'(x)\,\eta);\ (f(x), \eta) \in \Gamma\}. \tag{8.2.4}$$

In particular we have for every $u \in \mathscr{D}'(Y)$ satisfying (8.2.3)

$$WF(f^*u) \subset f^* WF(u).$$

Proof. Define $f^*u = u \circ f$ when $u \in C^\infty(Y)$. By Theorem 8.2.3 the theorem will be proved if we show that f^* maps sequences $u_j \in C^\infty$ converging in $\mathscr{D}'_\Gamma(Y)$ to sequences converging in $\mathscr{D}'_{f^*\Gamma}(X)$. First we shall just prove convergence in $\mathscr{D}'(X)$. If $u \in C_0^\infty(Y)$ and $\chi \in C_0^\infty(X)$ we have by Fourier's inversion formula applied to u

$$\langle f^*u, \chi \rangle = (2\pi)^{-n} \int \hat{u}(\eta)\, I_\chi(\eta)\, d\eta, \qquad I_\chi(\eta) = \int \chi(x)\, e^{i\langle f(x), \eta\rangle}\, dx. \tag{8.2.5}$$

Let $x_0 \in X$, set $y_0 = f(x_0)$, $\Gamma_{y_0} = \{\eta;\ (y_0, \eta) \in \Gamma\}$ and choose

a) a closed conic neighborhood V of Γ_{y_0} in $\mathbb{R}^n \setminus 0$ such that ${}^t f'(x_0)\,\eta \neq 0$, $\eta \in V$,

b) a compact neighborhood Y_0 of y_0 such that V is a neighborhood of Γ_y for every $y \in Y_0$; such a neighborhood exists since Γ is closed,

c) a compact neighborhood X_0 of x_0 with $f(X_0)$ in the interior of Y_0 and ${}^tf'(x)\eta \neq 0$ if $x \in X_0$ and $\eta \in V$.

Choose $\phi \in C_0^\infty(Y_0)$ equal to 1 on $f(X_0)$. Then (8.2.5) is valid when $\chi \in C_0^\infty(X_0)$ for every $u \in C^\infty(Y)$ if u is replaced by ϕu in the right-hand side. Since $d\langle f(x), \eta\rangle = \langle dx, {}^tf'(x)\eta\rangle$ and

$$|\eta| \leqq C|{}^tf'(x)\eta| \quad \text{if } x \in \operatorname{supp}\chi \text{ and } \eta \in V,$$

it follows from Theorem 7.7.1 that

$$|I_\chi(\eta)| \leqq C_{N,\chi}(1+|\eta|)^{-N}, \quad \eta \in V, \quad N=1,2,\ldots. \tag{8.2.6}$$

If $u_j \in C^\infty(Y)$ and $u_j \to u$ in $\mathscr{D}'_\Gamma(Y)$ we have

$$|\widehat{\phi u_j}(\eta)| \leqq C'_N(1+|\eta|)^{-N}, \quad \eta \notin V, \quad N=1,2,\ldots$$

and for some M (cf. the proof of Theorem 2.1.8)

$$|\widehat{\phi u_j}(\eta)| \leqq C(1+|\eta|)^M, \quad \eta \in \mathbb{R}^n.$$

Hence dominated convergence in V and in $\complement V$ gives

$$\langle f^*u_j, \chi\rangle \to (2\pi)^{-n} \int \widehat{\phi u}(\eta)\, I_\chi(\eta)\, d\eta,$$

so f^*u_j converges in $\mathscr{D}'$ to a limit independent of the sequence chosen. (See the remark after Theorem 2.2.4.) We denote the limit by f^*u.

In the proof of the continuity of $f^*\colon \mathscr{D}'_\Gamma(Y) \to \mathscr{D}'_{f^*\Gamma}(X)$ we set $\chi f^*u_j = v_j$. Then (8.2.5) with χ replaced by $\chi e^{-i\langle\cdot,\xi\rangle}$ gives

$$\hat{v}_j(\xi) = (2\pi)^{-n} \int \widehat{\phi u_j}(\eta)\, I_\chi(\eta,\xi)\, d\eta,$$
$$I_\chi(\eta,\xi) = \int \chi(x)\, e^{i(\langle f(x),\eta\rangle - \langle x,\xi\rangle)}\, dx.$$

Let W be an open conic neighborhood of ${}^tf'(x_0)\Gamma_{y_0} = (f^*\Gamma)_{x_0}$. We may assume the neighborhoods V and X_0 above chosen so that

$${}^tf'(x)\eta \in W \quad \text{if } x \in X_0 \text{ and } \eta \in V.$$

Then we have for some $\varepsilon > 0$

$$|{}^tf'(x)\eta - \xi| \geqq \varepsilon(|\xi| + |\eta|) \quad \text{if } x \in X_0, \quad \eta \in V \text{ and } \xi \notin W$$

for the left hand side cannot vanish when $|\xi| + |\eta| = 1$. Hence Theorem 7.7.1 gives for any N

$$|I_\chi(\eta,\xi)| \leqq C_N(1+|\xi|+|\eta|)^{-N} \quad \text{if } \xi \notin W \text{ and } \eta \in V. \tag{8.2.6$'$}$$

If $\eta \notin V$ we use another obvious consequence of Theorem 7.7.1,

$$|I_\chi(\eta,\xi)| \leqq C_N(1+|\eta|)^N(1+|\xi|)^{-N}. \tag{8.2.6$''$}$$

Summing up the preceding estimates we obtain when $\xi \notin W$

$$|\hat{v}_j(\xi)| \leqq C'_N\Big(\int_V (1+|\xi|+|\eta|)^{M-N} d\eta + (1+|\xi|)^{-N} \int_{\complement V} |\widehat{\phi u_j}(\eta)|(1+|\eta|)^N d\eta\Big).$$

Since ϕu_j satisfies condition (ii)′ (after Definition 8.2.2) in $\complement V$ it follows that

$$\sup_j \sup_{\xi \notin W} |\xi|^N |\widehat{\chi f^* u_j}(\xi)| < \infty, \qquad N = 1, 2, \ldots$$

if W is a conic neighborhood of $(f^*\Gamma)_{x_0}$ and $\operatorname{supp}\chi$ is sufficiently close to x_0. By a partition of unity it follows now that $f^*u_j \to f^*u$ in $\mathscr{D}'_{f^*\Gamma}$, which completes the proof.

If X is a C^∞ manifold and $u \in \mathscr{D}'(X)$ we can now define $WF(u) \subset T^*(X) \setminus 0$ so that the restriction to a coordinate patch X_κ is equal to $\kappa^* WF(u \circ \kappa^{-1})$. In fact, when f is a diffeomorphism between open sets in $\mathbb{R}^n$ it follows from Theorem 8.2.4 that $WF(f^*v)$ is the pullback of $WF(v)$ considered as a subset of the cotangent bundle. Hence the preceding definition is independent of the choice of local coordinates. It is clear that $WF(u)$ is a closed subset of $T^*(X) \setminus 0$ which is conic in the sense that the intersection with the vector space $T_x^*(X)$ is a cone for every $x \in X$. Indeed, these are local properties inherited from the local coordinate patches.

If E is a C^∞ vector bundle over X and $u \in \mathscr{D}'(X, E)$, we define $WF(u)$ locally as $\bigcup WF(u_i)$ where $(u_1, \ldots, u_N)$ are the components of u with respect to a local trivialization of E. Passage to another local trivialization only means that $(u_1, \ldots, u_N)$ is multiplied by an invertible C^∞ matrix so the definition is independent of the choice of local trivialization.

Example 8.2.5. If u is a C^∞ density on a C^∞ submanifold Y of the manifold X, then

$$WF(u) = \{(x, \xi) \in T^*(X);\, x \in \operatorname{supp} u,\, \xi \neq 0 \text{ and } \langle T_x(Y), \xi\rangle = 0\}.$$

In fact, with suitable local coordinates this is just Theorem 8.1.5. Thus the wave front set is the restriction to $\operatorname{supp} u$ of the normal bundle

$$N(Y) = \{(y, \xi),\, y \in Y,\, \langle T_y(Y), \xi\rangle = 0\}$$

with the zero section removed.

Example 8.2.6. For the distributions $(A \pm i0)^{(2-n)/2}$ in (6.2.1) we have

$$WF((A \pm i0)^{(2-n)/2}) = \{(x, t\,dA(x)),\, x \neq 0,\, A(x) = 0,\, t \gtrless 0\} \cup T_0^* \setminus 0. \tag{8.2.7}$$

In fact, since $WF((t \pm i0)^a) \subset \{(0, \tau), \tau \gtrless 0\}$ by Theorem 8.1.6, it follows from Theorem 8.2.4 that the left-hand side is contained in the right-hand side when $x \neq 0$, and this is trivial when $x=0$. Since

$$B(\partial)(A \pm i0)^{(2-n)/2} = c\,\delta_0,$$

it follows from (8.1.11) and the fact that $c \neq 0$, $WF(\delta_0) = T_0^* \setminus 0$ (by Theorem 8.1.5) that the left-hand side of (8.2.7) contains $T_0^* \setminus 0$. Now $(A \pm i0)^{(2-n)/2}$ is not in C^∞ at any $x \neq 0$ where $A(x)=0$ which proves that there is equality in (8.2.7). If we recall from the proof of Theorem 6.2.1 how $\chi_\pm^a$ is written as a linear combination of $(t \pm i0)^a$ when $a \notin \mathbb{Z}_+$, it follows that in $\mathbb{R}^n \setminus 0$

$$WF(A^* \chi_\pm^{(2-n)/2}) = \{(x, t\,dA(x));\ A(x)=0,\ x \neq 0,\ t \neq 0\}. \tag{8.2.8}$$

Since the wave front set is closed it must at the origin contain

$$\{(0, dA(x));\ A(x)=0,\ x \neq 0\} = \{(0,\xi);\ \xi \neq 0,\ B(\xi)=0\}, \tag{8.2.9}$$

and when n_+ (n_-) is odd the argument above shows that it contains $T_0^* \setminus 0$. It will follow from Theorem 8.3.1 that the wave front set at 0 is given by (8.2.9) when n_+ (n_-) is even. For the advanced fundamental solution of the wave operator the preceding argument gives at once that the wave front set is the normal bundle of the forward light cone with the origin removed, together with $T_0^* \setminus 0$.

In Chapter VI we could never pull a distribution back to a manifold of lower dimension. However, this is sometimes allowed by Theorem 8.2.4 and we list an important special case:

Corollary 8.2.7. *Let X be a manifold and Y a submanifold with normal bundle denoted by $N(Y)$. For every distribution u in X with $WF(u)$ disjoint with $N(Y)$ the restriction $u_{|Y}$ to Y is a well defined distribution on Y, the pullback by the inclusion $Y \hookrightarrow X$.*

Example 8.2.8. Let Z be another submanifold of X and u a C^∞ density on Z. Then $WF(u) \subset N(Z)$, and $N(Z) \cap N(Y)$ is contained in the zero section if and only if $x \in Z \cap Y$ and $\xi \in T_x^*$ orthogonal to $T_x(Z)$ and $T_x(Y)$ implies $\xi = 0$. This means that $T_x(Z) + T_x(Y) = T_x(X)$, that is, that Z and Y have a *transversal* intersection. The restriction of u to Y is then defined. It is a density in $Y \cap Z$. In fact, we may choose the coordinates locally so that $X = \mathbb{R}^n$, Y is defined by $x'=0$ and Z by $x''=0$ where x' denotes the first n' coordinates and x'' the next n'' coordinates. Write $x=(x', x'', x''')$ where x''' are thus coordinates in $Z \cap Y$. Then $u = a(x', x''')\,\delta(x'')$ which is the limit in $\mathscr{D}'_{N(Z)}(X)$ when $\varepsilon \to 0$ of $a(x', x''')\,\phi(x''/\varepsilon)\,\varepsilon^{-n''}$, if $\phi \in C_0^\infty(\mathbb{R}^{n''})$ and $\int \phi(x'')\,dx'' = 1$. The restriction to Y is the function $a(0, x''')\,\phi(x''/\varepsilon)\,\varepsilon^{-n''}$ which converges in $\mathscr{D}'(Y)$ to the density $a(0, x''')\,\delta(x'')$ on $Z \cap Y$ considered as a submanifold of Y.

We are now able to define the product of some pairs of distributions which have singularities at the same point. To do so we first observe that if u and v are functions in X then the product $u(x)v(x)$ is the restriction to the diagonal of the tensor product $u(x)v(y)$ defined for $(x, y)\in X\times X$. Thus we shall first examine the tensor product.

Theorem 8.2.9. *If* $u\in\mathscr{D}'(X)$, $v\in\mathscr{D}'(Y)$ *then*

$$WF(u\otimes v)\subset(WF(u)\times WF(v))\cup((\operatorname{supp} u\times\{0\})\times WF(v))\cup(WF(u)\times(\operatorname{supp} v\times\{0\})). \tag{8.2.10}$$

Proof. If $u\in\mathscr{E}'(\mathbb{R}^m)$ and $v\in\mathscr{E}'(\mathbb{R}^n)$ then the Fourier transform of $u\otimes v$ is $\hat{u}(\xi)\,\hat{v}(\eta)$. If $u\neq 0$ and $v\neq 0$ it is clear that

$$\Sigma(u\otimes v)\subset(\Sigma(u)\times\Sigma(v))\cup(\{0\}\times\Sigma(v))\cup(\Sigma(u)\times\{0\}).$$

To prove (8.2.10) we just have to apply this to ϕu and ψv where ϕ and ψ are in C^∞ with supports close to x and y respectively.

Theorem 8.2.10. *If* $u, v\in\mathscr{D}'(X)$ *then the product* uv *can be defined as the pullback of the tensor product* $u\otimes v$ *by the diagonal map* $\delta\colon X\to X\times X$ *unless* $(x,\xi)\in WF(u)$ *and* $(x,-\xi)\in WF(v)$ *for some* (x,ξ). *When the product is defined we have*

$$WF(uv)\subset\{(x,\xi+\eta);\ (x,\xi)\in WF(u) \text{ or } \xi=0,\ (x,\eta)\in WF(v) \text{ or } \eta=0\}. \tag{8.2.11}$$

Proof. For the diagonal map $\delta(x)=(x,x)$ from X to $X\times X$ we have

$$\delta'(x)\,t=(t,t) \quad \text{if } t\in T_x(X),$$
$${}^t\delta'(x)(\xi,\eta)=\xi+\eta \quad \text{if } \xi,\ \eta\in T_x^*(X).$$

The theorem is therefore an immediate consequence of Theorems 8.2.4, 8.2.9.

Example 8.2.11. If u and v are C^∞ densities on submanifolds Y and Z of X intersecting transversally, then uv is a C^∞ density on $Y\cap Z$. In fact, $u\otimes v$ is a C^∞ density on $Y\times Z$ which intersects the diagonal in $X\times X$ transversally so the statement follows from Example 8.2.8.

Since pullbacks have been defined here by continuous extension of composition of functions, it is clear that the preceding definition of multiplication extends the multiplication of distributions and smooth functions defined in Chapter II. Similarly all standard rules of calculus remain valid for the extended operations defined in this section; we leave for the reader to fill in these obvious details.

Finally we shall discuss $WF(\mathscr{K}u)$ when $\mathscr{K}$ is a linear transformation from $C_0^\infty(Y)$ to $\mathscr{D}'(X)$. For the sake of simplicity in statements we restrict ourselves to open sets X and Y in Euclidean spaces.

Theorem 8.2.12. *Let $X \subset \mathbb{R}^n$ and $Y \subset \mathbb{R}^m$ be open sets and let $K \in \mathscr{D}'(X \times Y)$. Denote the corresponding linear transformation from $C_0^\infty(Y)$ to $\mathscr{D}'(X)$ by $\mathscr{K}$. Then we have*

$$WF(\mathscr{K}u) \subset \{(x,\xi);\, (x,y,\xi,0) \in WF(K) \text{ for some } y \in \operatorname{supp} u\},\ u \in C_0^\infty(Y).$$

Proof. Let $x_0 \in X$, choose $\chi \in C_0^\infty(X)$ with $\chi(x_0)=1$ and set

$$K_1 = (\chi \otimes u)\, K \in \mathscr{E}'(X \times Y).$$

The Fourier transform of $\chi \mathscr{K} u$ is $\hat{K}_1(\xi, 0)$. Now Proposition 8.1.3 gives

$$\Sigma(K_1) \subset \{(\xi,\eta);\, (x,y,\xi,\eta) \in WF(K) \text{ for some } x \in \operatorname{supp}\chi,\ y \in \operatorname{supp} u\}.$$

Hence it follows that

$$\Sigma(\chi \mathscr{K} u) \subset \{\xi;\, (x,y,\xi,0) \in WF(K) \text{ for some } x \in \operatorname{supp}\chi \text{ and } y \in \operatorname{supp} u\}.$$

When $\operatorname{supp}\chi \to \{x_0\}$ the theorem follows.

The proof shows that $\mathscr{K}$ maps $C_0^\infty(M)$ continuously into $\mathscr{D}'_\Gamma(X)$ if M is a compact subset of Y and

$$\Gamma = \{(x,\xi);\, (x,y,\xi,0) \in WF(K) \text{ for some } y \in M\}.$$

For the union of all such sets we shall use the notation

$$WF(K)_X = \{(x,\xi);\, (x,y,\xi,0) \in WF(K) \text{ for some } y \in Y\}.$$

It is of course not necessarily a closed set. If it is empty then $\mathscr{K}$ is a continuous map from $C_0^\infty(Y)$ to $C^\infty(X)$.

The first part of the following theorem is essentially dual to Theorem 8.2.12.

Theorem 8.2.13. *There is a unique way of defining $\mathscr{K}u \in \mathscr{D}'(X)$ for every $u \in \mathscr{E}'(Y)$ with $WF(u) \cap WF'(K)_Y = \emptyset$, where*

$$WF'(K)_Y = \{(y,\eta);\, (x,y,0,-\eta) \in WF(K) \text{ for some } x \in X\}$$

so that the map $\mathscr{E}'(M) \cap \mathscr{D}'_\Gamma \ni u \to \mathscr{K}u \in \mathscr{D}'(X)$ is continuous for all compact sets $M \subset Y$ and all closed conic sets Γ disjoint with $WF'(K)_Y$. We have

$$WF(\mathscr{K}u) \subset WF(K)_X \cup WF'(K) \circ WF(u) \tag{8.2.12}$$

where

$$WF'(K) = \{(x,y,\xi,\eta);\, (x,y,\xi,-\eta) \in WF(K)\}$$

is considered as a relation mapping sets in $T^*(Y)\setminus 0$ *to sets in* $T^*(X)\setminus 0$.

Proof. Let $\psi\in C_0^\infty(Y)$ be equal to 1 in a neighborhood of M. When $u\in C_0^\infty$ and the support is contained in this neighborhood then

$$\mathscr{K}u=\mathscr{K}(u\psi)=\mathscr{K}_u\psi$$

where $\mathscr{K}_u$ has the kernel

$$K_u=K(1\otimes u).$$

If $u\in\mathscr{D}'(Y)$ we have by Theorem 8.2.9

$$WF(1\otimes u)=\{(x,y,0,\eta);\,(y,\eta)\in WF(u)\}.$$

The product K_u is therefore defined for every $u\in\mathscr{D}'_\Gamma(Y)$ if Γ is disjoint with $WF'(K)_Y$, and Theorem 8.2.10 also gives

$$WF(K_u)\subset\{(x,y,\xi,\eta+\eta');\,(y,\eta)\in WF(u) \text{ and } (x,y,\xi,\eta')\in WF(K)\}\\ \cup WF(K)\cup WF(1\otimes u).$$

It is clear that the map $\mathscr{D}'_\Gamma\ni u\to K_u\in\mathscr{D}'(X\times Y)$ is continuous. Setting $\mathscr{K}u=\mathscr{K}_u\psi$ therefore gives a continuous extension of $\mathscr{K}$ to all $u\in\mathscr{D}'_\Gamma$ with support close to M. The uniqueness of such an extension follows from Theorem 8.2.3. Using Theorem 8.2.12 we obtain

$$WF(\mathscr{K}u)\subset\{(x,\xi);\,(x,y,\xi,-\eta)\in WF(K)\\ \text{for some } (y,\eta)\in WF(u)\}\cup WF(K)_X,$$

which proves (8.2.12).

Let now $X\subset\mathbb{R}^n$, $Y\subset\mathbb{R}^m$ and $Z\subset\mathbb{R}^p$ be open sets and let $K_1\in\mathscr{D}'(X\times Y)$, $K_2\in\mathscr{D}'(Y\times Z)$. Assume that the projection

(8.2.13) $$\operatorname{supp}K_2\ni(y,z)\to z$$

is proper, that is, the inverse image of any compact set is compact. If $u\in C_0^\infty(Z)$ it follows then that $\mathscr{K}_2u\in\mathscr{E}'(Y)$, and by Theorem 8.2.12 we have

$$WF(\mathscr{K}_2u)\subset WF(K_2)_Y.$$

If we assume that

(8.2.14) $$WF'(K_1)_Y\cap WF(K_2)_Y=\emptyset$$

then the composition $\mathscr{K}_1\circ\mathscr{K}_2$ is defined as a continuous map $C_0^\infty(Z)\to\mathscr{D}'(X)$. Thus it has a Schwartz kernel $K\in\mathscr{D}'(X\times Z)$. When K_1 and K_2 are smooth then

$$K(x,z)=\int K_1(x,y)\,K_2(y,z)\,dy.$$

In general the kernel K is also obtained by pulling the tensor product $K_1 \otimes K_2 \in \mathscr{D}'(X \times Y \times Y \times Z)$ back to $X \times Y \times Z$ with the map $(x, y, z) \to (x, y, y, z)$. The normal set of this map is $\{(x, y, y, z;\ 0, \eta, -\eta, 0)\}$ so it follows from (8.2.14) that the pullback is defined. Finally the pullback is integrated with respect to y over Y, which gives K and an estimate for $WF(K)$ if one also recalls Theorem 8.2.12. This leads to the following theorem for which the reader should have no difficulty at all in supplying the missing and rather repetitive details of proof.

Theorem 8.2.14. *When* (8.2.14) *is valid and the projection* (8.2.13) *is proper then the composition* $\mathscr{K}_1 \circ \mathscr{K}_2$ *is defined and for the kernel K we have*

$$WF'(K) \subset WF'(K_1) \circ WF'(K_2) \cup (WF(K_1)_X \times Z \times \{0\}) \cup (X \times \{0\} \times WF'(K_2)_Z).$$

To illustrate the preceding results we shall consider convolution by a distribution $k \in \mathscr{D}'(\mathbb{R}^n)$. This has as kernel the distribution K obtained by pulling back k with the map

$$\mathbb{R}^n \times \mathbb{R}^n \ni (x, y) \to x - y \in \mathbb{R}^n.$$

Theorem 8.2.4 gives

$$WF(K) \subset \{(x, y, \xi, -\xi);\ (x-y, \xi) \in WF(k)\}.$$

For any constant c we have $k = f_c^* K$ where $f_c(x) = (x+c, c)$, thus ${}^t f_c'(x)(\xi, \eta) = \xi$. Hence Theorem 8.2.4 also gives

$$WF(k) \subset \{(x, \xi);\ (x+c, c, \xi, -\xi) \in WF(K)\}$$

so there is in fact equality,

$$WF(K) = \{(x, y, \xi, -\xi);\ (x-y, \xi) \in WF(k)\}. \tag{8.2.15}$$

Since the two frequency components vanish simultaneously it follows that convolution with k maps C_0^∞ into C^∞ and has a continuous extension to a map $\mathscr{E}' \to \mathscr{D}'$. Furthermore, we have

$$WF(k * u) \subset \{(x+y, \xi);\ (x, \xi) \in WF(k) \text{ and } (y, \xi) \in WF(u)\}, \quad u \in \mathscr{E}'. \tag{8.2.16}$$

This improves Theorem 4.2.5 a great deal. (A direct proof of (8.2.16) is easily obtained from Theorem 4.2.5 and the obvious fact that $\Sigma(k * u)$ is contained in $\Sigma(k) \cap \Sigma(u)$ when $k, u \in \mathscr{E}'$.)

8.3. The Wave Front Set of Solutions of Partial Differential Equations

A differential operator with C^∞ coefficients of order m in an open set $X \subset \mathbb{R}^n$ is of the form

(8.3.1) $$P = P(x, D) = \sum_{|\alpha| \leq m} a_\alpha(x) D^\alpha.$$

The principal part (or symbol) P_m is defined by

(8.3.2) $$P_m(x, \xi) = \sum_{|\alpha| = m} a_\alpha(x) \xi^\alpha.$$

Note that the definition differs from that in Section 6.4 by a factor i^m. Corresponding to (6.4.6)′ we have

(8.3.2)′ $$P_m(x, d\phi) = \lim_{t \to \infty} t^{-m} e^{-it\phi} P e^{it\phi}.$$

If X is a C^∞ manifold then a differential operator of order m on X is by definition an operator which has the form (8.3.1) in local coordinate systems. From (8.3.2)′ it follows that the principal symbol is invariantly defined in the cotangent bundle. We shall now prove a weak converse of (8.1.11).

Theorem 8.3.1. *If P is a differential operator of order m with C^∞ coefficients on a manifold X, then*

(8.3.3) $$WF(u) \subset \operatorname{Char} P \cup WF(Pu), \qquad u \in \mathscr{D}'(X),$$

where the characteristic set Char P *is defined by*

(8.3.4) $$\operatorname{Char} P = \{(x, \xi) \in T^*(X) \setminus 0, P_m(x, \xi) = 0\}.$$

Corollary 8.3.2. *If P is elliptic, that is, $P_m(x, \xi) \neq 0$ in $T^*(X) \setminus 0$, then*

$$WF(u) = WF(Pu), \qquad u \in \mathscr{D}'(X).$$

Hence

$$\operatorname{sing\,supp} u = \operatorname{sing\,supp} Pu, \qquad u \in \mathscr{D}'(X).$$

Proof of Theorem 8.3.1. We have stated the result for a manifold but it is purely local so we may assume that $X \subset \mathbb{R}^n$ in the proof. If $P_m(x_0, \xi_0) \neq 0$ we can choose a neighborhood $U \subset X$ of x_0 and an open cone $V \ni \xi_0$ such that

(8.3.5) $$|\xi|^m \leq C |P_m(x, \xi)| \quad \text{if } x \in U \text{ and } \xi \in V,$$

for some C. Later on another condition will be imposed on U and V. Choose a fixed $\phi \in C_0^\infty(U)$ with $\phi(x_0) = 1$. To estimate $\widehat{\phi u}(\xi)$ when $\xi \in V$

we first note that if

$$^tPv=\sum(-D)^\alpha(a_\alpha v),$$

that is, ^tP is the *formal adjoint* of P, then $Pu=f$ means that

$$\langle u, {}^tPv\rangle=\langle f, v\rangle, \qquad v\in C_0^\infty(X).$$

We would like to find v so that the left-hand side is $\widehat{\phi u}(\xi)$, that is,

$$^tPv(x)=\phi(x)\,e^{-i\langle x,\xi\rangle}.$$

For large ξ an approximate solution is $e^{-i\langle x,\xi\rangle}\phi(x)/P_m(x,\xi)$. To improve it we set

$$v(x)=w(x)\,e^{-i\langle x,\xi\rangle}/P_m(x,\xi)$$

which gives the equation for v the form

$$w-Rw=\phi.$$

Here $R=R_1+\ldots+R_m$ and $R_j|\xi|^j$ is a differential operator of order $\leqq j$ which is a homogeneous function of ξ of degree 0. In fact, to obtain a term in R which is homogeneous of degree $-j$ we must let $m-j$ derivatives fall on the exponential $e^{-i\langle x,\xi\rangle}$ and have no more than j left which can act on w. By (8.3.5) all x derivatives of the coefficients of $R_j|\xi|^j$ are bounded in $U\times V$. Formally the equation $w-Rw=\phi$ is satisfied by $w=\sum R^k\phi$. However, the sum is unlikely to converge, so we take instead a large partial sum

$$w_N=\sum_{k<N} R^k\phi.$$

Then we have

$$w_N-Rw_N=\phi-R^N\phi$$

and R^N is a sum of terms each containing a factor $|\xi|^{-k}$ for some $k\geqq N$. The preceding equation means that

$$^tP(x,D)(e^{-i\langle x,\xi\rangle}w_N(x)/P_m(x,\xi))=e^{-i\langle x,\xi\rangle}(\phi-R^N\phi).$$

Hence

$$\widehat{\phi u}(\xi)=u(e^{-i\langle\cdot,\xi\rangle}R^N\phi)+f(e^{-i\langle\cdot,\xi\rangle}w_N/P_m(\cdot,\xi)), \qquad \xi\in V. \tag{8.3.6}$$

If the distribution u is of order μ in a neighborhood of $\operatorname{supp}\phi$ then the first term on the right-hand side of (8.3.6) can be estimated by

$$C\sum_{|\alpha|\leqq\mu}\sup|D^\alpha(e^{-i\langle\cdot,\xi\rangle}R^N\phi)|\leqq C_N|\xi|^{\mu-N}, \qquad |\xi|\geqq 1.$$

Here $N-\mu$ is as large as we please. If $(x_0,\xi_0)\notin WF(f)$, it follows from (8.1.3)$'$ that we can choose the neighborhood U of x_0 and the conic neighborhood V of ξ_0 such that for some integer M and $k=1,2,\ldots$

$$\sup_V|\xi|^k|\widehat{\psi f}(\xi)|\leqq C_k\sum_{|\alpha|\leqq k+M}\sup|D^\alpha\psi|, \qquad \psi\in C_0^\infty(U).$$

Taking $\psi = w_N/P_m(.,\xi)$ we conclude that the second term on the right hand side of (8.3.6) is $O(|\xi|^{-k})$ as $\xi \to \infty$ in V. Hence

$$\widehat{\phi u}(\xi) = O(|\xi|^{-k}), \qquad \xi \in V, \quad k = 1, 2, \dots$$

which proves Theorem 8.3.1.

Theorem 8.3.1 allows us to complete the proof that the wave front set of $A^* \chi_\pm^{(2-n)/2}$ (defined in Theorem 6.2.1) is given by (8.2.9) at 0 when n_+ (n_-) is even. In fact, we know already that (8.2.9) is a lower bound, and since $B(D) A^* \chi_\pm^{(2-n)/2} = 0$ when n_+ (n_-) is even it follows that (8.2.9) is also an upper bound.

With the notation in Theorem 6.2.1 we note that

$$B(D) E_\pm = \delta, \qquad E_\pm = c_\pm (A \pm i0)^{(2-n)/2}$$

for a suitable choice of c_+, c_-. If we write $\xi = 2tAx$, that is, $x = (2t)^{-1} B\xi$ in (8.2.7) we have

$$WF(E_\pm) = \{(tB'(\xi), \xi);\, t \gtrless 0,\, \xi \neq 0,\, B(\xi) = 0\} \cup T_0^* \setminus 0. \tag{8.3.7}$$

The difference $E_+ - E_-$ satisfies the equation $B(D)(E_+ - E_-) = 0$ and

$$WF(E_+ - E_-) = \{(tB'(\xi), \xi);\, t \in \mathbb{R},\, \xi \neq 0,\, B(\xi) = 0\}.$$

In fact, this follows from (8.3.7) when $x \neq 0$, for the two terms have disjoint wave front sets then, and at 0 we can use the argument just given for $\chi_\pm^{(2-n)/2}$. By a translation we obtain solutions of the equation $B(D) u = 0$ with $WF(u)$ containing any desired point in $\operatorname{Char} B$. However, not every subset of $\operatorname{Char} B$ can be the wave front set of a solution. To prove this we first take $v \in \mathscr{E}'(\mathbb{R}^n)$ and set $B(D) v = g$. Then

$$v = E_+ * g$$

so (8.2.16) and (8.3.7) give

$$WF(v) \subset WF(g) \cup \{(x + tB'(\xi), \xi);$$
$$(x, \xi) \in WF(g),\, t > 0,\, \xi \neq 0,\, B(\xi) = 0\}.$$

Using E_- instead of E_+ gives the same inclusion with $t < 0$ instead. If

$$(x, \xi) \in WF(v) \setminus WF(g)$$

it follows from Theorem 8.3.1 that $B(\xi) = 0$, and the preceding inclusions show that one can find t_- and t_+ so that $t_- < 0 < t_+$ and $(x - t_\pm B'(\xi), \xi) \in WF(g)$. This leads to

Theorem 8.3.3. *Let B be a real non-singular quadratic form in $\mathbb{R}^n$, let X be an open set in $\mathbb{R}^n$ and $u \in \mathscr{D}'(X)$ a solution of the equation $B(D) u = f$.*

If $(x, \xi) \in WF(u) \setminus WF(f)$ *then* $B(\xi)=0$ *and*

$$I \times \{\xi\} \subset WF(u)$$

if $I \subset X$ *is a line segment containing* x *with direction* $B'(\xi)$ *such that* $I \times \{\xi\}$ *does not meet* $WF(f)$.

Thus singularities of u with frequency ξ propagate with fixed frequency in the direction $B'(\xi)$ in X until they meet the singularities of f.

Proof. That $B(\xi)=0$ follows from Theorem 8.3.1. Choose $\phi \in C_0^\infty(X)$ so that $\phi(x)=1$ and $L \cap \operatorname{supp} \phi \subset I$ if L is the line through I. Then $v = \phi u \in \mathscr{E}'$ and

$$B(D)v = \phi B(D)u + w = \phi f + w$$

where $\operatorname{supp} w \subset \operatorname{supp} d\phi$. Since

$$(L \times \{\xi\}) \cap WF(B(D)v) = (L \times \{\xi\}) \cap WF(w)$$

it follows from the discussion preceding the statement of the theorem that there are points $z_\pm \in L$ on either side of x such that $(z_\pm, \xi) \in WF(w)$, hence

$$z_\pm \in L \cap \operatorname{supp} d\phi \quad \text{and} \quad (z_\pm, \xi) \in WF(u).$$

If y_+ and y_- are arbitrary points in the interior of I on different sides of x we can choose ϕ so that $L \cap \operatorname{supp} d\phi$ is as close to $\{y_+, y_-\}$ as we wish. Hence $(y_\pm, \xi) \in WF(u)$ which proves the theorem.

In a moment we shall prove that Theorem 8.3.3 is valid for much more general differential operators with constant coefficients although we do not have quite so explicit fundamental solutions to work with then. However, we give first an example of a solution of the wave equation in $\mathbb{R}^4$ which indicates that Theorem 8.3.3 gives all conditions which $WF(u)$ must satisfy when $B(D)u=0$.

Example 8.3.4. There exists a solution $u \in \mathscr{D}'(\mathbb{R}^4)$ of the wave equation

$$\Box u = (c^{-2}\partial^2/\partial t^2 - \Delta_x)u = 0$$

such that for a given y with $|y|=1$

$$WF(u) = \{(t, cty; sc, -sy), t \in \mathbb{R}, s \neq 0\}.$$

To construct u we change notation and let E_+, E_- be the advanced and retarded fundamental solutions (see Section 6.2). These are proportional to $\delta(c^2t^2 - |x|^2)$ when $t \gtrless 0$, so for the solution $E_0 = E_+ - E_-$

of $\Box E_0=0$ we have by Example 8.2.5, Theorem 8.2.4 and Theorem 8.3.1

$$WF(E_0)\subset\{(t, ctx, sc, -sx); t\in\mathbb{R}, s\in\mathbb{R}\setminus 0, x\in\mathbb{R}^3, |x|=1\}.$$

Let f be a positive C_0^∞ density on the line L through 0 with direction $(1, cy)$, and set $u=E_0*f$. Example 8.2.5 gives

$$WF(f)\subset\{(t, cty; \tau, \xi); t\in\mathbb{R}, \tau+c\langle y, \xi\rangle=0\}.$$

Now the tangent plane $\tau+c\langle y, \xi\rangle=0$ of the characteristic cone $\tau^2-c^2|\xi|^2=0$ meets the cone only when (τ, ξ) is proportional to $(c, -y)$ so (8.2.16) gives

$$WF(u)\subset\{(t, cty; sc, -sy); t\in\mathbb{R}, s\neq 0\}. \tag{8.3.8}$$

By (4.2.2) we have $\{(t, x);\ ct=\langle x, y\rangle\}\cap\operatorname{supp} u\subset L$. If t is so large that $u=f*E_+$ in a neighborhood of $(t, cty)\in L$ it follows from (6.2.7) that the total mass of the measure u at distance $\leqq\delta$ from (t, cty) is at least $C\delta^2$ for some $C>0$. Hence $(t, cty)\in\operatorname{sing\,supp} u$. Since u is real valued the wave front set is symmetric with respect to the origin in the frequency variable. Hence there is equality in (8.3.8) for a missing point would make the left-hand side empty by Theorem 8.3.3.

By another convolution it would be easy to construct a solution with "one half" of the wave front set above. We leave this for the reader since a general result of this kind will be proved below (Theorem 8.3.8).

We shall now extend Theorem 8.3.3 to general differential operators with constant real coefficients and non-singular characteristic set:

Definition 8.3.5. A differential operator $P(D)$ with constant coefficients in $\mathbb{R}^n$ is said to be of real principal type if the principal symbol P_m is real and

$$P_m'(\xi)\neq 0 \quad \text{when } \xi\in\mathbb{R}^n\setminus 0. \tag{8.3.9}$$

Since $P_m'(\xi)=0$ implies $mP_m(\xi)=\langle P_m'(\xi), \xi\rangle=0$ it would be sufficient to assume (8.3.9) when $P_m(\xi)=0$.

By Theorem 7.1.23 the Fourier transform of the fundamental solution $E_\pm$ of $B(D)$ used in the proof of Theorem 8.3.3 is $(B(\xi)\mp i0)^{-1}$. According to Lemma 6.2.2 this is the limit as $\varepsilon\to\mp 0$ of $B(\xi+i\varepsilon v(\xi))^{-1}$ if v is a vector field with $\langle B'(\xi), v(\xi)\rangle>0$. We shall take this as a guide for the construction of fundamental solutions in the general case, but the presence of lower order terms will force us to go a finite distance into the complex domain. With P of real principal type we set

$$v(\xi)=P'_m(\xi)|\xi|^{1-m}.$$

This vector field is homogeneous of degree 0 with respect to ξ. In the following lemma we give a lower bound for P in the direction $iv(\xi)$ from ξ.

Lemma 8.3.6. *There exist positive constants t, C_1, C_2, C_3 such that*

$$\begin{aligned}(8.3.10)\quad \operatorname{Im} P(\xi+itv(\xi)+iV)\\ \geqq C_1(1+|\xi|)^{m-1}+\langle P'_m(\xi), V\rangle - C_2(|V|+1)|V|(|\xi|+|V|)^{m-2}\end{aligned}$$

if $\xi\in\mathbb{R}^n, |\xi|\geqq C_3, V\in\mathbb{R}^n$.

Proof. Taylor's formula gives

$$\begin{aligned}\operatorname{Im} P(\xi+itv(\xi)+iV)\geqq \langle P'_m(\xi), tv(\xi)+V\rangle - C(t+|V|)^2(|\xi|+t+|V|)^{m-2}\\ -C(|\xi|+t+|V|)^{m-1}.\end{aligned}$$

Here we have by (8.3.9) for some $c>0$

$$\langle P'_m(\xi), v(\xi)\rangle \geqq c|\xi|^{m-1}.$$

If t is fixed so that $tc>C$, we obtain (8.3.10) when $V=0$; an obvious estimate of the terms involving V gives (8.3.10) in general.

Theorem 8.3.7. *If $P(D)$ is of real principal type then one can find $E_\pm\in\mathscr{D}'(\mathbb{R}^n)$ and $\omega_\pm\in C^\infty(\mathbb{R}^n)$ such that $P(D)E_\pm=\delta+\omega_\pm$ and*

$$(8.3.11)\quad WF(E_\pm)\subset\{(tP'_m(\xi), \xi);\ t\gtrless 0, P_m(\xi)=0, \xi\neq 0\}\cup T^*_0\setminus 0.$$

Proof. It is no restriction to assume that $m>n+1$, for if $E_\pm$ has the properties stated in the theorem for $\Delta^k P(D)$, then $\Delta^k E_\pm$ has these properties for $P(D)$. (Δ is the Laplace operator.) Replacing P by $-P$ interchanges E_+ and E_- so it suffices to construct E_-. Guided by the second order case as indicated above we let Γ be the chain

$$(8.3.12)\qquad \mathbb{R}^n\ni\xi\to\xi+itv(\xi),\qquad |\xi|\geqq C_3,$$

where C_3 and t are given by Lemma 8.3.6. Noting that by (8.3.10) $\operatorname{Im} P(\zeta)\geqq C_1(1+|\operatorname{Re}\zeta|)^{n+1}$ on Γ we set

$$(8.3.13)\quad E_-(x)=(2\pi)^{-n}\int_\Gamma e^{i\langle x,\zeta\rangle}/P(\zeta)\,d\zeta_1\wedge\ldots\wedge d\zeta_n,\qquad x\in\mathbb{R}^n.$$

In terms of the parameters $\xi_1,\ldots,\xi_n$ on Γ we have explicitly

$$d\zeta_1\wedge\ldots\wedge d\zeta_n=J\,d\xi_1\wedge\ldots\wedge d\xi_n,$$
$$J=D(\xi_1+itv_1(\xi),\ldots,\xi_n+itv_n(\xi))/D(\xi_1,\ldots,\xi_n)\to 1 \text{ at } \infty,$$

so the integral (8.3.13) is locally absolutely and uniformly convergent. When $\phi \in C_0^\infty$ we have with $\psi = P(-D)\phi$

$$\langle P(D)E_-, \phi\rangle = \langle E_-, \psi\rangle = (2\pi)^{-n} \iint_\Gamma \psi(x) e^{i\langle x,\zeta\rangle}/P(\zeta)\, dx\, d\zeta_1 \wedge \ldots \wedge d\zeta_n.$$

Here we integrate first with respect to x and use that $\hat{\psi}(-\zeta) = P(\zeta)\hat{\phi}(-\zeta)$. Now $F(\zeta)\, d\zeta_1 \wedge \ldots \wedge d\zeta_n$ is a closed differential form for every analytic function F, since dF is a linear combination of $d\zeta_1, \ldots, d\zeta_n$. Thus $\hat{\phi}(-\zeta) d\zeta_1 \wedge \ldots \wedge d\zeta_n$ is a closed differential form which decreases rapidly at infinity, so Stokes' formula gives

$$\begin{aligned}\langle P(D)E_-, \phi\rangle &= (2\pi)^{-n} \int_\Gamma \hat{\phi}(-\zeta) d\zeta_1 \wedge \ldots \wedge d\zeta_n \\ &= (2\pi)^{-n} \int_{\mathbb{R}^n} \hat{\phi}(-\xi) d\xi - (2\pi)^{-n} \int_{\Gamma_0} \hat{\phi}(-\zeta) d\zeta_1 \wedge \ldots \wedge d\zeta_n.\end{aligned}$$

Here Γ_0 is the chain

$$\mathbb{R}^n \ni \xi \to \xi + itv_0(\xi), \quad |\xi| \leqq C_3,$$

where v_0 is a smooth extension of v from $|\xi| = C_3$ to $|\xi| \leqq C_3$, which makes $\Gamma \cup \Gamma_0$ homotopic to $\mathbb{R}^n$. Thus $P(D)E_- = \delta + \omega_-$ where

$$\omega_-(x) = -(2\pi)^{-n} \int_{\Gamma_0} e^{i\langle x,\zeta\rangle} d\zeta_1 \wedge \ldots \wedge d\zeta_n$$

is an entire analytic function since Γ_0 is compact. Hence Theorem 8.3.1 shows that $P_m(\xi) = 0$ if $(x, \xi) \in WF(E_-)$ and $x \neq 0$.

To complete the proof we must show that $(x_0, \xi_0) \notin WF(E_-)$ if $x_0 \notin \mathbb{R}_- P'_m(\xi_0)$. This condition means precisely that we can find $V \in \mathbb{R}^n$ with $|V| = 1$ and

$$\langle x_0, V\rangle > 0, \quad \langle P'_m(\xi_0), V\rangle > 0. \tag{8.3.14}$$

Choosing a conic neighborhood W of ξ_0 such that for some $c > 0$

$$\langle P'_m(\xi), V\rangle > c|\xi|^{m-1}, \quad \xi \in W,$$

we obtain from (8.3.10) when $\xi \in W$

$$\begin{aligned}&\operatorname{Im} P(\xi + itv(\xi) + isV) \\ &\geqq C_1(1+|\xi|)^{m-1} + c|\xi|^{m-1} s - C_2(s+1)s(|\xi|+s)^{m-2} \\ &\geqq C_1(1+|\xi|)^{m-1}\end{aligned}$$

if $0 < s < \varepsilon|\xi|$ and $|\xi|$ is large enough. Replacing V by εV we have

$$\begin{aligned}&\operatorname{Im} P(\xi + itv(\xi) + isV) \geqq C_1(1+|\xi|)^{m-1}, \\ &\xi \in W, \quad 0 \leqq s \leqq |\xi|, \quad |\xi| \geqq C'_3.\end{aligned} \tag{8.3.15}$$

Choose $\chi \in C^\infty(\mathbb{R}^n \setminus 0)$ homogeneous of degree 0 with support in W so that $0 \leqq \chi \leqq 1$, and $\chi = 1$ in a conic neighborhood W_0 of ξ_0. If

$\langle x, V\rangle > 0$, which is true in a neighborhood of x_0, we obtain using Stokes' formula

$$E_-(x) = (2\pi)^{-n} \int_{\Gamma' \cup \Gamma'_0} e^{i\langle x, \zeta\rangle}/P(\zeta)\, d\zeta_1 \wedge \ldots \wedge d\zeta_n \tag{8.3.13$'$}$$

where Γ' is the chain

$$\mathbb{R}^n \ni \xi \to \xi + itv(\xi) + i|\xi|\chi(\xi)V, \qquad |\xi| \geqq C'_3,$$

and Γ'_0 is the union of the part of Γ where $C_3 < |\xi| < C'_3$ with the chain

$$\{(\xi, s);\ |\xi| = C'_3, 0 < s < C'_3\} \to \xi + itv(\xi) + is\chi(\xi)V$$

with suitable orientations. The contribution to (8.3.13)$'$ when $\zeta \in \Gamma'_0$ or $\operatorname{Re}\zeta \in W_0$ is an analytic function of x when $\langle x, V\rangle > 0$. If M is a measurable conic set contained in a closed proper convex cone G, then the wave front set of the function

$$x \to \int_{\zeta \in \Gamma',\ \operatorname{Re}\zeta \in M} e^{i\langle x, \zeta\rangle}/P(\zeta)\, d\zeta_1 \wedge \ldots \wedge d\zeta_n, \qquad \langle x, V\rangle > 0,$$

is contained in $\{(x, \xi);\ \langle x, V\rangle > 0,\ \xi \in G\}$. This follows from Theorem 8.1.6. In fact, replacing x by $z = x + iy$ we obtain a bounded analytic function when $|x|$ is bounded, $\langle x, V\rangle > 0$, and y is in the interior of the dual cone of G, for

$$\operatorname{Re} i\langle z, \zeta\rangle = -\langle x, \operatorname{Im}\zeta\rangle - \langle y, \operatorname{Re}\zeta\rangle \leqq -t\langle x, v(\xi)\rangle < C|x|.$$

We can cover $\complement W_0$ with a finite number of such cones G which do not contain ξ_0, so it follows that $(x_0, \xi_0) \notin WF(E_-)$. The proof is complete.

Repetition of the proof of Theorem 8.3.3 gives now

Theorem 8.3.3$'$. *Let $P(D)$ be of real principal type. If $u \in \mathscr{D}'(X)$, $P(D)u = f$ and $(x, \xi) \in WF(u) \setminus WF(f)$, then $P_m(\xi) = 0$ and*

$$I \times \{\xi\} \subset WF(u)$$

if $I \subset X$ is a line segment containing x with direction $P'_m(\xi)$ such that $I \times \{\xi\}$ does not meet $WF(f)$.

Finally we shall give a general version of Example 8.3.4.

Theorem 8.3.8. *Let $P(D)$ be of real principal type, $0 \neq \xi \in \mathbb{R}^n$ and $P_m(\xi) = 0$. Then one can find $u \in C^m(\mathbb{R}^n)$ such that $P(D)u \in C^\infty(\mathbb{R}^n)$ and*

$$WF(u) = \{(tP'_m(\xi), s\xi);\ t \in \mathbb{R}, s > 0\}. \tag{8.3.16}$$

Proof. Set $L = \mathbb{R}P'_m(\xi)$ and let $\mathscr{F}$ be the set of all $u \in C^m(\mathbb{R}^n)$ with $Pu \in C^\infty(\mathbb{R}^n)$, $u \in C^\infty(\complement L)$ and $WF(u) \subset \mathbb{R}^n \times (\mathbb{R}_+\xi)$. The theorem states that there is an element $u\in\mathscr{F}$ which is not in C^∞, for $u\in\mathscr{F}$ implies

$$WF(u)\subset \mathbb{R}P'_m \times \mathbb{R}_+\xi$$

and by Theorem 8.3.3′ $u\in C^\infty$ if the inclusion is strict. Now $\mathscr{F}$ is a Fréchet space with the seminorms

(i) $\sup_K |D^\alpha u|$, $|\alpha|\leqq m$, K a compact subset of $\mathbb{R}^n$,

(ii) $\sup_K |D^\alpha u|$, α arbitrary, K a compact subset of $\complement L$,

(iii) $\sup_K |D^\alpha P(D)u|$, α arbitrary, K a compact subset of $\mathbb{R}^n$,

(iv) $\sup_{\complement\Gamma_N} |\eta|^N |\widehat{\phi u}(\eta)|$, $N=1, 2, \ldots$, $\phi\in C_0^\infty(\mathbb{R}^n)$.

Here Γ_N is a sequence of conic neighborhoods of ξ in $\mathbb{R}^n$ shrinking to $\mathbb{R}_+\xi$. We need only use a countable number of compact sets K and functions ϕ since the semi-norms (iv) can be estimated by the corresponding ones with ϕ replaced by a function ψ which is 1 in $\operatorname{supp}\phi$. (See the proof of Lemma 8.1.1.) The proof of completeness is an exercise for the reader. If $\mathscr{F}\subset C^{m+1}$ then the closed graph theorem shows that the inclusion $\mathscr{F}\hookrightarrow C^{m+1}$ is continuous. Thus one can find N, $\phi\in C_0^\infty(\mathbb{R}^n)$, $K_1\Subset\mathbb{R}^n$ and $K_2\Subset\complement L$ so that

$$\sum_{|\alpha|=m+1} |D^\alpha u(0)| \leqq C\{\sum_{|\alpha|\leqq m} \sup_{K_1} |D^\alpha u| + \sum_{|\alpha|\leqq N} \sup_{K_2} |D^\alpha u| + \sum_{|\alpha|\leqq N} \sup_{K_1} |D^\alpha P(D)u| + \sup_{\complement\Gamma_N} (1+|\eta|)^N |\widehat{\phi u}(\eta)|, \quad u\in\mathscr{F}. \tag{8.3.17}$$

To show that (8.3.17) is not valid we need to construct approximate solutions of the equation $Pu=0$ concentrated close to L, thus away from K_2. To make the last term small the Fourier transform of u should be concentrated close to the direction ξ. It is therefore natural to set for $t>0$

$$u_t(x)=e^{it\langle x,\xi\rangle} v_t(x).$$

Then

$$P(D)u_t(x)=e^{it\langle x,\xi\rangle} P(D+t\xi)v_t(x)$$
$$=t^{m-1} e^{it\langle x,\xi\rangle} \left(\sum_1^n P_m^{(j)}(\xi) D_j v_t + P_{m-1}(\xi) v_t + \ldots\right)$$

where terms indicated by dots contain a negative power of t, and $P_m^{(j)} =\partial_j P_m$. A formal solution

$$v_t=v_0+t^{-1}v_1+\ldots$$

may be found by solving the first order equation

$$Lv_0=\sum_1^n P_m^{(j)}(\xi) D_j v_0 + P_{m-1}(\xi) v_0 = 0 \tag{8.3.18}$$

and then successively equations

$$Lv_j = f_j \tag{8.3.19}$$

where f_j is determined by $v_0, \dots, v_{j-1}$. The support of v_0 is a cylinder with the axis in the direction $P_m'(\xi)$; we can choose v_0 with $v_0(0)=1$ and support close to L by prescribing such values on a plane Σ orthogonal to $P_m'(\xi)$. If the other functions v_j are determined by the boundary condition $v_j=0$ on Σ, it is clear that $\operatorname{supp} v_j \subset \operatorname{supp} v_0$ for $j \neq 0$. For

$$v_t = \sum_{j<M} v_j t^{-j}$$

the third sum on the right-hand side of (8.3.17) is $O(t^{m-1-M+N})$. The last term is rapidly decreasing when $t \to \infty$ since

$$\widehat{\phi u_t}(\eta) = \sum_{j<M} t^{-j} (\widehat{\phi v_j})(\eta - t\xi)$$

and $t + |\eta| \leqq C|\eta - t\xi|$ when $\eta \notin \Gamma_N$. The first sum on the right-hand side of (8.3.17) is $O(t^m)$, the second sum is 0 for an appropriate choice of v_0, but the left-hand side grows as t^{m+1} since $\xi \neq 0$. If we take $M=N$ this is a contradiction which completes the proof.

The preceding argument can of course be given a more constructive look by summing a very lacunary sequence of the functions u_t as in the proof of Theorem 8.1.4. However, making the proof of the closed graph theorem explicit in this way tends to hide the idea of the proof. – In Chapter X we shall prove that the equation $P(D)v=f$ has a solution $v \in C^\infty(\mathbb{R}^n)$ for every $f \in C^\infty(\mathbb{R}^n)$. If we take $f = P(D)u$ and replace u by $u-v$ we obtain a solution of the homogeneous equation $P(D)u=0$ satisfying (8.3.16).

The fact that we have restricted ourselves to discussing operators of real principal type with constant coefficients does not mean that results such as Theorem 8.3.3′ and Theorem 8.3.8 are not available when the coefficients are variable. We shall return to these matters in Chapter XXVI after introducing the appropriate tools.

8.4. The Wave Front Set with Respect to C^L

Let L_k be an increasing sequence of positive numbers such that $L_0 = 1$ and

$$k \leqq L_k, \quad L_{k+1} \leqq C L_k \tag{8.4.1}$$

for some constant C. If $X \subset \mathbb{R}^n$ is an open set we shall denote by $C^L(X)$ the set of all $u \in C^\infty(X)$ such that for every compact set $K \subset X$ there is a constant C_K with

$$|D^\alpha u(x)| \leqq C_K (C_K L_{|\alpha|})^{|\alpha|}, \quad x \in K, \tag{8.4.2}$$

for all multi-indices α. (This notation, used from now on, differs slightly from the standard one used in Section 1.3.) When $L_k = k+1$ this means that $C^L(X)$ is the set of real analytic functions in X. The class C^L with $L_k = (k+1)^a$, $a > 1$, is called the Gevrey class of order a. It occurs quite frequently in the theory of partial differential equations.

Proposition 8.4.1. *$C^L(X)$ is a ring which is closed under differentiation. If $f\colon Y \to X$ is an analytic map from the open set $Y \subset \mathbb{R}^m$ to the open set $X \subset \mathbb{R}^n$, then composition with f defines a map $f^*\colon C^L(X) \to C^L(Y)$.*

Proof. Since L_k is increasing we obtain by Leibniz' rule

$$\sup_K |D^\alpha(uv)| \leqq C_K^2 (2 C_K L_{|\alpha|})^{|\alpha|}$$

if u and v satisfy (8.4.2). Thus C^L is a ring. That C^L is closed under differentiation follows from the inequality

$$(L_{j+1})^{j+1} \leqq (C L_j)^{j+1} \leqq C^{2j+1} L_j^j \tag{8.4.3}$$

which is a consequence of the second part of (8.4.1). To prove the last statement we note that the derivatives of f^*u at y of order k are the same as the derivatives of

$$z \to \sum_{|\alpha| \leqq k} (D^\alpha u)(f(y))(if(z) - if(y))^\alpha/\alpha!$$

when $z = y$. The right-hand side is an analytic function of z when $y \in K$ and z is complex with $|y - z| < r$ sufficiently small. It can then be estimated by

$$C \sum_{|\alpha| \leqq k} (C L_k)^{|\alpha|} (C' r)^{|\alpha|}/\alpha! = C \sum_0^k (n C C' r L_k)^j / j!.$$

Now $L_k^j / j! \leqq L_k^k / k!$ by the first part of (8.4.1), so this sum can be estimated by

$$C L_k^k / k! \sum_0^k (n C C' r)^j < 2 C L_k^k / k!$$

if $r \leqq 1/(2nCC')$. By Cauchy's inequalities we conclude that

$$|D^\alpha f^* u(y)| \leqq 2 C r^{-k} L_k^k \quad \text{if } |\alpha| = k, \quad y \in K,$$

which completes the proof.

Proposition 8.4.1 shows that we can define $C^L(X)$ by means of local coordinate systems when X is a real analytic manifold. (This means that an atlas for X is given such that the maps (6.3.1) are all real analytic.)

For any distribution $u \in \mathscr{D}'(X)$ we define $\operatorname{sing\,supp}_L u$ to be the smallest closed subset of X such that u is in C^L in the complement. (When C^L is the real analytic class we shall use the notation $\operatorname{sing\,supp}_A u$.) The purpose of this section is to show how one can make a spectral analysis of this set parallel to Sections 8.1 and 8.2. A new difficulty occurs when

(8.4.4) $$\sum_1^\infty 1/L_k = \infty$$

for then the class C^L is quasi-analytic by the Denjoy-Carleman theorem (Theorem 1.3.8) so one cannot choose cutoff functions in C^L. (Multiplication by C^∞ functions not in C^L may of course increase $\operatorname{sing\,supp}_L u$.) However, this difficulty can be circumvented by using Theorem 1.4.2 to choose test functions with adequate bounds for derivatives up to a certain order only. This leads to a description of $\operatorname{sing\,supp}_L$ in terms of Fourier transforms:

Proposition 8.4.2. *Let $x_0 \in X \subset \mathbb{R}^n$ and $u \in \mathscr{D}'(X)$. Then $u \in C^L$ in a neighborhood of x_0 if and only if for some neighborhood U of x_0 there is a bounded sequence $u_N \in \mathscr{E}'(X)$ which is equal to u in U and satisfies*

(8.4.5) $$|\hat{u}_N(\xi)| \leqq C(CL_N/|\xi|)^N, \qquad N=1,2,\ldots$$

for some constant C.

Proof. a) Necessity. Let $u \in C^L$ when $|x-x_0|<3r$ and choose $\chi_N \in C_0^\infty$ so that $\chi_N=1$ when $|x-x_0|<r$, $\chi_N=0$ when $|x-x_0|>2r$ and

(8.4.6) $$|D^\alpha \chi_N| \leqq (C_1 N)^{|\alpha|}, \qquad |\alpha| \leqq N,$$

where C_1 does not depend on N. This is possible by Theorem 1.4.2 with $d_k=r/2N$ for $k \leqq N$. With $u_N=\chi_N u$ we obtain from (8.4.2) and (8.4.6) since $N \leqq L_N$ and $L_{|\alpha|} \leqq L_N$ when $|\alpha| \leqq N$

$$|D^\alpha u_N| \leqq C_K(C_K+C_1)^N L_N^N, \qquad |\alpha|=N.$$

Hence

$$|\xi^\alpha \hat{u}_N(\xi)| \leqq C_2^{N+1} L_N^N, \qquad |\alpha|=N,$$

and since $|\xi| \leqq n^{\frac{1}{2}} \max |\xi_j|$ this proves (8.4.5).

b) Sufficiency. In U we have

(8.4.7) $$D^\alpha u(x)=(2\pi)^{-n} \int \xi^\alpha \hat{u}_N(\xi) e^{i\langle x,\xi\rangle} d\xi$$

when $N=|\alpha|+n+1$, for $\xi^\alpha \hat{u}_N(\xi)$ is then integrable by (8.4.5). That u_N is bounded in $\mathscr{E}'$ implies by the Banach-Steinhaus theorem that

$$|\hat{u}_N(\xi)| \leqq C(1+|\xi|)^M, \qquad N=1,2,\ldots$$

where C and M are independent of N. (See the proof of Theorem 2.1.8.) We use this to estimate the integrand in (8.4.7) when $|\xi|<L_N$ but use the estimate (8.4.5) when $|\xi|>L_N$. This gives that in U

$$|D^\alpha u(x)| \leqq C_3 (C_3 L_{|\alpha|+n+1})^{|\alpha|+n+M}$$

and repeated use of (8.4.3) shows now that $u \in C^L(U)$.

Proposition 8.4.2 suggests the following definition:

Definition 8.4.3. If $X \subset \mathbb{R}^n$ and $u \in \mathscr{D}'(X)$ we denote by $WF_L(u)$ the complement in $X \times (\mathbb{R}^n \setminus 0)$ of the set of (x_0, ξ_0) such that there is a neighborhood $U \subset X$ of x_0, a conic neighborhood Γ of ξ_0 and a bounded sequence $u_N \in \mathscr{E}'(X)$ which is equal to u in U and satisfies (8.4.5) when $\xi \in \Gamma$. When C^L is the analytic class we use the notation $WF_A(u)$.

By definition $WF_L(u)$ is a closed subset of $X \times (\mathbb{R}^n \setminus 0)$. The following lemma shows that u_N can always be chosen as products of u and suitable cutoff functions, obtained by regularizing those in the proof of Proposition 8.4.2.

Lemma 8.4.4. *Let $u \in \mathscr{D}'(X)$ and let K be a compact subset of X, F a closed cone $\subset \mathbb{R}^n$ such that $WF_L(u) \cap (K \times F) = \emptyset$. If $\chi_N \in C_0^\infty(K)$ and for all α*

$$|D^{\alpha+\beta}\chi_N| \leqq C_\alpha (C_\alpha L_N)^{|\beta|}, \qquad |\beta| \leqq N=1,2,\ldots, \tag{8.4.6$'$}$$

it follows that $\chi_N u$ is bounded in $\mathscr{E}'^M$ if u is of order M in a neighborhood of K, and we have

$$|\widehat{\chi_N u}(\xi)| \leqq C(CL_N/|\xi|)^N, \qquad N=1,2,\ldots,\ \xi \in F. \tag{8.4.5$'$}$$

Proof. The boundedness of $\chi_N u$ is obvious since χ_N is bounded in C_0^∞. Let $x_0 \in K$, $\xi_0 \in F \setminus \{0\}$ and choose U, Γ, u_N according to Definition 8.4.3. If $\operatorname{supp} \chi_N \subset U$ then $\chi_N u = \chi_N u_N$. By hypothesis u_N satisfies (8.4.5) in Γ, and $|\hat{u}_N(\xi)| \leqq C(1+|\xi|)^M$ for fixed C, $M \geqq 0$. From (8.4.6)$'$ it follows that

$$|\hat{\chi}_N(\eta)| \leqq C^{N+1}(L_N/(|\eta|+L_N))^N (1+|\eta|)^{-n-1-M}.$$

If we apply (8.1.3)$'$ with $v=u_N$ and $\phi=\chi_N$, the estimate (8.4.5)$'$ is proved if F is replaced by a closed conic neighborhood of ξ_0 contained in the interior of Γ apart from the origin. Since F can be covered by a finite number of such neighborhoods it follows that (8.4.5)′ is valid if $\operatorname{supp}\chi_N \subset U$ for a sufficiently small neighborhood U of x_0. We can cover K by such neighborhoods U_j, $j=1, \dots, J$, and choose $\chi_{N,j} \in C_0^\infty(U_j)$ so that $\sum \chi_{N,j}=1$ in K and $\chi_{N,j}$ satisfies (8.4.6)′ for $j=1, \dots, J$. To do so we just have to regularize any partition of unity by a function ψ_N chosen according to Theorem 1.4.2 so that it satisfies (8.4.6). If $\chi_N \in C_0^\infty(K)$ satisfies (8.4.6)′ it is clear that $\chi_{N,j}\chi_N$ also satisfies (8.4.6)′ with some other constants. Hence (8.4.5)′ is valid with χ_N replaced by $\chi_{N,j}\chi_N$. Since $\sum \chi_{N,j}\chi_N=\chi_N$ this completes the proof.

Lemma 8.4.4 and Proposition 8.4.2 give easily

Theorem 8.4.5. *The projection of* $WF_L(u)$ *in* X *is equal to* $\operatorname{sing\,supp}_L u$ *if* $u \in \mathscr{D}'(X)$.

Proof. a) If $u \in C^L$ in a neighborhood of x_0 it follows from Proposition 8.4.2 that $(x_0, \xi_0) \notin WF_L(u)$, $\xi_0 \in \mathbb{R}^n \setminus \{0\}$. b) Assume that $(x_0, \xi_0) \notin WF_L(u)$ for all $\xi_0 \in \mathbb{R}^n \setminus \{0\}$. Then we can choose a compact neighborhood K of x_0 so that $WF_L(u) \cap (K \times \mathbb{R}^n)=\emptyset$. By Lemma 8.4.4 there is a sequence $\chi_N \in C_0^\infty(K)$ which is equal to 1 in a neighborhood U of x_0 such that $\chi_N u$ is bounded in $\mathscr{E}'$ and satisfies (8.4.5). Hence $x_0 \notin \operatorname{sing\,supp}_L u$ by Proposition 8.4.2.

The condition (8.4.6)′ is satisfied by any fixed function in C^L with support in K. If C^L is non-quasianalytic we can therefore simplify Definition 8.4.3 to the existence of a fixed distribution v which is equal to u in a neighborhood of x_0 and has a Fourier transform satisfying (8.4.5) in a conic neighborhood of ξ_0. This is parallel to Definition 8.1.2, so we obtain

Theorem 8.4.6. *For all* u *and* L *we have* $WF(u) \subset WF_L(u) \subset WF_A(u)$; *moreover, if* $L'_j \leqq L''_j$ *then* $WF_{L''}(u) \subset WF_{L'}(u)$.

The conditions (8.4.6)′ remain valid if we multiply all χ_N by the same function in $C^L(X)$. This gives

Theorem 8.4.7. $WF_L(au) \subset WF_L(u)$ *if* $a \in C^L(X)$ *and* $u \in \mathscr{D}'(X)$.

It is obvious that

$$WF_L(\partial u/\partial x_j) \subset WF_L(u)$$

for (8.4.5) implies in view of (8.4.3)

$$|\xi_j \hat{u}_{N+1}(\xi)| \leqq C|\xi|(CL_{N+1}/|\xi|)^{N+1} \leqq C'(C'L_N/|\xi|)^N.$$

If we combine this with Theorem 8.4.7 we obtain

$$WF_L(P(x,D)u) \subset WF_L(u),$$

if $u \in \mathscr{D}'(X)$ and

$$P(x,D)u = \sum_{|\alpha| \leqq m} a_\alpha(x) D^\alpha$$

is a differential operator with coefficients in $C^L(X)$.

We could now proceed to study $WF_L(u)$ using arguments completely analogous to those in Sections 8.1 to 8.3. However, to avoid such boring repetitions we shall use an alternative approach which has the advantage that it is also applicable in the study of hyperfunctions in Chapter IX. The first step is just an improvement of Theorem 8.1.6.

Theorem 8.4.8. *If the hypotheses of Theorem 3.1.15 are fulfilled, then*

$$WF_A(f_0) \subset X \times (\Gamma^\circ \setminus \{0\}) \tag{8.4.8}$$

where Γ° is the dual cone of Γ.

Proof. Let $X_1 \Subset X_0 \Subset X$ be open sets and choose using Theorem 1.4.2 a sequence $\phi_\nu \in C_0^\infty(X_0)$ with $\phi_\nu = 1$ in X_1 such that

$$|D^\alpha \phi_\nu| \leqq (C_1(\nu+1))^{|\alpha|}, \quad |\alpha| \leqq \nu+1.$$

As in (3.1.18) we set

$$\Phi_\nu(x,y) = \sum_{|\alpha| \leqq \nu} \partial^\alpha \phi_\nu(x)(iy)^\alpha/\alpha!$$

and have by (8.1.15) for a fixed $Y \in \Gamma$ with $|Y| < \gamma$

$$\widehat{\phi_\nu f_0}(\xi) = \int \Phi_\nu(x,Y) f(x+iY) e^{-i\langle x+iY,\xi\rangle} dx$$
$$+(\nu+1) \iint_{0<t<1} f(x+itY) e^{-i\langle x+itY,\xi\rangle} \sum_{|\alpha|=\nu+1} \partial^\alpha \phi_\nu(x)(iY)^\alpha/\alpha!\, t^\nu \, dx\, dt.$$

For $\mu \leqq \nu+1$ we have with $|Y|_1 = \sum |Y_j|$

$$\Big|\sum_{|\alpha|=\mu} \partial^\alpha \phi_\nu(x)(iY)^\alpha/\alpha!\Big| \leqq C_1^\mu (\nu+1)^\mu |Y|_1^\mu/\mu!$$

so it follows with $C_2 = 2\, e^{C_1|Y|_1}$ that

$$|\Phi_\nu(x,Y)| \leqq C_2^{\nu+1}, \quad \Big|(\nu+1) \sum_{|\alpha|=\nu+1} \partial^\alpha \phi_\nu(x)(iY)^\alpha/\alpha!\Big| \leqq C_2^{\nu+1}.$$

The estimate (8.1.16) is therefore replaced by

$$|\widehat{\phi_\nu f_0}(\xi)| \leqq C_3^{\nu+1}(e^{\langle Y,\xi\rangle} + (\nu-N)!\,\langle -Y,\xi\rangle^{N-\nu-1}), \quad \langle Y,\xi\rangle < 0.$$

Set $f_\nu = \phi_{N+\nu-1} f_0$. When $\langle Y, \xi\rangle < -c|\xi|$ for a fixed c we obtain for some C_4

$$|\hat{f}_\nu(\xi)| \leqq C_4^{\nu+1} \nu! |\xi|^{-\nu}, \tag{8.4.9}$$

for $e^{-c|\xi|} \leqq \nu! (c|\xi|)^{-\nu}$.

Now $f_0 \in \mathscr{D}'^{N+1}$ by Theorem 3.1.15. If we choose the sequence ϕ_ν bounded in C_0^{N+1} it follows that f_ν is bounded in $\mathscr{E}'^{N+1}$. Thus we have proved that

$$WF_A(f_0) \subset X \times \{\xi; \langle Y, \xi\rangle \geqq 0\},$$

and since Y has an arbitrary direction in Γ, this proves the theorem.

We shall now show that it is possible to associate with any distribution $u \in \mathscr{S}'(\mathbb{R}^n)$ an analytic function U in the convex tube

$$\Omega = \{z \in \mathbb{C}^n; |\operatorname{Im} z| < 1\}$$

(Euclidean norm) so that

$$u = \int_{|\omega|=1} U(.+i\omega)\, d\omega. \tag{8.4.10}$$

The interest of such a decomposition is clear if we note that for the boundary value $U(.+i\omega) = \lim_{r \nearrow 1} U(.+ir\omega)$ we must have

$$WF_A(U(.+i\omega)) \subset \mathbb{R}^n \times \mathbb{R}_- \omega,$$

by Theorem 8.4.8, so one should be able to determine if $(x, -\omega) \in WF_A(u)$ by just looking at the behavior of U near $x + i\omega$.

Assuming that U is well behaved at infinity we denote by $\hat{U}(\xi, y)$ the Fourier transform of $U(x+iy)$ with respect to x and observe that the Cauchy-Riemann equations $\partial U/\partial x_j + i\partial U/\partial y_j = 0$ give

$$\xi_j \hat{U} + \partial \hat{U}/\partial y_j = 0, \quad j = 1, \dots, n.$$

Thus $\hat{U}(\xi, y) = U_0(\xi) e^{-\langle y, \xi\rangle}$, and (8.4.10) leads to the condition $\hat{u}(\xi) = I(\xi) U_0(\xi)$,

$$I(\xi) = \int_{|\omega|=1} e^{-\langle \omega, \xi\rangle}\, d\omega. \tag{8.4.11}$$

When $n=1$ we have $I(\xi) = 2\cosh \xi$, and when $n>1$ we have $I(\xi) = I_0(\langle \xi, \xi\rangle^{\frac{1}{2}})$

$$I_0(\rho) = c_{n-1} \int_{-1}^{1} (1-t^2)^{(n-3)/2} e^{t\rho}\, dt \tag{8.4.11$'$}$$

where c_{n-1} is the area of S^{n-2}. This is a multiple of the Bessel function $J_{(n-2)/2}(i\rho)$ divided by $\rho^{(n-2)/2}$ so the asymptotic behavior is well known. For the convenience of the reader we give a direct proof of what we need.

Lemma 8.4.9. *I_0 is an even analytic function in $\mathbb{C}$ such that for every $\varepsilon>0$*

(8.4.12) $$I_0(\rho)=(2\pi)^{(n-1)/2}e^\rho\rho^{-(n-1)/2}(1+O(1/\rho))$$
$$\text{if } \rho\to\infty, \quad |\arg\rho|<\pi/2-\varepsilon.$$

There is a constant C such that for all $\rho\in\mathbb{C}$

(8.4.12)′ $$|I_0(\rho)|\leqq C(1+|\rho|)^{-(n-1)/2}e^{|\operatorname{Re}\rho|}.$$

Proof. Since $I_0(\bar\rho)=\overline{I_0(\rho)}$ we may assume that $0\leqq\arg\rho<\pi/2$ when proving (8.4.12) and (8.4.12)′. To prove (8.4.12) we take the integration from -1 to $+1$ in (8.4.11)′ along the short sides of a right triangle with hypothenuse $(-1,1)$ and $(t-1)\rho<0$ on the side through 1. Writing $1-t^2=(1-t)(1+t)$ and taking $s=(1-t)\rho$ as integration variable on this side we obtain since $1+t=2-s/\rho$

$$e^{-\rho}I_0(\rho)\rho^{(n-1)/2}=c_{n-1}\int_0^\infty(2s)^{(n-3)/2}e^{-s}\,ds+O(1/\rho).$$

(8.4.12) follows in view of (3.4.2). If we integrate along the half lines $(\pm1-t)\,\rho<0$ instead, the estimate (8.4.12)′ is obtained.

Now we introduce the function K correspondending to decomposition of $u=\delta$,

(8.4.13) $$K(z)=(2\pi)^{-n}\int e^{i\langle z,\xi\rangle}/I(\xi)\,d\xi, \qquad z\in\Omega.$$

It follows at once from (8.4.12) that the integral converges and defines an analytic function in Ω.

Lemma 8.4.10. *$K(z)$ is an analytic function in the connected open set*

$$\tilde\Omega=\{z\in\mathbb{C}^n;\ \langle z,z\rangle\notin(-\infty,-1]\}\supset\Omega.$$

For any closed cone $\Gamma\subset\tilde\Omega$ such that $\langle z,z\rangle$ is never $\leqq 0$ when $z\in\Gamma\setminus\{0\}$ there is some $c>0$ such that $K(z)=O(e^{-c|z|})$ when $z\to\infty$ in Γ. We have for real x and y

(8.4.14) $$|K(x+iy)|\leqq K(iy)$$
$$=(n-1)!\,(2\pi)^{-n}(1-|y|)^{-n}(1+O(1-|y|)), \qquad |y|\nearrow 1.$$

Proof. Introducing polar coordinates we obtain $K(z)=K_0(\langle z,z\rangle^{\frac12})$ where K_0 is the even analytic function

$$K_0(w)=(2\pi)^{-n}\iint\exp(i\rho\omega_1 w)/I_0(\rho)\,\rho^{n-1}\,d\omega\,d\rho$$
$$=(2\pi)^{-n}\int_0^\infty I_0(i\rho w)/I_0(\rho)\rho^{n-1}\,d\rho, \qquad |\operatorname{Im} w|<1, \quad w\in\mathbb{C}.$$

When $w=-it$, $1/2<t<1$ then

$$K_0(-it)=(2\pi)^{-n}\int_0^\infty I_0(\rho t)/I_0(\rho)\rho^{n-1}d\rho$$
$$=(2\pi)^{-n}\int_0^\infty e^{-\rho(1-t)}(1+O(1/(\rho+1)))t^{-(n-1)/2}\rho^{n-1}d\rho$$
$$=(n-1)!\,(2\pi)^{-n}(1-t)^{-n}(1+O(1-t))$$

which proves (8.4.14). To study the analyticity of K we note that if $w=it$, $-1<t<1$, then Cauchy's integral formula gives

$$K_0(w)=(2\pi)^{-n}\int_0^\infty I_0(i\rho w(1+is))/I_0(\rho(1+is))(1+is)^n\rho^{n-1}d\rho$$

for any real s. In fact, $I_0\neq 0$ in the right half plane since the Bessel function J_ν has only real zeros when $\nu>-1$ (see Hurwitz [1]). The right-hand side is an analytic function $K_s(w)$ in the set

$$Z_s=\{w;\ |\mathrm{Re}\,(iw(1+is))|<1\}=\{w;\ |s\,\mathrm{Re}\,w+\mathrm{Im}\,w|<1\}.$$

Z_s is a strip with boundary lines passing through $\pm i$ and slope $-s$. We have $K_s(w)=K_{s'}(w)$ in $Z_s\cap Z_{s'}$ since this is a convex set containing an interval on the imaginary axis where we know that the equality is valid. Hence the functions K_s define together an analytic extension of K_0 to $\mathbb{C}\setminus\{it;\ t\in\mathbb{R},\ |t|\geqq 1\}$, so K is analytic in $\tilde{\Omega}$. The set $\tilde{\Omega}$ is connected for the component of $\mathbb{R}^n$ contains all $x+iy$ with $\langle x,y\rangle\neq 0$ since $x+ity\in\tilde{\Omega}$ when $0\leqq t\leqq 1$, and $\tilde{\Omega}$ is contained in the closure of this set.

It remains to prove that $K_0(w)$ is exponentially decreasing when $|w|<C\,\mathrm{Re}\,w$. When $\mathrm{Re}\,w>1$ we have

$$|K_0(w)|=\left|(2\pi)^{-n}\int_0^\infty I_0(i\rho|w|^2)/I_0(\rho\bar{w})\bar{w}^n\rho^{n-1}d\rho\right|$$
$$\leqq C_1\int_0^\infty e^{-\rho\,\mathrm{Re}\,w}((1+\rho|w|)/(1+\rho|w|^2))^{(n-1)/2}|w|^n\rho^{n-1}d\rho\leqq C_2|w|^n.$$

Furthermore, K_0 is exponentially decreasing on $\mathbb{R}$ since

$$K(x)e^{\langle x,\eta\rangle}=(2\pi)^{-n}\int e^{i\langle x,\xi\rangle}/I(\xi+i\eta)d\xi$$

is bounded for $x\in\mathbb{R}^n$ and small $|\eta|$. Hence the Phragmén-Lindelöf theorem gives $|K_0(w)|\leqq C_3e^{-c\,\mathrm{Re}\,w}|w|^n$ if $\mathrm{Re}\,w>1$. The proof is complete.

Theorem 8.4.11. *If* $u\in\mathscr{S}'(\mathbb{R}^n)$ *and* $U=K*u$, *where* K *is defined by* (8.4.13), *then* U *is analytic in* $\Omega=\{z;\ |\mathrm{Im}\,z|<1\}$ *and for some* C, a, b

(8.4.15) $$|U(z)|\leqq C(1+|z|)^a(1-|\mathrm{Im}\,z|)^{-b},\qquad z\in\Omega.$$

The boundary values $U(.+i\omega)$ *are continuous functions of* $\omega \in S^{n-1}$ *with values in* $\mathscr{S}'(\mathbb{R}^n)$, *and*

$$\langle u, \phi \rangle = \int \langle U(.+i\omega), \phi \rangle d\omega, \quad \phi \in \mathscr{S}. \tag{8.4.16}$$

Conversely, if U *is given satisfying* (8.4.15) *then* (8.4.16) *defines a distribution* $u \in \mathscr{S}'$ *with* $U = K * u$. *We have for any* L

$$\begin{aligned}(\mathbb{R}^n \times S^{n-1}) \cap WF_L(u) \\ = \{(x, \omega); |\omega| = 1, U \text{ is not in } C^L \text{ at } x - i\omega\},\end{aligned} \tag{8.4.17}$$

and an analogous description is valid for $WF(u)$.

That U is in C^L at $x - i\omega$ means of course that for some neighborhood V of $x - i\omega$ and some constant C we have

$$|\partial_z^\alpha U(z)| \leqq C^{1+|\alpha|} L_{|\alpha|}^{|\alpha|} \qquad \text{if } z \in V \text{ and } |\operatorname{Im} z| < 1.$$

For the real analytic class this means that U can be continued analytically to a full neighborhood of $x - i\omega$.

Proof of Theorem 8.4.11. That $u \in \mathscr{S}'$ means that for some a, b

$$|u(\phi)| \leqq C \sum_{|\alpha| \leqq a, |\beta| \leqq b} \sup |x^\alpha D^\beta \phi|, \qquad \phi \in \mathscr{S}.$$

By Cauchy's inequalities and (8.4.14) we have for every β

$$|D^\beta K(x+iy)| \leqq C_\beta (1 - |y|)^{-n-|\beta|} e^{-c|x|}, \qquad |y| < 1,$$

which gives (8.4.15) with b replaced by $b + n$. The Fourier transform of $U(x+iy)$ is $e^{-\langle y, \xi \rangle} \hat{u}(\xi)/I(\xi)$, hence continuous when $|y| \leqq 1$ with values in $\mathscr{S}'$, and we obtain (8.4.16) by the definition of I.

Conversely, assume given an analytic function U satisfying (8.4.15). Set $U_y = U(.+iy)$ when $|y| < 1$. By the proof of Theorem 3.1.15 the limit U_ω of U_y when $y \to \omega$, $|\omega| = 1$, exists in $\mathscr{S}'$. If b is an integer $\geqq 0$ then

$$\begin{aligned}U_\omega(\phi) = \int U(x) \sum_{|\beta| \leqq b} \partial^\beta \phi(x)(-i\omega)^\beta/\beta! \, dx \\ + (b+1) \iint_{0<t<1} U(x + i(1-t)\omega) \sum_{|\beta| = b+1} \partial^\beta \phi(x)(-i\omega)^\beta/\beta! \, t^b \, dx \, dt, \\ \phi \in \mathscr{S}.\end{aligned}$$

Thus U_ω is a continuous function of ω with values in $\mathscr{S}'$, and

$$|U_\omega(\phi)| \leqq C \sum_{|\beta| \leqq b+1} \int (1 + |x|)^a |\partial^\beta \phi(x)| \, dx.$$

It follows that (8.4.16) defines a distribution $u \in \mathscr{S}'$ with a similar estimate. We have $\hat{U}_\omega = e^{-\langle \cdot, \omega \rangle} \hat{U}_0$, hence $\hat{u} = I \hat{U}_0$, so $U = u * K$ as claimed.

To prove (8.4.17) we first assume that $(x_0, \omega_0) \notin WF_L(u)$ and that $|\omega_0|=1$. We must show that $U=K*u \in C^L$ at $x_0 - i\omega_0$. By hypothesis we can find $r>0$, a bounded sequence $u_N \in \mathscr{E}'(\mathbb{R}^n)$ equal to u when $|x-x_0|<r$ and a conic neighborhood Γ of ω_0 such that

$$|\hat{u}_N(\xi)| \leqq C(CL_N/|\xi|)^N, \quad N=1,2,\ldots, \xi \in \Gamma. \tag{8.4.18}$$

Recall that the boundedness of u_N implies that for fixed C and $M \geqq 0$

$$|\hat{u}_N(\xi)| \leqq C(1+|\xi|)^M, \quad \xi \in \mathbb{R}^n. \tag{8.4.19}$$

Set $u=u_N+v_N$. Then $U=K*u_N+K*v_N$ and

$$K*v_N(z)=\langle K(z-.), v_N\rangle.$$

Now $K(x+iy-t)$ is well defined when $|y|^2<1+|x-t|^2$, so it is well defined and rapidly decreasing with all derivatives when $|t-x_0| \geqq r$ if

$$|y|^2<1+(r-|x-x_0|)^2, \quad |x-x_0|<r. \tag{8.4.20}$$

It follows that $K*v_N$ is analytic and uniformly bounded in compact subsets of the set defined by (8.4.20), which is a neighborhood of $x_0 - i\omega_0$.

The Fourier transform of $K*u_N(.+iy)$ is $e^{-\langle y,\xi\rangle}\hat{u}_N(\xi)/I(\xi)$, so (8.4.12) gives

$$|D^\alpha K*u_N(x+iy)| \leqq C\int e^{-\langle y,\xi\rangle-|\xi|}(1+|\xi|)^{(n-1)/2}|\xi^\alpha|\,|\hat{u}_N(\xi)|\,d\xi.$$

Using (8.4.18) we can estimate the integral when $\xi \in \Gamma$ and $|\xi|>L_N$ by

$$C^{N+1}L_N^N \int_{|\xi|>L_N} |\xi|^{|\alpha|+(n-1)/2-N}d\xi \leqq C_0^{N+1}L_N^N \quad \text{if } |\alpha| \leqq N-2n, |y|<1.$$

The integral when $\xi \in \Gamma$ and $|\xi| \leqq L_N$ has the bound

$$CL_N^{|\alpha|+(n-1)/2+M+n} \leqq C_1^{N+1}L_N^N, \quad |\alpha| \leqq N-M-2n,$$

in view of (8.4.19). It remains to examine

$$C\int_{\xi \notin \Gamma} e^{-\langle y,\xi\rangle-|\xi|}(1+|\xi|)^{M+(n-1)/2}|\xi|^{|\alpha|}d\xi.$$

Choose $\varepsilon>0$ so that $\langle \omega_0, \xi\rangle<(1-2\varepsilon)|\xi|$ when $\xi \notin \Gamma$. Then

$$-\langle y,\xi\rangle-|\xi|<-\varepsilon|\xi| \quad \text{if } \xi \notin \Gamma \quad \text{and} \quad |y+\omega_0|<\varepsilon. \tag{8.4.21}$$

Hence we obtain if $N=|\alpha|+2n+M$

$$C\int_{\xi \notin \Gamma} e^{-\langle y,\xi\rangle-|\xi|}(1+|\xi|)^{M+(n-1)/2}|\xi|^{|\alpha|}d\xi \leqq C_0^N N! \leqq C_0^N L_N^N,$$
$$|y+\omega_0|<\varepsilon.$$

Summing up, it follows by repeated use of (8.4.3) that

$$|D^\alpha U(z)| \leqq C_1^{|\alpha|+2n+M+1} L_{|\alpha|+2n+M}^{|\alpha|+2n+M} \leqq C_2^{|\alpha|+1} L_{|\alpha|}^{|\alpha|}$$

in the intersection of Ω and a neighborhood of $x_0 - i\omega_0$.

The other inclusion in (8.4.17) follows from the next lemma.

Lemma 8.4.12. *Let $d\mu$ be a measure on S^{n-1} and Γ an open convex cone such that*

$$\langle y, \omega\rangle < 0 \quad \text{when } 0 \neq y \in \bar{\Gamma}, \qquad \omega \in \operatorname{supp} d\mu.$$

If U is analytic in Ω and satisfies (8.4.15) *then*

$$F(z) = \int U(z + i\omega)\, d\mu(\omega)$$

is analytic and $|F(z)| \leqq C'(1+|\operatorname{Re} z|)^a |\operatorname{Im} z|^{-b}$ when $\operatorname{Im} z \in \Gamma$ and $|\operatorname{Im} z|$ is small enough. For every measure $d\mu$ on S^{n-1} we have

(8.4.22) $\quad WF_L(U_\mu) \subset \{(x, \xi);\ -\xi/|\xi| \in \operatorname{supp} d\mu$ *and* $U \notin C^L$ *at* $x - i\xi/|\xi|\}$,

(8.4.22)' $\quad WF(U_\mu) \subset \{(x, \xi);\ -\xi/|\xi| \in \operatorname{supp} d\mu$ *and* $U \notin C^\infty$ *at* $x - i\xi/|\xi|\}$.

Here $U_\mu = \int U(.+i\omega)\, d\mu(\omega)$.

Proof. If $\omega \in \operatorname{supp} d\mu$ and $\operatorname{Im} z \in \Gamma$, we have for some $c > 0$

$$\begin{aligned} |\operatorname{Im}(z+i\omega)|^2 &= 1 + 2\langle \omega, \operatorname{Im} z\rangle + |\operatorname{Im} z|^2 \\ &< 1 - 2c|\operatorname{Im} z| + |\operatorname{Im} z|^2 < 1 - c|\operatorname{Im} z| \end{aligned}$$

if $|\operatorname{Im} z| < c$. Hence $1 - |\operatorname{Im}(z+i\omega)| > c|\operatorname{Im} z|/2$ when $|\operatorname{Im} z| < c$, which proves the first statement. By Theorem 8.4.8 it follows that

$$WF_A(U_\mu) \subset \mathbb{R}^n \times \Gamma^\circ$$

where Γ° is the dual cone of Γ, and it is obvious that

$$\operatorname{sing\,supp}_L U_\mu \subset \{x;\ U \text{ is not in } C_L \text{ at } x + i\omega \text{ for some } \omega \in \operatorname{supp} d\mu\}.$$

To prove (8.4.22) we write $d\mu = \sum d\mu_j$ where $\operatorname{supp} d\mu_j$ is contained in the intersection of $\operatorname{supp} d\mu$ and a narrow open convex cone V_j. Applying the result just proved with $d\mu$ replaced by $d\mu_j$ and Γ replaced by the interior of the dual cone $-V_j^\circ$ we obtain

$$WF_L(U_\mu) \subset \bigcup_j \{(x, \xi);\ -\xi/|\xi| \in \bar{V}_j,\ \ U \notin C^L \text{ at } x + i\omega \text{ for some } \omega \in V_j\}.$$

If $-\xi/|\xi| \notin \operatorname{supp} d\mu$ or $U \in C^L$ at $x - i\xi/|\xi|$ we can choose the covering so that $-\xi/|\xi| \notin \bar{V}_j$ for every j or for all $j \neq 1$ while $U \in C^L$ at $x + i\omega$ for every $\omega \in V_1$. In both cases it follows that $(x, \xi) \notin WF_L(U_\mu)$ which proves (8.4.22). The proof of (8.4.22)′ is of course identical. This completes the proof of Lemma 8.4.12 and of Theorem 8.4.11.

Corollary 8.4.13. *Let $\Gamma_1, \dots, \Gamma_j$ be closed cones in $\mathbb{R}^n \setminus 0$ with union $\mathbb{R}^n \setminus 0$. Any $u \in \mathscr{S}'(\mathbb{R}^n)$ can then be written $u = \sum u_j$ where $u_j \in \mathscr{S}'$ and*

$$WF_L(u_j) \subset WF_L(u) \cap (\mathbb{R}^n \times \Gamma_j).$$

If $u = \sum u'_j$ is another such decomposition then $u'_j = u_j + \sum_k u_{jk}$ where $u_{jk} \in \mathscr{S}'$, $u_{jk} = -u_{kj}$ and

$$WF_L(u_{jk}) \subset WF_L(u) \cap (\mathbb{R}^n \times (\Gamma_j \cap \Gamma_k)).$$

Proof. If ϕ_j is the characteristic function of $\Gamma_j \setminus (\Gamma_j \cap (\Gamma_1 \cup \dots \cup \Gamma_{j-1}))$ we have $\sum \phi_j = 1$. With $U = K * u$ and $U_j = K * (u'_j - u_j)$ we have $\sum U_j = 0$, and it follows from Theorem 8.4.11 and Lemma 8.4.12 that we can take

$$u_j = \int U(. - i\omega)\, \phi_j(\omega)\, d\omega,$$
$$u_{jk} = \int U_j(. - i\omega)\, \phi_k(\omega)\, d\omega - \int U_k(. - i\omega)\, \phi_j(\omega)\, d\omega.$$

Next we improve Theorem 8.1.4.

Theorem 8.4.14. *If $X \subset \mathbb{R}^n$ is open and S is a closed conic set in $X \times (\mathbb{R}^n \setminus 0)$ then one can find $u \in \mathscr{D}'(X)$ with $WF(u) = WF_L(u) = S$ for every L.*

Proof. It is sufficient to prove the statement when $X = \mathbb{R}^n$, and we only have to verify for the chosen u that

$$WF(u) = WF_A(u) = S.$$

Let (x_k, θ_k) be a sequence without repetitions which is dense in $\{(x, \theta) \in S;\ |\theta| = 1\}$. With K defined by (8.4.13) we set

$$U(z) = \sum_1^\infty 3^{-k} K((z - x_k - i\theta_k)/2), \qquad |\operatorname{Im} z| < 1.$$

Since

$$|K((z - x_k - i\theta_k)/2)| \leqq K(i(\operatorname{Im} z - \theta_k)/2) \leqq C(1 - |\operatorname{Im} z|)^{-n}$$

it is clear that U is an analytic function satisfying (8.4.15). Noting that $\sum_{N+1}^\infty 3^{-k} = 3^{-N}/2$ and that $|t\theta_k + \theta| \leqq t + 1$ if $|\theta| = 1$, $t > 0$, we obtain

$$|U(x_k - it\theta_k)| > 3^{-k} |K(-i(t+1)\,\theta_k/2)|/2 - |\sum_{j<k} 3^{-j} K((x_k - x_j - i(t\theta_k + \theta_j))/2)| \to \infty, \qquad t \nearrow 1.$$

Hence U is not even bounded in Ω near any point in

$$S' = \{(x, -\theta); (x, \theta) \in S \text{ and } |\theta| = 1\}.$$

On the other hand, it is clear that U is analytic near any point in $(\mathbb{R}^n \times S^{n-1}) \setminus S'$. By Theorem 8.4.11 this completes the proof of the theorem.

Next we prove a converse of Theorems 8.1.6 and 8.4.8.

Theorem 8.4.15. *Let $u \in \mathscr{D}'(X)$, $X \subset \mathbb{R}^n$, and assume that $WF_L(u) \subset X \times \Gamma^\circ$ (resp. $WF(u) \subset X \times \Gamma^\circ$) where Γ° is the dual of an open convex cone Γ. If $X_1 \Subset X$ and Γ_1 is an open convex cone with closure $\subset \Gamma \cup \{0\}$, then one can find a function F analytic in $\{x + iy;\, x \in X_1,\, y \in \Gamma_1,\, |y| < \gamma\}$, such that*

$$|F(x+iy)| < C|y|^{-N}, \quad y \in \Gamma_1, \quad |y| < \gamma, \quad x \in X_1,$$

and the limit of $F(.+iy)$ when $y \to 0$ in Γ_1 differs from u by an element in $C^L(X_1)$ (resp. $C^\infty(X_1)$).

Proof. Set $v = \chi u$ where $\chi \in C_0^\infty(X)$ is equal to 1 in X_1. If $V = K * v$ is defined as in Theorem 8.4.11 we have $V \in C^L$ at every point in $X_1 + i(S^{n-1} \cap \complement(-\Gamma^\circ))$. Choose $M \subset S^{n-1}$ open with $\Gamma^\circ \cap S^{n-1} \subset M$ and $\overline{M}$ in the interior of Γ_1°. Then $v = v_1 + v_2$ where

$$v_1 = \int_{-\omega \notin M} V(. + i\omega)\, d\omega$$

belongs to C^L in X_1 and v_2 is the boundary value of the analytic function

$$F(z) = \int_{-\omega \in M} V(z + i\omega)\, d\omega, \qquad \operatorname{Im} z \in \Gamma_1, \quad |\operatorname{Im} z| < \gamma.$$

Lemma 8.4.12 completes the proof.

Remark. It follows from Corollary 8.4.13, Theorem 8.4.15 and Theorem 8.4.8 that $WF_L(u)$ (resp. $WF(u)$) is the intersection of $WF_A(u - u_1)$ for all $u_1 \in C^L$ (resp. C^∞). Thus the notions $WF_L(u)$ and $WF(u)$ can be derived from $WF_A(u)$. – For the analytic case the statement of Theorem 8.4.15 can be simplified: the restriction of u to X_1 is the boundary value of an analytic function F with the stated properties.

Corollary 8.4.16. *If $u \in \mathscr{D}'(X)$ where X is an interval on $\mathbb{R}$ and if $x_0 \in X$ is a boundary point of* $\operatorname{supp} u$, *then $(x_0, \pm 1) \in WF_A(u)$.*

Proof. Assume for example that $(x_0, -1) \notin WF_A(u)$. Then we can find F analytic in $\Omega = \{z;\, \operatorname{Im} z > 0,\, |z - x_0| < r\}$ with boundary value u. There is an interval $I \subset (x_0 - r, x_0 + r)$ where $u = 0$. By Theorem 3.1.12 (and Theorem 4.4.1) F can be extended analytically across I so that $F = 0$ below I. Thus the uniqueness of analytic continuation gives $F = 0$, hence $u = 0$ in $(x_0 - r, x_0 + r)$. This contradicts that x_0 is a boundary point of $\operatorname{supp} u$ and proves the corollary. (See also Theorem 3.1.15 for a more general form of the proof.)

Note that the corollary can be phrased as a uniqueness theorem: If we know that $WF_A(u)$ does not contain $T_x^*(\mathbb{R}) \setminus 0$ for any $x \in X$ then u must vanish identically if u vanishes in an open set. We shall extend this statement to several variables in the next section.

Finally we prove an analogue of Lemma 8.1.7 and of Theorem 8.1.8.

Lemma 8.4.17. *If $u \in \mathscr{S}'$ then $WF_A(u) \subset \mathbb{R}^n \times F$ where F is the limit cone of $\operatorname{supp} \hat{u}$ defined in Lemma 8.1.7.*

Proof. The Fourier transform of $u * K$ is $\hat{u}(\xi)/I(\xi)$. If Γ is an open cone with $\bar{\Gamma} \cap F = \{0\}$ we can choose a closed cone F' with $F \setminus \{0\}$ in its interior and $\Gamma \cap F' = \{0\}$. Then we have for some $c < 1$

$$\langle y, \xi \rangle \leqq c|y|\,|\xi| \quad \text{if } y \in \Gamma, \quad \xi \in F'.$$

Hence Lemma 8.4.9 shows that $e^{-\langle y, \xi \rangle}/I(\xi)$ and all its ξ derivatives are bounded if $y \in -\Gamma$, $|y| < 2/(1+c)$, and $\xi \in F'$. Since $\operatorname{supp} \hat{u}$ is contained in the union of F' and a compact set, it follows that $\hat{u}(\xi) e^{-\langle y, \xi \rangle}/I(\xi)$ is in $\mathscr{S}'$ when $y \in -\Gamma$, $|y| < 2/(1+c)$. By Theorem 7.4.2 it follows that $u * K$ has an analytic continuation to $\{z;\, \operatorname{Im} z \in -\Gamma,\ |\operatorname{Im} z| < 2/(1+c)\}$. Hence $WF_A(u) \subset \complement \Gamma$ by Theorem 8.4.11 which proves the lemma.

Theorem 8.4.18. *If $u \in \mathscr{S}'(\mathbb{R}^n)$ is homogeneous in $\mathbb{R}^n \setminus 0$ then*

$$(x, \xi) \in WF_L(u) \Leftrightarrow (\xi, -x) \in WF_L(\hat{u}) \quad \text{if } \xi \neq 0, \quad x \neq 0, \tag{8.4.23}$$

$$x \in \operatorname{supp} u \Leftrightarrow (0, -x) \in WF_L(\hat{u}), \qquad x \neq 0, \tag{8.4.24}$$

$$\xi \in \operatorname{supp} \hat{u} \Leftrightarrow (0, \xi) \in WF_L(u), \qquad \xi \neq 0. \tag{8.4.25}$$

Proof. Since $\hat{\hat{u}} = (2\pi)^n \check{u}$ it follows from Lemma 8.4.17 that

$$x \notin \operatorname{supp} u \Rightarrow (0, -x) \notin WF_L(\hat{u}), \quad \text{if } x \neq 0.$$

On the other hand, Theorem 8.1.8 gives

$$(0, -x) \notin WF_L(\hat{u}) \Rightarrow (0, -x) \notin WF(\hat{u}) \Rightarrow x \notin \operatorname{supp} u$$

which proves (8.4.24). If u is homogeneous then (8.4.25) is (8.4.24) applied to $\hat{u}$. Otherwise $\operatorname{supp} \hat{u} = \mathbb{R}^n$ and $(0, \xi) \in WF(u) \subset WF_L(u)$ for every ξ as in the proof of (8.1.19). Hence (8.4.25) is true. Arguing exactly as in the proof of (8.1.17) we also conclude that (8.4.23) follows if we prove that

$$(x_0, \xi_0) \notin WF_L(u) \Rightarrow (\xi_0, -x_0) \notin WF_L(\hat{u}) \tag{8.4.23$'$}$$

if $x_0 \neq 0$, $\xi_0 \neq 0$ and u is homogeneous in $\mathbb{R}^n$. The proof will essentially be a repetition of that of (8.1.17)$'$, with cutoff functions chosen more carefully.

Choose compact neighborhoods K and $\hat{K}$ in $\mathbb{R}^n \setminus 0$ of x_0 and ξ_0 such that

$$(K \times \hat{K}) \cap WF_L(u) = \emptyset. \tag{8.4.26}$$

By Theorem 1.4.2 we can find a sequence $\chi_N \in C_0^\infty(\hat{K})$ equal to 1 in a fixed neighborhood of ξ_0 such that (8.4.6)$'$ is valid for every α. We shall estimate the Fourier transform of $v_N = \chi_N \hat{u}$ in a conic neighborhood of $-x_0$. The homogeneity of u gives as in the proof of (8.1.17)$'$

$$\hat{v}_N(-tx) = t^{a+n} \langle u, \hat{\chi}_N(t(. - x))\rangle.$$

Choose $r > 0$ and $\psi_N \in C_0^\infty(K)$ so that $\psi_N(x) = 1$ when $|x - x_0| < 2r$ and ψ_N satisfies (8.4.6)$'$, and set $u_{0N} = \psi_N u$, $u_{1N} = (1 - \psi_N) u$. Then

$$I_0 = \langle u_{0N}, \hat{\chi}_N(t(. - x))\rangle = \int \hat{u}_{0N}(\xi) \chi_N(\xi/t) e^{i\langle x, \xi\rangle} d\xi/t^n,$$

and by Lemma 8.4.4 and (8.4.26)

$$|\hat{u}_{0N}(t\xi)| \leqq C(CL_N/t)^N, \qquad \xi \in \operatorname{supp} \chi_N.$$

Hence

$$|I_0| \leqq C'(CL_N/t)^N.$$

Moreover,

$$I_1 = \langle u_{1N}, \hat{\chi}_N(t(. - x))\rangle = \langle u, (1 - \psi_N) \hat{\chi}_N(t(. - x))\rangle,$$

and since $u \in \mathcal{S}'$ it follows when $|x - x_0| < r$ that for some C, C' and μ

$$\begin{aligned} |I_1| &\leqq C \sum_{|\alpha+\beta| \leqq \mu} \sup |y^\alpha D^\beta (1 - \psi_N(y)) \hat{\chi}_N(t(y - x))| \\ &\leqq C' \sum_{|\alpha+\beta| \leqq \mu} \sup_{|y| > tr} t^{|\beta| - |\alpha|} |y^\alpha D^\beta \hat{\chi}_N(y)|. \end{aligned}$$

Now (8.4.6)$'$ implies that for some constant C

$$|y|^N |D^\beta \hat{\chi}_N(y)| \leqq C(CL_N)^N, \qquad |\beta| \leqq \mu.$$

Hence we obtain the estimate

$$|I_1| \leqq C_1 t^\mu (C_1 L_N/t)^N.$$

If μ' is an integer $\geqq 0$ and $\geqq \mu+a+n$ it follows in view of (8.4.3) that

$$|\hat{v}_N(x)| \leqq C_2(C_2 L_{N-\mu'}/|x|)^{N-\mu'}$$

if x is in the cone generated by $\{x; |x+x_0|<r\}$. This means that $(\xi_0, -x_0) \notin WF_L(\hat{u})$, and (8.4.23)′ is proved.

8.5. Rules of Computation for WF_L

We shall now prove analogues for WF_L of the results on WF in Section 8.2, starting with an analogue of the basic Theorem 8.2.4.

Theorem 8.5.1. *Let X and Y be open subsets of $\mathbb{R}^n$ and $\mathbb{R}^m$ respectively and let $f: X \to Y$ be a real analytic map with normal set N_f. Then we have*

(8.5.1) $$WF_L(f^*u) \subset f^* WF_L(u), \quad \text{if} \quad u \in \mathscr{D}'(Y), \quad N_f \cap WF_L(u) = \emptyset.$$

Proof. First assume that there is an analytic function Φ in

$$\Omega = \{y' + iy''; \, y' \in Y, \, y'' \in \Gamma, \, |y''| < \gamma\},$$

where Γ is an open convex cone, such that

$$|\Phi(y'+iy'')| \leqq C|y''|^{-N} \quad \text{in } \Omega,$$

$$u = \lim_{\Gamma \ni y \to 0} \Phi(.+iy).$$

This implies that $WF_A(u) \subset Y \times \Gamma^\circ$. Let $x_0 \in X$ and assume that ${}^t f'(x_0)\eta \neq 0$, $\eta \in \Gamma^\circ \setminus \{0\}$. Then ${}^t f'(x_0)\Gamma^\circ$ is a closed, convex, proper cone. We claim that

(8.5.2) $$WF_A(f^*u)|_{x_0} \subset \{(x_0, {}^t f'(x_0)\eta), \, \eta \in \Gamma^\circ \setminus \{0\}\}.$$

To give another expression for the right-hand side we let Γ_1 be an open convex cone with closure contained in $\Gamma \cup \{0\}$ and ${}^t f'(x_0)\eta \neq 0$, $\eta \in \Gamma_1^\circ \setminus \{0\}$. Then ${}^t f'(x_0)\Gamma_1^\circ$ is a closed convex cone with dual cone

$$\{h \in \mathbb{R}^n; \, \langle h, {}^t f'(x_0)\eta \rangle \geqq 0, \, \eta \in \Gamma_1^\circ\} = \{h; \, f'(x_0)h \in \bar{\Gamma}_1\},$$

which implies that

$${}^t f'(x_0)\Gamma_1^\circ = \{\xi; \, \langle h, \xi \rangle \geqq 0 \text{ if } f'(x_0)h \in \bar{\Gamma}_1\}.$$

Since $\bar{\Gamma}_1 \subset \Gamma \cup \{0\}$ we obtain when $\Gamma_1 \nearrow \Gamma$ that

$${}^t f'(x_0)\Gamma^\circ = \{\xi; \, \langle h, \xi \rangle \geqq 0 \text{ if } f'(x_0)h \in \Gamma\}.$$

Thus let $h\in\mathbb{R}^n$ and $f'(x_0)h\in\Gamma$. Then $\operatorname{Im} f(x+i\varepsilon h)\in\Gamma$ for small $\varepsilon>0$ if x is in a sufficiently small neighborhood X_0 of x_0. In X_0 we have

$$f^*u=\lim_{\varepsilon\to+0}\Phi(f(.+i\varepsilon h)).$$

In fact, the proof of Theorem 3.1.15 shows that $\Phi(f(.+i\varepsilon h)+iy)$ is a continuous function of $(\varepsilon,y)\in\mathbb{R}_+\times\bar{\Gamma}$, near 0, with values in $\mathscr{D}'$. Letting $\varepsilon\to 0$ first we obtain the left-hand side of the equation and letting $y\to 0$ first we obtain the right-hand side. By Theorem 8.4.8 we now obtain

$$WF_A(f^*u)|_{x_0}\subset\{(x_0,\xi);\langle h,\xi\rangle\geqq 0\}$$

which proves (8.5.2).

Using Corollary 8.4.13 and Theorem 8.4.15 we can write a general u as a finite sum $\sum u_j$ where each term is either a C^L function in a neighborhood of $f(x_0)$ or else satisfies the hypotheses above with some cone Γ_j such that Γ_j° is small and intersects $WF_L(u)|_{f(x_0)}$. By hypothesis ${}^tf'(x_0)\eta\neq 0$ when $(f(x_0),\eta)\in WF_L(u)$ so we conclude that

$$WF_L(f^*u)|_{x_0}\subset\{(x_0,{}^tf'(x_0)\eta),\eta\in\bigcup\Gamma_j^\circ\},$$

and this implies (8.5.1).

Remark. Theorem 8.5.1 shows in particular that $WF_L(u)$ can be defined as a subset of $T^*(X)\smallsetminus 0$ if X a real analytic manifold.

Theorem 8.5.2. *Theorem 8.2.9 remains valid with WF replaced by WF_L.*

The proof is an obvious modification of that of Theorem 8.2.9. In particular, Theorems 8.5.1 and 8.5.2 show that for a non-zero analytic density u on a real analytic submanifold $WF_A(u)$ is the normal bundle. Example 8.2.6 is also immediately obtained with WF_A in the left hand side of (8.2.7). Combining Theorems 8.5.1 and 8.5.2 we conclude:

Theorem 8.5.3. *If $u,v\in\mathscr{D}'(X)$ and $(x,\xi)\in WF_L(u)$ implies $(x,-\xi)\notin WF_L(v)$ then the product uv is defined and*

$$\begin{aligned}WF_L(uv)\subset\{(x,\xi+\eta);\ &(x,\xi)\in WF_L(u)\\ &\text{or } \xi=0,\ (x,\eta)\in WF_L(v) \text{ or } \eta=0\}.\end{aligned}$$

When proving an analogue of Theorem 8.2.12 we begin with a special case which fits the notations in Theorem 8.4.11 well.

Theorem 8.5.4. *Let $u\in\mathscr{E}'(\mathbb{R}^n)$, split the coordinates in $\mathbb{R}^n$ into two groups $x'=(x_1,\dots,x_{n'})$ and $x''=(x_{n'+1},\dots,x_n)$, and set*

$$u_1(x')=\int u(x',x'')\,dx''$$

in the sense defined in Section 5.2. Then

$$WF_L(u_1)\subset\{(x',\xi');\ (x',x'',\xi',0)\in WF_L(u)\ \text{for some}\ x''\}.$$

Proof. With the notation in Theorem 8.4.11 we have

$$\langle u,\phi\otimes\psi\rangle=\int_{|\omega|=1}\langle U(.+i\omega),\phi\otimes\psi\rangle\,d\omega,$$
$$\phi\in C_0^\infty(\mathbb{R}^{n'}),\quad \psi\in C_0^\infty(\mathbb{R}^{n-n'}).$$

Take $\psi(x'')=\chi(\delta x'')$ where $\chi=1$ in the unit ball, and let $\delta\to 0$. Since U is exponentially decreasing at infinity it follows then that

$$\langle u_1,\phi\rangle=\int_{|\omega|=1}\langle U(.+i\omega),\phi\otimes 1\rangle\,d\omega=\int_{|\omega|=1}\langle U_1(.+i\omega'),\phi\rangle\,d\omega$$

where

$$U_1(z')=\int U(z',x'')\,dx''=\int U(z',x''+iy'')\,dx'',\qquad |\operatorname{Im} z'|^2+|y''|^2<1,$$

is an analytic function when $|\operatorname{Im} z'|<1$ which is bounded by $C(1-|\operatorname{Im} z'|)^{-N}$. If $|\omega_0'|=1$ and $(x',x'',\omega_0')\notin WF_L(u)$ for every $x''\in\mathbb{R}^{n-n'}$ then $U_1\in C^L$ at $x'-i\omega_0'$. Hence Lemma 8.4.12 implies that $(x',\omega_0')\notin WF_L(u_1)$.

The following statement is parallel to Theorem 8.2.12 but essentially equivalent to Theorem 8.5.4.

Theorem 8.5.4′. *Let $X\subset\mathbb{R}^n$, $Y\subset\mathbb{R}^m$ be open sets and $K\in\mathscr{D}'(X\times Y)$ be a distribution such that the projection $\operatorname{supp} K\to X$ is proper. If $u\in C^L(Y)$ then*

$$WF_L(\mathscr{K}u)\subset\{(x,\xi);\ (x,y,\xi,0)\in WF_L(K)\ \text{for some}\ y\in\operatorname{supp} u\}.$$

Here $\mathscr{K}$ is the linear operator with kernel K.

Proof. Replacing K by $K(1\otimes u)$ we may assume that $u=1$. Without changing K over a given compact subset of X we may replace K by a distribution of compact support, and then the statement is identical to Theorem 8.5.4.

The following is an analogue of Theorem 8.2.13 and the notations employed are obvious modifications of those used in Theorem 8.2.13.

Theorem 8.5.5. *If* $u \in \mathscr{E}'(Y)$ *and* $WF_L(u) \cap WF'_L(K)_Y = \emptyset$ *then*

$$WF_L(\mathscr{K} u) \subset WF_L(K)_X \cup (WF'_L(K) \circ WF_L(u)).$$

Proof. By Theorem 8.5.2 we have

$$WF_L(1 \otimes u) \subset \{(x, y, 0, \eta);\ (y, \eta) \in WF_L(u)\}.$$

If $K_u = K(1 \otimes u)$ it follows Theorem 8.5.3 that

$$WF_L(K_u) \subset \{(x, y, \xi, \eta + \eta');\ (y, \eta) \in WF_L(u), (x, y, \xi, \eta') \in WF_L(K)\} \\ \cup WF_L(K) \cup WF_L(1 \otimes u).$$

Since $\mathscr{K} u$ is the integral of K_u over Y, an application of Theorem 8.5.4 completes the proof.

The applications to convolutions indicated at the end of Section 8.2 obviously carry over to WF_L with no change. We shall not dwell on this any longer but shall instead prove an n dimensional analogue of Corollary 8.4.16 concerning the uniqueness of analytic continuation.

Theorem 8.5.6. *Let* $u \in \mathscr{D}'(X)$, $X \subset \mathbb{R}^n$, *and assume that* f *is a real valued real analytic function in* X *and* x^0 *a point in* $\operatorname{supp} u$ *such that*

$$df(x^0) \neq 0, \quad f(x) \leqq f(x^0) \quad \text{if } x \in \operatorname{supp} u. \tag{8.5.3}$$

Then it follows that

$$(x^0, \pm df(x^0)) \in WF_A(u). \tag{8.5.4}$$

Proof. Replacing f by $f(x) - |x - x^0|^2$ we may assume that

$$f(x) < f(x^0) \quad \text{if } x^0 \neq x \in \operatorname{supp} u.$$

Since $df(x^0) \neq 0$ we may take f as a coordinate locally, so we may assume that $f(x) = x_n$ and that $x^0 = 0$. Choose a neighborhood Y of $0 \in \mathbb{R}^{n-1}$ so that $Y \times \{0\} \Subset X$. Since $\operatorname{supp} u \cap (\bar{Y} \times \{0\}) = \{0\}$ we can choose an open interval $I \subset \mathbb{R}$ with $0 \in I$ so that

$$Y \times I \Subset X \quad \text{and} \quad (\partial Y \times I) \cap \operatorname{supp} u = \emptyset.$$

If $a(x')$ is an entire analytic function of $x' = (x_1, \ldots, x_{n-1})$ then Theorem 8.5.4′ (with $X \times Y$, x and y replaced by $I \times Y$, x_n and x') gives that

$$U_a(x_n) = \int_Y u(x) a(x') dx'$$

is well defined as a distribution in I and that

$$WF_A(U_a) \subset \{(x_n, \xi_n);\ (x', x_n, 0, \xi_n) \in WF_A(u) \text{ for some } x' \in Y\}.$$

Here (x', x_n) must be close to 0 if x_n is small. If, say $(0, e_n) \notin WF_A(u)$, $e_n = (0, \dots, 0, 1)$, then we can choose I so that $(x, e_n) \notin WF_A(u)$ if $x \in Y \times I$. Hence $(x_n, 1) \notin WF_A(U_a)$ if $x_n \in I$, so Corollary 8.4.16 gives that $U_a = 0$ in I, because $U_a = 0$ when $x_n > 0$. Thus, if u_1 is u restricted to $Y \times I$,

$$\langle u_1, a \otimes \phi \rangle = 0$$

for all real analytic a and all $\phi \in C_0^\infty(I)$. Since a is free to vary in a dense subset of $C^\infty(\mathbb{R}^{n-1})$ it follows from Theorem 5.1.1 that $u = 0$ in $Y \times I$. This contradiction proves (8.5.4).

Theorem 8.5.6 obtains a more suggestive form if one introduces the normal set of any closed set $F \subset X$:

Definition 8.5.7. If F is a closed subset of a C^2 manifold X then the exterior normal set $N_e(F) \subset T^*(X) \setminus 0$ is defined as the set of all (x^0, ξ^0) such that $x^0 \in F$ and there is a real valued function $f \in C^2(X)$ with $df(x^0) = \xi^0 \neq 0$ and

$$f(x) \leqq f(x^0) \quad \text{when } x \in F. \tag{8.5.3)$'$}$$

It would be sufficient to require f to be defined in a neighborhood U of x^0, for if $0 \leqq \phi \in C_0^2(U)$ is equal to 1 near x^0 we can replace f by $\phi(x) f(x) + (1 - \phi(x)) f(x^0)$. Thus the definition of $N_e(F)$ is entirely local. In local coordinates we can always replace f by $f_2(x) - |x - x^0|^2$ where f_2 is the second order Taylor expansion at x^0, so f can always be taken analytic and strictly smaller than $f(x^0)$ when $x^0 \neq x \in F$. The following proposition shows that there is always a large normal set if $\partial F \neq \emptyset$.

Proposition 8.5.8. *For every closed subset F of the C^2 manifold X the projection of $N_e(F)$ in X is dense in ∂F. If $x^0 \in F$, $f \in C^1(X)$, $df(x^0) = \xi^0 \neq 0$ and $f(x) \leqq f(x^0)$ when $x \in F$, then $(x^0, \xi^0) \in \overline{N_e(F)}$. If $X \subset \mathbb{R}^n$ and Y is a convex open set $\subset X \setminus F$, $x^0 \in F \cap \partial Y$, we have $(x^0, \xi^0) \in N_e(F)$ for some ξ^0 with $\langle x - x^0, \xi^0 \rangle > 0$, $x \in Y$.*

Proof. Since the assertions are local we may assume that $X \subset \mathbb{R}^n$. If $y \in X \setminus F$ and $z \in F$ has minimal Euclidean distance to y, thus

$$|z - y|^2 - |x - y|^2 \leqq 0, \qquad x \in F,$$

then $(z, y - z) \in N_e(F)$. If $x^0 \in \partial F$ we can choose a sequence $y_\nu \in X \setminus F$ converging to x^0 and obtain a sequence $(z_\nu, \zeta_\nu) \in N_e(F)$ with $z_\nu \to x^0$, which proves the first statement. If $f \in C^1$ and $f(x) \leqq f(x^0)$ when $x \in F$, we choose $y_\nu = x^0 + f'(x^0)/\nu$ on the outer normal and obtain if $z_\nu = x^0 + w_\nu$

$$|f'(x^0)/\nu|^2 \geqq |w_\nu - f'(x^0)/\nu|^2, \qquad f(x^0 + w_\nu) \leqq f(x^0).$$

Thus $\nu |w_\nu|^2/2 \leqq \langle w_\nu, f'(x^0)\rangle \leqq o(|w_\nu|)$, so $\nu w_\nu \to 0$ as $\nu \to \infty$ and

$$\nu(y_\nu - z_\nu) = f'(x^0) - \nu w_\nu \to f'(x^0)$$

which proves that $(x^0, f'(x^0)) \in \overline{N_e(F)}$.

Next assume that Y is convex and that $Y \Subset X$, $F \cap \partial Y = \{x^0\}$, $F \cap Y = \emptyset$. Let $0 \in Y$. Then the homogeneous function f of degree 1 such that $Y = \{x; f(x) < 1\}$ is convex, for if $f(x) + f(y) = M$ then $f(x+y) \leqq M$ since

$$(x+y)/M = (x/f(x))\, f(x)/M + (y/f(y))\, f(y)/M \in \bar{Y}.$$

Let $0 \leqq \chi \in C_0^\infty$, $\int \chi\, dx = 1$, and set $f_\varepsilon = f * \chi_\varepsilon$ where $\chi_\varepsilon(x) = \varepsilon^{-n}\chi(x/\varepsilon)$. Then we have $|f - f_\varepsilon| < C\varepsilon$ since f is Lipschitz continuous. If we note that $1 - f \leqq 0$ in F with equality only at x^0 we conclude that the maximum of $1 - f_\varepsilon$ in F is $\leqq C\varepsilon$ and is attained at a point x_ε such that $x_\varepsilon \to x^0$ when $\varepsilon \to 0$. Since f_ε is convex we have if $\xi_\varepsilon = -f_\varepsilon'(x_\varepsilon)/|f_\varepsilon'(x_\varepsilon)|$

$$(x_\varepsilon, \xi_\varepsilon) \in N_e(F); \qquad \langle x - x_\varepsilon, \xi_\varepsilon\rangle \geqq 0 \qquad \text{when } f_\varepsilon(x) < f_\varepsilon(x_\varepsilon).$$

If ξ^0 is a limit point of ξ_ε when $\varepsilon \to 0$ it follows that $(x^0, \xi^0) \in \overline{N_e(F)}$ and that $\langle x - x^0, \xi^0\rangle > 0$ if $x \in Y$.

If Y is any convex open set $\subset X \setminus F$ with $x^0 \in F \cap \partial Y$ we can apply the preceding result to the interior of the convex hull of x^0 and any compact subset K of Y with interior points. Then the last statement in the proposition follows when $K \nearrow Y$.

In what follows we also use the notation

$$N_i(F) = \{(x, \xi);\ (x, -\xi) \in N_e(F)\}$$

for the *interior normal set* of F and $N(F) = N_e(F) \cup N_i(F)$ for the whole normal set. The closure in $T^*(X) \setminus 0$ will be denoted by $\bar{N}(F)$. Theorem 8.5.6 can now be restated as follows:

Theorem 8.5.6′. *For every $u \in \mathscr{D}'(X)$ we have*

$$\bar{N}(\operatorname{supp} u) \subset WF_A(u). \tag{8.5.5}$$

The importance of Theorem 8.5.6′ will be enhanced in Section 8.6 where we prove that if u satisfies a differential equation $P(x, D)u = 0$ with analytic coefficients, then $WF_A(u)$ is contained in the characteristic set of P. Thus the principal symbol $p(x, \xi)$ vanishes on $WF_A(u)$, so it must vanish on $\bar{N}(\operatorname{supp} u)$ by (8.5.5). We shall now examine the purely geometrical consequences of having such a function. Recall (see

Section 6.4) that if p is a real valued function in $C^\infty(T^*(X)\setminus 0)$ and $p(x^0, \xi^0)=0$, then the Hamilton equations

$$dx/dt=\partial p(x,\xi)/\partial \xi, \qquad d\xi/dt=-\partial p(x,\xi)/\partial x$$

with initial data $(x,\xi)=(x^0,\xi^0)$ when $t=0$ define a curve through (x^0,ξ^0) on which p remains equal to 0. It is called a *bicharacteristic (strip)*; the projection in X is called a *bicharacteristic curve*. It is non-singular if $\partial p(x,\xi)/\partial\xi \neq 0$.

Theorem 8.5.9. *Let F be a closed subset of X and assume that $p\in C^\infty(T^*(X)\setminus 0)$ is real valued and vanishes on $N_e(F)$. If $(x^0,\xi^0)\in N_e(F)$ it follows that a neighborhood of (x^0,ξ^0) on the bicharacteristic strip $t\to(x(t),\xi(t))$ for p through (x^0,ξ^0) remains in $N_e(F)$. If $p'_\xi(x^0,\xi^0)\neq 0$ there is a function $\Phi\in C^\infty(X)$ such that for some $\varepsilon>0$*

$$\Phi(x(t))=0, \qquad d\Phi(x(t))=\xi(t) \qquad \textit{if } |t|<\varepsilon,$$

and $\Phi(x)<0$ when x is in a neighborhood of $\Gamma=\{x(t), |t|<\varepsilon\}$ in F but $x\notin\Gamma$.

Proof. We may assume that $X\subset\mathbb{R}^n$ since the result is local. Choose $f\in C^\infty$ with $f(x^0)=0$, $df(x^0)=\xi^0$, and $f(x)<0$ if $x^0\neq x\in F$. Using Theorem 6.4.5 we can find a solution ϕ of the Cauchy problem

$$\partial\phi/\partial t+p(x,\phi'_x)=0, \qquad \phi(0,x)=f(x), \tag{8.5.6}$$

for $|t|<\delta$ and x in a compact convex neighborhood W of x^0 where $df(x)\neq 0$. We have $\phi'_x\neq 0$ in $(-\delta,\delta)\times W$, and $\phi'_x(t,x)=\xi$, $\phi'_t(t,x)=-p(x,\xi)$ on the curves

$$dx/dt=\partial p(x,\xi)/\partial\xi, \qquad d\xi/dt=-\partial p(x,\xi)/\partial x; \qquad (x,\xi)=(y,f'(y)), \quad t=0. \tag{8.5.7}$$

When $y=x^0$ we have in particular $\phi'_x(t,x(t))=\xi(t)$, and $\phi(t,x(t))$ is independent of the choice of f.

If δ is small enough then

$$M(t)=\max_{F\cap W}\phi(t,x)>\max_{F\cap\partial W}\phi(t,x), \qquad |t|<\delta,$$

for this is true when $t=0$. Thus the maximum $M(t)$ is attained at a point $x_t\in F$ in the interior of W. We claim that $M(t)=0$, $|t|<\delta$, which is true by hypothesis when $t=0$. To prove this we observe that since $|\sup u-\sup v|\leqq\sup|u-v|$ for all bounded functions u, v, we have

$$|M(t)-M(s)|\leqq\max_{F\cap W}|\phi(t,x)-\phi(s,x)|.$$

We may replace $F \cap W$ by $\{x_t, x_s\}$ here for this does not change the maximum values $M(t)$ and $M(s)$. Now $(x_t, \phi'_x(t, x_t)) \in N_e(F)$ and $(x_s, \phi'_x(s, x_s)) \in N_e(F)$, hence $p(x_t, \phi'_x(t, x_t)) = p(x_s, \phi'_x(s, x_s)) = 0$. Thus it follows from (8.5.6) that $\phi'_t(t, x_t) = \phi'_s(s, x_s) = 0$, so Taylor's formula gives

$$\phi(t, x_t) - \phi(s, x_t) = O((t-s)^2), \qquad \phi(t, x_s) - \phi(s, x_s) = O((t-s)^2).$$

This means that $M(t) - M(s) = O((t-s)^2)$ so $M'(t) = 0$ and $M(t) = M(0) = 0$.

Now replace $f(x)$ by $g(x) = f(x) - |x - x^0|^2$ in the preceding argument. Thus we let ψ be the solution of the equation

$$\partial\psi/\partial t + p(x, \psi'_x) = 0, \qquad \psi(0, x) = g(x).$$

For sufficiently small δ and W the results proved for ϕ are also valid for ψ, and $\phi(t, x) - \psi(t, x)$ is a strictly convex function of $x \in W$. When $x = x(t)$ we know that $\phi(t, x(t)) = \psi(t, x(t))$ and that $\phi'_x(t, x(t)) = \psi'_x(t, x(t)) = \xi(t)$, so the convexity gives $\phi(t, x) > \psi(t, x)$ when $x \neq x(t)$. Since $\phi(t, x)$ and $\psi(t, x)$ both have the maximum value zero in $W \cap F$, it must be attained at $x(t)$ and only there in the case of $\psi(t, x)$. Hence $(x(t), \xi(t)) \in N_e(F)$ and $\phi(t, x(t)) = 0$.

Assume now that $p'_\xi(x^0, \xi^0) \neq 0$. The equation $\phi'_t(T, x) = 0$ defines a C^∞ function $T(x)$ in a neighborhood of x^0 with $T(x^0) = 0$ unless

$$\begin{aligned} 0 = \phi''_{tt}(0, x^0) &= -\langle p'_\xi(x^0, \xi^0), \phi''_{tx}(0, x^0)\rangle \\ &= \langle p'_\xi, p'_x\rangle + \langle f''_{xx} p'_\xi, p'_\xi\rangle. \end{aligned}$$

In that case we just replace f by g. Note that $\phi'_t(t, x(t)) = -p(x(t), \xi(t)) = 0$ because $(x(t), \xi(t)) \in N_e(F)$, so $T(x(t)) = t$. We may assume that $\phi(t, x) < 0$ if $x \in W \cap F$ and $x \neq x(t)$. Then $\Phi(x) = \phi(T(x), x)$ has the required properties.

Corollary 8.5.10. *Let F be a closed subset of X and set*

$$\mathcal{N}_F = \{p \in C^\infty(T^*(X) \setminus 0);\ p = 0 \text{ on } N_e(F)\}.$$

Then $\mathcal{N}_F$ is an ideal in $C^\infty(T^*(X) \setminus 0)$ which is closed under Poisson brackets, that is, if p and q are in $\mathcal{N}_F$ then $\mathcal{N}_F$ contains the Poisson bracket

$$\{p, q\} = \sum (\partial p/\partial\xi_j \partial q/\partial x_j - \partial p/\partial x_j \partial q/\partial\xi_j).$$

Proof. Only the last assertion needs verification. We may assume that p is real valued. If $(x^0, \xi^0) \in N_e(F)$ then the bicharacteristic strip $t \to (x(t), \xi(t))$ for p through (x^0, ξ^0) is in $N_e(F)$ for small t by Theorem 8.5.9, hence

$$0 = dq(x(t), \xi(t))/dt = \{p, q\}(x(t), \xi(t))$$

which proves the corollary.

If $p_1, \dots, p_k \in \mathcal{N}_F$ and $dp_1, \dots, dp_k$ are linearly independent at $(x^0, \xi^0) \in N_e(F)$, then repeated use of Theorem 8.5.9 gives a k dimensional manifold through (x^0, ξ^0) contained in $N_e(F)$. The restriction of the one form $\langle \xi, dx \rangle$ and therefore of the symplectic form to any manifold $\Sigma \subset N_e(F)$ is equal to 0. In fact, if $(x^0, \xi^0) \in N_e(F)$ we can choose f with $f \leqq f(x^0)$ in F so that $\langle \xi^0, dx \rangle = df(x)$ in the tangent space of Σ at (x^0, ξ^0). This is zero since f has a maximum in Σ at (x^0, ξ^0). Hence $k \leqq n$, and since N_e is conic it follows that $k < n$ if $\partial p_1/\partial \xi, \dots, \partial p_k/\partial \xi$ are linearly independent.

We shall now discuss the dual objects of $N_e(F)$ briefly. This is not really related to the main theme of the chapter, but the proofs contain arguments similar to the proof of Proposition 8.5.8 and throw more light on the set $N_e(F)$. Moreover, the result will be important in Chapter XXVI. It confirms the geometrically plausible fact that an integral curve of a vector field cannot leave a closed set without pointing out of it at some boundary point.

Theorem 8.5.11. *Let v be a C^1 (or just Lipschitz continuous) vector field in X, and let F be a closed subset of X. Then the following conditions are equivalent:*

(i) *Every integral curve of $dx/dt = v(x(t))$, $0 \leqq t \leqq T$, with $x(0) \in F$ is contained in F.*

(ii) $$\langle v(x), \xi \rangle \leqq 0 \quad \text{for all } (x, \xi) \in N_e(F).$$

Proof. (i) ⇒ (ii). Let $f \in C^2$, $df(x) = \xi$ and assume that f restricted to F has a local maximum at x. Locally we can solve the equation $dy/dt = v(y(t))$ with $y(0) = x$. Since (i) gives that $f(x) \geqq f(y(t))$ for small $t > 0$, the derivative at $t = 0$ of $f(y(t))$ must be $\leqq 0$, that is, $\langle v(x), \xi \rangle \leqq 0$. To prove that (ii) ⇒ (i) we may assume that $X = \mathbb{R}^n$ and begin by proving an elementary lemma:

Lemma 8.5.12. *Let F be a closed set in $\mathbb{R}^n$ and set*

$$f(x) = \min_{z \in F} |x - z|^2$$

where $|\ |$ is the Euclidean norm. Then we have

$$f(x+y) = f(x) + f'(x, y) + o(|y|),$$
$$f'(x, y) = \min \{\langle 2y, x - z \rangle;\ z \in F, |x - z|^2 = f(x)\}.$$

Proof. We may assume in the proof that $x = 0$. Set

$$q_\varepsilon(y) = \inf\{-2\langle y, z \rangle;\ z \in F, |z| \leqq (f(0))^{\frac{1}{2}} + \varepsilon\}.$$

q_ε is homogeneous of degree 1, and $q_\varepsilon \nearrow q_0$ as $\varepsilon \downarrow 0$. The limit is therefore uniform on the unit sphere, so

$$q_0(y) \geqq q_\varepsilon(y) \geqq q_0(y) - c_\varepsilon |y|, \qquad c_\varepsilon \to 0 \text{ as } \varepsilon \to 0.$$

Now

$$|y-z|^2 = |z|^2 - 2\langle y, z\rangle + |y|^2$$

which gives immediately

$$f(y) \leqq f(0) + q_0(y) + |y|^2.$$

On the other hand, when $|y| \leqq \varepsilon/2$ the minimum in the definition of $f(y)$ is assumed for some z with $|z| \leqq f(0)^{\frac{1}{2}} + \varepsilon/2$, so

$$f(y) \geqq f(0) + q_\varepsilon(y) + |y|^2, \qquad |y| \leqq \varepsilon/2.$$

The lemma is proved.

Proof of Theorem 8.5.11. With the notation in (i) and Lemma 8.5.12 we have if $t<T$

$$\lim_{s \to t+0} (f(x(s)) - f(x(t)))/(s-t) = f'(x(t), v(x(t))).$$

Since the result to be proved is local we may assume that for all x and y

$$|v(x) - v(y)| \leqq C|x-y|.$$

When $z \in F$ and $|x(t)-z|^2 = f(x(t))$ we have

$$2\langle v(x(t)), x(t)-z\rangle = 2\langle v(z), x(t)-z\rangle - 2\langle v(z) - v(x(t)), x(t)-z\rangle.$$

The last term is $\leqq 2Cf(x(t))$. From the proof of Proposition 8.5.8 we recall that if $f(x(t))>0$ then $(z, x(t)-z) \in N_e(F)$ for every z such that $|x(t)-z|^2 = f(x(t))$. Hence the first term on the right is $\leqq 0$ by condition (ii) so the right-hand derivative of $f(x(t))$ is $\leqq 2\,C f(x(t))$. Thus the right-hand derivative of $f(x(t))\, e^{-2Ct}$ is $\leqq 0$ so this is a decreasing function by a simple modification of the proof of Theorem 1.1.1. (Note that f is continuous.) If $f(x(0))=0$ we obtain $f(x(t))=0$, $0<t<T$, as claimed.

8.6. WF_L for Solutions of Partial Differential Equations

If

$$P(x, D) = \sum_{|\alpha| \leqq m} a_\alpha(x) D^\alpha$$

is a differential operator with coefficients in $C^L(X)$, we have proved that

$$WF_L(P(x, D)u) \subset WF_L(u), \qquad u \in \mathscr{D}'(X).$$

When a_α are real analytic this is also an easy consequence of Theorem 8.4.8, Lemma 8.4.2 and Theorem 8.4.5. Making that assumption we shall now prove a converse similar to Theorem 8.3.1.

Theorem 8.6.1. *If* $P(x,D)$ *is a differential operator of order* m *with real analytic coefficients in* X, *then*

$$WF_L(u) \subset \operatorname{Char} P \cup WF_L(Pu), \quad u \in \mathscr{D}'(X), \tag{8.6.1}$$

where the characteristic set $\operatorname{Char} P$ *is defined by* (8.3.4).

Proof. We shall repeat the proof of Theorem 8.3.1 but make a more careful choice of cutoff functions. We must prove that if (x_0, ξ_0) does not belong to the right-hand side of (8.6.1) and $\xi_0 \neq 0$, then $(x_0, \xi_0) \notin WF_L(u)$. The hypothesis means that we can choose a compact neighborhood K of x_0 and a closed conic neighborhood V of ξ_0 in $\mathbb{R}^n \setminus 0$ such that

$$P_m(x,\xi) \neq 0 \quad \text{in } K \times V \tag{8.6.2}$$

$$(K \times V) \cap WF_L(Pu) = \emptyset. \tag{8.6.3}$$

Using Theorem 1.4.2 we now choose a sequence $\chi_N \in C_0^\infty(K)$ equal to 1 in a fixed neighborhood U of x_0 such that for every α

$$|D^{\alpha+\beta} \chi_N| \leqq C_\alpha (C_\alpha N)^{|\beta|}, \quad |\beta| \leqq N. \tag{8.6.4}$$

Then the sequence $u_N = \chi_{2N} u$ is bounded in $\mathscr{E}'$ and equal to u in U. The theorem will be proved if we show that (8.4.5) is valid in V when $|\xi| > N$, for (8.4.5) follows from the boundedness of u_N when $|\xi| \leqq N \leqq L_N$.

To estimate $\hat{u}_N(\xi)$ in V we must solve the equation

$${}^tP v(x) = \chi_{2N}(x) e^{-i\langle x, \xi\rangle} \tag{8.6.5}$$

approximately. Writting $v = e^{-i\langle x,\xi\rangle} w / P_m(x,\xi)$ and noting that the principal symbol of ${}^tP$ is $P_m(x, -\xi)$, we obtain instead of (8.6.5) an equation of the form

$$w - Rw = \chi_{2N}, \quad R = R_1 + R_2 + \ldots + R_m \tag{8.6.6}$$

where $R_j |\xi|^j$ is a differential operator of order less than or equal to j with analytic coefficients which are homogeneous of degree 0 with respect to ξ when $\xi \in V$ and $x \in K$. Formally a solution would be given by

$$\sum_0^\infty R^k \chi_{2N}.$$

However, we must not introduce derivatives of very high order so we set

$$w_N = \sum_{j_1+\ldots+j_k \leqq N-m} R_{j_1} \ldots R_{j_k} \chi_{2N}.$$

A simple calculation gives

$$w_N - R w_N = \chi_{2N} - \sum_{j_1+\ldots+j_k > N-m \geqq j_2+\ldots+j_k} R_{j_1} \ldots R_{j_k} \chi_{2N} = \chi_{2N} - e_N.$$

This means that

$${}^tP(x, D)(e^{-i\langle x,\xi\rangle} w_N(x, \xi)/P_m(x, \xi)) = e^{-i\langle x,\xi\rangle}(\chi_{2N}(x) - e_N(x, \xi)).$$

With integrals denoting action of distributions we obtain

$$(8.6.7)\quad \int u(x)\chi_{2N}(x) e^{-i\langle x,\xi\rangle} dx = \int u(x) e_N(x, \xi) e^{-i\langle x,\xi\rangle} dx + \int f(x) e^{-i\langle x,\xi\rangle} w_N(x, \xi)/P_m(x,\xi)\, dx.$$

Here $f = P(x, D)u$. To estimate the right-hand side of (8.6.7) we first prove a simple lemma.

Lemma 8.6.2. *There is a constant C' such that, if $j = j_1 + \ldots + j_k$ and $j + |\beta| \leqq 2N$,*

$$(8.6.8)\qquad |D^\beta R_{j_1} \ldots R_{j_k} \chi_{2N}| \leqq C'^{N+1} N^{j+|\beta|} |\xi|^{-j}, \qquad \xi \in V.$$

Proof. By the homogeneity it suffices to prove the lemma when $|\xi| = 1$. All coefficients occurring in some R_j when $|\xi| = 1$, $\xi \in V$, have a fixed bound in a fixed complex neighborhood of K, so the lemma is a consequence of the following one.

Lemma 8.6.3. *Let K be a compact set in $\mathbb{R}^n$ and K' a neighborhood of K in $\mathbb{C}^n$. If $a_1, \ldots, a_{j-1}$ are analytic and $|a_1| < 1, \ldots, |a_{j-1}| < 1$ in K', $j \leqq N$, we have*

$$(8.6.9)\qquad |D_{i_1} a_1 D_{i_2} \ldots a_{j-1} D_{i_j} \chi_N| \leqq C'^{N+1} N^j.$$

Proof. By Cauchy's inequalities we have for some $r > 0$

$$|D^\alpha a_l| \leqq |\alpha|!\, r^{-|\alpha|} \quad \text{in } K,$$

and (8.6.4) gives

$$|D^\alpha \chi_N| \leqq C_0 (C_0 N)^{|\alpha|} \leqq C_0 e^N C_0^{|\alpha|} |\alpha|! \quad \text{if } |\alpha| \leqq N,$$

for $N^{|\alpha|}/|\alpha|! \leqq e^N$. Now it is clear that $D_{i_1} a_1 \ldots a_{j-1} D_{i_j} \chi_N$ is a sum of terms of the form $(D^{\alpha_1} a_1) \ldots (D^{\alpha_{j-1}} a_{j-1}) D^{\alpha_j} \chi_N$ where

$$|\alpha_1| + \ldots + |\alpha_j| = j.$$

If there are $C_{k_1 \dots k_j}$ terms with $|\alpha_1|=k_1, \dots, |\alpha_j|=k_j$, the left-hand side of (8.6.9) can be estimated by

$$C_0 e^N (\max(C_0, 1/r))^N \sum C_{k_1 \dots k_j} k_1! \dots k_j!.$$

Since the derivative D_{i_k} in (8.6.9) operates on all the following factors, it is easy to see that

$$\sum C_{k_1 \dots k_j} x_1^{k_1} \dots x_j^{k_j} = (x_1 + \dots + x_j)(x_2 + \dots + x_j) \dots x_j.$$

It follows that

$$\sum C_{k_1 \dots k_j} k_1! \dots k_j! = \int_0^\infty \cdots \int_0^\infty (x_1 + \dots + x_j)(x_2 + \dots + x_j) \dots x_j e^{-(x_1 + \dots + x_j)} dx$$
$$= (2j-1)!! \leqq (2N)^j.$$

(The integral is computed by taking $x_1 + \dots + x_j$, $x_2 + \dots + x_j$, ... as new variables.) This completes the proof of the lemma.

End of the proof of Theorem 8.6.1. If M is the order of u in a neighborhood of K, we can estimate the first term on the right-hand side of (8.6.7) for large N and $|\xi| > N$, $\xi \in V$, by

$$C \sum_{|\alpha| \leqq M} (1+|\xi|)^{M-|\alpha|} \sup_x |D^\alpha e_N(x, \xi)|.$$

The number of terms in e_N cannot exceed 2^N, and each term can be estimated by means of (8.6.8), which gives the bound

$$C_1 |\xi|^{M+m-N} C'^{N+1} N^{N+M} 2^N.$$

If N is replaced by $N+m+M$ this is an estimate of the desired form (8.4.5) even for the analytic class. To estimate the last term in (8.6.7) we observe that (8.6.8) gives

$$|D^\beta w_N| \leqq C_1^{N+1} N^{|\beta|}, \quad |\beta| \leqq N, \ \xi \in V, \ |\xi| > N. \tag{8.6.10}$$

We have a similar bound for $w_N |\xi|^m / P_m(x, \xi)$. The proof is therefore completed by the following lemma.

Lemma 8.6.4. *Let $f \in \mathscr{D}'(X)$, let K be a compact subset of X and V a closed cone $\subset \mathbb{R}^n \setminus 0$ such that*

$$WF_L(f) \cap (K \times V) = \emptyset.$$

If $w_N \in C_0^\infty(K)$ and (8.6.10) is fulfilled, then

$$|\widehat{w_N f}(\xi)| \leqq C_2^{N+1} (L_{N-M-n}/|\xi|)^{N-M-n} \quad \text{if } \xi \in V, \ |\xi| > N, \ N > M+n. \tag{8.6.11}$$

Here M is the order of f in a neighborhood of K.

Proof. By Lemma 8.4.4 we can find a sequence f_N which is bounded in $\mathscr{E}'^M$ and equal to f in a neighborhood of K so that

$$|\hat{f}_N(\eta)| \leqq C(CL_N/|\eta|)^N, \qquad \eta \in W,$$

where W is a conic neighborhood of V. Then $w_N f = w_N f_{N'}$, $N' = N - M - n$. Since

$$|\hat{w}_N(\eta)| \leqq C_2^{N+1}(N/(N+|\eta|))^N$$

by (8.6.10), it follows from (8.1.3) that

$$|\widehat{w_N f}(\xi)| \leqq C_3^{N+1}((L_{N'}/|\xi|)^{N'} + N^N |\xi|^{n+M-N}), \qquad \xi \in V, \ |\xi| > N.$$

Since $N' \leqq L_{N'}$ this proves (8.6.11).

Combination of Theorems 8.6.1 and 8.5.6′ gives

Theorem 8.6.5 (Holmgren's uniqueness theorem). *If $u \in \mathscr{D}'(X)$ is a solution of a differential equation $P(x, D)u = 0$ with analytic coefficients, then the principal symbol $P_m(x, \xi)$ must vanish on $N(\operatorname{supp} u)$. Thus $u = 0$ in a neighborhood of a non-characteristic C^1 surface if this is true on one side.*

The last statement follows in view of Proposition 8.5.8. (Recall from Section 6.4 that a C^1 surface with normal ξ at x is non-characteristic at x if $P_m(x, \xi) \neq 0$.) If P is elliptic (cf. Corollary 8.3.2) then the theorem states that $N(\operatorname{supp} u)$ is empty, so $\operatorname{supp} u$ has no boundary point in X. If X is connected and $u = 0$ near a point in X, it follows that $u = 0$ in X. A stronger unique continuation theorem is obtained if we use Corollary 8.5.10 also:

Theorem 8.6.6. *Let $P(x, D)$ be a differential operator with analytic coefficients and let $\mathscr{C}$ be the smallest subset of $C^\infty(T^*X \setminus 0)$ which contains all C^∞ functions vanishing on* Char P *and is closed under Poisson brackets. If $u \in \mathscr{D}'(X)$ and $P(x, D)u = 0$ it follows then that all functions in $\mathscr{C}$ must vanish on $N(\operatorname{supp} u)$.*

In particular, if the functions in $\mathscr{C}$ have no common zeros then we conclude that u vanishes identically if X is connected and u vanishes in an open set. If u vanishes on one side of a C^1 surface with normal ξ at x, then u vanishes in a neighborhood of x unless all functions in $\mathscr{C}$ vanish at (x, ξ). This is an improvement of the classical uniqueness theorem of Holmgren as the following example shows:

Example 1. If $P(x,\xi)=\xi_1^2+x_1^2\xi_2^2+\ldots+x_{n-1}^2\xi_n^2$ then $\xi_1, x_1\xi_2, \ldots, x_{n-1}\xi_n$ vanish on Char P. Taking Poisson brackets we obtain

$$\{\xi_1, x_1\xi_2\}=\xi_2,\ \{\xi_2, x_2\xi_3\}=\xi_3, \ldots, \{\xi_{n-1}, x_{n-1}\xi_n\}=\xi_n$$

so the functions in $\mathscr{C}$ have no common zeros.

Example 2. If $P(x,\xi)=x_2^2\xi_1^2+\xi_2^2+\xi_3^2$ then $\mathscr{C}$ contains ξ_2, ξ_3, x_2, and since $\{\xi_2, x_2\}=1\in\mathscr{C}$ there are no common zeros. However, the solutions of $P(x,D)u=0$ need not be analytic. In fact,

$$u_\tau(x)=\exp(\tau x_3+ix_1\tau^2-x_2^2\tau^2/2)$$

is a solution for every τ. Hence

$$u(x)=\int_0^\infty u_\tau(x)e^{-\tau}d\tau$$

is a C^∞ solution when $|x_3|<1$, but u is not real analytic since

$$D_1^k u(0)=\int_0^\infty \tau^{2k}e^{-\tau}d\tau=(2k)!.$$

For differential operators with constant coefficients forming Poisson brackets is of no use, for the Poisson bracket of any two functions of ξ is 0. The following is then a partial converse of Theorem 8.6.5.

Theorem 8.6.7. *Let the plane $\langle x, N\rangle=0$, $N\in\mathbb{R}^n$, be characteristic with respect to the differential operator $P(D)$, that is, $P_m(N)=0$. Then there exists a solution u of the equation $P(D)u=0$ such that $u\in C^\infty(\mathbb{R}^n)$ and* $\operatorname{supp} u=\{x;\langle x,N\rangle\leqq 0\}$.

Proof. Let $P=P_m+P_{m-1}+\ldots+P_0$ where P_j is homogeneous of degree j and $P_m\neq 0$. With a fixed vector ξ such that $P_m(\xi)\neq 0$ we shall study the solutions of the equation

$$P(sN+t\xi)=0 \tag{8.6.12}$$

for large s. To do so we set $t=ws$ which reduces (8.6.12) to an algebraic equation in w and $1/s$,

$$P_m(N+w\xi)+\ldots+(1/s)^{m-k}P_k(N+w\xi)+\ldots=0.$$

When $1/s=0$ this algebraic equation in w is not identically satisfied since $P_m(\xi)\neq 0$ but it is true for $w=0$ since $P_m(N)=0$. Hence it follows from Lemma A.1.3 in the appendix to Volume II that for some integer p the equation (8.6.12) has a solution which is an analytic function of $(1/s)^{1/p}$ in a neighborhood of the origin and vanishes at the origin. This means that (8.6.12) has a solution

$$t(s)=s\sum_1^\infty c_j(s^{-1/p})^j \tag{8.6.13}$$

analytic for $|s^{1/p}|>M$ where M is a constant. Thus we have with a constant C

$$|t(s)|\leqq C|s|^{1-1/p}, \qquad |s|>(2M)^p. \tag{8.6.14}$$

Now choose a number ρ such that $1-1/p<\rho<1$ and set with $\tau>(2M)^p$

$$u(x)=\int_{i\tau-\infty}^{i\tau+\infty} e^{i\langle x,\, sN+t(s)\xi\rangle}e^{-(s/i)^\rho}\,ds. \tag{8.6.15}$$

Here we define $(s/i)^\rho$ so that it is real and positive when s is on the positive imaginary axis, and we choose a fixed branch of $s^{1/p}$ in the upper half plane. The integral is convergent and independent of τ, for when x is in a fixed bounded set we have in view of (8.6.14)

$$\begin{aligned}&\operatorname{Re}(i\langle x, sN+t(s)\xi\rangle-(s/i)^\rho)\\ &\leqq -\tau\langle x,N\rangle+C|x|\,|\xi|\,|s|^{1-1/p}-|s|^\rho\cos(\pi\rho/2)\\ &\leqq -\tau\langle x,N\rangle-c|s|^\rho\end{aligned} \tag{8.6.16}$$

if $0<c<\cos(\pi\rho/2)$ and $|s|$ is large. This estimate also shows that, when x is in a compact set, the integral (8.6.15) is uniformly convergent even after an arbitrary number of differentiations with respect to x. Hence $u\in C^\infty$ and using (8.6.12) we conclude that $P(D)u=0$. From (8.6.16) we also obtain

$$|u(x)|\leqq e^{-\tau\langle x,N\rangle}\int_{-\infty}^{\infty} e^{-c|\sigma|^\rho}\,d\sigma.$$

Hence it follows when $\tau\to+\infty$ that $u(x)=0$ if $\langle x,N\rangle>0$. (Compare the proof of Theorem 7.3.1.)

When $\langle x,N\rangle<0$ we can replace the integration contour in (8.6.15) by the negatively oriented boundary of the set

$$\{s;\ |s|<(2M)^p \text{ or } \operatorname{Im} s<0,\ |\operatorname{Re} s|<(2M)^p\}.$$

The integrand is then exponentially decreasing and remains so for all complex x with $\langle \operatorname{Re} x, N\rangle<0$. Hence u is analytic in the half space $\{x\in\mathbb{R}^n;\ \langle x,N\rangle<0\}$ and does not vanish identically by Fourier's inversion formula. (Note that $u(x)$ is the Fourier transform of $\exp(-(s/i)^\rho)$ when $\langle x,N\rangle=t$ and $\langle x,\xi\rangle=0$.) This proves that $\operatorname{supp} u$ is the closure of the half space. The theorem is proved.

Remark. Separating the integration in (8.6.15) for $\operatorname{Re} s>0$ and for $\operatorname{Re} s<0$ we can write $u(x)=u_+(x)+u_-(x)$ where $u_+(u_-)$ is the boundary value of a function analytic when $\langle \operatorname{Im} z, N\rangle>0$ (resp. $\langle \operatorname{Im} z, N\rangle<0$). This proves that $WF_A(u)=\{(x,tN); \langle x,N\rangle=0\}$ is the normal bundle of the boundary of the support.

The following theorem gives a useful summary of the results in the constant coefficient case.

Theorem 8.6.8. *Let X_1 and X_2 be open convex sets in $\mathbb{R}^n$ such that $X_1\subset X_2$, and let $P(D)$ be a differential operator with constant coefficients. Then the following conditions are equivalent:*

(i) *Every $u\in\mathscr{D}'(X_2)$ satisfying the equation $P(D)u=0$ in X_2 and vanishing in X_1 must also vanish in X_2.*

(ii) *Every hyperplane which is characteristic with respect to P and intersects X_2 also intersects X_1.*

Proof. (i) $\Rightarrow$ (ii) Assume that π is a characteristic hyperplane which does not intersect X_1. Let H be the half space bounded by π which does not intersect X_1. By Theorem 8.6.7 we can find a solution u of the equation $P(D)u=0$ with $\operatorname{supp} u=H$, so (i) shows that $H\cap X_2=\emptyset$, hence $\pi\cap X_2=\emptyset$.

(ii) $\Rightarrow$ (i). Let y_2 be a point in X_2. Choose a point $y_1\in X_1$ and denote by I the line segment between y_1 and y_2. We can find an open convex set $X\Subset X_1$ such that every characteristic plane intersecting I also meets X. In fact, if $x_0\in I$ and $\xi_0\in\mathbb{R}^n$, $P_m(\xi_0)=0$, $|\xi_0|=1$, we can choose an open ball $\Sigma\Subset X_1$ which meets the plane $\langle x-x_0,\xi_0\rangle=0$ and consequently meets every characteristic plane with normal close to ξ_0 passing through a point near x_0. By the Borel-Lebesgue lemma a set X with the required properties can therefore be constructed by taking the convex hull of a finite number of open balls $\Sigma\Subset X_1$.

Let Y_t be the interior of the convex hull of X and $y_t=y_1+t(y_2-y_1)$, $0\leqq t\leqq 1$. For small t we have $y_t\in X_1$, hence $Y_t\subset X_1$ and $u=0$ in Y_t. Let T be the supremum of all $t\in[0,1]$ such that $u=0$ in Y_t. Then $u=0$ in Y_T and $y_T\notin X_1$. If π is a supporting plane of Y_T then π is non-characteristic if $y_T\in\pi$ since π intersects I but not X, and if $y_T\notin\pi$ then $\pi\cap\bar{Y}_T\subset\bar{X}\Subset X_1$. Hence it follows from Proposition 8.5.8 and Theorem 8.6.5 that $\partial Y_T\cap\operatorname{supp} u=\emptyset$. Hence $T=1$, and $u=0$ in a neighborhood of the arbitrarily chosen point $y_2\in X_2$. This proves condition (i).

Corollary 8.6.9. *If the support of a solution $u\in\mathscr{D}'(\mathbb{R}^n)$ of the equation $P(D)u=0$ is contained in a half space with non-characteristic boundary, then $u=0$.*

Proof. Every characteristic plane intersects the half space.

Corollary 8.6.10. *Let N_1 and N_2 be real vectors such that*

$$P_m(\tau_1 N_1 + \tau_2 N_2) \neq 0 \quad \text{when } \tau_1 > 0 \text{ and } \tau_2 \geqq 0. \tag{8.6.17}$$

Set $X_{a_1,a_2} = \{x; \langle x, N_j\rangle < a_j, j=1,2\}$, where a_1 and a_2 are real numbers or $+\infty$. If $u \in \mathscr{D}'(X_{a_1,a_2})$ satisfies the equation $P(D)u=0$ and if $u=0$ in X_{c,a_2} for some $c<a_1$ then $u=0$ in X_{a_1,a_2}.

Proof. The normal of a plane which does not intersect X_{c,a_2} must be a linear combination with non-negative coefficients of N_1 and N_2. If the plane is characteristic, the normal must therefore be proportional to N_2 by (8.6.17). Hence (ii) in Theorem 8.6.8 is fulfilled.

Corollary 8.6.11. *Let X be an open proper convex cone with vertex y such that no hyperplane through y which is characteristic with respect to $P(D)$ intersects $\bar{X}$ only at y. Every $u \in \mathscr{D}'(X)$ satisfying the equation $P(D)u=0$ and vanishing outside a bounded subset of X must then vanish in all of X.*

Proof. Since X is proper we can find a plane π through y which meets $\bar{X}$ only at y. We can apply Theorem 8.6.8 with $X_2 = X$ and X_1 equal to the intersection of X and a suitable half space with boundary parallel to π. The hypothesis means that in every characteristic plane containing a point in X_2 there is a half ray which lies entirely in X_2. Hence it meets X_1 so the corollary follows from Theorem 8.6.8.

In the proof of Theorem 8.3.7 we actually saw that ω_- was an entire analytic function, and if the reference to Theorem 8.1.6 in the proof is replaced by a reference to Theorem 8.4.8, we obtain

Theorem 8.6.12. *If $P(D)$ is of real principal type, then one can find $E_\pm \in \mathscr{D}'(\mathbb{R}^n)$ and analytic $\omega_\pm$ such that $P(D)E_\pm = \delta + \omega_\pm$ and*

$$WF_A(E_\pm) \subset \{(tP'_m(\xi), \xi);\, t \gtrless 0,\, P_m(\xi)=0,\, \xi \neq 0\} \cup T_0^* \setminus \{0\}. \tag{8.6.18}$$

We can now prove an analogue of Theorem 8.3.3′.

Theorem 8.6.13. *Let $P(D)$ be of real principal type. If $u \in \mathscr{D}'(X)$, $P(D)u = f$ and $(x,\xi) \in WF_L(u) \setminus WF_L(f)$, then $P_m(\xi)=0$ and*

$$I \times \{\xi\} \subset WF_L(u)$$

if $I \subset X$ is a line segment with direction $P'_m(\xi)$ containing x such that

$$(I \times \{\xi\}) \cap WF_L(f) = \emptyset.$$

Proof. Without restricting the generality we may of course assume that $u \in \mathscr{E}'$. Let U and F be the analytic functions in $\{z; |\operatorname{Im} z| < 1\}$ corresponding to u and f as in Theorem 8.4.11. Since $U = K * u$ and $F = K * f$ we have $P(D)\, U = F$. In particular, if $|\omega| = 1$ then

$$P(D)\, U(. - i\omega) = F(. - i\omega). \tag{8.6.19}$$

If there is some $y \in I$ such that $(y, \xi) \notin WF_L(u)$ then there is a neighborhood W of $-\xi/|\xi|$ in S^{n-1} such that

$$u_1 = \int_W U(. + i\omega)\, d\omega \in C^L(V), \quad f_1 = \int_W F(. + i\omega)\, d\omega \in C^L(V_1)$$

where V is a neighborhood of y and V_1 is a neighborhood of I. We have $(x, \xi) \in WF_L(u_1)$ since $(x, \xi) \notin WF_L(u - u_1)$, and $P(D)\, u_1 = f_1$.

We may assume that $x - y = t P_m'(\xi)$ for some $t > 0$. Set

$$I_- = \{x\} + \mathbb{R}_- P_m'(\xi).$$

Now choose a cutoff function $\chi \in C_0^\infty$ which is 1 near the segment $[y, x]$ so that $I_- \cap \operatorname{supp} d\chi \subset V$. We can choose χ so that $\chi(x) = \psi(\langle x, \eta \rangle)$, $x \in V$, for some $\psi \in C^\infty$ and some η with $\langle P_m'(\xi), \eta \rangle \neq 0$, thus η is linearly independent of ξ. By Theorem 8.5.1 this implies that

$$WF_A(\chi)|_V \subset V \times \mathbb{R}\eta.$$

Hence $v = \chi u_1 \in \mathscr{E}'$ and

$$WF_L(v)|_V \subset V \times \mathbb{R}\eta,$$

$$WF_L(P(D)\, v)|_{I_-} \subset V \times \mathbb{R}\eta.$$

It follows from Theorem 8.5.5 that the analogue for WF_L of (8.2.16) is valid, so for $v = E_+ * P(D)\, v - \omega_+ * v$ we have

$$(x, \xi) \notin WF_L(v).$$

Since $v = u_1$ in a neighborhood of x this is a contradiction proving that $(y, \xi) \in WF_L(u)$. The proof is complete.

Corollary 8.6.14. *Let $P(D)$ be of real principal type, $u \in \mathscr{D}'(X)$ and $P(D)\, u = 0$. If $(x, \xi) \in \bar{N}(\operatorname{supp} u)$ and $I \subset X$ is an interval containing x on the line through x with direction $P_m'(\xi)$, then $I \subset \operatorname{supp} u$.*

Proof. By Theorem 8.5.6′ we have $(x,\xi)\in WF_A(u)$ so $I\times\{\xi\}\subset WF_A(u)$ by Theorem 8.6.13. This proves the statement.

Corollary 8.6.14 should be compared with Theorem 8.5.9 which is far more general but only local. Theorem 8.6.13 and Corollary 8.6.14 can be extended to operators with real analytic coefficients, but the proofs require additional technical tools then. Instead we shall prove an extension of Theorem 8.6.1 to convolution equations which will be useful in Section 12.9.

If $\mu\in\mathscr{S}'(\mathbb{R}^n)$ we can define a characteristic set as follows. First we let Γ be the set of all $\xi_0\in\mathbb{R}^n\setminus\{0\}$ such that there is a complex conic neighborhood V of ξ_0 and an analytic function Φ in $V_C=\{\zeta\in V, |\zeta|>C\}$ for some C such that $\Phi\hat{\mu}=1$ in $V_C\cap\mathbb{R}^n$ and

$$|\Phi(\zeta)|\leqq C_1|\zeta|^N, \quad \zeta\in V_C, \tag{8.6.20}$$

for some N and C_1. We shall denote by $\operatorname{Char}\mu$ the complement of Γ in $\mathbb{R}^n\setminus\{0\}$.

Theorem 8.6.15. *If $\mu\in\mathscr{S}'(\mathbb{R}^n)$ and $u\in\mathscr{E}'(\mathbb{R}^n)$, then*

$$WF_A(u)\subset WF_A(\mu*u)\cup(\mathbb{R}^n\times\operatorname{Char}\mu). \tag{8.6.21}$$

Proof. We shall use the interpretation of WF_A in Theorem 8.4.11. With the notation in that theorem we must show that $u*K(z)$ is analytic at $x_0-i\xi_0$ if $\xi_0\notin\operatorname{Char}\mu$, $|\xi_0|=1$ and $(x_0,\xi_0)\notin WF_A(f)$, $f=\mu*u$. Choose V and Φ as above so that (8.6.20) is valid and $\Phi\hat{\mu}=1$ in $V_C\cap\mathbb{R}^n$. Let W' and W'' be closed conic neighborhoods of ξ_0 in $\mathbb{R}^n\setminus\{0\}$ such that W'' is contained in the interior of W' and $W'\subset V$. Choose $\chi\in C^\infty$ with $0\leqq\chi\leqq 1$ equal to 1 in a neighborhood of W''_{3C} and $\operatorname{supp}\chi\subset W'_{2C}$ so that χ is homogeneous of degree 0 when $|\xi|>3C$. Then the Fourier transform of $u*K(.+iy)$, $|y|<1$, can be decomposed as follows

$$\hat{u}e^{-\langle y,\xi\rangle}/I(\xi)=\hat{u}(1-\chi(\xi))e^{-\langle y,\xi\rangle}/I(\xi)+\hat{f}\Phi(\xi)\chi(\xi)\,e^{-\langle y,\xi\rangle}/I(\xi).$$

If we introduce the inverse Fourier transforms

$$K_1(z)=(2\pi)^{-n}\int(1-\chi(\xi))e^{i\langle z,\xi\rangle}/I(\xi)\,d\xi,$$

$$K_2(z)=(2\pi)^{-n}\int\chi(\xi)\Phi(\xi)e^{i\langle z,\xi\rangle}/I(\xi)\,d\xi$$

which are rapidly decreasing when $\operatorname{Re}z\to\infty$, $|\operatorname{Im}z|<1$, it follows that

$$K*u(z)=K_1*u(z)+K_2*f(z), \quad |\mathrm{Im}\, z|<1. \tag{8.6.22}$$

It is clear that K_1 remains analytic when $|\mathrm{Im}\, z+\xi_0|$ is sufficiently small, so $K_1*u(z)$ is analytic at $x_0-i\xi_0$. To study the properties of K_2 we shall follow the proof of Lemma 8.4.10 although we must now work in all variables and apply Stokes' formula. Let $\chi_1(\xi)$ be a C^∞ function with support in W''_{3C} which is 1 in W'''_{4C} for another conic neighborhood W''' of ξ_0 and is homogeneous of degree 0 for $|\xi|>4C$. We want to move the integration to the cycle ($x=\mathrm{Re}\, z$)

$$\mathbb{R}^n\ni\xi\to\xi+i\delta\chi_1(\xi)|\xi|x(1+|x|^2)^{-\frac{1}{2}},$$

where $0<\delta\leqq 1$ is chosen so small that we do not leave V_C when $\xi\in\mathrm{supp}\,\chi_1$. To estimate the integrand we shall use Lemma 8.4.9 and the inequality

$$\begin{aligned}&\mathrm{Re}(i\langle x+iy,\xi+i\eta\rangle-\langle\xi+i\eta,\xi+i\eta\rangle^{\frac{1}{2}})\\&\leqq-\langle x,\eta\rangle-\langle y,\xi\rangle-(|\xi|^2-|\eta|^2)^{\frac{1}{2}}\end{aligned}$$

valid when $|\eta|<|\xi|$. (This follows from the fact that $\mathrm{Re}\, w^2\leqq(\mathrm{Re}\, w)^2$.) When $\eta=\rho\,|\xi|\,x(1+|x|^2)^{-\frac{1}{2}}$, $0\leqq\rho\leqq 1$, we obtain the estimate

$$|\xi|(-\rho|x|^2/(1+|x|^2)^{\frac{1}{2}}-(1-\rho^2|x|^2/(1+|x|^2))^{\frac{1}{2}})-\langle y,\xi\rangle.$$

The parenthesis is a convex function of ρ which is -1 for $\rho=0$ and $-(1+|x|^2)^{\frac{1}{2}}$ when $\rho=1$. Hence

$$\begin{aligned}&\mathrm{Re}(i\langle x+iy,\xi+i\eta\rangle-\langle\xi+i\eta,\xi+i\eta\rangle^{\frac{1}{2}})\\&\leqq-|\xi|(1-\rho+\rho(1+|x|^2)^{\frac{1}{2}})-\langle y,\xi\rangle.\end{aligned}$$

Using Stokes' formula as just indicated it follows that for some $\delta>0$ there is an analytic continuation of K_2 to

$$\{z;|\mathrm{Im}\, z|<1-\delta+\delta(1+|\mathrm{Re}\, z|^2)^{\frac{1}{2}},|\mathrm{Im}\, z+\xi_0|<\delta\}$$

where the second restriction as in the discussion of K_1 comes from the set where the integration contour has not been deformed. An integration by parts shows that K_2 is rapidly decreasing at infinity in this set.

The properties of K_2 show that the boundary value $K_2*f(.-i\xi_0)$ is equal to the convolution of f and the boundary values $K_2(.-i\xi_0)$ which are analytic except at 0. Write $f=f_1+f_2$ where $f_1\in\mathscr{E}'$ and f_2 vanishes when $|x-x_0|<r$, say. Then

$$K_2*f_2(z)=f_2(K_2(.-z))$$

is analytic when z is so close to $x_0-i\xi_0$ that $K(.-z)$ is uniformly in $\mathscr{S}$ when $|x-x_0|\geqq r$. By Theorem 8.4.8 we have $WF_A(K_2(.-i\xi_0))$ $\subset\{(0,t\xi_0),t>0\}$. (Cf. Lemma 8.4.12.) If we recall that $(x_0,\xi_0)\notin WF_A(f_1)$ it follows by the analogue of (8.2.16) for WF_A that

$$x_0\notin \operatorname{sing\,supp}_A f_1*K_2(.-i\xi_0).$$

Hence $K*u$ is analytic at $x_0-i\xi_0$ which completes the proof.

Remark. The theorem remains valid if $u\in\mathscr{D}'(\mathbb{R}^n)$ and

$$\operatorname{supp}\mu\times\operatorname{supp}u\ni(x,y)\to x+y$$

is proper. In fact, $\mu*u$ is then defined (see Section 4.2) and for any x we have $\mu*u=\mu*(\phi u)$, $u=\phi u$ in a neighborhood of x if $\phi\in C_0^\infty$ is equal to 1 in a sufficiently large set. If we apply (8.6.21) to ϕu we conclude that (8.6.21) is valid for the fiber at x.

8.7. Microhyperbolicity

If F is a real valued real analytic function in the open set $X\subset\mathbb{R}^n$ and θ is a real vector such that $\langle\theta,F'(x)\rangle>0$ when $F(x)=0$, then

$$F_\theta^{-1}=\lim_{\varepsilon\to+0}1/F(.+i\varepsilon\theta)$$

is a well defined distribution with

$$WF_A(F_\theta^{-1})=\{(x,tF'(x));\ F(x)=0,\ t>0\}.$$

In fact, if Γ is an open convex cone such that

$$\langle y,F'(x)\rangle>0,\qquad 0\neq y\in\bar{\Gamma},\ x\in\bar{X}_0\Subset X,$$

then it follows from Taylor's formula that

(8.7.1) $\quad |y|\leqq C|F(x+iy)|\quad$ if $x\in X_0$, $y\in\Gamma$ and $|y|$ is small.

The statement is therefore a consequence of Theorem 8.4.8. Now a weaker form of (8.7.1) may be valid also when F has critical points. A typical example is the Lorentz form $F(x)=x_1^2-x_2^2-\ldots-x_n^2$. By (7.4.8)

$$F(y)\leqq|F(x+iy)|;\qquad x,y\in\mathbb{R}^n;$$

so (8.7.1) is valid with $|y|$ replaced by $|y|^2$ if $\bar{\Gamma}\setminus\{0\}$ is contained in the open light cone. The following terminology is motivated by this example.

Definition 8.7.1. A real analytic function F in the open set $X \subset \mathbb{R}^n$ is called microhyperbolic with respect to $\theta \in \mathbb{R}^n$ if there is a positive continuous function $t(x)$ in X such that

$$F(x+it\theta) \neq 0, \quad \text{if } 0<t<t(x),\ x \in X. \tag{8.7.2}$$

In the following discussion of the local properties of F we may shrink X so that t is bounded from below in X by a positive constant and then replace θ by a multiple to achieve that

$$F(x+it\theta) \neq 0 \quad \text{if } 0<t\leqq 1,\ x \in X. \tag{8.7.2)'$$

To simplify notation we also assume that $0 \in X$ and study F near 0.

Lemma 8.7.2. *If F satisfies (8.7.2)′ and $F(t\theta)$ has a zero of order m exactly when $t=0$, then*

$$F(x)=F_0(x)+O(|x|^{m+1})$$

where F_0 is a homogeneous polynomial of degree m and

$$F_0(\theta) \neq 0, \quad F_0(x+it\theta) \neq 0 \quad \textit{if } 0 \neq t \in \mathbb{R},\ x \in \mathbb{R}^n. \tag{8.7.3}$$

Proof. Let y be a fixed real vector and set

$$g(t,s)=F(t\theta+sy).$$

Then $g(t,0)=ct^m+O(t^{m+1})$, $c \neq 0$, and we claim that

$$g(t,s)=O(|t|+|s|)^m \quad \text{at } (0,0).$$

If this is not true then the largest λ such that

$$g(t,s)=O(|t|+|s|^\lambda)^m \quad \text{at } (0,0)$$

is a rational number with $1/m \leqq \lambda < 1$. Write $\lambda = p/q$ where p and q are positive relatively prime integers, and consider the limits

$$g_0^\pm(w)=\lim_{s\to\pm 0} g(w|s|^\lambda, s)/|s|^{m\lambda}.$$

If at^js^k is a term in $g(t,s)$ with $j+k/\lambda=m$ then q divides $m-j$. Hence

$$g_0^\pm(w)=cw^m+(\pm 1)^p c_1 w^{m-q}+(\pm 1)^{2p} c_2 w^{m-2q}+\dots$$

where all c_j are not equal to 0. From (8.7.2)′ it follows that $\operatorname{Im} w \leqq 0$ for the zeros of $g_0^\pm(w)$, for if $g(w|s|^\lambda, s)=F(\operatorname{Re} w|s|^\lambda\theta+sy+i\operatorname{Im} w|s|^\lambda\theta)=0$ and s is small it follows that $\operatorname{Im} w \leqq 0$. Now we can find a number $z \neq 0$ such that $g_0^\pm(w)=0$ if $w^q=(\pm 1)^p z$. All such w cannot lie in a half plane unless $q=2$ and p is even, which contradicts that $1 \leqq p < q$. Hence $\lambda=1$, and since y is arbitrary we conclude that $F(x)$

$=O(|x|^m)$, $x\to 0$. Now
$$F_0(x)=\lim_{\varepsilon\to 0} F(\varepsilon x)/\varepsilon^m$$
is a homogeneous polynomial of degree m. By the argument above it follows from (8.7.2)′ that
$$F_0(x+w\theta)\neq 0 \quad \text{if } x\in\mathbb{R}^n \text{ and } \operatorname{Im} w>0.$$
Hence $F_0(x+w\theta)=(-1)^m F_0(-x-w\theta)\neq 0$ if $x\in\mathbb{R}^n$ and $\operatorname{Im} w<0$, which proves (8.7.3).

Remark. In the proof that $g(t,s)=O(|t|+|s|)^m$ it would have been sufficient to assume that for small real s and small $|t|$
$$g(t,s)=0 \Rightarrow \operatorname{Im} t<C|s|.$$
In fact, for the zeros of $g(w|s|^\lambda, s)/|s|^{m\lambda}$ we have $\operatorname{Im} w|s|^\lambda \leqq C|s|$ then. Since $\lambda<1$ it follows that $\operatorname{Im} w\leqq 0$ if w is a zero of $g_0^\pm$. This observation will be useful in Chapter XII.

Lemma 8.7.3. *Let F_0 be a homogeneous polynomial satisfying* (8.7.3). *Then the component Γ of θ in $\{x\in\mathbb{R}^n;\ F_0(x)\neq 0\}$ is a convex cone. The zeros t of $F_0(x+ty)$ are real if $x\in\mathbb{R}^n$ and $y\in\Gamma$; they are then negative if and only if x is also in Γ. The coefficients of $F_0(x)/F_0(\theta)$ are real.*

Proof. a) The zeros t of $F_0(x+t\theta)$ are all real for every real x, for if $F_0(x+t\theta)=0$ then $F_0(x+\operatorname{Re} t\theta+i\operatorname{Im} t\theta)=0$, hence $\operatorname{Im} t=0$ by (8.7.3). This implies that the quotient $F_0(x)/F_0(\theta)$ of the lowest and the highest coefficient in this polynomial in t is real.

b) Set
$$\Gamma_\theta=\{x\in\mathbb{R}^n;\ F_0(x+t\theta)=0 \Rightarrow t<0\}.$$
Then Γ_θ is open, and $\theta\in\Gamma_\theta$ since the zeros are -1 when $x=\theta$. If x_0 is in the closure of Γ_θ then $F_0(x_0+t\theta)=0 \Rightarrow t\leqq 0$, so $x_0\in\Gamma_\theta$ if $F_0(x_0)\neq 0$. Thus Γ_θ is open and closed in $\{x\in\mathbb{R}^n, F_0(x)\neq 0\}$, so Γ_θ contains the component Γ of θ there.

c) If $x\in\Gamma_\theta$ then
$$F_0(\varepsilon x+(1-\varepsilon)\theta)=\varepsilon^m F_0(x+(1-\varepsilon)\varepsilon^{-1}\theta)\neq 0 \quad \text{if } 0<\varepsilon\leqq 1.$$
Hence $F_0\neq 0$ on the line segment between x and θ. In particular $\Gamma_\theta\subset\Gamma$ so these cones are identical.

d) If $y\in\Gamma$ and $\varepsilon>0$ is fixed, then
$$E_y=\{x\in\mathbb{R}^n;\ F_0(x+i\varepsilon\theta+isy)=0 \Rightarrow \operatorname{Re} s<0\}$$
is open, and $0\in E_y$ since $F_0(i\varepsilon\theta+isy)=(is)^m F_0(\varepsilon\theta/s+y)=0$ implies $s<0$. If x is in the closure of E_y then $F_0(x+i\varepsilon\theta+isy)=0$ implies

$\operatorname{Re} s \leqq 0$, and $\operatorname{Re} s=0$ would contradict (8.7.3). Hence $x \in E_y$ so $E_y=\mathbb{R}^n$ and

$$F_0(x+i(\varepsilon\theta+y)) \neq 0 \quad \text{if } x \in \mathbb{R}^n,\ y \in \Gamma \text{ and } \varepsilon>0.$$

Since Γ is open it follows that

$$F_0(x+iy) \neq 0 \quad \text{if } x \in \mathbb{R}^n,\ y \in \Gamma.$$

Thus $F_0(x+ty)$ has only real zeros then (see a)). Since the component of y in $\{x \in \mathbb{R}^n;\ F_0(x) \neq 0\}$ is equal to Γ, it follows from b), c) that the zeros are negative if and only if $x \in \Gamma$, and then the line segment between x and y also lies in Γ. The proof is complete.

Homogeneous polynomials satisfying (8.7.3) are called *hyperbolic* with respect to θ. We shall resume their study in Section 12.4. However, what we need to prove now is that F has essentially the properties proved for F_0. The proof will be based on the idea in part d) of the proof of Lemma 8.7.3.

Lemma 8.7.4. *If Γ_1 is a closed cone contained in $\Gamma \cup \{0\}$ then one can find $\delta>0$ such that*

$$\delta|y|^m \leqq |F(x+iy)| \quad \text{if } y \in \Gamma_1,\ x \in \mathbb{R}^n,\ |y|<\delta,\ |x|<\delta. \tag{8.7.4}$$

Proof. Let K be a convex compact subset of Γ containing θ, which generates a cone containing Γ_1 and θ. Since

$$F(rz)/r^m \to F_0(z) \quad \text{if } r \to 0$$

and $F_0 \neq 0$ in K, we can choose $r>0$ so that

$$F(ty) \neq 0 \quad \text{if } t \in \mathbb{C},\ 0<|t| \leqq r,\ y \in K. \tag{8.7.5}$$

Note that $t=0$ is a zero of order m. With $y \in K$ consider the equation

$$F(x+i\varepsilon\theta+isy)=0.$$

When $(x,\varepsilon) \in \mathbb{R}^{n+1}$ and $|x|+|\varepsilon|$ is small enough there are exactly m roots s with $|s|<r$, for there is none with $|s|=r$ and when $x=\varepsilon=0$ there are m roots $s=0$. If $x=0$ and $y=\theta$ the roots are $s=-\varepsilon$ so they are negative if $\varepsilon>0$, as we now assume. If s is a root with $\operatorname{Re} s=0$ then $F(x-y\operatorname{Im} s+i\varepsilon\theta)=0$ which contradicts our hypothesis (8.7.2)'. It follows that for small $|x|+\varepsilon$, $\varepsilon>0$, and $y \in K$, we have m roots with $\operatorname{Re} s<0$, $|s|<r$. Letting $\varepsilon \to 0$ we conclude that $F(x+isy)$ has m zeros with $\operatorname{Re} s \leqq 0$ and $|s|<r$.

If $f(z)$ is an analytic function in $\{z \in \mathbb{C};\ |z| \leqq r\}$ and has m zeros $z_1, \ldots, z_m$ in the disc, then

$$\left|\prod_1^m (z-z_j)\right| \Big/ |f(z)| \leqq (2r)^m \sup_{|w|=r} 1/|f(w)|$$

by the maximum principle. When $\operatorname{Re} z_j \leqq 0$, $j=1, \ldots, m$, it follows that

$$1/|f(z)| \leqq (2r/\operatorname{Re} z)^m \sup_{|w|=r} |1/f(w)|, \qquad \text{if } |z|<r, \ \operatorname{Re} z>0.$$

If we apply this to $F(x+izy)$ it follows for small $x\in\mathbb{R}^n$ and $y\in K$ that

$$1/|F(x+isy)| \leqq (2r/s)^m \sup_{|w|=r} 1/|F(x+iwy)|, \qquad 0<s<r,$$

and this proves (8.7.4).

We are now ready to prove the main result of the section:

Theorem 8.7.5. *Let F be real analytic in the open set $X\subset\mathbb{R}^n$ and microhyperbolic with respect to $\theta\in\mathbb{R}^n$. If $x\in X$ we denote by $F_x(y)$ the lowest homogeneous part in the Taylor expansion of $y\to F(x+y)$. Then $F_x(\theta)\neq 0$ and the component Γ_x of θ in $\{y\in\mathbb{R}^n;\ F_x(y)\neq 0\}$ is an open convex cone. If Γ_x° is the dual cone then*

$$\Gamma^\circ = \{(x,\xi);\ x\in X, \xi\in\Gamma_x^\circ\} \subset T^*(X)$$

is closed. The limit

$$F_\theta^{-1} = \lim_{\varepsilon\to+0} F(.+i\varepsilon\theta)^{-1}$$

exists in $\mathscr{D}'(X)$, and

(8.7.6)
$$WF_A(F_\theta^{-1}) \subset \Gamma^\circ \setminus \{0\}.$$

The canonical one form $\omega=\langle\xi, dx\rangle$ vanishes in Γ° in the sense that if $t\to(x(t),\xi(t))\in\Gamma^\circ$ is a C^∞ curve then $\langle\xi(t), x'(t)\rangle=0$.

Proof. The existence of the limit and the inclusion (8.7.6) are consequences of Theorem 8.4.8 and Lemma 8.7.4. If $(x_0,\xi_0)\notin\Gamma^\circ$ then we can find $y_0\in\Gamma_{x_0}$ so that $\langle y_0,\xi_0\rangle<0$. By Lemma 8.7.4 F is also microhyperbolic with respect to y_0 in a neighborhood U of x_0, hence $F_x(y_0)\neq 0$ when $x\in U$ in view of Lemma 8.7.2. But this implies that $(x,\xi)\notin\Gamma^\circ$ if $x\in U$ and $\langle y_0,\xi\rangle<0$, so Γ° is closed.

If I is an open interval on $\mathbb{R}$ and $I\ni t\mapsto(x(t),\xi(t))$ is a C^∞ curve contained in Γ° then the degree of $F_{x(t)}(y)$ is locally constant in an open everywhere dense set. In fact, if J is an open interval $\subset I$ and m is the minimum of the degree of $F_{x(t)}$ when $t\in J$, then the degree of $F_{x(t)}$ is equal to m for all t in an open interval $J'\subset J$ because it is upper semi-continuous. By Taylor's formula we have for small y

$$F(x(t)+y) = \sum_{|\alpha|=m} R_\alpha(t,y)\, y^\alpha, \qquad t\in J',$$

where $R_\alpha\in C^\infty$. Differentiation with respect to t gives

$$\langle F'(x(t)+y), x'(t)\rangle = \sum_{|\alpha|=m} \partial R_\alpha(t,y)/\partial t\, y^\alpha.$$

Hence $\langle \partial F_{x(t)}(y)/\partial y, x'(t)\rangle \equiv 0$, for the Taylor expansion of the left-hand side starts with this polynomial of order $m-1$. This means that $\Gamma_{x(t)} + \mathbb{R}x'(t) = \Gamma_{x(t)}$, so $\langle \xi, x'(t)\rangle = 0$ if $\xi \in \Gamma^{\circ}_{x(t)}$, which proves that $\langle \xi(t), x'(t)\rangle = 0$, $t \in J'$. Hence $\langle \xi(t), x'(t)\rangle = 0$, $t \in I$, for this is proved in a dense subset.

Remark. It is not possible to have an inclusion (8.7.6) where Γ° has smaller *convex* fibers. In fact, assume that Γ_0 is a closed convex proper conic neighborhood of the fiber of $WF_A(F_\theta^{-1})$ at x_0. Then it follows from Theorem 8.4.15 and the remark after its proof that there is a function G with boundary value F_θ^{-1} which is analytic in

$$\{x+iy;\ |x-x_0|<\delta,\ |y|<\delta,\ y\in\Gamma_0^{\circ}\} \tag{8.7.7}$$

for some $\delta>0$. Now continuous analytic functions are uniquely determined by their boundary values, and Theorem 3.1.15 shows that this is still true when the boundary values are assumed in the distribution sense. Thus G is an analytic continuation of $1/F$, so $F \neq 0$ in the set (8.7.7). Hence

$$F_{x_0}(z) = \lim_{\varepsilon\to 0} F(x_0+\varepsilon z)/\varepsilon^m$$

has no zero with $\operatorname{Im} z$ in the interior of Γ_0°, so $\Gamma_0^{\circ} \subset \bar{\Gamma}_{x_0}$ and $\Gamma_0 \supset \Gamma_{x_0}^{\circ}$.

We shall now discuss the example

$$F(x) = x_1^2 - x_2^2 - \ldots - x_n^2, \qquad \theta = (1, 0, \ldots, 0),$$

mentioned at the beginning of the section. Then Theorem 8.7.5 gives

$$WF_A(F_\theta^{-1}) \subset \{(x, tF'(x)),\ F(x)=0,\ tx_1>0\} \cup \{(0,y);\ y_1 \geqq 0,\ F(y) \geqq 0\}.$$

On the other hand, $WF_A(F_\theta^{-1})$ must contain the first set on the right-hand side since $\operatorname{sing\,supp} F_\theta^{-1}$ is the set of zeros of F. Hence it also contains the closure which is the boundary of the second set apart from 0. However, when $n=4$ there is nothing else in $WF_A(F_\theta^{-1})$. To prove this we observe that by (7.4.7) the Fourier transform of F_θ^{-1} is a multiple of $\delta(\xi_1^2-\xi_2^2-\ldots-\xi_4^2)$ restricted to $\xi_1>0$. Hence it follows from Theorem 8.4.18 that $(0,y)\in WF_A(F_\theta^{-1})$, $y \neq 0$, is equivalent to $y_1>0$ and $F(y)=0$ which proves the assertion.

Notes

That singularities should be classified according to their spectrum was recognized independently and from different points of view by several mathematicians around 1970. The first was perhaps Sato [3, 4] (see also Sato-Kawai-Kashiwara [1]) who introduced and studied for hyperfunctions u a set $SS(u)$ (called the singular support) which is our $WF_A(u)$ in the case of distributions. As proved by Bony [3] it is also equal to the essential support of Bros and Iagolnitzer (see Section 9.6 and Iagolnitzer [1]). $WF(u)$ was first defined by Hörmander [25] by means of pseudo-differential operators. This definition, which will be given in Section 18.1 below, was in fact more or less implicit in standard methods for localization by means of such operators. The equivalent definition of $WF(u)$ used here comes from Hörmander [26] where the results of Section 8.2 were also proved. In Section 8.4 we start with the definition of $WF_L(u)$ given in Hörmander [27] but shift to equivalent definitions closely related to those of Sato by means of an analytic decomposition of the δ function. This is quite similar to the decomposition of δ used in Sato-Kawai-Kashiwara [1, p. 473] and Bony [3], but the analyticity of the decomposition is an essential advantage. This was pointed out to us by Louis Boutet de Monvel; see also the related exposition by his student Lebeau [1] and the survey by Schapira [2].

The wave front set was introduced by Hörmander [25] to simplify the study of the propagation of singularities. Note that results like Theorem 8.3.3′ on the wave front set are entirely local and therefore easier to prove than the corresponding weaker results on sing supp u. Indeed, these state in the simplest form that if $P(D)u \in C^\infty$ and $0 \in \operatorname{sing\,supp} u$ then $\mathbb{R}P'_m(\xi) \subset \operatorname{sing\,supp} u$ for some ξ with $P_m(\xi)=0$. This was first proved by Grušin [1] who constructed a fundamental solution with singular support contained in any "half" of the bicharacteristic cone obtained by projecting Char P in $\mathbb{R}^n$. (The method was extended to the analytic case by Andersson [1].) The fundamental solution must be adapted to the distribution u being studied. Here on the other hand we have just needed two natural fundamental solutions $E_\pm$ (with properties more or less classical in quantum electrodynamics in the case of the Klein-Gordon equation). Particularly in the analytic case and for variable coefficients this eliminates considerable technical difficulties. Conceptually it is of course a great advantage that one knows unambiguously in which direction a singularity described by a point in $WF(u)$ is going to travel. For the sources of Example 8.3.4 and Theorem 8.3.8 see Zerner [1, 2] and Hörmander [24].

The results on differential operators in Sections 8.3 and 8.6 are merely intended as examples. The third part of this book will mainly be devoted to the study of $WF(u)$ for solutions of (pseudo-)differential equations. In the analytic case there is also a vast theory of $WF_A(u)$, usually even for hyperfunction solutions. We refer the reader to Sato-

Kawai-Kashiwara [1], Kashiwara [1], Sjöstrand [1, 2] and the references given there.

The Holmgren uniqueness theorem (Theorem 8.6.5) was proved by Holmgren [1] in a special case and by John [1] in full generality for classical solutions. The key to the proof is a result on analyticity of integrals over non-characteristic surfaces depending on a parameter for solutions of differential equations. This was used by John [1] to prove analyticity of solutions of elliptic equations and related regularity results. As observed in Hörmander [27] and independently by Kawai (see Sato-Kawai-Kashiwara [1, 470–473]) one can now reverse the order and deduce uniqueness theorems from microlocal regularity theorems. The purpose of this was to prove uniqueness theorems related to Theorem 8.6.13 in the case of characteristic boundaries. Unique continuation across a surface Σ at a characteristic point where Σ is strictly convex with respect to the corresponding tangential bicharacteristic was proved in the predecessor of this book by geometrical arguments combined with the Holmgren theorem. Successively refined geometrical arguments were then given by Bony [1, 2] and Hörmander [28]. They are now superseded by Theorem 8.5.9 which is due to Sjöstrand [1]. One of the original results of Bony is presented as Theorem 8.6.6. The construction in Example 2 following it is due to Baouendi and Goulaouic [1]. (Hypoellipticity of such operators will be proved in Chapter XXII where further references are given.) Theorem 8.6.7 is from Hörmander [1]; it was proved in Hörmander [9] that the null solutions are dense in all solutions in $C^\infty(\{x;\langle x,N\rangle>0\})$ if $P(D)$ has no non-characteristic factor. Theorem 8.6.8 - Corollary 8.6.11 are close to results of John [1] and were proved as stated here in the predecessor of this book. Further relations between $\operatorname{supp} u$ and $WF_A(u)$ will be discussed in Section 9.6.

K.G. Andersson [1] introduced the notion of local hyperbolicity with respect to θ which is the conjunction of microhyperbolicity with respect to θ and $-\theta$, and Gårding [5] continued his study. Microhyperbolicity was defined as here by Kashiwara and Kawai [1] who used the local Bochner tube theorem (see Komatsu [1]) to prove the crucial Lemma 8.7.4. The reader is referred to Chapter XII for further information in this context.

Chapter IX. Hyperfunctions

Summary

We defined $\mathscr{D}'(X)$ as the space of continuous linear forms on $C_0^\infty(X)$. This is by no means the most general concept of its kind, for a larger space of distributions is obtained if $C_0^\infty(X)$ is replaced by a dense subspace with a stronger topology. An example is the space of elements of compact support in C^L (defined in Section 8.4) provided that it does not contain just the 0 function, that is,

$$\sum 1/L_k < \infty.$$

The study of the dual space of distributions is then fairly similar to that of $\mathscr{D}'(X)$.

The situation is rather different in the quasi-analytic case where

$$\sum 1/L_k = \infty.$$

No analogue of $C_0^\infty(X)$ is then available but we may regard $C^L(X)$ as a substitute for $C^\infty(X)$. The dual space of $C^L(X)$ can be taken as the elements of compact support in a distribution theory preserving many of the features of $\mathscr{D}'(X)$ but differing in some respects. The largest space of distributions is obtained when C^L is the real analytic class. It was introduced in a different way by Sato who coined the term hyperfunction for its elements. In this chapter we shall give an introduction to the theory of hyperfunctions in a manner which follows Schwartz distribution theory as closely as possible.

Section 9.1 is devoted to the study of hyperfunctions of compact support. In particular we give an elementary proof of the crucial and non-trivial fact that there is a good notion of support. The general definition of hyperfunctions can then be given in Section 9.2 along lines first proposed by Martineau. Section 9.3 is devoted to the wave front set with respect to analytic functions of a hyperfunction and the definition of operations such as multiplication. This is done rather quickly for most proofs in Sections 8.4 and 8.5 were chosen so that they are applicable to hyperfunctions after a few basic facts have been established.

Section 9.4 is devoted to the existence of analytic solutions of analytic differential equations. In addition to the classical Cauchy-Kovalevsky theorem precise information on bounds and existence domains is given. These are applied in Section 9.5 to prove some basic facts on hyperfunction solutions of analytic differential equations. In Section 9.6 finally we present the Bros-Iagolnitzer definition of $WF_A(u)$ and prove as an application a theorem of Kashiwara on the relation between $\operatorname{supp} u$ and $WF_A(u)$ similar to Holmgren's uniqueness theorem.

9.1. Analytic Functionals

If K is a compact set in $\mathbb{R}^n$ then a distribution $u \in \mathscr{E}'(K)$ is a linear form on $C^\infty(\mathbb{R}^n)$ such that, if ω is a neighborhood of K,

$$|u(\phi)| \leqq C_\omega \sum_{|\alpha| \leqq k} \sup_\omega |D^\alpha \phi|, \qquad \phi \in C^\infty(\mathbb{R}^n).$$

One can extend $u(\phi)$ by continuity to all $\phi \in C^\infty(\omega)$. (See Theorem 2.3.1 and the remarks after its statement.) Since the derivatives of an analytic function can be estimated in a compact set by the maximum of the modulus in a neighborhood, the following definition is quite analogous:

Definition 9.1.1. If $K \subset \mathbb{C}^n$ is a compact set, then $A'(K)$, the space of analytic functionals carried by K, is the space of linear forms u on the space A of entire analytic functions in $\mathbb{C}^n$ such that for every neighborhood ω of K

$$|u(\phi)| \leqq C_\omega \sup_\omega |\phi|, \qquad \phi \in A. \tag{9.1.1}$$

Example. $u(\phi) = \sum_\alpha a_\alpha D^\alpha \phi(0)/\alpha!$ is an analytic functional carried by 0 if and only if $|a_\alpha| \leqq C_\varepsilon \varepsilon^{|\alpha|}$ for every $\varepsilon > 0$. (The sufficiency follows from Cauchy's inequalities and the necessity by taking $\phi(z) = z^\alpha$.) u is not a distribution unless the sum is finite.

It would suffice to consider only polynomials ϕ in the definition, for every entire analytic function is locally uniformly the sum of its Taylor series. Note that $A'(K)$ is a Fréchet space with the best constants $C_\omega(u)$ as semi-norms.

In contrast to what one might expect from the analogy with $\mathscr{E}'(K)$ it is not always true that $u \in A'(K_1) \cap A'(K_2)$ implies $u \in A'(K_1 \cap K_2)$. For example,

$$u(\phi)=\int_0^1 \phi(z)\,dz, \qquad \phi \in A(\mathbb{C}^1),$$

has any C^1 curve from 0 to 1 as a minimal carrier. However, we shall prove that this is true if $K_1, K_2 \subset \mathbb{R}^n$, which is the case we are mainly interested in. As a first step in this direction we shall prove that if $u \in A'(K)$ and $K \subset \mathbb{R}^n$, then $u(\phi)$ can be defined for every ϕ which is analytic just in a neighborhood of K.

Proposition 9.1.2. *Let $K \subset \mathbb{R}^n$ be a compact set, and set for $\varepsilon > 0$*

$$K_\varepsilon = \{z \in \mathbb{C}^n;\ |\operatorname{Re} z - x| + 2|\operatorname{Im} z| \leqq \varepsilon \text{ for some } x \in K\}.$$

For every ϕ which is analytic in a neighborhood V of K_ε one can then find a sequence $\phi_j \in A$ such that

$$\sup_{K_\varepsilon} |\phi_j - \phi| \to 0, \qquad j \to \infty.$$

Proof. Choose $\chi \in C_0^\infty(V \cap \mathbb{R}^n)$ equal to 1 near $K_\varepsilon \cap \mathbb{R}^n$, and set

$$\phi_j(z) = \int E_j(z-x)\,\chi(x)\,\phi(x)\,dx, \qquad z \in \mathbb{C}^n,$$

where E_j is the normalized Gaussian function

$$E_j(x) = (j/\pi)^{n/2} \exp(-j\langle x, x\rangle).$$

Since $E_j \in A$ it is clear that $\phi_j \in A$. The set K_ε is defined by

$$K_\varepsilon = \{z;\ 2|\operatorname{Im} z| \leqq F(\operatorname{Re} z)\}, \qquad F(y) = \varepsilon - \min_{x \in K} |x-y|.$$

By the triangle inequality $|F(y) - F(y')| \leqq |y - y'|$. If $z_0 = x_0 + i y_0 \in K_\varepsilon$ it follows that K_ε contains the chains

$$\Gamma(z_0, t):\ x \to x + i t y_0 (1 - |x - x_0|/2|y_0|), \qquad |x - x_0| \leqq 2|y_0|,$$

when $0 \leqq t \leqq 1$, and they have the same boundary. Since the form $E_j(z-\zeta)\,\phi(\zeta)\,d\zeta_1 \wedge \ldots \wedge d\zeta_n$ is closed in K_ε we obtain by Stokes' formula

$$\phi_j(z) = \int_{|x-x_0| > 2|y_0|} E_j(z-x)\,\chi(x)\,\phi(x)\,dx + \int_{\Gamma(z_0, 1)} E_j(z-\zeta)\,\phi(\zeta)\,d\zeta_1 \wedge \ldots \wedge d\zeta_n.$$

We take $z=z_0$ and observe that

$$\operatorname{Re} -\langle z_0-x, z_0-x\rangle = -|x_0-x|^2+|y_0|^2 \leqq -3|x_0-x|^2/4 \quad \text{if } |x-x_0|>2|y_0|,$$

$$\operatorname{Re} -\langle z_0-\zeta, z_0-\zeta\rangle = -|x_0-x|^2+|x-x_0|^2/4 = -3|x-x_0|^2/4 \quad \text{if } \zeta\in\Gamma(z_0,1),\ \operatorname{Re}\zeta=x.$$

Since $|\phi(\zeta)-\phi(z_0)|\leqq C|x-x_0|$ in the integrals we conclude that

$$|\phi_j(z_0)-\phi(z_0)\textstyle\int E_j(z_0-x)\chi(x)\,dx| \leqq C_0\int (j/\pi)^{n/2} e^{-3j|x-x_0|^2/4}|x-x_0|\,dx \leqq C_1 j^{-\frac{1}{2}}.$$

Now

$$1-\int E_j(z_0-x)\chi(x)\,dx = \int (1-\chi(x))\,E_j(z_0-x)\,dx$$

is exponentially decreasing as $j\to\infty$ since $\operatorname{Re}\langle z_0-x, z_0-x\rangle$ has a positive lower bound when $z_0\in K_\varepsilon$ and $x\in\operatorname{supp}(1-\chi)$. This completes the proof.

With the notation in the proposition we have by (9.1.1) if $u\in A'(K)$

$$|u(\phi_j)-u(\phi_k)|\leqq C\sup_{K_\varepsilon}|\phi_j-\phi_k|\to 0 \quad \text{as } j,k\to\infty$$

so we can define

$$u(\phi)=\lim_{j\to\infty} u(\phi_j).$$

Since the sets K_ε form a fundamental system of neighborhoods of K, it is clear that we have now obtained a unique definition of $u(\phi)$ for every ϕ which is analytic in a neighborhood ω of K, and (9.1.1) remains valid for all ϕ analytic in ω.

We shall now associate with every $u\in A'(K)$ a regularization from which u can be conveniently reconstructed. Let E be the fundamental solution of the Laplace operator in $\mathbb{R}^{n+1}$ given in Theorem 3.3.2 and set

$$P=\partial E/\partial x_{n+1}=x_{n+1}|X|^{-n-1}/c_{n+1}.$$

Here $X=(x,x_{n+1})\in\mathbb{R}^{n+1}$. If $u\in C_0^0(K)$, K compact in $\mathbb{R}^n$, then

$$U(X)=P*(u\otimes\delta)(X)=x_{n+1}\int u(y)(|x-y|^2+x_{n+1}^2)^{-(n+1)/2}\,dy/c_{n+1}$$

is odd as a function of x_{n+1}, harmonic outside $K\times\{0\}$ and converges to $\pm u/2$ as $x_{n+1}\to\pm 0$. Thus u is the jump across the plane $x_{n+1}=0$ of the harmonic function U, which is also clear from the fact that

$$\Delta U=\Delta P*(u\otimes\delta)=u\otimes\partial_{n+1}\delta.$$

Very little has to be changed in this discussion if $u\in A'(K)$:

Proposition 9.1.3. *If K is a compact set in $\mathbb{R}^n$ and $u \in A'(K)$, then*

$$U(X) = u_y P(X-(y,0)) \tag{9.1.2}$$

is a harmonic function in $\mathbb{R}^{n+1} \setminus (K \times \{0\})$ which is odd as a function of x_{n+1}. If Φ is a harmonic function in $\mathbb{R}^{n+1}$ and $\chi \in C_0^\infty(\mathbb{R}^{n+1})$ is equal to 1 in a neighborhood of $K \times \{0\}$, then

$$u(\partial_{n+1} \Phi|_{x_{n+1}=0}) = -\int U \Delta(\chi\Phi)\, dX. \tag{9.1.3}$$

Proof. $U(X)$ is defined when $X \notin K \times \{0\}$ for $y \to P(X-(y,0))$ is then analytic in a neighborhood of K. The continuity of u implies that U is continuous and that we may compute the derivatives of U by differentiating on P in (9.1.2). Since P is harmonic outside 0 it follows that U is harmonic. To prove (9.1.3) we note that if $Y \in \mathbb{R}^{n+1}$ and $\chi = 1$ near Y then

$$\begin{aligned}\int P(X-Y)\, \Delta(\chi\Phi)(X)\, dX &= \langle \Delta P(X-Y), \chi\Phi(X)\rangle = \langle \partial_{n+1}\delta_Y, \chi\Phi\rangle \\ &= -\partial_{n+1}\Phi(Y).\end{aligned}$$

By the uniqueness of analytic continuation this remains true for all Y in a complex neighborhood of $K \times \{0\}$ in $\mathbb{C}^{n+1}$, and the left-hand side is then the uniform limit of the corresponding Riemann sums. Setting $Y=(y,0)$ and letting u operate on each term in the Riemann sum, we obtain (9.1.3).

(9.1.3) determines u completely, for we have

Lemma 9.1.4. *For every entire analytic function ϕ in $\mathbb{C}^n$ there is a unique entire analytic function Φ in $\mathbb{C}^{n+1}$ such that*

$$\sum_1^{n+1} \partial^2 \Phi / \partial z_j^2 = 0, \quad \text{and} \quad \Phi = 0, \partial_{n+1}\Phi = \phi \quad \text{when} \quad z_{n+1} = 0. \tag{9.1.4}$$

For every $R > 1$ there is a constant C_R such that

$$|\Phi(z, z_{n+1})| \leq C_R |z_{n+1}| \sup_{|\zeta - z| < R|z_{n+1}|} |\phi(\zeta)|. \tag{9.1.5}$$

Proof. If Φ satisfies (9.1.4) then

$$u(t,x) = \Phi(z + i z_{n+1} x, z_{n+1} t); \quad t \in \mathbb{R},\ x \in \mathbb{R}^n;$$

satisfies the wave equation $\partial^2 u/\partial t^2 = \Delta_x u$, and

$$u = 0, \quad \partial u/\partial t = z_{n+1}\phi(z + i z_{n+1} x) \quad \text{when } t = 0.$$

Hence it follows from Theorem 6.2.4 that

$$\Phi(z, z_{n+1}) = z_{n+1} \langle E_+(1), \phi(z + i z_{n+1} \cdot)\rangle$$

where $E_+(1)$ is a distribution with support in $\{x\in\mathbb{R}^n; |x|\leqq 1\}$, given explicitly by (6.2.4)'. This proves the uniqueness and (9.1.5). On the other hand, if we define Φ by this formula we obtain an entire analytic function which satisfies (9.1.4) when $z_{n+1}>0$ and z/i is real. But the entire function $\sum\partial^2\Phi/\partial z_j^2$ must vanish identically if it vanishes in this set, so the lemma is proved.

Remark. A direct proof can also be made by estimating the terms in the power series expansion

$$\Phi(z, z_{n+1})=\sum_0^\infty z_{n+1}^{2k+1}(-\Delta)^k\phi(z)/(2k+1)!.$$

As already pointed out Lemma 9.1.4 implies that the map from analytic functionals to harmonic functions defined by (9.1.2) is injective. Using Lemma 9.1.4 we shall now prove that it is also surjective.

Proposition 9.1.5. *If K is a compact set in $\mathbb{R}^n$ and U is a harmonic function in $\mathbb{R}^{n+1}\setminus(K\times\{0\})$ which is odd as a function of x_{n+1}, then there is a unique $u\in A'(K)$ such that (9.1.3) is valid when $\chi\in C_0^\infty(\mathbb{R}^{n+1})$, $\chi=1$ near $K\times\{0\}$ and Φ is any harmonic function in $\mathbb{R}^{n+1}$. We have*

$$u_y P(.-(y,0))-U=H$$

where H is harmonic in $\mathbb{R}^{n+1}$, and H vanishes identically if and only if $U\to 0$ at ∞.

Proof. The right-hand side of (9.1.3) is independent of the choice of χ, for it is equal to 0 if $\chi\in C_0^\infty(\mathbb{R}^{n+1}\setminus(K\times\{0\}))$. For any $\delta>0$ we may therefore choose χ so that every point in $\operatorname{supp}\chi$ has distance $<\delta$ to $K\times\{0\}$, and we can always take χ even as a function of x_{n+1}. Then (9.1.3) is automatically true if Φ is even as a function of x_{n+1}. When ϕ is a polynomial in $\mathbb{C}^n$ we now define

$$u(\phi)=-\int U\Delta(\chi\Phi)\,dX \tag{9.1.6}$$

where Φ is given by Lemma 9.1.4. Taking $R=4/3$ in the lemma we obtain

$$|\Delta(\chi\Phi)|\leqq C_\delta\sup_{K_{7\delta}}|\phi|,$$

for if $|x-y|^2+x_{n+1}^2<\delta^2$ for some $y\in K$ then $|z-x|\leqq 4|x_{n+1}|/3$ implies $|z-y|<7\delta/3$, hence $z\in K_{7\delta}$. This proves that $u\in A'(K)$. For reasons of continuity it follows now that (9.1.3) is valid for every entire harmonic Φ which is odd with respect to x_{n+1}, hence for all entire harmonic Φ.

For the harmonic function in $\mathbb{R}^{n+1} \setminus (K \times \{0\})$ defined by

$$U_1(X) = u_y(P(X-(y,0)))$$

we know from Proposition 9.1.3 that (9.1.3) is valid with U replaced by U_1. Writing $H = U_1 - U$ we obtain

$$\int H \Delta(\chi \Phi)\, dX = 0$$

for every entire harmonic Φ. Now choose $\chi_1 \in C_0^\infty(\mathbb{R}^{n+1})$ so that $\chi_1 = 1$ in a neighborhood of $K \times \{0\}$ and $\chi = 1$ in $\operatorname{supp} \chi_1$. Then $(1-\chi_1)H = H_1 \in C^\infty$ and for all exponential solutions Φ of Δ in $\mathbb{R}^{n+1}$ we have

$$0 = \int H_1 \Delta(\chi \Phi)\, dX = \int (\Delta H_1) \Phi\, dX.$$

Hence it follows from Theorem 7.3.2 and Lemma 7.3.7 that $\Delta H_1 = \Delta f$ for some $f \in C_0^\infty$. This means that $H_1 - f$ is a harmonic function which is equal to H outside a compact set, and therefore outside $K \times \{0\}$. Thus H has been extended to a function which is harmonic in $\mathbb{R}^{n+1}$. Since $H = 0$ is equivalent to $H \to 0$ at ∞ by the maximum principle, and since $U_1 \to 0$ at ∞, the last statement in the proposition follows. We obtain (9.1.3) for every harmonic function in $\mathbb{R}^{n+1}$ since this is true with U replaced by U_1 and since

$$\int H \Delta(\chi \Phi)\, dX = 0.$$

We are now ready to prove some important facts on the elements of

$$A'(\mathbb{R}^n) = \bigcup_{K \Subset \mathbb{R}^n} A'(K).$$

Theorem 9.1.6. *If $u \in A'(\mathbb{R}^n)$ then there is a smallest compact set $K \subset \mathbb{R}^n$ such that $u \in A'(K)$; it is called the support of u.*

Proof. Let K be the intersection of all compact sets $K' \subset \mathbb{R}^n$ such that $u \in A'(K')$. By Proposition 9.1.3 a harmonic function in $\mathbb{R}^{n+1} \setminus (K' \times \{0\})$ is defined by (9.1.2). It is uniquely determined by its restriction to the complement of the plane $x_{n+1} = 0$. The functions obtained for different choices of K' must therefore agree in their common domain of definition and give together a harmonic function in $\mathbb{R}^{n+1} \setminus (K \times \{0\})$. Hence $u \in A'(K)$ by Proposition 9.1.5.

Next we shall prove a completeness theorem for analytic functionals. In order to prepare for the construction of boundary values of analytic functions in Section 9.3 we shall then consider some analytic functionals which are carried by compact sets close to $\mathbb{R}^n$ but not contained in $\mathbb{R}^n$. This requires another look at Propositions 9.1.3 and 9.1.5. If $u \in A'(K_\varepsilon)$ where K_ε is defined in Proposition 9.1.2, then (9.1.2) defines a harmonic function U in the complement of

$$\tilde{K}_\varepsilon = \{X \in \mathbb{R}^{n+1};\ |x-y|^2 + x_{n+1}^2 \leqq \varepsilon^2 \text{ for some } y \in K\}.$$

This will follow if we just show that

$$\text{(9.1.7)} \qquad \operatorname{Re}\langle x-z, x-z\rangle + x_{n+1}^2 > 0 \quad \text{if } X \notin \tilde{K}_\varepsilon \text{ and } z \in K_\varepsilon.$$

The left-hand side is equal to

$$|x - \operatorname{Re} z|^2 + x_{n+1}^2 - |\operatorname{Im} z|^2$$

and for some $y \in K$ we have

$$|\operatorname{Re} z - y| + 2|\operatorname{Im} z| \leqq \varepsilon.$$

Since $X \notin \tilde{K}_\varepsilon$ it follows that

$$\varepsilon^2 < |x-y|^2 + x_{n+1}^2,$$

so the triangle inequality gives

$$(|x - \operatorname{Re} z|^2 + x_{n+1}^2)^{\frac{1}{2}} > \varepsilon - |\operatorname{Re} z - y| \geqq |\operatorname{Im} z|$$

which proves (9.1.7). On the other hand, we actually saw in the proof of Proposition 9.1.5 that U harmonic outside $\tilde{K}_\varepsilon$ implies $u \in A'(K_{7\varepsilon})$. We are now ready to prove the crucial completeness result:

Theorem 9.1.7. *Let K_0 and K be compact sets with $K_0 \subset K \subset \mathbb{R}^n$, let $u_j \in A'(\mathbb{C}^n)$ and assume that*

(i) *For any compact neighborhood V of K in $\mathbb{C}^n$ we have $u_j \in A'(V)$ for large j.*

(ii) *For any compact neighborhood V_0 of K_0 in $\mathbb{C}^n$ we have*

$$u_j - u_k \in A'(V_0) \quad \textit{for large } j, k.$$

Then one can find $u \in A'(K)$ so that for any compact neighborhood V_0 of K_0

$$\text{(iii)} \qquad u - u_j \in A'(V_0) \quad \textit{for large } j.$$

Condition (iii) *determines u uniquely* mod $A'(K_0)$.

Proof. If (iii) is fulfilled with u replaced by v also, then $u - v \in A'(V_0)$ for every V_0 so $u - v \in A'(K_0)$ as claimed. Thus we only have to prove the existence of u. Choose a sequence $\varepsilon_j \to 0$ so that with the notation in Proposition 9.1.2

$$\text{(i)}' \qquad u_j \in A'(K_{\varepsilon_j})$$

$$\text{(ii)}' \qquad u_j - u_k \in A'(K_{0,\varepsilon_j}) \quad \text{if } k \geqq j.$$

As we have just seen it follows from (i)′ that

$$U_j(X) = u_{jy} P(X - (y, 0))$$

is a harmonic function outside $\tilde{K}_{\varepsilon_j}$, and from (ii)′ that $U_j - U_k$ has a harmonic extension to the complement of $\tilde{K}_{0,\varepsilon_j}$ if $k>j$. By Runge's approximation theorem (Theorem 4.4.5) we can approximate $U_{j+1} - U_j$ in $\complement \tilde{K}_{0,\varepsilon_j}$ by functions harmonic in the complement Ω of $K_0 \times \{0\}$. In fact, $\tilde{K}_{0,\varepsilon_j}$ is not the union of two disjoint non-empty compact sets one of which is disjoint with $K_0 \times \{0\}$, for it is a union of balls with center in $K_0 \times \{0\}$. Let

$$M_j = \{X \in \mathbb{R}^{n+1};\ |X| \leqq j,\ |x-y|^2 + x_{n+1}^2 \geqq 2\varepsilon_j^2 \text{ for all } y \in K_0\}.$$

This is a compact subset of the complement of $\tilde{K}_{0,\varepsilon_j}$ and it increases to Ω when $j \to \infty$. We can therefore choose V_j harmonic in Ω so that

$$|U_{j+1} - U_j - V_j| \leqq 2^{-j} \quad \text{in } M_j. \tag{9.1.8}$$

(Strictly speaking $U_{j+1} - U_j$ should be replaced by the harmonic extension to the complement of $\tilde{K}_{0,\varepsilon_j}$.) Since V_j can be replaced by $(V_j(x, x_{n+1}) - V_j(x, -x_{n+1}))/2$ in (9.1.8), we can take V_j odd as a function of x_{n+1}.

It follows from (9.1.8) that the limit

$$\begin{aligned} U &= \lim (U_j - V_1 - \ldots - V_{j-1}) \\ &= U_j - V_1 - \ldots - V_{j-1} + \sum_j^\infty (U_{k+1} - U_k - V_k) \end{aligned}$$

exists and is harmonic outside $K \times \{0\}$, for the sum is harmonic outside $\tilde{K}_{0,\varepsilon_j}$ by (9.1.8), and the other terms are harmonic outside $\tilde{K}_{\varepsilon_j}$. Let u be the corresponding element in $A'(K)$. Since $U - U_j$ is harmonic outside $\tilde{K}_{0,\varepsilon_j}$ we have $u - u_j \in A'(K_{0,7\varepsilon_j})$ which proves (iii).

The following theorem is a substitute for the existence of partitions of unity:

Theorem 9.1.8. *If* $K_1, \ldots, K_r$ *are compact subsets of* $\mathbb{R}^n$ *and* $u \in A'(K_1 \cup \ldots \cup K_r)$, *then one can find* $u_j \in A'(K_j)$ *so that*

$$u = u_1 + \ldots + u_r.$$

Proof. It is sufficient to prove the statement when $r=2$. The function U defined by (9.1.2) is harmonic outside $\tilde{K}_1 \cup \tilde{K}_2$ where $\tilde{K}_j = K_j \times \{0\}$. The theorem will be proved if we can split U into a sum

$$U = U_1 + U_2$$

where U_j is harmonic outside $\tilde{K}_j$. To do so we choose using Corollary 1.4.11 a function $\phi \in C^\infty(\mathbb{R}^{n+1} \setminus (\tilde{K}_1 \cap \tilde{K}_2))$ which vanishes for large

$|X|$ and near $\tilde{K}_2 \setminus (\tilde{K}_1 \cap \tilde{K}_2)$ while $\phi = 1$ near $\tilde{K}_1 \setminus (\tilde{K}_1 \cap \tilde{K}_2)$. Then

$$U_1 = \phi U - v, \qquad U_2 = (1 - \phi) U + v$$

have the required properties if $v \in C^\infty(\mathbb{R}^{n+1} \setminus (\tilde{K}_1 \cap \tilde{K}_2))$ and

$$\Delta v = \Delta(\phi U).$$

Here we define $\phi U = 0$ near $\tilde{K}_2 \setminus (\tilde{K}_1 \cap \tilde{K}_2)$ and $(1-\phi) U = 0$ near $\tilde{K}_1 \setminus (\tilde{K}_1 \cap \tilde{K}_2)$. Since $\Delta(\phi U)$ vanishes near $(\tilde{K}_1 \cup \tilde{K}_2) \setminus (\tilde{K}_1 \cap \tilde{K}_2)$ it is a C^∞ function outside $\tilde{K}_1 \cap \tilde{K}_2$. The existence of v is therefore a consequence of Theorem 4.4.6. (Note that this is based on another application of Runge's approximation theorem.)

Finally we note that if $u \in \mathscr{E}'(\mathbb{R}^n)$ then u defines an element in $A'(\mathbb{R}^n)$ with the same support. In fact, the harmonic function $U(x, x_{n+1})$ defined by (9.1.2) has the $\mathscr{D}'$ limit $\pm u/2$ as $x_{n+1} \to \pm 0$, so continuation of U as a harmonic function is only possible outside $\operatorname{supp} u \times \{0\}$. Thus we have an injection preserving supports

$$\mathscr{E}'(\mathbb{R}^n) \hookrightarrow A'(\mathbb{R}^n).$$

The operations defined for distributions in Chapters III to VII carry over easily to $A'(\mathbb{R}^n)$. We shall just recall them briefly and leave all details for the reader.

a) If $u \in A'(\mathbb{R}^n)$ then $\partial_j u \in A'(\mathbb{R}^n)$ can be defined by

$$(\partial_j u)(\phi) = -u(\partial_j \phi)$$

when ϕ is an entire function, for $\sup_\omega |\partial_j \phi|$ can be estimated by the supremum of $|\phi|$ over a neighborhood of $\bar{\omega}$.

b) If $u \in A'(K)$, $K \subset \mathbb{R}^n$, and f is analytic in a neighborhood ω of K, then we define the product fu by

$$(fu)(\phi) = u(f\phi)$$

when ϕ is analytic in ω. Here it is of course important that Proposition 9.1.2 allowed us to extend u to all functions analytic in ω.

c) If $u \in A'(\mathbb{R}^n)$ and $v \in A'(\mathbb{R}^m)$ then $u \otimes v \in A'(\mathbb{R}^{n+m})$ is defined by

$$(u \otimes v)(\phi) = u_x(v_y(\phi(x, y))) = v_y(u_x(\phi(x, y)))$$

when ϕ is a polynomial in $\mathbb{C}^{n+m}$. The second equality is obvious then and the first is a definition; it is clear that we obtain an analytic functional supported by $\operatorname{supp} u \times \operatorname{supp} v$.

d) If $K \in A'(\mathbb{R}^n \times \mathbb{R}^m)$ and v is analytic in a neighborhood of the projection of $\operatorname{supp} K$ in $\mathbb{R}^m$, then $\mathscr{K} v \in A'(\mathbb{R}^n)$ is defined by

$$(\mathscr{K} v)(\phi) = K(\phi \otimes v)$$

for every entire function ϕ in $\mathbb{C}^n$.

e) If $u \in A'(K)$, where K is a compact set in $\mathbb{R}^n$, and if f is a real analytic bijection of an open set $\omega \subset \mathbb{R}^n$ on a neighborhood of K with inverse h, then the pullback $f^* u \in A'(f^* K)$ is defined by

$$(f^* u)(\phi) = u((\phi \circ h)\,|\det h'|)$$

when ϕ is analytic in ω.

f) If $u, v \in A'(\mathbb{R}^n)$ then c), d), e) allow us to define $u * v$ by letting the pullback of $u \otimes v$ by the map $(x, y) \to (y, x-y)$ act on the function 1.

g) If $u \in A'(K)$, K compact in $\mathbb{R}^n$, then the Fourier-Laplace transform

$$\hat{u}(\zeta) = u(\exp -i\langle ., \zeta\rangle)$$

is an entire analytic function such that for every $\varepsilon > 0$

$$|\hat{u}(\zeta)| \leqq C_\varepsilon \exp(H_K(\operatorname{Im}\zeta) + \varepsilon|\zeta|), \qquad \zeta \in \mathbb{C}^n.$$

This follows at once from the definition of $A'(K)$. Conversely, every entire function satisfying these estimates is the Fourier-Laplace transform of a unique element in $A'(K)$. The uniqueness follows from the fact that

$$u(P) = P(-D)\,\hat{u}(0)$$

for every polynomial P. The existence proof will be given as Theorem 15.1.5, and the result will not be used in the meantime.

9.2. General Hyperfunctions

We want to define hyperfunctions in $\mathbb{R}^n$ in such a way that they are locally equivalent to analytic functionals with compact support in $\mathbb{R}^n$. This will be done in two steps.

Definition 9.2.1. If $X \subset \mathbb{R}^n$ is open and bounded we define the space of hyperfunctions $B(X)$ in X by

$$B(X) = A'(\bar{X})/A'(\partial X). \tag{9.2.1}$$

The reader might object here that this does not give the desired result in the case of distributions, for

$$\mathscr{E}'(\bar{X})/\mathscr{E}'(\partial X) \hookrightarrow \mathscr{D}'^F(X).$$

However, the definition will be justified in a moment when we prove that the analogue of the Localization Theorem 2.2.4 is valid.

If $u, v \in A'(\bar{X})$ and $u - v \in A'(\partial X)$, then $X \cap \operatorname{supp} u = X \cap \operatorname{supp} v$ since $\operatorname{supp} u \subset \operatorname{supp} v \cup \partial X$ and $\operatorname{supp} v \subset \operatorname{supp} u \cup \partial X$. Thus it is legitimate to define the support of the class $u^\cdot$ of u in $B(X)$ by

$$\operatorname{supp} u^\cdot = X \cap \operatorname{supp} u.$$

If $Y \subset X$ and X, Y are open and bounded, we can for every $u \in A'(\bar{X})$ find $v \in A'(\bar{Y})$ so that $Y \cap \operatorname{supp}(u-v) = \emptyset$. This follows from Theorem 9.1.8 applied to u and the compact sets $\bar{Y}$ and $\overline{(X \setminus Y)}$. The class $v^\cdot$ of v in $B(Y)$ is uniquely determined by the class $u^\cdot$ of u in $B(X)$ and is called the *restriction* of $u^\cdot$ to Y. Note that the restriction of $u^\cdot$ to Y is 0 if and only if $Y \cap \operatorname{supp} u = \emptyset$. As in the case of distributions the support of a hyperfunction is therefore the smallest closed set such that the restriction of the hyperfunction to the complement is equal to 0. The definition of $\operatorname{sing\,supp} u$ is also extended to hyperfunctions with no change.

We can now state and prove the localization theorem.

Theorem 9.2.2. *Let X_j be open sets in $\mathbb{R}^n$ with bounded union X. If $u_j \in B(X_j)$ and for all i, j we have $u_i = u_j$ in $X_i \cap X_j$ (that is, the restrictions are equal) then there is a unique $u \in B(X)$ such that the restriction of u to X_j is equal to u_j for every j.*

Proof. The uniqueness is clear for if v has the same property as u then the support of $u - v$ is empty so $u - v = 0$. To prove the existence we begin with the case of just *two open sets* X_1 and X_2. Choose $U_j \in A'(\bar{X}_j)$ defining u_j for $j = 1, 2$. The support of $U_1 - U_2$ is contained in

$$(\bar{X}_1 \cup \bar{X}_2) \setminus (X_1 \cap X_2) \subset (\complement X_1 \cap \bar{X}_2) \cup (\bar{X}_1 \cap \complement X_2),$$

so Theorem 9.1.8 gives a decomposition

$$U_1 - U_2 = V_1 - V_2, \quad V_1 \in A'(\complement X_1 \cap \bar{X}_2), \quad V_2 \in A'(\bar{X}_1 \cap \complement X_2).$$

Now

$$U = U_1 - V_1 = U_2 - V_2 \in A'(\overline{X_1 \cup X_2})$$

defines an element in $B(X_1 \cup X_2)$ which restricts to u_j in X_j for $j = 1, 2$.

Next we assume that we have countably many sets X_j, $j = 1, 2, \dots$. Repeated use of the special case just proved gives a sequence v_j in $B(X_1 \cup \dots \cup X_j)$ with restriction u_i to X_i for $i \leqq j$. Let $V_j \in A'(\overline{X_1 \cup \dots \cup X_j})$ be in the class of v_j. Since $\operatorname{supp}(V_j - V_k) \subset \bar{X} \setminus (X_1 \cup \dots \cup X_j)$ when $k > j$, it follows from Theorem 9.1.7 that there is an element $V \in A'(\bar{X})$ such that $\operatorname{supp}(V - V_j) \subset \bar{X} \setminus (X_1 \cup \dots \cup X_j)$ for every j. Hence the class u of V has the desired restrictions.

If we have more than countably many X_j we just choose countably many of them with the same union and then a corresponding u. The uniqueness established at the beginning of the proof shows that the restriction of u to X_j is then u_j for every j.

It follows from Theorem 2.2.4 and the remarks at the end of Section 9.1 that we have an injection $\mathscr{D}'(X) \to B(X)$. Let us also note here that the elements with compact support in $B(X)$ can be identified with the elements in $A'(\mathbb{R}^n)$ having support in X. In fact, let $u \in A'(\bar{X})$ and assume that the class $u^{\bullet}$ has compact support $K \subset X$. Then $\operatorname{supp} u \subset K \cup \partial X$ so Theorem 9.1.8 gives a decomposition

$$u = u_1 + u_2, \qquad u_1 \in A'(K),\ u_2 \in A'(\partial X)$$

which is unique since K and ∂X are disjoint. This means that $u^{\bullet} = u_1^{\bullet}$ for a unique $u_1 \in A'(K)$.

It is easy to extend the operations on $A'(\mathbb{R}^n)$ discussed at the end of Section 9.1 to operations on $B(X)$. First it is clear that if X and Y are bounded open sets in $\mathbb{R}^n$ and f is a real analytic diffeomorphism of a neighborhood of $\bar{Y}$ on a neighborhood of $\bar{X}$, then we obtain a bijection

$$f^*\colon B(X) \to B(Y)$$

from the bijections

$$f^*\colon A'(\bar{X}) \to A'(\bar{Y}) \quad \text{and} \quad f^*\colon A'(\partial X) \to A'(\partial Y).$$

The easy proof that

$$(fg)^* = g^* f^*$$

is left for the reader.

We can now define $B(X)$ for any real analytic manifold X. First we choose an atlas $\mathscr{F}$ of analytic diffeomorphisms of coordinate patches $X_\kappa \Subset X$ on open sets $\tilde{X}_\kappa \Subset \mathbb{R}^n$ such that κ has an analytic extension to a neighborhood of the closures. Then

$$\kappa\kappa'^{-1}\colon \kappa'(X_\kappa \cap X_{\kappa'}) \to \kappa(X_\kappa \cap X_{\kappa'}), \qquad \kappa, \kappa' \in \mathscr{F},$$

has an analytic extension to a neighborhood of the closures, so

$$(\kappa\kappa'^{-1})^*\colon B(\kappa(X_\kappa \cap X_{\kappa'})) \to B(\kappa'(X_\kappa \cap X_{\kappa'}))$$

is defined. We can therefore define a hyperfunction $u \in B(X)$ as a collection of hyperfunctions $u_\kappa \in B(\tilde{X}_\kappa)$ for every $\kappa \in \mathscr{F}$ such that (6.3.3) is valid. The easy but tedious proof that $B(X)$ is independent of the choice of atlas and that it agrees with our previous definition when $X \Subset \mathbb{R}^n$ is left for the conscientious reader.

The notion of support and restriction carry over immediately to the general case. A final justification of Definition 9.2.1 is given by

Theorem 9.2.3. *If X is a real analytic manifold and Y an open subset then every $u \in B(Y)$ is the restriction to Y of a hyperfunction $v \in B(X)$ with support in $\bar{Y}$.*

Proof. Let $\kappa: X_\kappa \to \tilde{X}_\kappa$ be a coordinate system $\in \mathscr{F}$ on X. Then $u_\kappa \in B(\kappa(Y \cap X_\kappa))$ is the class of an element $U \in A'(\overline{\kappa(Y \cap X_\kappa)})$ which also defines a hyperfunction $V \in B(\tilde{X}_\kappa)$ since $\kappa(Y \cap X_\kappa) \subset \tilde{X}_\kappa$. The restriction of V to $\kappa(Y \cap X_\kappa)$ is equal to u_κ. The desired extension of u to $Y \cup X_\kappa$ is now obtained immediately if to an atlas for Y with coordinate patches $\Subset Y$ we add the coordinate system κ with the hyperfunction V. Continuing in this way we can successively extend u to all of X. (If X is not countable one should use Zorn's lemma but we are not interested in such generality.)

The extension of Theorem 9.2.2 to a real analytic manifold X with open subsets X_j is immediate. So is the definition of the product fu of a hyperfunction $u \in B(X)$ by a function f which is real analytic in a neighborhood of $\operatorname{supp} u$, as well as the definition of the tensor product.

9.3. The Analytic Wave Front Set of a Hyperfunction

Definitions 8.1.2 and 8.4.3 of WF and WF_L make no sense for hyperfunctions but it is possible to use the equivalent characterization in Theorem 8.4.11. For the sake of brevity we shall only discuss WF_A. With K still denoting the analytic function in $\{z;\ |\operatorname{Im} z|^2 < 1 + |\operatorname{Re} z|^2\}$ constructed in Lemma 8.4.10 we first prove an analogue of a part of Theorem 8.4.11.

Proposition 9.3.1. *If $u \in A'(\mathbb{R}^n)$ then*

$$U(z) = K * u(z) = u_t K(z-t)$$

is an analytic function in

$$Z = \{z;\ |\operatorname{Im} z|^2 < 1 + |\operatorname{Re} z - t|^2,\ t \in \operatorname{supp} u\}.$$

If X is a bounded open neighborhood of $\operatorname{supp} u$ in $\mathbb{R}^n$ then

$$u(\phi) = \lim_{r \to 1} \int_{|\omega|=1} \int_X U(x + ir\omega)\, \phi(x)\, dx\, d\omega, \qquad \phi \in A. \tag{9.3.1}$$

For any function U which is analytic when $|\operatorname{Im} z| < 1$ and any bounded open set X let $\Sigma(U, X)$ be the set of all $y \in \mathbb{R}^n$ with $|y| = 1$ such that U

is analytic at $x+iy$ *for every* $x\in\partial X$. *Then*

$$U_y^X(\phi)=\int_X U(x+iy)\,\phi(x)\,dx, \qquad \phi\in A,$$

is in $A'(\bar{X})$ *if* $y\in\mathbb{R}^n$, $|y|<1$, *and* U_y^X *can be defined for all* $y\in\Sigma(U,X)$ *so that* U_y^X *remains a continuous function of* y *with values in* $A'(\bar{X})$. *Thus*

$$U_\mu^X(\phi)=\int U_\omega^X(\phi)\,d\mu(\omega)=\lim_{r\to 1}\int\int_X U(x+ir\omega)\,\phi(x)\,dx\,d\mu(\omega), \qquad \phi\in A,$$

defines an element of $A'(\bar{X})$ *for every measure* $d\mu$ *with support in* $\Sigma(U,X)$.

Proof. $K(z-t)$ is an analytic function of t in a complex neighborhood of $\operatorname{supp} u$ if $z\in Z$. Hence U is defined in Z, and U is analytic since derivatives of U can be put on K. If $\phi\in A$ and $r<1$ then

$$\int_{|\omega|=1}\int_X U(x+ir\omega)\,\phi(x)\,dx\,d\omega=u(\Phi_r)$$

$$\begin{aligned}\Phi_r(t)&=\int_{|\omega|=1}\int_X K(x+ir\omega-t)\,\phi(x)\,dx\,d\omega\\ &=\int_{|\omega|=1}\int_X K(t+ir\omega-x)\,\phi(x)\,dx\,d\omega.\end{aligned}$$

(Recall that K is even.) By Theorem 8.4.11 $\Phi_r\to\phi$ in $\mathscr{D}'(X)$ as $r\to 1$. Choose $\chi\in C_0^\infty(X)$ with $0\leqq\chi\leqq 1$ so that $\chi=1$ in a neighborhood of $\operatorname{supp} u$. Then

$$\begin{aligned}&\int_X K(t+ir\omega-x)\,\phi(x)\,dx\\ &\quad=\int_{\gamma(\omega,\varepsilon)} K(t+ir\omega-z)\,\phi(z)\,dz_1\wedge\ldots\wedge dz_n, \qquad 0<\varepsilon<1,\end{aligned}$$

by Stokes' formula, $\gamma(\omega,\varepsilon)$ denoting the chain

$$X\ni x\to x+i\varepsilon\chi(x)\,\omega.$$

Letting $r\to 1$ we conclude that $\Phi_r(t)$ has an analytic limit in a complex neighborhood of $\operatorname{supp} u$. The limit is necessarily equal to $\phi(t)$, which proves (9.3.1).

With the same notation we obtain in the second part of the proposition

$$U_y^X(\phi)=\int_{\gamma(\omega,\varepsilon)} U(z+iy)\,\phi(z)\,dz_1\wedge\ldots\wedge dz_n$$

if $|y|<1$ and $|y+\varepsilon\omega|<1$. This proves the asserted continuity in

$$\{y;\ |y|=1,\ \langle y,\omega\rangle<-1/2,\ U \text{ is analytic at } x+iy \text{ if } x\in X,\ \chi(x)=0\}.$$

Since $\omega \in S^{n-1}$ is arbitrary and χ can be chosen equal to 1 on any compact subset of X, the proof is complete.

Suppose with the notation in the first part of the proposition that there is a point $x_0 \in \operatorname{supp} u$ such that U is analytic at $x_0 + i\omega$ when $|\omega|=1$. Then there is a compact neighborhood $M \subset X$ of x_0 such that U is analytic at $M + iS^{n-1}$. Hence

$$u(\phi) = U_{d\omega}^{X \setminus M}(\phi) + \int_{|\omega|=1} \int_M U(x+i\omega)\,\phi(x)\,dx\,d\omega.$$

The restriction of u to the interior of M is therefore the analytic function

$$\int_{|\omega|=1} U(x+i\omega)\,d\omega, \qquad x \in M.$$

Definition 9.3.2. If $u \in A'(\mathbb{R}^n)$ then $WF_A(u)$ is the set of all $(x, \xi) \in \mathbb{R}^n \times (\mathbb{R}^n \setminus 0)$ such that $U = K * u$ is not analytic at $x - i\xi/|\xi|$.

We have just proved that the projection of $WF_A(u)$ in $\mathbb{R}^n$ is $\operatorname{sing\,supp}_A u$. Since $WF_A(u)$ is determined at x by the local properties of u, the definition is immediately extended to general hyperfunctions in an open subset of $\mathbb{R}^n$. To prove that the results of Sections 8.4 to 8.6 can be extended to hyperfunctions we shall work consistently with boundary values of functions analytic in tube domains. The following is an analogue of Theorem 3.1.15 and Theorem 8.4.8.

Theorem 9.3.3. *Let X be an open set in $\mathbb{R}^n$, Γ a connected open cone in $\mathbb{R}^n \setminus \{0\}$ and f an analytic function in an open set $Z \subset \mathbb{C}^n$ such that for every open set $X_1 \Subset X$ and closed convex cone $\Gamma_1 \subset \Gamma \cup \{0\}$ we have for some $\gamma > 0$*

$$Z \supset \{z \in \mathbb{C}^n;\ \operatorname{Re} z \in X_1,\ \operatorname{Im} z \in \Gamma_1,\ 0 < |\operatorname{Im} z| < \gamma\}.$$

Then

(i) *there is an element $f_{X_1} \in A'(\bar{X}_1)$ independent of Γ_1 and uniquely determined* $\operatorname{mod} A'(\partial X_1)$ *such that the analytic functional*

$$A \ni \phi \to f_{X_1}(\phi) - \int_{X_1} f(x+iy)\,\phi(x+iy)\,dx$$

is carried by any given neighborhood of ∂X_1 in $\mathbb{C}^n$ when $y \in \Gamma_1$ and $|y|$ is small enough. Thus f_{X_1} defines uniquely a hyperfunction in $B(X_1)$.

(ii) *If $X_2 \subset X_1$ is another open set then $f_{X_1} - f_{X_2} \in A'(\bar{X}_1 \setminus X_2)$ so there is a unique $f_X \in B(X)$ such that f_X and f_{X_1} have the same restriction to X_1 for every X_1.*

(iii) *If $|f(x+iy)| \leq C|y|^{-N}$, $x \in X_1$, $y \in \Gamma_1$, $|y| < \gamma$, as in Theorem 3.1.15 then f_X restricted to X_1 is equal to the distribution limit which exists by Theorem 3.1.15.*

(iv) *If f can be continued analytically to a neighborhood of ∂X_1 then*

$$\phi \to \int_{X_1} f(x+iy)\,\phi(x)\,dx, \qquad \phi \in A,$$

converges in $A'(\bar{X}_1)$, when $\Gamma_1 \ni y \to 0$, to an element satisfying the condition (i) *above.*

(v) *If $f_X = 0$ in some non-empty $X_1 \subset X$ and Z is connected, then $f = f_X = 0$.*

(vi) $WF_A(f_X) \subset X \times (\Gamma^\circ \setminus \{0\})$.

Proof. (i) The analytic functional

$$\phi \to \int_{X_1} f(x+iy_1)\,\phi(x+iy_1)\,dx - \int_{X_1} f(x+iy_2)\,\phi(x+iy_2)\,dx$$

is carried by $\partial X_1 + i[y_1, y_2]$. In fact, if ∂X_1 is smooth it follows from Stokes' formula that the difference is equal to the integral of the closed form $f(z)\,\phi(z)\,dz_1 \wedge \ldots \wedge dz_n$ over the chain $\partial X_1 + i[y_2, y_1]$, and we can approximate X_1 arbitrarily closely by an open set with C^∞ boundary. (i) is thus a consequence of Theorem 9.1.7. To prove (ii) we just have to observe that the analytic functional

$$\int_{X_1 \setminus X_2} f(x+iy)\,\phi(x+iy)\,dx$$

is carried by $\bar{X}_1 \setminus X_2 + iy$, for this implies that $f_{X_1} - f_{X_2}$ is carried by any neighborhood of $\bar{X}_1 \setminus X_2 \supset \partial X_1 \cup \partial X_2$. (iii) Let f_0 be the distribution limit in $\mathscr{D}'(X_1)$. If $\psi \in A$ then ψf_0 is the distribution limit of ψf, so (3.1.20) gives

$$\langle \phi f_0, \psi \rangle = \int \Phi(x,y)(\psi f)(x+iy)\,dx$$
$$+ (N+1) \iint_{0<t<1} (\psi f)(x+ity) \sum_{|\alpha|=N+1} \partial^\alpha \phi(x)(iy)^\alpha/\alpha!\, t^N\,dx\,dt.$$

Here $\phi \in C_0^\infty(X_1)$, Φ is defined by (3.1.18) and $y \in \Gamma_1$, $|y| < \gamma$. The formula (3.1.20) extends f_0 to a distribution in $\mathscr{E}'(\bar{X}_1)$ if we integrate only over X_1 for arbitrary $\phi \in C_0^\infty(\mathbb{R}^n)$. With $\phi \in C_0^\infty(X_1)$ equal to 1 in a large compact subset of X_1 it follows then from the preceding formula that

$$\psi \to \langle f_0, \psi \rangle - \int_{X_1} f(x+iy)\,\psi(x+iy)\,dx$$

is carried by $\partial X_1 + i[0, y]$. Hence the uniqueness in (i) gives $f_0 - f_{X_1} \in A'(\partial X_1)$ so $f_X = f_0$ in X_1. (iv) The existence of the limit f_0 in $A'(\bar{X}_1)$ follows exactly as in the proof of Proposition 9.3.1 so we leave this for the reader. Since

$$\int_{X_1} f(x+iy)\,\phi(x+iy)\,dx \to f_0(\phi), \qquad \phi \in A,$$

it follows from (i) that $f_{X_1} - f_0$ is carried by ∂X_1.

To prove (v) and (vi) we denote by R_y the difference in (i) which is carried by any neighborhood of ∂X_1 when $|y|$ is small. Then

$$K * f_{X_1}(z) - \int_{X_1} K(z-t-iy)\, f(t+iy)\, dt = R_y K(z-.)$$

when $|\operatorname{Im} z| + |y| < 1$, $y \in \Gamma_1$, $|y| < \gamma$. Here $R_y K(z-.)$ is analytic in any compact subset of $\{z; |\operatorname{Im} z|^2 < 1 + |\operatorname{Re} z - x|^2,\ x \in \partial X_1\}$ when $|y|$ is small. If $f_{X_1} \in A'(\partial X_1)$ and $X_2 \Subset X_1$ it follows that

$$F_y(z) = \int_{X_1} K(z-t-iy)\, f(t+iy)\, dt$$

is analytic in a fixed complex neighborhood of $\bar{X}_2 + iS^{n-1}$ when $y \in \Gamma_1$ and $|y|$ is small. Thus

$$f(x+iy) = \int_{|\omega|=1} F_y(x+iy+i\omega)\, d\omega$$

is analytic for x in a fixed complex neighborhood of $\bar{X}_2$ when $|y|$ is small. Hence f can be analytically extended to a neighborhood of $\bar{X}_2$. This extension is identically 0 in X_2 so $f=0$ in Z if Z is connected, which proves (v). To prove (vi) we observe that

$$K * f_{X_1}(z) = \int_{X_1} K(z-t-iy)\, f(t+iy)\, dt + R_y K(z-.)$$

is analytic at $\bar{X}_2 + i\omega$ if $|\omega - y| < 1$. Since

$$|\omega - y|^2 = 1 - 2\langle \omega, y\rangle + |y|^2$$

this is true if y is replaced by εy for some small $\varepsilon > 0$ and $\langle \omega, y\rangle > 0$. Hence $K * f_{X_1}(z)$ is analytic at $\bar{X}_2 + i\omega$ unless $\langle \omega, y\rangle \leqq 0$ for every $y \in \Gamma$, that is, $-\omega \in \Gamma^\circ$. This completes the proof of (vi) and of the theorem.

The hyperfunction $f_X \in B(X)$ will be called the boundary value of f from Γ. Occasionally the notation f_0 of Theorem 3.1.15 will be used, but we shall use the notation $b_\Gamma f$ when we want to emphasize that the limit is taken from Γ. This notation assumes tacitly that f is analytic in a set Z with the properties listed in Theorem 9.3.3. Then f is called Γ analytic at X.

There is a converse of (vi) in Theorem 9.3.3 (cf. Theorem 8.4.15):

Theorem 9.3.4. *Let X be an open set in $\mathbb{R}^n$ and let $u \in B(X)$. If*

$$WF_A(u) \subset X \times (\Gamma^\circ \setminus 0)$$

where Γ is an open convex cone in $\mathbb{R}^n \setminus 0$, then there is a Γ analytic function f such that $u = b_\Gamma f$.

Proof. If $X_1 \Subset X$ we can choose $v \in A'(\bar{X}_1)$ defining u in X_1. Set $V = K * v$, let Γ_1 be a closed cone $\subset \Gamma \cup \{0\}$ and choose $M \subset S^{n-1}$ open with $\Gamma^\circ \cap S^{n-1} \subset M$ so that $\bar{M}$ is in the interior of Γ_1°. If $X_1 \Subset X_0 \Subset X$ it follows from Proposition 9.3.1 that we can write $v = v_1 + v_2$ where

$$v_1(\phi) = \int_{\omega \notin -M} V_\omega^{X_0}(\phi)\, d\omega, \qquad \phi \in A,$$

is analytic in X_1 and v_2 is the boundary value of the analytic function

$$V_2(z) = \int_{-M} V(z + i\omega)\, d\omega, \qquad |\mathrm{Im}\, z| < \gamma, \ \mathrm{Im}\, z \in \Gamma_1, \ \mathrm{Re}\, z \in X_0.$$

If $X_2 \Subset X_1$ it follows that the restriction of u to X_2 is the boundary value of a function f analytic in $\{z;\ \mathrm{Re}\, z \in X_2,\ \mathrm{Im}\, z \in \Gamma_1,\ |\mathrm{Im}\, z| < \gamma_1\}$ for some $\gamma_1 > 0$. The uniqueness of f implied by Theorem 9.3.3(v) shows that the functions f obtained for different Γ_1 and X_2 together define a Γ analytic function at X.

An immediate consequence of Theorem 9.3.3 is the classical "edge of the wedge" theorem:

Theorem 9.3.5. *Let $f^\pm$ be analytic in*

$$Z^\pm = \{x + iy;\ x \in X,\ \pm y \in \Gamma, |y| < \gamma\}$$

where $\gamma > 0$ and Γ is an open convex cone. If $f_0 = b_\Gamma f^+ = b_{-\Gamma} f^-$ then f_0 is an analytic function which extends both f^+ and f^-.

Proof. If $\xi \in \Gamma^\circ \cap (-\Gamma)^\circ$ then $\langle y, \xi \rangle = 0$ when $y \in \Gamma$ so $\xi = 0$. Hence it follows from Theorem 9.3.3(vi) that $WF_A(f_0)$ is empty, which means that f_0 is real analytic. If we subtract the analytic continuation of f_0 from $f^\pm$ it follows from Theorem 9.3.3(v) that the difference vanishes in $Z^\pm$ when $|\mathrm{Im}\, z|$ is small, and this proves the theorem.

We can supplement Proposition 9.3.1 with an analogue of Lemma 8.4.12:

Lemma 9.3.6. *With the notation in Proposition 9.3.1 we have*

$$WF_A(U_\mu^X) \subset \{(x, \xi);\ x \in X,\ -\xi/|\xi| \in \operatorname{supp} d\mu \text{ and } U \text{ is not analytic at } x - i\xi/|\xi|\} \cup \partial X \times (\mathbb{R}^n \setminus 0).$$

Proof. Replace the reference to Theorem 8.4.8 in the proof of Lemma 8.4.12 by a reference to Theorem 9.3.3(vi).

We come now to the important decomposition theorem corresponding to Corollary 8.4.13:

Theorem 9.3.7. *Let $\Gamma_1, \dots, \Gamma_J$ be closed cones in $\mathbb{R}^n \setminus \{0\}$ with union $\mathbb{R}^n \setminus \{0\}$. If X is a bounded open set in $\mathbb{R}^n$ and $u \in B(X)$ then $u = \sum u_j$ where $u_j \in B(X)$ and $WF_A(u_j) \subset WF_A(u) \cap (X \times \Gamma_j)$, $j = 1, \dots, J$. If $u = \sum u'_j$ is another such decomposition then $u'_j = u_j + \sum_k u_{jk}$ with $u_{jk} \in B(X)$, $u_{jk} = -u_{kj}$ and $WF_A(u_{jk}) \subset X \times (\Gamma_j \cap \Gamma_k)$.*

Proof. Extend u to $v \in A'(\bar{X})$, set $V = K * v$ and define $v_j \in A'(\bar{Y})$ by

$$v_j(\psi) = \int V^Y_\omega(\psi)\, \phi_j(\omega)\, d\omega, \qquad \psi \in A.$$

Here Y is a bounded open neighborhood of $\bar{X}$, V^Y_ω is defined in Proposition 9.3.1 and ϕ_j is the partition of unity in the proof of Corollary 8.4.13. Then $\sum v_j = v$, and if u_j is the restriction of v_j to X we obtain

$$WF(u_j) \subset WF_A(u) \cap (X \times \Gamma_j)$$

in view of Lemma 9.3.6. To prove the second part we let $w_j \in A'(\bar{X})$ be an extension of $u'_j - u_j$, thus $w = \sum w_j \in A'(\partial X)$. Replacing w_1 by $w_1 - w$ we may assume that $\sum w_j = 0$. Set $W_j = K * w_j$ and

$$w_{jk}(\psi) = \int W^Y_{j\omega}(\psi)\, \phi_k(\omega)\, d\omega - \int W^Y_{k\omega}(\psi)\, \phi_j(\omega)\, d\omega.$$

Since

$$\sum_k w_{jk}(\psi) - w_j(\psi) = 0$$

we obtain for the restrictions u_{jk} of w_{jk} to X

$$\sum_k u_{jk} = u'_j - u_j,$$

which completes the proof.

The last part of the theorem is called Martineau's edge of the wedge theorem. Its significance is perhaps more clear if we take Γ_j convex and proper and denote by G_j the interior of the dual cone. Then it follows from Theorem 9.3.4 that $u_j = b_{G_j} f_j$ and $u'_j = b_{G_j} f'_j$ for some G_j analytic f_j, f'_j. Moreover, u_{jk} is the limit from $G_j + G_k$ of a function $f_{jk}, f_{jk} = -f_{kj}$, thus

$$b_{G_j} f_{jk} = b_{G_j + G_k} f_{jk} = b_{G_k} f_{jk} = -b_{G_k} f_{kj}$$

which confirms that

$$\sum_j u'_j - \sum_j u_j = \sum_{j,k} b_{G_j} f_{jk} = 0.$$

It is therefore possible to define $B(X)$ as the set of all $(f_1, \dots, f_J)$ such that f_j is G_j analytic at X, identifying with 0 the set of all

$$\left(\sum_k f_{1k}, \dots, \sum_k f_{Jk}\right)$$

where $f_{jk}=-f_{kj}$ is G_j+G_k analytic. (Two G_j analytic functions are called equal if they have identical G_j analytical restrictions.) We can also use this observation to extend operations on hyperfunctions as we did in Section 8.5. Assume for example that $X\subset\mathbb{R}^n$, $Y\subset\mathbb{R}^m$ are open sets and h a real analytic map $X\to Y$. Set

$$N_h=\{(h(x),\eta)\in Y\times\mathbb{R}^m;\ {}^th'(x)\eta=0\}.$$

If $u\in B(Y)$ and $N_h\cap WF_A(u)=\emptyset$ we shall define h^*u so that as before

$$WF_A(h^*u)\subset h^*WF_A(u). \tag{9.3.2}$$

To do so we take an arbitrary $x_0\in X$ and write u in a neighborhood of $h(x_0)$ in the form

$$u=f_0+\sum b_{G_j}f_j$$

where f_0 is real analytic and ${}^th'(x_0)\eta\neq 0$ if $0\neq\eta\in G_j^\circ$. Then the proof of Theorem 8.5.1 shows that $f_j\circ h$ is G_j' analytic near x_0 for any G_j' with closure contained in $h'(x_0)^{-1}G_j$, which is the interior of the dual cone of ${}^th'(x_0)G_j^\circ$. Thus we can define

$$h^*u=f_0\circ h+\sum b_{G_j'}(f_j\circ h)$$

in a neighborhood of x_0. That this is independent of the decomposition of u follows from the remarks above. (Note that if one has two different coverings $\mathbb{R}^m\setminus\{0\}=\bigcup\Gamma_j$ then one can pass to the covering consisting of all Γ_j occurring in one of them.) The proof of Theorem 8.5.1 gives (9.3.2) without any change. In particular $WF_A(u)\subset T^*(X)\setminus 0$ is now well defined if u is a hyperfunction on a real analytic manifold X. (We leave as an exercise to verify that h^*u as defined above agrees with h^*u as defined in Sections 9.1 and 9.2 when h is an analytic diffeomorphism.)

If $u\in B(X)$ and $v\in B(Y)$ then $u\otimes v\in B(X\times Y)$ is defined and

$$\begin{aligned}WF_A(u\otimes v)\subset{}&(WF_A(u)\times WF_A(v))\\&\cup(WF_A(u)\times(\operatorname{supp}v\times\{0\}))\cup((\operatorname{supp}u\times\{0\})\times WF_A(v)).\end{aligned} \tag{9.3.3}$$

It is sufficient to prove this when $X\subset\mathbb{R}^n$ and $Y\subset\mathbb{R}^m$. We can then use Theorem 9.3.7 to decompose u and v. The statement follows if one notes that when f and g are G_f and G_g analytic respectively, then $f\otimes g$ is $G_f\times G_g$ analytic. The dual cone of $G_f\times G_g$ is $G_f^\circ\times G_g^\circ$, which reduces to $G_f^\circ\times\{0\}$ if $G_g=\mathbb{R}^m$ and to $\{0\}\times G_g^\circ$ if $G_f=\mathbb{R}^n$.

The rest of Section 8.5 can now be extended without any change of the proofs to the case of hyperfunctions. We leave this repetition to the reader.

9.4. The Analytic Cauchy Problem

Already in Lemma 9.1.4 we solved an analytic Cauchy problem for the Laplacean. The extension of Theorem 8.6.1 to hyperfunctions in Section 9.5 will require a general existence theorem for the analytic Cauchy problem with precise information on the existence domain. Such results will be proved in this section.

Points in $\mathbb{C}^n$ will be denoted by $z=(z_1, \dots, z_n)$, and

$$\Omega_R=\{z;\ |z_j|<R,\ j=1, \dots, n\}$$

is the polydisc of radius R with center at 0. In this section only we shall use the notation D^α for the operator $(\partial/\partial z_1)^{\alpha_1} \dots (\partial/\partial z_n)^{\alpha_n}$ acting on analytic functions. We start by solving the simple differential equation

$$D^\beta u=f \tag{9.4.1}$$

with the boundary conditions

$$D_j^k u=0 \quad \text{when } z_j=0 \text{ and } 0 \leqq k<\beta_j; \quad j=1, \dots, n. \tag{9.4.2}$$

Lemma 9.4.1. *For any f which is analytic in Ω_R there is a unique solution of* (9.4.1) *which is analytic in Ω_R and satisfies* (9.4.2). *We have*

$$\sup_{\Omega_R} |u| \leqq R^{|\beta|} \sup_{\Omega_R} |f|/\beta!. \tag{9.4.3}$$

Proof. When $n=1$ we obtain from Taylor's formula the unique solution

$$u(z)=\int_0^z f(t)(z-t)^{\beta-1}\, dt/(\beta-1)!=z^\beta \int_0^1 f(t z)(1-t)^{\beta-1}/(\beta-1)!\, dt.$$

Taking care of one variable at a time we find for any n that if $\beta_1>0, \dots, \beta_k>0$, $\beta_{k+1}=\dots=\beta_n=0$, then

$$u(z)=z^\beta \int_0^1 \dots \int_0^1 f(t_1 z_1, \dots, t_k z_k, z_{k+1}, \dots, z_n) \prod_1^k (1-t_j)^{\beta_j-1}/(\beta_j-1)!\, dt$$

is the unique solution of the boundary problem. It is obvious that (9.4.3) follows.

A slightly weaker existence theorem is valid for small perturbations of (9.4.1). We take $R=1$ for the sake of simplicity.

Theorem 9.4.2. *Let β be a fixed multi-index, $|\beta|=m$, and let a^α, $|\alpha| \leqq m$, and f be bounded analytic functions in Ω_1 with*

$$A=(2^n e)^m \sup_{\Omega_1} \sum |a^\alpha|<1.$$

Then the equation

(9.4.1)′
$$D^\beta u = \sum_{|\alpha| \leqq m} a^\alpha D^\alpha u + f$$

has a unique solution satisfying (9.4.2) *in* $\Omega_{\frac{1}{2}}$, *and*

(9.4.3)′
$$\sup_{\Omega_{\frac{1}{2}}} |u| \leqq (1-A)^{-1} \sup_{\Omega_1} |f|/\beta!.$$

The proof requires two elementary lemmas.

Lemma 9.4.3. *If* v *is an analytic function of one complex variable* ζ *when* $|\zeta|<1$, *such that*

$$|v'(\zeta)| \leqq C(1-|\zeta|)^{-a}, \quad |\zeta|<1, \text{ and } v(0)=0,$$

where $a \geqq 1$, *then it follows that with the same* C

$$|v(\zeta)| \leqq C a^{-1} (1-|\zeta|)^{-a}, \quad |\zeta|<1.$$

Proof. If $a>1$ the statement follows from the fact that

$$(1-r)^a \int_0^r (1-t)^{-a}\, dt = ((1-r)-(1-r)^a)/(a-1)$$

takes its maximum when $a(1-r)^{a-1}=1$, for the maximum value is $(1-r)/a<1/a$. Letting $a \to 1$ we obtain the statement when $a=1$.

Lemma 9.4.4. *If* v *is analytic when* $|\zeta|<1$ *and*

$$|v(\zeta)| \leqq C(1-|\zeta|)^{-a}, \quad |\zeta|<1,$$

where $a \geqq 0$, *it follows with the same* C *that*

$$|v'(\zeta)| \leqq Ce(1+a)(1-|\zeta|)^{-a-1}, \quad |\zeta|<1.$$

Proof. If $0<\varepsilon<\rho=1-|\zeta|$ we have $|v(\zeta_1)| \leqq C(\rho-\varepsilon)^{-a}$ when $|\zeta_1-\zeta|<\varepsilon$. Hence Cauchy's inequality gives $|v'(\zeta)| \leqq C\varepsilon^{-1}(\rho-\varepsilon)^{-a}$. Choosing $\varepsilon = \rho/(a+1)$ we minimize the right hand side and obtain

$$|v'(\zeta)| \leqq C(a+1)(1+1/a)^a \rho^{-a-1} < Ce(a+1)\rho^{-a-1}.$$

Proof of Theorem 9.4.2. We shall pass from (9.4.1) to (9.4.1)′ by using Lemma 9.4.1 to solve recursively the equations

(9.4.1)″
$$D^\beta u_{\nu+1} = \sum_{|\alpha| \leqq m} a^\alpha D^\alpha u_\nu + f$$

starting with $u_0=0$. By Lemma 9.4.1 we have

$$M = \sup_{\Omega_1} |u_1| \leqq \sup_{\Omega_1} |f|/\beta!.$$

Subtracting two successive equations (9.4.1)″ and writing $v_\nu = u_{\nu+1} - u_\nu$ we obtain $v_0 = u_1$ and

(9.4.4) $$D^\beta v_{\nu+1} = \sum_{|\alpha| \leqq m} a^\alpha D^\alpha v_\nu, \qquad \nu = 0, 1, 2, \ldots.$$

We claim that with $C = A/2^{nm}$

(9.4.5) $$|v_\nu(z)| \leqq C^\nu M \prod_1^n (1 - |z_j|)^{-m\nu}, \qquad z \in \Omega_1, \ \nu = 0, 1, \ldots.$$

This is just the definition of M when $\nu = 0$. If (9.4.5) holds for one value of ν, it follows from (9.4.4) and Lemma 9.4.4 that

$$|D^\beta v_{\nu+1}(z)| \leqq A(2^n e)^{-m} C^\nu M (em(\nu+1))^m \prod_1^n (1 - |z_j|)^{-m(\nu+1)}, \qquad z \in \Omega_1.$$

Since $v_{\nu+1}$ satisfies (9.4.2) we may apply Lemma 9.4.3 m times which gives

$$|v_{\nu+1}(z)| \leqq A(2^n e)^{-m} C^\nu M e^m \prod_1^n (1 - |z_j|)^{-m(\nu+1)}, \qquad z \in \Omega_1.$$

Thus we have verified (9.4.5) with ν replaced by $\nu+1$. When $z \in \Omega_{\frac{1}{2}}$ it follows from (9.4.5) that

$$|v_\nu(z)| \leqq A^\nu M.$$

Hence $u = \lim u_\nu = \sum v_\nu$ exists and is analytic in $\Omega_{\frac{1}{2}}$, and $|u| \leqq M/(1-A)$ there which proves (9.4.3)′. Since $\lim D^\alpha u_\nu = D^\alpha u$ for every α, letting $\nu \to \infty$ in (9.4.1)″ gives (9.4.1)′.

If u is a bounded solution in Ω_1 of (9.4.1)′ with $f = 0$ and if (9.4.2) is fulfilled, then $v_\nu = u$ satisfies (9.4.4). By the preceding proof $v_\nu \to 0$ in a neighborhood of 0, hence $u = 0$. If u is just a solution in a neighborhood of 0 we can apply this conclusion to $u(Rz)$ for some $R > 0$, replacing a_α by $a_\alpha(Rz) R^{m-|\alpha|}$. This completes the proof.

Theorem 9.4.2 can be used to solve some mixed problems, but we specialize now to the Cauchy problem which is all we shall need here. Since the variable z_n will play a distinguished role we introduce the polydiscs with unequal radii

$$\Omega_{R,\delta} = \{z \in \mathbb{C}^n; |z_j| < R \text{ when } j < n, \ |z_n| < \delta R\}.$$

Theorem 9.4.5 (Cauchy-Kovalevsky). *Assume that the coefficients in the differential equation*

(9.4.6) $$\sum_{|\alpha| \leqq m} a^\alpha D^\alpha u = f$$

are analytic in $\Omega_{R,\delta}$ *and that the coefficient* a^β, $\beta=(0,\dots,0,m)$, *of* $D_n^m u$ *is equal to* 1. *If*

(9.4.7) $$2(2^n e)^m \sum_{\alpha\neq\beta} R^{m-|\alpha|}\delta^{m-\alpha_n}|a^\alpha(z)|\leqq 1, \qquad z\in\Omega_{R,\delta},$$

and f *is bounded and analytic in* $\Omega_{R,\delta}$, *then* (9.4.6) *has a unique analytic solution in* $\Omega_{R/2,\delta}$ *satisfying the Cauchy boundary conditions*

(9.4.8) $$D_n^j u=0 \quad \text{when } z_n=0,\ j<m.$$

For u *we have the estimate*

(9.4.9) $$\sup_{\Omega_{R/2,\delta}} |u|\leqq 2(R\delta)^m \sup_{\Omega_{R,\delta}}|f|.$$

Proof. When $\delta=R=1$ the statement follows from Theorem 9.4.2 where $A\leqq 1/2$. With $z'=(z_1,\dots,z_{n-1})$ the equation (9.4.6) can be written

$$\sum a^\alpha(Rz',R\delta z_n)R^{-|\alpha|}\delta^{-\alpha_n}D^\alpha u(Rz',R\delta z_n)=f(Rz',R\delta z_n).$$

After multiplication by $R^m\delta^m$ we have reduced the statement to $\delta=R=1$, which completes the proof.

Since $|\alpha|\leqq m$ and $\alpha_n<m$ in (9.4.7), the condition is automatically fulfilled for small R and δ if a^α are analytic at 0. Hence we have in particular a local existence theorem for the homogeneous Cauchy problem. Instead of the homogeneous Cauchy conditions (9.4.8) we can pose inhomogeneous ones

(9.4.8)′ $$D_n^j(u-\phi)=0 \quad \text{when } z_n=0,\ j<m,$$

where ϕ is a given analytic function in $\Omega_{R,\delta}$. Writing $u=\phi+v$ we get homogeneous Cauchy conditions for v and the equation

$$\sum a^\alpha D^\alpha v=f-\sum a^\alpha D^\alpha\phi.$$

Thus we are back in the situation discussed in Theorem 9.4.5. Only the analyticity of $D_n^j\phi|_{z_n=0}$, $j<m$, is important for we can replace ϕ by

$$\sum_{j<m} D_n^j\phi(z',0)\,z_n^j/j!.$$

We shall now give global versions of Theorem 9.4.5. Set

$$P(z,D)=\sum_{|\alpha|\leqq m} a^\alpha(z)D^\alpha$$

and let

$$P_m(z,\zeta)=\sum_{|\alpha|=m} a^\alpha(z)\zeta^\alpha$$

be the principal symbol. If ϕ is an analytic function with $\phi=0$ and

$$\partial\phi/\partial z=(\partial\phi/\partial z_1,\dots,\partial\phi/\partial z_n)\neq 0 \quad \text{at } z_0,$$

then the equation $\phi(z)=0$ defines an analytic hypersurface at z_0. It is called non-characteristic if

$$P_m(z,\partial\phi/\partial z)\neq 0.$$

If ψ is a real valued C^1 function with $\psi=0$ and $d\psi\neq 0$ at z_0, then the equation $\psi=0$ defines a real C^1 hypersurface Σ at z_0. The tangent plane

$$d\psi=\partial\psi+\bar\partial\psi=0;$$

$$\partial\psi=\sum\partial\psi/\partial z_j\,dz_j, \quad \partial\psi/\partial z_j=(\partial/\partial x_j-i\partial/\partial y_j)/2$$

contains a unique analytic hyperplane defined by $\partial\psi=0$. We call Σ non-characteristic if this hyperplane is non-characteristic, that is,

$$P_m(z,\partial\psi/\partial z)\neq 0.$$

Formally this condition looks just as in the analytic case.

Theorem 9.4.6. *Let $P(z,D)$ be an analytic differential operator of order m in the open set $Z\subset\mathbb{C}^n$. If $S\subset Z$ is an analytic non-characteristic hypersurface then the Cauchy problem*

$$P(z,D)u=f, \quad D^\alpha(u-\phi)=0 \text{ on } S \quad \text{when } |\alpha|<m,$$

has a unique analytic solution u in a neighborhood of S for arbitrary f and ϕ which are analytic in Z.

Proof. At any point $z_0\in S$ we can make an analytic change of coordinates so that z_0 becomes the origin and S is defined by $z_n=0$. Then we have the situation in Theorem 9.4.5 after dividing by a^β. (Note that $a^\beta\neq 0$ at 0, because S is non-characteristic.) Thus we obtain local solutions, and since they are unique they fit together to a solution with the properties stated in the theorem.

The set in which the solution exists can often be enlarged by a continuity method:

Theorem 9.4.7. *Let f be analytic in the open subset Z of $\mathbb{C}^n$ and let u be an analytic solution of the equation $P(z,D)u=f$ in the open set $Z_0\subset Z$. If $z_0\in Z\cap\partial Z_0$ and Z_0 has a C^1 non-characteristic boundary at z_0, then u can be continued analytically as a solution of the equation $P(z,D)u=f$ in a neighborhood of z_0.*

Proof. Changing coordinates near z_0 we may assume that $z_0=0$ and that Z_0 is defined near 0 by $\operatorname{Re} z_n<\psi(z_1,\dots,z_{n-1},\operatorname{Im} z_n)$ where $\psi\in C^1$,

$\psi(0)=0$ and $d\psi(0)=0$. The Cauchy problem

$$P(z,D)v=f, \quad D_n^j v=D_n^j u \quad \text{when } z_n=-\varepsilon$$

satisfies the hypothesis of Theorem 9.4.5 in $\Omega_{R,\delta}+(0,\dots,0,-\varepsilon)$ if δ and ε are small and

$$-\varepsilon<\psi(z',0), \quad |z'|<R. \tag{9.4.10}$$

Hence v gives an analytic continuation of u to a neighborhood of 0 if $R\delta/2>\varepsilon$. With δ fixed and $R=3\varepsilon/\delta$ the condition (9.4.10) is fulfilled for small ε since $\psi(z',0)=o(|z'|)$, and this proves the theorem.

We shall now prove a general global existence theorem similar to Lemma 9.1.4 which will be important in Section 9.5. As before we write $z'=(z_1,\dots,z_{n-1})$ and we use the Euclidean norm,

$$|z'|^2=|z_1|^2+\dots+|z_{n-1}|^2.$$

Theorem 9.4.8. *Let Z be an open convex set in $\mathbb{C}^n$ such that*

$$\{(z',0);|z'|<R\}\subset Z\subset\{z;|z_n|/\varepsilon+|z'|<R\}, \tag{9.4.11}$$

and let $P(z,D)$ be an analytical differential operator of order m in Z such that

$$P_m(z,\zeta)\neq 0 \quad \text{when } z\in Z, \quad |\zeta'|<\varepsilon|\zeta_n|. \tag{9.4.12}$$

Then the Cauchy problem

$$P(z,D)u=f \text{ in } Z, \quad D_n^j u=0 \quad \text{when } z_n=0,\ j<m, \tag{9.4.13}$$

has a unique analytic solution in Z for every f which is analytic in Z.

Proof. We can find $r>0$ so that $(0,z_n)\in Z$ if $|z_n|<r$. Then the convexity of Z gives that $z\in Z$ if $|z_n|/r+|z'|/R<1$. If we apply Theorem 9.4.5 to smaller polydiscs with center in the plane $z_n=0$, it follows that there exists a solution of the Cauchy problem in a neighborhood $\hat{Z}$ of $\{(z',0);|z'|<R\}$ in Z. We have to show that it can be continued analytically to an arbitrary $w=(w',w_n)\in Z$ along the straight line from $(w',0)$. In doing so we may assume that $w_n>0$. Replacing Z by the intersection with a smaller set of the form $\{z;|z_n|/\varepsilon+|z'-w'|<a\}$ where $a+|w'|<R$, we may even assume that $w'=0$ and that $(0,w_n)\in Z$ when $0\leqq w_n<\varepsilon R$.

Thus we assume now that $(0,z_n)\in Z$ when $0\leqq z_n<\varepsilon R$. Fix a small $c>0$, then a large M and set for $0\leqq t\leqq\varepsilon$

$$Z_t=\{z;0<\operatorname{Re} z_n<t(R-(|z'|^2+|M\operatorname{Im} z_n|^2+c)^{\frac{1}{2}})\}.$$

In Z_t we have $|\operatorname{Im} z_n|<R/M$, $0<\operatorname{Re} z_n<\varepsilon(R-(|z'|^2+c)^{\frac{1}{2}})$, and since Z contains all z with $0<\operatorname{Re} z_n<\varepsilon(R-|z'|)$, $\operatorname{Im} z_n=0$, it follows that $\bar{Z}_t\subset Z$

if M is large enough. Then we also have $Z_t \subset \hat{Z}$ for small t, and $\{z \in \partial Z_t;\ \operatorname{Re} z_n = 0\} \subset \hat{Z}$, $0 < t \leqq \varepsilon$. On the other part of ∂Z_t the analytic tangent plane is defined by $\langle dz, \zeta \rangle = 0$ where

$$\zeta' = t\bar{z}'(|z'|^2 + |M \operatorname{Im} z_n|^2 + c)^{-\frac{1}{2}}, \quad \operatorname{Re} \zeta_n = 1. \tag{9.4.14}$$

Hence $|\zeta'| < t|\zeta_n| \leqq \varepsilon|\zeta_n|$, so $P_m(z, \zeta) \neq 0$ by (9.4.12). In view of Theorem 9.4.7 it follows that the set of $t \in [0, \varepsilon]$ such that u can be continued analytically to Z_t is open. However, it is obviously closed and non-void so it must be equal to $[0, \varepsilon]$. Continuation is therefore possible to Z_ε for every $c > 0$, and this completes the proof that analytic continuation is possible from 0 to $(0, z_n)$ if $0 < z_n < \varepsilon R$.

Corollary 9.4.9. *Let Γ be an open convex cone in $\mathbb{C}^n$ such that*

$$\operatorname{Im} z_n > a|z'|,\ \operatorname{Re} z_n = 0 \Rightarrow (z', z_n) \in \Gamma. \tag{9.4.15}$$

Here $a > 0$. Let W be an open neighborhood of 0 and $P(z, D)$ an analytic differential operator of order m in W such that

$$P_m(0, \zeta) \neq 0 \quad \text{when } |\zeta'| \leqq a|\zeta_n| \neq 0. \tag{9.4.16}$$

For every analytic function f in $W \cap \Gamma$ it is then possible to find an analytic solution of the equation $P(z, D)u = f$ in $W' \cap \Gamma$ for some other neighborhood W' of 0 independent of f. If u is any such solution and f is analytic in a full neighborhood of 0 then u has an analytic continuation to a neighborhood of 0.

Proof. We can choose $\varepsilon > a$ and a neighborhood W_0 of 0 such that

$$P_m(z, \zeta) \neq 0 \quad \text{when } z \in W_0 \text{ and } |\zeta'| < \varepsilon|\zeta_n|. \tag{9.4.16)$'$}$$

Let $T = (0, \dots, 0, aR)$ and

$$Z = \{z \in \mathbb{C}^n; |z_n|/\varepsilon + |z'| < R\} \cap (\Gamma - iT).$$

Z satisfies (9.4.11) since $(z', iaR) \in \Gamma$ if $|z'| < R$, by (9.4.15). If R is sufficiently small then $P(z + iT, D)$ is analytic in Z and satisfies (9.4.12) there by (9.4.16)$'$. Also $f(z + iT)$ is analytic in Z. Hence it follows from Theorem 9.4.8 that the Cauchy problem with homogeneous Cauchy data for the equation $P(z + iT, D)u(z) = f(z + iT)$ has an analytic solution in Z. Then $P(z, D)u(z - iT) = f(z)$ in $Z + iT$, which contains the intersection of Γ and a neighborhood of 0 since $\varepsilon > a$. The first statement is now proved.

To prove the second statement we take

$$Z = \{z; |z_n|/\varepsilon + |z'| < R\}$$

and solve the equation

$$P(z+iT, D)\, v = f(z+iT), \qquad z \in Z,$$

with Cauchy data

$$D_n^j v(z', 0) = D_n^j u(z', iaR); \quad j < m, \ |z'| < R.$$

If R is sufficiently small this is a Cauchy problem for

$$w(z) = v(z) - \sum_0^{m-1} z_n^j D_n^j u(z', iaR)/j!$$

of the form studied in Theorem 9.4.8. Hence a solution exists in Z, and $v(z-iT)$ is then an analytic continuation of u to a neighborhood of 0.

9.5. Hyperfunction Solutions of Partial Differential Equations

If $P(x, D)$ is a differential operator with real analytic coefficients in the open set $X \subset \mathbb{R}^n$ then

$$WF_A(P(x, D)\, u) \subset WF_A(u), \qquad u \in B(X). \tag{9.5.1}$$

This is an immediate consequence of Theorems 9.3.7 and 9.3.4, for if $u = b_\Gamma f$ then $P(x, D)\, u = b_\Gamma P(z, D) f$. We shall now prove a converse (cf. Theorem 8.6.1):

Theorem 9.5.1. *If $P(x, D)$ is a differential operator of order m with real analytic coefficients in $X \subset \mathbb{R}^n$ then*

$$WF_A(u) \subset \operatorname{Char} P \cup WF_A(P(x, D)\, u), \qquad u \in B(X). \tag{9.5.2}$$

Proof. We must show that if $(x_0, \xi_0) \notin WF_A(P(x, D)\, u)$ and $P_m(x_0, \xi_0) \neq 0$ then $(x_0, \xi_0) \notin WF_A(u)$. In doing so we may assume that $x_0 = 0$ and that $\xi_0 = (0, \dots, 0, 1)$. Choose $a > 0$ so that

$$(0, \xi) \notin WF_A(P(x, D)\, u) \quad \text{if } \xi_n = 1, \quad |\xi'| \leqq a, \tag{9.5.3}$$

$$P_m(0, \zeta) \neq 0 \quad \text{when } |\zeta'| \leqq a, \quad \zeta_n = 1. \tag{9.5.4}$$

Let $\mathbb{R}^n \setminus 0 = \bigcup_0^J \Gamma_j$ where Γ_j are proper, closed, convex cones, $|\xi'| < a\xi_n$ in Γ_0 and $\xi_0 \notin \Gamma_j$ for $j \neq 0$. By the first part of Theorem 9.3.7 and (9.5.3) we can then write in a neighborhood Y of 0

$$P(x,D)\,u=f=\sum_1^J f_j, \qquad WF_A(f_j)\subset Y\times\Gamma_j;$$

$$u=\sum_0^J u_j, \qquad WF_A(u_j)\subset Y\times\Gamma_j.$$

Hence

$$P(x,D)\,u_0+\sum_1^J (P(x,D)\,u_j-f_j)=0.$$

By the second part of Theorem 9.3.7 it follows that

$$P(x,D)\,u_0=\sum_1^J f_{0j}, \qquad WF_A(f_{0j})\subset Y\times(\Gamma_0\cap\Gamma_j).$$

The closure of $G_0=\{y\in\mathbb{R}^n;\, y_n>a|y'|\}$ in $\mathbb{R}^n\setminus 0$ is in the interior of the dual cone of Γ_0. We can choose open convex cones $G_j, j\neq 0$, with closure in the interior of the dual cone of Γ_j and $\xi_0\notin G_j^\circ$, for $\xi_0\notin\Gamma_j$ when $j\neq 0$. By Theorem 9.3.4

$$f_{0j}=b_{G_0+G_j}F_j$$

where F_j is analytic in the intersection of $\mathbb{R}^n+i(G_0+G_j)=\hat{\Gamma}_j$ and a neighborhood of 0. Now $\hat{\Gamma}_j$ satisfies (9.4.15) since $\mathbb{R}^n+iG_0$ does, and (9.4.16) is valid by (9.5.4). Hence it follows from the first part of Corollary 9.4.9 that we can find U_j analytic in the intersection of $\hat{\Gamma}_j$ and a neighborhood of 0 so that $P(z,D)\,U_j=F_j$. We also have $u_0=b_{G_0}U_0$ where U_0 is analytic in the intersection of $\mathbb{R}^n+iG_0$ and a neighborhood of 0. Thus Theorem 9.3.3(v) gives in $\mathbb{R}^n+iG_0$ near 0

$$P(z,D)\left(U_0-\sum_1^J U_j\right)=0.$$

But then the second part of Corollary 9.4.9 proves that $U_0-\sum_1^J U_j$ can be continued analytically to a neighborhood of 0. Hence

$$u_0-\sum_1^J b_{G_j}U_j$$

is analytic at 0. Since $(0,\xi_0)\notin WF_A(b_{G_j}U_j)$ by Theorem 9.3.3(vi), and $(0,\xi_0)\notin WF_A(u_j)$, when $j\neq 0$, we obtain $(0,\xi_0)\notin WF_A(u)$. The proof is complete.

The proof gives also a microlocal existence theorem:

Theorem 9.5.2. *Let $(x_0,\xi_0)\in T^*(X)\setminus 0$ be non-characteristic with respect to the differential operator $P(x,D)$ with real analytic coefficients. For* *every $f \in B(X)$ one can then find $u \in B(X)$ with*

$$(x_0, \xi_0) \notin WF_A(P(x,D)u - f).$$

Proof. With Γ_j defined as in the proof of Theorem 9.5.1 we write

$$f = \sum f_j, \quad WF_A(f_j) \subset X \times \Gamma_j.$$

Then $f_0 = b_{G_0} F_0$ where F_0 is analytic in the intersection of $\mathbb{R}^n + iG_0$ and a neighborhood of x_0. By Corollary 9.4.9 we can find a solution U of the equation $P(z,D)U = F_0$ in the intersection of $\mathbb{R}^n + iG_0$ and another neighborhood of x_0. Taking $u = b_{G_0} U$ in a neighborhood of x_0 we have proved the theorem.

In view of Theorem 9.5.1 the Holmgren uniqueness theorem and its refinements proved in Section 8.6 for distribution solutions of the equation $P(x,D)u = 0$ remain valid for hyperfunction solutions. We shall prove next that Cauchy data can also be defined on an arbitrary non-characteristic analytic surface. For the sake of simplicity we shall assume that it is a plane.

Thus let X be an open subset of $\mathbb{R}^n$ and set

$$X_\pm = \{x \in X, x_n \gtrless 0\}, \quad X_0 = \{x \in X; x_n = 0\}.$$

If $u \in B(X_+)$ satisfies the analytic differential equation $P(x,D)u = 0$ in X_+ and X_0 is non-characteristic with respect to P, we shall prove that $D_n^j u|_{x_n=0} \in B(X_0')$ can be defined in a natural way. (Here X_0' is X_0 considered as a subset of $\mathbb{R}^{n-1}$, that is, $X_0 = X_0' \times \{0\}$.) First we let u_0 be any hyperfunction in X which is equal to u in X_+ and 0 in X_-. Then $f = P(x,D)u_0$ has support in X_0. We can replace u_0 by $u_0 - v$ for any $v \in B(X)$ with $\operatorname{supp} v \subset X_0$, and this changes f to $f - P(x,D)v$. There is natural best choice of v:

Theorem 9.5.3. *Assume that $P(x,D)$ has real analytic coefficients and that X_0 is non-characteristic with respect to $P(x,D)$. Then there is for every $f \in B(X)$ with $\operatorname{supp} f \subset X_0$ a unique decomposition*

$$f = P(x,D)v + \sum_{j<m} v_j \otimes D_n^j \delta_0(x_n) \tag{9.5.5}$$

where $v \in B(X)$, $\operatorname{supp} v \subset X_0$, *and $v_j \in B(X_0')$ for $0 \leqq j < m$.*

Proof. Assume first that we have some f, v, v_j satisfying (9.5.5) with $\operatorname{supp} f$, $\operatorname{supp} v$ and $\operatorname{supp} v_j \times \{0\}$ contained in a compact subset K of X_0. Then (9.5.5) is equivalent to

$$f(\phi) = v({}^tP(x,D)\phi) + \sum_{j<m} v_j((-D_n)^j \phi(\cdot, 0)), \tag{9.5.6}$$

if ϕ is analytic in a neighborhood of K. By Theorem 9.4.6 we can choose ϕ so that in a neighborhood of K

$$(9.5.7) \qquad {}^tP(z,D)\,\phi=\psi; \qquad (-D_n)^j\,\phi(.,0)=\psi_j, \quad j=0,\dots,m-1,$$

if ψ and ψ_j are arbitrary entire functions of n and $n-1$ variables respectively. Hence $f=0$ implies $v=0$ and $v_j=0$ for every j. To prove existence we assume given $f\in B(X)$ with $\operatorname{supp} f=K\subset X_0$ compact and define

$$(9.5.8) \qquad v(\psi)+\sum_{j<m} v_j(\psi_j)=f(\phi).$$

If Ω is a neighborhood of K in $\mathbb{C}^n$ and $\Omega'\subset\mathbb{C}^{n-1}$ is a neighborhood of $K_0=\{x'\in\mathbb{R}^{n-1},(x',0)\in K\}$ it follows from Theorem 9.4.5 that there is a neighborhood ω of K in $\mathbb{C}^n$ where (9.5.7) has a solution ϕ such that

$$\sup_\omega|\phi|\leqq C(\sup_\Omega|\psi|+\sum\sup_{\Omega'}|\psi_j|).$$

Hence

$$|v(\psi)+\sum_{j<m} v_j(\psi_j)|\leqq C'(\sup_\Omega|\psi|+\sum\sup_{\Omega'}|\psi_j|)$$

which proves that (9.5.8) defines $v\in A'(K)$ and $v_j\in A'(K_0)$ satisfying (9.5.5). In view of the uniqueness proved first we conclude that

$$(9.5.9) \qquad \operatorname{supp} f=\operatorname{supp} v\cup\bigcup_j \operatorname{supp} v_j\times\{0\}$$

whenever (9.5.5) holds with f, v, v_j of compact support, $\operatorname{supp} v\cup\operatorname{supp} f\subset X_0$.

For an arbitrary $f\in B(X)$ with $\operatorname{supp} f\subset X_0$ and any open $Y\Subset X_0'$ we can choose $f_Y\in B(X)$ with $\operatorname{supp} f_Y\subset\bar{Y}\times\{0\}$ and $f_Y=f$ in $X\cap(Y\times\mathbb{R})$. In fact, we just have to extend the hyperfunction which is equal to f in $X\cap(Y\times\mathbb{R})$ and 0 in $X_+\cup X_-\cup(X\cap(\complement\bar{Y}\times\mathbb{R}))$. Writing

$$f_Y=P(x,D)\,u_Y+\sum_{j<m} u_{Yj}\otimes D_n^j\,\delta_0(x_n),$$

by the first part of the proof, we conclude from (9.5.9) that

$$\operatorname{supp}(u_Y-u_Z)\cup\bigcup_j \operatorname{supp}(u_{Yj}-u_{Zj})\times\{0\}\subset\operatorname{supp}(f_Y-f_Z)$$

if $Z\Subset X_0'$. Hence u_Y and u_{Yj} define when Y increases a global solution of (9.5.5). The uniqueness follows similarly from (9.5.9) if we start from a solution of (9.5.5) with $f=0$ and "cut v, v_j off" outside a neighborhood of some point in X_0. The proof is complete.

Corollary 9.5.4. *If $P(x,D)$ has real analytic coefficients in X and X_0 is non-characteristic with respect to $P(x,D)$ then any $u\in B(X_+)$ satisfying* *the equation* $P(x,D)u=0$ *has a unique extension* u_0 *to* X *vanishing in* X_- *such that for some* $v_j \in B(X'_0)$

$$P(x,D)u_0 = \sum_{j<m} v_j \otimes D_n^j \delta_0(x_n). \tag{9.5.10}$$

If all v_j *vanish then* $u=0$ *in a neighborhood of* X_0.

Proof. The extension satisfying (9.5.10) is obtained if we first take an arbitrary extension u_0 vanishing in X_-, apply Theorem 9.5.3 to $f = P(x,D)u_0$ and subtract v from u_0. The last statement follows from Holmgren's uniqueness theorem.

To give another interpretation of (9.5.10) we assume for a moment that $u \in C^m(\bar{X}_+)$ and set $u_0 = u$ in $\bar{X}_+$, $u_0 = 0$ in X_-. Then we obtain

$$D_n^k u_0 = (D_n^k u)_0 - i \sum_{j<k} D_n^{k-j-1} u(.,0) \otimes D_n^j \delta_0(x_n)$$

for $k \leqq m$ by induction. With $P_j(x,\zeta)$ denoting the polynomial part of $P(x,\zeta)/\zeta_n^{j+1}$ it follows if $P(x,D)u=0$ that

$$P(x,D)u_0 = -i \sum_{j<m} P_j(x,D)u|_{x_n=0} \otimes D_n^j \delta_0(x_n).$$

Thus the extension u_0 is the same as the one in the Corollary, and

$$-iP_j(x,D)u|_{x_n=0} = v_j; \quad j=0,\dots,m-1. \tag{9.5.11}$$

These equations can be considered as a system of differential equations for $u_j = D_n^j u|_{x_n=0}$, $j<m$. If $a(x)$ is the coefficient of D_n^m in $P(x,D)$, which is $\neq 0$ in X_0 by assumption, then the jth equation contains a term $-iau_{m-j-1}$ and otherwise only u_k with $k<m-1-j$. Starting with the Equation (9.5.11) with $j=m-1$ we can therefore determine $u_0, u_1, \dots, u_{m-1}$ successively in terms of $v_{m-1}, \dots, v_0$. This works for hyperfunctions as well, so under the hypotheses in Corollary 9.5.4 we obtain well defined "normal derivatives" $u_j \in B(X'_0)$. The definition agrees with that in Theorem 4.4.8′ when both are applicable, for the discussion above in the C^m case remains valid then by the remarks following Theorem 4.4.8′.

Thus we have now what is needed to state boundary conditions for differential equations involving hyperfunctions. The correspondence between analytic functionals and harmonic functions in Section 9.1 is just the special case of the Dirichlet problem in a half space. However, we must stop at this point in our brief introduction to hyperfunction theory.

9.6. The Analytic Wave Front Set and the Support

We shall begin by giving another equivalent definition of $WF_A(u)$ due to Bros and Iagolnitzer. If $u \in A'(\mathbb{R}^n)$ we define an entire function $T_\lambda u$ depending on a positive parameter λ by the Gaussian convolution

$$T_\lambda u(z) = u_y \exp(-\lambda(z-y)^2/2), \qquad z \in \mathbb{C}^n. \tag{9.6.1}$$

Here $z^2 = z_1^2 + \ldots + z_n^2$. Since

$$\operatorname{Re} -(z-y)^2 = (\operatorname{Im}(z-y))^2 - (\operatorname{Re}(z-y))^2 \leqq (\operatorname{Im}(z-y))^2; \quad z, y \in \mathbb{C}^n; \tag{9.6.2}$$

we have for every $\varepsilon > 0$

$$|T_\lambda u(z)| \leqq C_\varepsilon \exp(\lambda(|\operatorname{Im} z| + \varepsilon)^2/2). \tag{9.6.3}$$

T_λ is closely related to the Fourier transformation for

$$\exp(\lambda(i\langle x, \xi\rangle - \xi^2/2))\, T_\lambda u(x + i\xi) = u_y(\exp(-\lambda(x-y)^2/2 + i\lambda\langle y, \xi\rangle))$$

is the Fourier transform at $-\lambda\xi$ of u multiplied by $e^{-\lambda|x-y|^2/2}$. This factor localizes at x in very much the same way as the cutoff functions used in Lemma 8.4.4. It is therefore natural to expect that $(x_0, \xi_0) \notin WF_A(u)$ if and only if the right-hand side is $O(e^{-c\lambda})$ for some $c > 0$ in a neighborhood of $(x_0, -\xi_0)$ as $\lambda \to \infty$. To prove this we shall begin by estimating $T_\lambda u$ when u is the boundary value of an analytic function. We shall use the notation $d(x, A) = \inf_{y \in A} |x - y|$ for the Euclidean distance from $x \in \mathbb{R}^n$ to $A \subset \mathbb{R}^n$.

Proposition 9.6.1. *Let $\Omega \subset \mathbb{R}^n$ be an open convex set with $0 \in \bar{\Omega}$, let $\Gamma = \mathbb{R}_+ \Omega$ be the convex cone generated by Ω and let $X \subset \mathbb{R}^n$ be an open set. If f is an analytic function in $X + i\Omega$ and $u \in A'(\mathbb{R}^n)$ is equal to $b_\Gamma f$ in X, then*

$$|T_\lambda u(z)| \leqq C \exp \lambda \Phi(z), \qquad z \in K, \ \lambda > 0, \tag{9.6.4}$$

for every compact set $K \subset \mathbb{C}^n$ and every continuous function Φ on K such that

$$\Phi(z) > (|\operatorname{Im} z|^2 - d(\operatorname{Re} z, \complement X)^2)/2, \qquad z \in K, \tag{9.6.5}$$

$$\Phi(z) > d(\operatorname{Im} z, \Omega)^2/2 \quad \text{if } \operatorname{Re} z \in X, \qquad z \in K. \tag{9.6.6}$$

If $u = 0$ in X then (9.6.6) *can be omitted.*

Proof. We may assume that X is bounded since (9.6.5) remains valid if X is replaced by the intersection with a sufficiently large ball. The estimate (9.6.4) follows from (9.6.2) and (9.6.5) if $\operatorname{Re} z$ is in a neighborhood of $\complement X$, so we may assume in the proof that $\operatorname{Re} K \subset X$. Write $u = u_1 + u_0$ where $\operatorname{supp} u_1 \subset \bar{X}$ and $\operatorname{supp} u_0 \subset \complement X$. By (9.6.5) and (9.6.2)

$$\operatorname{Re} -(z-y)^2/2 < \Phi(z), \qquad z \in K,$$

for all y in a complex neighborhood of $\operatorname{supp} u_0$, so (9.6.4) is valid for $T_\lambda u_0$. Let $X_0 \Subset X$. For every $\gamma \in \Omega$ the analytic functional

$$A \ni \phi \to u_1(\phi) - \int_{X_0} f(x+i\gamma)\,\phi(x+i\gamma)\,dx$$

is carried by $(\bar{X} \setminus X_0) \cup (\partial X_0 + [0, i\gamma])$. Hence u_1 is carried by

$$M = (\bar{X} \setminus X_0) \cup (\partial X_0 + [0, i\gamma]) \cup (\bar{X}_0 + \{i\gamma\}).$$

For every $z_0 \in K$ we can choose $\gamma \in \Omega$ so that

$$|\operatorname{Im} z_0 - \gamma|^2/2 < \Phi(z_0).$$

When $y \in \bar{X} \setminus X_0$ we have by (9.6.5) if X_0 is large enough

$$(|\operatorname{Im} z|^2 - |\operatorname{Re} z - y|^2)/2 < \Phi(z), \qquad z \in K.$$

Hence we obtain in view of the convexity of $|\operatorname{Im}(z-y)|^2$

$$\operatorname{Re}(-(z-y)^2/2) < \Phi(z)$$

for z in a neighborhood of z_0 in K and y in a neighborhood of M. The estimate (9.6.4) follows in a neighborhood of z_0 and by the Borel-Lebesgue lemma in all of K.

Proposition 9.6.1 combined with the decomposition provided by Theorem 9.3.7 will easily give that $T_\lambda u(x+i\xi)$ grows more slowly than $\exp \lambda \xi^2/2$ if $(x, -\xi) \notin WF_A(u)$. To prove a converse result we must study how u can be reconstructed from $T_\lambda u$. We start from Fourier's inversion formula

$$\phi(0) = \lim_{\varepsilon \to 0} (2\pi)^{-n} \iint e^{i\langle y, \xi\rangle - \varepsilon|\xi|}\,\phi(y)\,d\xi\,dy, \qquad \phi \in C_0^\infty(\mathbb{R}^n),$$

where the convergence factor $e^{-\varepsilon|\xi|}$ guarantees absolute convergence. If $\zeta = \xi + i\eta$ then

$$\zeta^2 = \xi^2 - \eta^2 + 2i\langle \xi, \eta\rangle$$

has an analytic square root with positive real part when $|\eta| < |\xi|$. When $a > 0$ is so small that $a|y| < 1$ when $y \in \operatorname{supp} \phi$, we can shift the integration with respect to ξ to the cycle

$$\xi \to \zeta = \xi + i a y |\xi|.$$

Note that $i\langle y, \zeta\rangle = i\langle y, \xi\rangle - a y^2 |\xi|$ has real part $\leqq 0$ and that

$$d\zeta_1 \wedge \ldots \wedge d\zeta_n = (1 + i a \langle y, \xi/|\xi|\rangle)\,d\xi_1 \wedge \ldots \wedge d\xi_n$$

since $d|\xi| = \sum \xi_j/|\xi|\, d\xi_j$ and $d|\xi| \wedge d|\xi| = 0$. Hence

$$\phi(0) = \lim_{\varepsilon \to 0} (2\pi)^{-n} \int d\xi \int \exp(i\langle y, \xi\rangle - a\, y^2 |\xi| - \varepsilon \sqrt{\zeta^2}) \cdot (1 + i a \langle y, \xi/|\xi|\rangle)\, \phi(y)\, dy.$$

Here $y \to (\langle y, \xi\rangle + i a\, y^2 |\xi| + i\varepsilon \sqrt{\zeta^2})/|\xi|$ has imaginary part $\geqq 0$, is bounded in C^∞ and has a lower bound independent of ε and ξ for the norm of its differential with respect to y. By Theorem 7.7.1 the inner integral is therefore a rapidly decreasing function of ξ, uniformly with respect to ε, thus

(9.6.7) $$\phi(0) = (2\pi)^{-n} \int d\xi \int e^{i\langle y, \xi\rangle - a y^2 |\xi|} (1 + i a \langle y, \xi/|\xi|\rangle)\, \phi(y)\, dy, \quad \phi \in C_0^\infty,$$

when $a > 0$ is small enough. However, the right-hand side is an analytic function of a when $\operatorname{Re} a > 0$ so the formula is valid for all $a > 0$.

With a fixed $r > 0$ we set $\xi = -\lambda r \omega$ where $\omega \in S^{n-1}$ and $\lambda > 0$. Replacing ϕ by $\phi(x + .)$ we obtain

(9.6.7)′ $$\phi(x) = (2\pi)^{-n} \int_0^\infty r^n \lambda^{n-1}\, d\lambda \int_{|\omega|=1} d\omega \int \exp \lambda(i\langle x - y, r\omega\rangle - a r (x-y)^2)\, (1 + i a \langle x - y, \omega\rangle)\, \phi(y)\, dy.$$

If we choose $a = 1/2r$ then the exponent becomes λ times

$$E(x-y, r\omega) = i\langle x - y, r\omega\rangle - (x-y)^2/2 = -(x - y - i r\omega)^2/2 - r^2/2,$$

and it follows that

$$e^{-\lambda r^2/2}\, T_\lambda \phi(x - i r\omega) = \int e^{\lambda E(x-y, r\omega)}\, \phi(y)\, dy,$$

$$e^{-\lambda r^2/2} \langle \omega, D\rangle\, T_\lambda \phi(x - i r\omega) = \lambda \int e^{\lambda E(x-y, r\omega)} (r + i\langle x - y, \omega\rangle)\, \phi(y)\, dy.$$

Hence we have for every $\phi \in C_0^\infty$

(9.6.8) $$\phi(x) = 2^{-1} (2\pi)^{-n} \int_0^\infty r^n \lambda^{n-1}\, d\lambda \cdot \int_{|\omega|=1} e^{-\lambda r^2/2} (1 + \langle \omega, D/r\lambda\rangle)\, T_\lambda \phi(x - i r\omega)\, d\omega$$

where the integrand is $O(\lambda^{-N})$ for every N as $\lambda \to \infty$ and is bounded as $\lambda \to 0$. The extension to analytic functionals is straightforward:

Proposition 9.6.2. *If* $u \in A'(\mathbb{R}^n)$ *then*

$$F(z)=2^{-1}(2\pi)^{-n}\int_0^\infty e^{-\lambda/2}\lambda^{n-1}\,T_\lambda u(z)\,d\lambda,$$

$$F_j(z)=2^{-1}(2\pi)^{-n}\int_0^\infty e^{-\lambda/2}\lambda^{n-2}\,\partial/\partial z_j\,T_\lambda u(z)\,d\lambda$$

are analytic functions in $\{z\in\mathbb{C}^n;\ |\operatorname{Im} z|<1\}$ *which remain analytic at* $x_0+i\omega_0, |\omega_0|=1$, *if for some* $C, c>0$

$$|T_\lambda u(x+i\xi)|\leqq C e^{\lambda/2-c\lambda}, \quad |x-x_0|+|\xi-\omega_0|<c,\ \lambda>0. \tag{9.6.9}$$

By Proposition 9.6.1 this is true for every ω_0 *if* $x_0\notin \operatorname{supp} u$. *If* X *is a bounded open neighborhood of* supp u *in* $\mathbb{R}^n$ *we have with the notation of Proposition 9.3.1*

$$u(\phi)=\int_{|\omega|=1} F^X_\omega(\phi)\,d\omega+i\sum\int_{|\omega|=1} F^X_{j\omega}(\phi)\,\omega_j\,d\omega, \qquad \phi\in A. \tag{9.6.8$'$}$$

Proof. The stated analyticity of F and F_j follows immediately from (9.6.3) and (9.6.9). To prove (9.6.8)′ we let $\phi_0\in C_0^\infty(X)$ be equal to ϕ in a neighborhood of supp u. Let $R(\phi_0)$ be the right-hand side of (9.6.8)′ with ϕ replaced by ϕ_0. Then $R(\phi_0)$ is the limit when $r\to 1$ of

$$2^{-1}(2\pi)^{-n}\int\phi_0(x)\,dx\int_0^\infty e^{-\lambda/2}\lambda^{n-1}\,d\lambda\int_{|\omega|=1}(1-\langle\omega,D/\lambda\rangle)\,T_\lambda u(x+ir\omega)\,d\omega.$$

Now the definition of T_λ gives

$$\int\phi_0(x)\,T_\lambda u(x+ir\omega)\,dx=u(T_\lambda\phi_0(\,.\,-ir\omega)),$$
$$\int\phi_0(x)\langle-\omega,D/\lambda\rangle\,T_\lambda u(x+ir\omega)\,dx=u(\langle\omega,D/\lambda\rangle\,T_\lambda\phi_0(\,.\,-ir\omega)).$$

Thus $R(\phi_0)=\lim u(\phi_{0,r})$ where

$$\phi_{0,r}(x)=2^{-1}(2\pi)^{-n}\int_0^\infty e^{-\lambda/2}\lambda^{n-1}\,d\lambda\int_{|\omega|=1}(1+\langle\omega,D/\lambda\rangle)\,T_\lambda\phi_0(x-ir\omega)\,d\omega.$$

It follows from (9.6.8) that $\phi_{0,r}\to\phi_0$ uniformly when $r\to 1$, for the integrand is uniformly rapidly decreasing as $\lambda\to\infty$ when $r\leqq 1$. Since ϕ_0 is analytic in a neighborhood of supp u we have by Prop. 9.6.1

$$|T_\lambda\phi_0(z-ir\omega)|\leqq C e^{\lambda/2-c\lambda}, \qquad 0<r\leqq 1$$

for some $c>0$ and all z in a complex neighborhood of supp u. Hence $\phi_{0,r}(z)$ is uniformly convergent in such a neighborhood of supp u, and the limit must be equal to ϕ by the uniqueness of analytic continuation. This implies that $u(\phi_{0,r})\to u(\phi)$, that is, $R(\phi_0)=u(\phi)$. Thus $R=0$ outside supp u and (9.6.8)′ is valid.

Combination of Propositions 9.6.1 and 9.6.2 gives the Bros-Iagolnitzer definition of $WF_A(u)$ (the essential support in their terminology):

Theorem 9.6.3. *Let* $u\in A'(\mathbb{R}^n)$ *and* $(x_0,\xi_0)\in T^*(\mathbb{R}^n)\setminus 0$. *Then* $(x_0,\xi_0)\notin WF_A(u)$ *if and only if there is a neighborhood* V *of* $x_0-i\xi_0$ *and positive constants* C, c *such that*

$$|T_\lambda u(z)|\leqq Ce^{\lambda(|\xi_0|^2/2-c)},\quad z\in V,\ \lambda>0. \tag{9.6.10}$$

Proof. Assume first that (9.6.10) is valid. If $0\neq t\in\mathbb{R}$ and $M_t(x)=tx$, $x\in\mathbb{R}^n$, then

$$t^n\, T_{\lambda t^2}(M_t^* u)(z)=T_\lambda u(tz)$$

for

$$T_\lambda u(tz)=u(e^{-\lambda(tz-.)^2})=t^n(M_t^* u)(e^{-\lambda t^2(z-.)^2}).$$

Taking $t=|\xi_0|$ we reduce the proof to the case $|\xi_0|=1$. Then we have the representation (9.6.8)′ of u where F and F_j are analytic at $x_0-i\xi_0$, so Lemma 9.3.6 gives that $(x_0,\xi_0)\notin WF_A(u)$.

Now assume that $(x_0,\xi_0)\notin WF_A(u)$. Choose closed convex proper conex $\Gamma_1,\dots,\Gamma_j$ which cover $\mathbb{R}^n\setminus\{0\}$ so that $\xi_0\notin\Gamma_j$, $j\neq 1$, and $\Gamma_1\cap WF_A(u)_{x_0}=\emptyset$. By Theorems 9.3.7 and 9.3.4 we can choose a neighborhood X of x_0 and $u_j\in A'(\bar X)$ so that $u=\sum u_j$ in X, u_1 is analytic in X and $u_j=b_{G_j}f_j$, $j\neq 1$, where G_j is the interior of the dual cone of Γ_j and f_j is analytic in $X+i\Omega_j$ where Ω_j is a convex set generating G_j. We shall apply Proposition 9.6.1 to each u_j and to $u-\sum u_j$. If $\xi\in\mathbb{R}^n$ we have

$$d(\xi,\Omega_j)^2<|\xi|^2$$

unless 0 is the point in $\bar\Omega_j$ closest to ξ. This implies that for every $y\in G_j$

$$|\xi-\varepsilon y|^2\geqq|\xi|^2$$

if $\varepsilon>0$ is small. Hence $\langle\xi,y\rangle\leqq 0$ which means that $-\xi\in\Gamma_j$. Since $\xi_0\notin\Gamma_j$ when $j\neq 1$ it follows that there is a neighborhood K of $x_0-i\xi_0$ and some $c>0$ such that for $z\in K$

$$|\xi_0|^2/2-c>(|\operatorname{Im} z|^2-d(\operatorname{Re} z,\complement X)^2)/2,$$

$$|\xi_0|^2/2-c>d(\operatorname{Im} z,\Omega_j)^2/2,\quad j\neq 1.$$

(9.6.10) is now a consequence of Proposition 9.6.1.

If we make a change of scales the distinction between Ω_j and G_j in the preceding proof is suppressed and we obtain a very precise supplement to Proposition 9.6.1.

Proposition 9.6.4. *Let* $u\in A'(\mathbb{R}^n)$ *and* $W_0=\{\xi;(0,\xi)\in WF_A(u)\}$. *Set* $u_\delta=M_\delta^* u=u(\delta.)$. *If* $K\subset\mathbb{C}^n$ *is a compact set and* Φ *a continuous function* *on K such that*

$$\Phi(z) > (|\operatorname{Im} z|^2 - d(\operatorname{Im} z, -W_0)^2)/2, \quad z \in K, \tag{9.6.11}$$

it follows for small δ that

$$|T_\lambda u_\delta(z)| \leqq C_\delta e^{\lambda\Phi(z)}, \quad z \in K, \ \lambda > 0. \tag{9.6.12}$$

Proof. Take any covering of $\mathbb{R}^n \setminus \{0\}$ with small convex proper cones as in the proof of Theorem 9.6.3. Then $u = u_0 + \sum u_j$ where $u_0 = 0$ in a neighborhood X of 0 and $u_j = b_{G_j} f_j$ in X with f_j analytic in $X + i\Omega_j$. Here Ω_j is a neighborhood of 0 if $\Gamma_j \cap W_0 = \emptyset$, and Ω_j generates the open convex cone G_j with dual Γ_j otherwise. We have $u_\delta = \sum u_{j\delta}$ where $u_{0\delta}$ vanishes in X/δ and $u_{j\delta}$ restricted to X/δ is the boundary value of $f_j(\delta z)$, $j \neq 0$, which is analytic in $X/\delta + i\Omega_j/\delta$. For small δ the estimate (9.6.12) follows from Proposition 9.6.1 if

$$\Phi(z) > d(\operatorname{Im} z, \Omega_j/\delta)^2/2, \quad z \in K, \ j \neq 0, \ \Gamma_j \cap W_0 \neq \emptyset.$$

Since $d(\operatorname{Im} z, \Omega_j/\delta) \searrow d(\operatorname{Im} z, G_j)$ when $\delta \to 0$, this is true for small δ if

$$\Phi(z) > d(\operatorname{Im} z, G_j)^2/2, \quad z \in K, j \neq 0, \Gamma_j \cap W_0 \neq \emptyset. \tag{9.6.13}$$

Now we have a kind of Pythagorean theorem

$$d(\xi, G_j)^2 + d(\xi, -\Gamma_j)^2 = |\xi|^2.$$

Since $\bar{G}_j \cap (-\Gamma_j) = \{0\}$ and the relation between these cones is symmetric it suffices to prove this when $\xi \notin \bar{G}_j$. If ξ^* is the point in $\bar{G}_j$ closest to ξ then $\langle \xi^* - \xi, \eta - \xi^* \rangle \geqq 0$, $\eta \in G_j$, hence $\xi^* - \xi \in \Gamma_j$ since G_j is a cone, and $\langle \xi^* - \xi, \xi^* \rangle = 0 \geqq \langle \theta, \xi^* \rangle$, $\theta \in -\Gamma_j$. Thus $\xi - \xi^*$ is the point in $-\Gamma_j$ closest to ξ and (see Fig. 5).

$$d(\xi, G_j)^2 + d(\xi, -\Gamma_j)^2 = |\xi - \xi^*|^2 + |\xi^*|^2 = |\xi|^2.$$

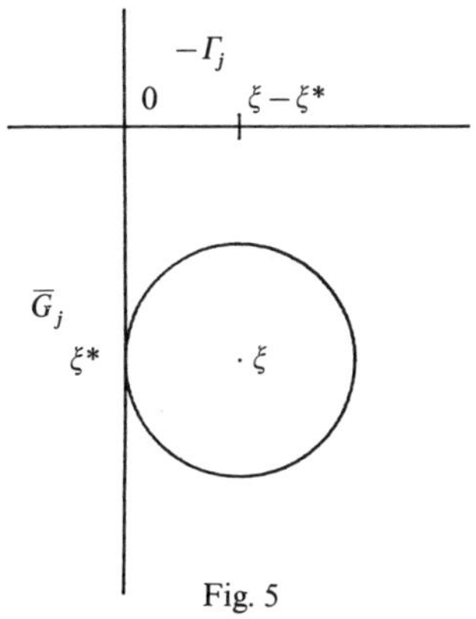

Fig. 5

Hence (9.6.13) means that

$$(9.6.13)' \qquad \Phi(z) > (|\operatorname{Im} z|^2 - d(\operatorname{Im} z, -\Gamma)^2)/2$$

where Γ is the union of the cones Γ_j with $\Gamma_j \cap W_0 \neq \emptyset$. Thus Γ is close to W_0 if all Γ_j are small, so (9.6.13)′ follows from (9.6.11). The proof is complete.

An important feature of (9.6.11) is that the right-hand side is linear along any outer normal of $-W_0$. To exploit this we need an elementary lemma.

Lemma 9.6.5. *Let a and b be positive numbers and u a subharmonic function in $R = \{z \in \mathbb{C};\, 0 < \operatorname{Re} z < a,\ |\operatorname{Im} z| < b\}$ such that for some $\varepsilon > 0$*

$$u(z) < (\max(0, -\operatorname{Im} z))^2 \quad \text{when } z \in R,$$
$$u(z) < -b^2/3 \quad \text{when } |\operatorname{Im} z| < b \text{ and } 0 < \operatorname{Re} z < \varepsilon.$$

With $A(x) = (b/3)\sinh(\pi(a-x)/b)/\sinh(\pi a/b)$ it follows that

$$(9.6.14) \quad u(x+iy) < -A(x)(2y + A(x)) \quad \text{if } |y| < A(x)/2, \quad 0 < x < a.$$

Proof. With $0 < \delta < b$ we apply the maximum principle in the rectangle $R_\delta = \{z;\, 0 < \operatorname{Re} z < a,\ -\delta < \operatorname{Im} z < b - \delta\}$ to

$$v(x+iy) = u(x+iy) - \delta^2 + bA(x)\sin(\pi(y+\delta)/b).$$

v is subharmonic since we have added a harmonic function, and $v < 0$ near the boundary. Hence $v < 0$ in R_δ. Now $\sin\theta > 2\theta/\pi$ for $0 < \theta < \pi/2$ so if $0 < y + \delta < b/2$ we obtain

$$u(x+iy) < \delta^2 - 2A(x)(y+\delta).$$

We minimize by taking $\delta = A(x)$ which is legitimate since $0 < y + A(x) < 3A(x)/2 \leqq b/2$ when $|y| < A(x)/2$. This proves (9.6.14).

The important point in (9.6.14) is that the upper bound is negative when $y = 0$; the good estimate at one end of the rectangle propagates with some decay along the line $\operatorname{Im} z = 0$.

We are now ready to state and prove the main result of this section relating $\operatorname{supp} u$ at a point x_0 to $WF_A(u)$ at x_0. It is due to Kashiwara but sometimes called the co-Holmgren theorem to underline the analogy with Theorem 8.5.6′. In the statement we need the notion of tangent cone of a set. If $x_0 \in M \subset \mathbb{R}^n$ then $T_{x_0}(M)$ is defined as the set of limits of sequences $t_j(x_j - x_0)$ when $t_j \to +\infty$ and $x_j \in M$. (See also the analogous limiting cone at infinity in Lemma 8.1.7.) It is clear that T_{x_0} is a closed cone. If $\psi \in C^1(\mathbb{R}^n, \mathbb{R}^m)$ then we have

$t_j(\psi(x_j)-\psi(x_0))\to\psi'(x_0)\,T$ if $t_j(x_j-x_0)\to T$. Hence

$$\psi'(x_0)\,T_{x_0}(M)\subset T_{\psi(x_0)}(\psi(M))$$

which shows in particular that $T_{x_0}(M)\subset T_{x_0}(X)$ is invariantly defined if M is a subset of a C^1 manifold X instead of $\mathbb{R}^n$.

Theorem 9.6.6. *If $u\in B(X)$ and $x_0\in X$ then*

$$\bar{N}(W_0)\subset\partial W_0\times T_{x_0}(\operatorname{supp}u) \tag{9.6.15}$$

if $W_0=\{\xi\in T^*_{x_0}(X);\,(x_0,\xi)\in WF_A(u)\}$ considered as a subset of the vector space $T^*_{x_0}(X)$ with the origin removed.

Proof. The statement is local so we may assume that $X=\mathbb{R}^n$, $x_0=0$ and $u\in A'(\mathbb{R}^n)$. Let $(\xi_0,t_0)\in N_e(W_0)$. According to Definition 8.5.7 this means that $\xi_0\in W_0$ and that there is a C^∞ function $f(\xi)$ such that $0\neq f'(\xi_0)=t_0$ and $f(\xi)\leqq f(\xi_0)=0$ when $\xi\in W_0$. If we prove that $t_0\notin T_{x_0}(\operatorname{supp}u)$ or $-t_0\notin T_{x_0}(\operatorname{supp}u)$ implies $\xi_0\notin W_0$ we shall have a contradiction proving (9.6.15).

Choose a compact neighborhood K of the interval $I=[-t_0,t_0]-i\xi_0$ in $\mathbb{C}^n$. For every fixed $\varepsilon>0$ we have by Proposition 9.6.4 for $u_\delta=u(\delta.)$ if δ is small enough

$$v_\delta(z)=2\lambda^{-1}\log|T_\lambda u_\delta(z)/C_\delta|<\varepsilon+|\operatorname{Im}z|^2-d(\operatorname{Im}z,-W_0)^2,\quad z\in K. \tag{9.6.16}$$

Choose a compact neighborhood K_0 of $t_0-i\xi_0$ or $-t_0-i\xi_0$ such that $\operatorname{Re}K_0\cap T_{x_0}(\operatorname{supp}u)=\emptyset$. Then it follows from Proposition 9.6.1 that there is a positive constant c such that for small δ

$$v_\delta(z)\leqq|\operatorname{Im}z|^2-c,\qquad z\in K_0, \tag{9.6.17}$$

if C_δ is chosen large enough.

Take any real x,ξ with $|x|+|\xi-\xi_0|$ small, $f(\xi)=0$, and consider the subharmonic function

$$V(w)=v_\delta(x-i\xi-wf'(\xi)),\qquad w\in\mathbb{C}.$$

If $R=\{w;\,|\operatorname{Re}w|<1,\,|\operatorname{Im}w|<b\}$ and $b+|x|+|\xi-\xi_0|$ is sufficiently small we have $x-i\xi-wf'(\xi)\in K$ when $w\in R$. The distance from $\operatorname{Im}wf'(\xi)+\xi$ to the hypersurface $f=0$ is $\pm\operatorname{Im}w\,|f'(\xi)|$ if b is sufficiently small. Hence the distance to W_0 is at least $\operatorname{Im}w\,|f'(\xi)|$, and we obtain from (9.6.16) when $w\in R$

$$\begin{aligned}V(w)&<\varepsilon+(\operatorname{Im}wf'(\xi)+\xi)^2-(\operatorname{Im}wf'(\xi))^2\\&=\varepsilon+\xi^2+2\operatorname{Im}w\langle f'(\xi),\xi\rangle,\qquad\operatorname{Im}w>0,\\V(w)&<\varepsilon+\xi^2+2\operatorname{Im}w\langle f'(\xi),\xi\rangle+|\operatorname{Im}w|^2\,|f'(\xi)|^2,\qquad\operatorname{Im}w\leqq0.\end{aligned}$$

When $w \in R$ and $\operatorname{Re} w$ is near 1 or -1 we obtain from (9.6.17)

$$V(w) < -c/2 + \xi^2 + 2 \operatorname{Im} w \langle f'(\xi), \xi \rangle$$

again if $|x| + |\xi - \xi_0| + b$ is sufficiently small. If we apply Lemma 9.6.5 to $(V(w) - \varepsilon - \xi^2 - 2 \operatorname{Im} w \langle f'(\xi), \xi \rangle)/|f'(\xi)|^2$ as a function of $1 \pm w$ it follows that for some $c_0 > 0$ and $r > 0$ independent of ε we have

$$V(w) \leqq -c_0 + \varepsilon + \xi^2 \quad \text{if } |w| < r \text{ and } |x| + |\xi - \xi_0| < r, \quad f(\xi) = 0.$$

Taking $\varepsilon = c_0/2$ we obtain

$$|T_\lambda u_\delta(x - i\xi - w f'(\xi))| \leqq C_\delta \exp(\lambda(-c_0/2 + \xi^2)/2)$$
$$\text{if } |w| < r \text{ and } |x| + |\xi - \xi_0| < r, \quad f(\xi) = 0.$$

This means that $|T_\lambda u_\delta(z)| \leqq C' \exp(\lambda(\xi_0^2 - c_0/3)/2)$ if $|z + i\xi_0|$ is sufficiently small, and by Theorem 9.6.3 it follows that $(0, \xi_0) \notin WF_A(u_\delta)$. The proof is complete.

All the arguments in Section 8.6 based on Holmgren's theorem have obvious analogues with the characteristic set replaced by the tangent cone of the support. In particular, the proof of Theorem 8.6.8 gives without change.

Corollary 9.6.7. *Let $W_1 \subset W_2$ be open convex sets in $T^*_{x_0}(X) \setminus \{0\}$ such that*

(i) $W_1 \cap WF_A(u)_{x_0} = \emptyset$

(ii) *every hyperplane with normal in $T_{x_0}(\operatorname{supp} u) \cap (-T_{x_0}(\operatorname{supp} u))$ which intersects W_2 also intersects W_1.*

Then $W_2 \cap WF_A(u)_{x_0} = \emptyset$.

Corollary 9.6.8. *If $(x_0, \xi_0) \in N_e(\operatorname{supp} u)$ then $(x_0, \xi) \in WF_A(u)$ implies that $(x_0, \xi + t\xi_0) \in WF_A(u)$ for every $t \in \mathbb{R}$.*

Proof. If ξ is proportional to ξ_0 the statement follows from Theorem 8.5.6′ (which conversely follows from Corollary 9.6.8 and the fact that $WF_A(u)_{x_0}$ cannot be empty). Otherwise, assuming that $\xi + t\xi_0 \notin WF_A(u)_{x_0}$, we take a convex open neighborhood W_1 of $\xi + t\xi_0$ such that $W_1 \cap WF_A(u)_{x_0} = \emptyset$ and $W_2 = W_1 + \mathbb{R}\xi_0$ does not contain 0. Since $T_{x_0}(\operatorname{supp} u) \cap (-T_{x_0}(\operatorname{supp} u))$ is in the orthogonal plane of ξ_0 and every hyperplane with normal orthogonal to ξ_0 intersecting W_2 must also meet W_1, the statement follows from Corollary 9.6.7.

Theorem 9.6.6 and Corollary 9.6.8 may be considered as the 0 and $n-1$ dimensional cases respectively of the following

Theorem 9.6.6′. *Let $u \in B(X)$, $x_0 \in X$, V a linear subspace of $T_{x_0}(X)$, and denote by $T_{x_0,V}(\operatorname{supp} u)$ the closure of the image of $T_{x_0}(\operatorname{supp} u)$ in*

$T_{x_0}(X)/V$. Then

(9.6.15)′ $\bar{N}(W_0 \cap (V' + \{\xi\})) \subset \partial(W_0 \cap (V' + \{\xi\})) \times T_{x_0, V}(\operatorname{supp} u)$

if $W_0 = WF_A(u)_{x_0}$ and $V' + \{\xi\}$ is any affine subspace of $T^*_{x_0}(X)$ parallel to the orthogonal space V' of V. Here $W_0 \cap (V' + \{\xi\})$ is regarded as a subset of $V' + \{\xi\}$, with the origin removed if $\xi \in V'$, so the normals belong to the dual space $T_{x_0}(X)/V$ of V'.

Proof. We may assume that $X = \mathbb{R}^n$, $x_0 = 0$, and that V' is defined by $\xi'' = (\xi_{k+1}, \dots, \xi_n) = 0$. Assume that $\xi_0 \neq 0$ and that ξ_0 has a compact neighborhood K such that $\xi_{01} = f(\xi_0')$ and

(9.6.18) $$\xi_0 \neq \xi \in K, \; \xi_1 \geqq f(\xi'), \; \xi'' = \xi_0'' \Rightarrow \xi \notin W_0,$$

where $\xi' = (\xi_2, \dots, \xi_k)$ and $f \in C^\infty(\mathbb{R}^{k-1})$. We must show that if either $(1, -\partial f/\partial \xi')$ or $(-1, \partial f/\partial \xi')$ is not in $T_{x_0, V}(\operatorname{supp} u)$ at ξ_0' then $\xi_0 \notin W_0$. Assume for example that

(9.6.19) $$(-1, \partial f/\partial \xi') \notin T_{x_0, V}(\operatorname{supp} u), \qquad \xi \in K.$$

Choose M so large that

(9.6.20) $$\xi \in \partial K, \; \xi_1 \geqq f(\xi') + M|\xi'' - \xi_0''|^2 \Rightarrow \xi \notin W_0.$$

This is possible since we have a compact subset of ∂K shrinking to

$$\{\xi \in \partial K; \xi_1 \geqq f(\xi'), \xi'' = \xi_0''\} \subset \complement W_0$$

by (9.6.18). Let t be the smallest number $\geqq 0$ such that

$$\xi \in K, \qquad \xi_1 > f(\xi') + t + M|\xi'' - \xi_0''|^2 \Rightarrow \xi \notin W_0.$$

If $\xi \in K$ and $\xi_1 = f(\xi') + t + M|\xi'' - \xi_0''|^2$ then $\xi \notin W_0$ by (9.6.20) if $\xi \in \partial K$, and if $\xi \in K \setminus \partial K$ this follows from Theorem 9.6.6 since

$$(-1, \partial f/\partial \xi', 2M(\xi'' - \xi_0'')) \notin T_{x_0}(\operatorname{supp} u)$$

by (9.6.19). Thus $t = 0$ and $\xi_0 \notin W_0$ which completes the proof.

We shall end with an application to the regularity of solutions of boundary problems. As in Section 9.5 we consider a plane boundary. Thus let X be an open neighborhood of $0 \in \mathbb{R}^n$ and

$$X_\pm = \{x \in X; x_n \gtrless 0\}, \qquad X_0 = \{x \in X; x_n = 0\}.$$

Let $P(x, D)$ be a differential operator of order m in X with analytic coefficients such that X_0 is non-characteristic. Set $x' = (x_1, \dots, x_{n-1})$.

Theorem 9.6.9. *If* $u \in B(X_+)$ *satisfies the equation* $P(x, D)u = 0$ *and*

$$(x'_0, \xi'_0) \notin WF_A(D_n^j u|_{x_n=0}), \quad j = 0, \ldots, m-1,$$

then there is an $\varepsilon > 0$ *such that*

$$(x, \xi) \notin WF_A(u) \quad \text{if } 0 < x_n < \varepsilon, \quad |x' - x'_0| + |\xi' - \xi'_0| < \varepsilon.$$

Note that there is no condition on ξ_n.

Proof. According to Corollary 9.5.4 we can extend u to $u_0 \in B(X)$ so that u_0 vanishes in X_- and

$$P(x, D)u_0 = \sum_{j<m} v_j \otimes D_n^j \delta_0(x_n) = f.$$

v_j is a linear combination of derivatives of the boundary values of $D_n^k u$ so $(x'_0, \xi'_0) \notin WF_A(v_j)$. Thus there is a positive ε_0 such that

$$(x, \xi) \notin WF_A(f) \quad \text{if } |x' - x'_0| + |\xi' - \xi'_0| < \varepsilon_0 \text{ and } x \in X.$$

If $\varepsilon_1 < \varepsilon_0$ is sufficiently small and M is sufficiently large we also have

$$P_m(x, \xi) \neq 0 \quad \text{when } |x' - x'_0| + |\xi' - \xi'_0| < \varepsilon_1, \quad |x_n| < \varepsilon_1, |\xi_n| > M,$$

for X_0 is non-characteristic. Hence $(x, \xi) \notin WF_A(u_0)$ for such x, ξ by Theorem 9.5.1. When $x_n = 0$ this remains true for all ξ_n by Corollary 9.6.8. Since $WF_A(u_0)$ is closed we can find $\varepsilon < \varepsilon_1$ so small that

$$(x, \xi) \notin WF_A(u_0) \quad \text{if } |x' - x'_0| + |\xi' - \xi'_0| < \varepsilon, \quad |x_n| < \varepsilon, |\xi_n| \leqq M.$$

This completes the proof since $u_0 = u$ in X_+.

Remark. The analogue of Theorem 9.6.9 with WF_A replaced by WF is false even when P has constant coefficients. In fact, if $P(D) = D_1 D_2 + D_3^2 + D_4^2$ we can by Example 8.3.4 choose $u_0 \in C^2$ so that $P(D)u_0 = 0$ and $\operatorname{sing\,supp} u_0$ is the x_1 axis. Then $u = \sum_1^\infty a_j u_0(x_1, x_2, x_3, x_4 - 1/j)$ is in $C^2(\mathbb{R}^4)$, u and $D_4 u$ are in $C^\infty(\mathbb{R}^3)$ when $x_4 = 0$, and $\operatorname{sing\,supp} u = \{(x_1, 0, 0, x_4);\ x_4 = 0 \text{ or } 1/x_4 \in \mathbb{Z}_+\}$ if a_j converges to 0 sufficiently rapidly. Since $P(D)u = 0$ we have a counter-example to the C^∞ analogue of Theorem 9.6.9. (A very general version of this example can be obtained from Theorem 11.3.1.)

Notes

The Fourier transform of a function f on $\mathbb{R}$ is the sum of boundary values of the analytic functions F_+ and F_- in the upper and lower half planes defined by the Fourier-Laplace transforms of f on $\mathbb{R}_-$ and $\mathbb{R}_+$ respectively. (See also a discussion after Theorem 7.1.5.) These are well defined if say $|f(x)|=O(e^{\varepsilon|x|})$ for every $\varepsilon>0$. Carleman [2] used this observation to define generalized Fourier transforms. However, taking boundary values in a classical sense requires additional limitations on the growth of f (see Beurling [2, 3]). One is therefore led to define boundary values in an abstract sense by introducing the space of pairs (F_+, F_-) of functions $F_\pm$ analytic when $\operatorname{Im} z \gtrless 0$ modulo the pairs $(F, -F)$ where F is entire. This is how Sato [1] first defined hyperfunctions on $\mathbb{R}$, and Sato [2] later extended the idea to $\mathbb{R}^n$. This is technically cumbersome since an invariant setup involves a relative cohomology group of degree n. It was pointed out by Martineau [1] that a fairly elementary presentation is obtained if one starts instead from the notion of analytic functional. The crucial existence of a unique support for a functional carried by a compact set in $\mathbb{R}^n$ was proved by Martineau using quite elementary facts on the cohomology of the sheaf of germs of holomorphic functions. While we follow Martineau on the whole in Sections 9.1 and 9.2 we have eliminated even these prerequisites by using the simple observation that solving a Dirichlet problem gives an isomorphism between analytic functionals and certain harmonic functions. The latter can be studied with the methods developed in the course of the preceding chapters. The advantage is that we just have to study one equation rather than the overdetermined Cauchy-Riemann system in several complex variables.

After the basic definitions in Sections 9.1 and 9.2 we study $WF_A(u)$ for hyperfunctions in Section 9.3. As already mentioned in the notes to Chapter VIII this notion was first introduced by Sato [3]. The proofs in Sections 8.4 and 8.5 were chosen so that they can be used with small modifications in the case of hyperfunctions. A crucial technical tool is Theorem 9.3.7, the second part of which is the edge of the wedge theorem of Martineau [2].

In Section 9.4 we leave hyperfunctions momentarily to study the analytic Cauchy problem. After a proof of a local existence theorem along the same lines as in the predecessor of this book, we give refinements concerning the existence domain taken from Leray [2], Zerner [3] and particularly Bony and Schapira [1]. This prepares for the proof by Bony and Schapira [1] given in Section 9.5 of the non-characteristic regularity theorem due to Sato [3]. We also define Cauchy data of hyperfunction solutions of differential equations on non-characteristic surfaces following Komatsu [2] and Schapira [1].

Section 9.6 presents the Bros-Iagolnitzer definition of $WF_A(u)$ in the spirit of Sjöstrand [1]. The proof of the Kashiwara theorem (Theorem 9.6.6) is also essentially taken from Sjöstrand [1]. For the application in Theorem 9.6.9 see also Schapira [3], Sjöstrand [2]. A very broad survey of analytic regularity theory can be found in Sjöstrand [1, 2].

The aim of this chapter has just been to give an introduction to hyperfunction theory which follows Schwartz distribution theory as closely as possible. The reader who wants to study the subject in depth should of course turn to the basic paper by Sato-Kawai-Kashiwara [1]. It may then be useful to consult also the introductions given by Kashiwara [1] and Cerezo-Chazarain-Piriou [1].

Exercises

In the following exercises some standard notation is used without explanation. In particular, H denotes the Heaviside function, the characteristic function of the positive real axis, and δ_a denotes the Dirac measure at a. The Fourier transform of u normalized as in Section 7.1 is denoted by $\hat{u}$.

Most of the exercises are intended to train the student in the routine use of the tools developed in the text. A few are extensions of the theory presented there. As a rule a rather complete though brief discussion is then given in the answers and hints following the exercises.

Chapter I

Exercise 1.1. Let $f \in C^\infty(\mathbf{R})$ be an even function. Prove that there is a function $g \in C^\infty(\mathbf{R})$ such that $f(x) = g(x^2)$.

Exercise 1.2. Show that every $f \in C^\infty(\mathbf{R})$ can be written in the form

$$f(x) = g_0(x^2) + x g_1(x^2)$$

with g_0 and g_1 in $C^\infty(\mathbf{R})$.

Exercise 1.3. Show that when $f \in C^\infty(\mathbf{R}^n)$ one can find g_0, g_1 in $C^\infty(\mathbf{R}^n)$ such that

$$f(x) = g_0(x_1^2, x_2, \dots, x_n) + x_1 g_1(x_1^2, x_2, \dots, x_n).$$

Exercise 1.4. Show that when $f \in C^\infty(\mathbf{R}^n)$ one can find a decomposition

$$f(x) = \sum_{1 \le i_1 < \dots < i_k \le n} x_{i_1} \cdots x_{i_k} g_{i_1 \dots i_k}(x_1^2, \dots, x_n^2)$$

with $g_{i_1 \dots i_k} \in C^\infty(\mathbf{R}^n)$ and k even (odd) if f is even (odd).

Exercise 1.5. Show that when $f \in C^\infty(\mathbf{R}^2)$ and $f(x_1, x_2) \equiv f(x_2, x_1)$, then one can find $g \in C^\infty(\mathbf{R}^2)$ such that

$$f(x_1, x_2) = g(x_1 + x_2, x_1 x_2).$$

Exercise 1.6. Show that there exist numbers a_k and b_k, $k = 0, 1, \dots$ such that

$$\text{(i)} \quad b_k < 0 \qquad \text{(ii)} \quad \sum_{k=0}^{\infty} |a_k b_k^n| < \infty, \quad n = 0, 1, \dots$$

$$\text{(iii)} \quad \sum_{k=0}^{\infty} a_k b_k^n = 1, \quad n = 0, 1, \dots \qquad \text{(iv)} \quad b_k \to -\infty \text{ as } k \to \infty.$$

Show that if $f \in C^\infty(\overline{\mathbf{R}}_+)$ and $g \in C^\infty(\mathbf{R})$ is chosen equal to 1 in $(-\infty, 1)$, equal to 0 in $(2, \infty)$, then

$$f(t) = \sum_{0}^{\infty} a_k g(b_k t) f(b_k t), \quad t < 0,$$

gives an extension of f which is in $C^\infty(\mathbf{R})$. Show that the extension is in $C^\nu(\mathbf{R})$ if $f \in C^\nu(\overline{\mathbf{R}}_+)$.

Exercise 1.7. f is a locally bounded real valued function on $\mathbf{R}$ with $f(x) = 0$ when $x < 0$ and $f(x) = O(x^N)$ for every $N > 0$ as $x \to 0$. Show that there is a function $F \in C^\infty(\mathbf{R})$ such that $F(x) = 0$, $x < 0$, and $f(x) \le F(x)$, $x \in \mathbf{R}$.

Chapter II

Exercise 2.1. For which $b \in \mathbf{C}$ does there exist a distribution $u \in \mathscr{D}'^k(\mathbf{R})$ with restriction $x \mapsto x^b$ to $\mathbf{R}_+$?

Exercise 2.2. Does there exist a distribution u on $\mathbf{R}$ with the restriction $x \mapsto e^{1/x}$ to $\mathbf{R}_+$?

Exercise 2.3. For which $a > 0$ and $b \in \mathbf{C}$ does there exist a distribution $u \in \mathscr{D}'^k(\mathbf{R})$ with restriction $x \mapsto e^{ix^{-a}} x^b$ to $\mathbf{R}_+$?

Exercise 2.4. Show that for every $f \in C^1(\mathbf{R}_+)$ one can find a real valued function $g \in C^1(\mathbf{R}_+)$ and a distribution $u \in \mathscr{D}'^1(\mathbf{R})$ with restriction $f e^{ig}$ to $\mathbf{R}_+$!

Exercise 2.5. Determine $\lim_{t\to+\infty} f_t$ in $\mathscr{D}'(\mathbf{R})$ when

$$\text{a)}\quad f_t(x) = t/(1+t^2x^2); \qquad \text{b)}\quad f_t(x) = t^{-\frac{1}{2}}e^{-x^2/4t}.$$

Exercise 2.6. Determine the following limits in $\mathscr{D}'(\mathbf{R})$:

$$\text{a)}\ \lim_{t\to\infty} t^2 x \cos tx \quad \text{b)}\ \lim_{t\to\infty} t^2|x| \cos tx \quad \text{c)}\ \lim_{t\to+\infty} x^{-1}\sin tx.$$

Exercise 2.7. Find the limit of $f_t(x) = te^{itx}\log|x|$ in $\mathscr{D}'(\mathbf{R})$ as $t \to +\infty$.

Exercise 2.8. For which values of $a \in \mathbf{R}$ is it true that the functions $f_t(x) = t^a \sin(tx)$ converge to 0 as $t \to +\infty$ a) as C^k functions; b) as distributions?

Exercise 2.9. Let $u_j \in C^1(X)$ where X is an open set in $\mathbf{R}^n$, and assume that for every compact set $K \subset X$ there is a constant C_K such that $|u_j'| \le C_K$ on K for every j. Show that if $u_j \to u$ in $\mathscr{D}'(X)$ then $u \in C(X)$ and $u_j \to u$ uniformly on every compact set in X.

Exercise 2.10. Determine a number a and a distribution $u \ne 0$ such that

$$u_\alpha(t) = (1-\cos t)^\alpha, \quad t \in \mathbf{R},$$

is locally integrable when $\alpha > a$ and $(\alpha - a)u_\alpha \to u$ in $\mathscr{D}'(\mathbf{R})$ when $\alpha \to a+0$. Show that the limit

$$v = \lim_{\alpha\to+\infty} 2^{-\alpha}\alpha^{-a}u_\alpha$$

exists and determine it.

Exercise 2.11. Prove that if $f(x) = (x_1\cdots x_n + i\varepsilon)^{-1} + (x_1\cdots x_n - i\varepsilon)^{-1}$, $x \in \mathbf{R}^n$, then $\lim_{\varepsilon\to 0} f_\varepsilon$ exists in $\mathscr{D}'(\mathbf{R}^n)$.

Exercise 2.12. Put $f_w(x) = 1/(x^4+w)$, $x \in \mathbf{R}$, where w belongs to the open right half plane H in $\mathbf{C}$. Determine homogeneous functions $a_j(w)$ in H, $0 \le j \le 2$, such that

$$f_w - \sum_0^2 a_j(w)\delta^{(j)}$$

has a limit in $\mathscr{D}'(\mathbf{R})$ as $H \ni w \to 0$.

Exercise 2.13. Put $\chi_\varepsilon(x) = \chi(x/\varepsilon)/\varepsilon$ where $\chi \in C_0^\infty(\mathbf{R})$ and $\int \chi\,dx = 1$. Determine a constant C and a distribution u such that the distribution $K_\varepsilon = \chi_\varepsilon(xy) + C\log\varepsilon\,\delta_0(x,y) \to u$ when $\varepsilon \to +0$.

Exercise 2.14. Define $u_{a,\varepsilon}(x) = \varepsilon/(\varepsilon^2 + (x - a\sqrt{\varepsilon})^2)$, $x \in \mathbf{R}$, where $a \in \mathbf{R}$ and $\varepsilon > 0$. Determine $\lim_{\varepsilon\to 0} u_{a,\varepsilon}$ and $\lim_{\varepsilon\to 0} u_{a,\varepsilon}u_{b,\varepsilon}$ when they exist in $\mathscr{D}'(\mathbf{R})$.

Exercise 2.15. Determine the limit in $\mathscr{D}'(\mathbf{R}^2 \setminus 0)$ as $t \to +\infty$ of

$$f_t(x) = t \sin(t||x|^2 - 1|), \quad x \in \mathbf{R}^2.$$

Does the limit exist in $\mathscr{D}'(\mathbf{R}^2)$?

Exercise 2.16. Set $f_{t,\varepsilon}(x) = e^{-itx}(x + i\varepsilon)^{-1}$, $x \in \mathbf{R}$, and determine the following limits in $\mathscr{D}'(\mathbf{R})$:

$$\text{a)} \quad \lim_{\varepsilon\to+0}\big(\lim_{t\to+\infty} f_{t,\varepsilon}\big) \qquad \text{b)} \quad \lim_{t\to+\infty}\big(\lim_{\varepsilon\to+0} f_{t,\varepsilon}\big).$$

Section 3.1

Exercise 3.1.1. Let $f \in \mathscr{D}'(I)$ where I is an open interval on $\mathbf{R}$. Show that there is a solution $u \in \mathscr{D}'(I)$ of the differential equation $u' = f$, and that the difference between two such primitive distributions is a constant.

Exercise 3.1.2. Let $u \in \mathscr{D}'(I)$, where I is an open interval $\subset \mathbf{R}$. Show that if u has order $k > 0$, then u' has order $k + 1$.

Exercise 3.1.3. Let $u \in \mathscr{D}'(I)$ where I is a finite open interval $\subset \mathbf{R}$. Show that if u is the restriction to I of a distribution of order k in a neighborhood of $\overline{I}$, then

$$|u(\varphi)| \le C \sum_{j \le k} \sup |\varphi^{(j)}|, \quad \varphi \in C_0^\infty(I),$$

and that conversely this estimate implies that there is a measure $d\mu$ on $\mathbf{R}$ with support in $\overline{I}$ such that u is the restriction to I of its kth derivative.

Exercise 3.1.4. Show that if f is a measurable function in $(-1, 1)$ and

$$\int_{-1}^{1} (1 - x^2)^m |f(x)|\, dx < \infty,$$

where m is a positive integer, then there is a distribution $F \in \mathscr{E}'^m([-1, 1])$ with restriction f to $(-1, 1)$.

Exercise 3.1.5. Does there exist a distribution $u \in \mathscr{D}'(\mathbf{R})$ which restricts to the function $x \mapsto e^{1/x} \exp(ie^{1/x})$, $x > 0$?

Exercise 3.1.6. For which $a \in \mathbf{R}$ does there exist a distribution $u \in \mathscr{D}'^k(\mathbf{R})$ with the restriction $x \mapsto e^{1/x} \exp(ie^{a/x})$ to $\mathbf{R}_+$?

Exercise 3.1.7. Prove that the limit

$$\langle \mathrm{vp}(1/x), \varphi \rangle = \lim_{\varepsilon \to 0} \int_{|x|>\varepsilon} \varphi(x)\, dx/x, \quad \varphi \in C_0^\infty(\mathbf{R}),$$

exists. What is the order of the distribution $\mathrm{vp}(1/x)$?

Exercise 3.1.8. f is an odd locally integrable function on $\mathbf{R}$ such that $xf(x) \to 1$ as $x \to \infty$. Prove that $g_t(x) = t^2 f'(tx)$ has a limit in $\mathscr{D}'(\mathbf{R})$ and determine it.

Exercise 3.1.9. Determine real numbers a_1, a_2 such that the integral

$$u(\varphi) = \int_0^\infty \big(a_1(\varphi(x) - \varphi(-x)) + a_2(\varphi(2x) - \varphi(-2x))\big)\, dx/x^3$$

exists when $\varphi \in C_0^\infty(\mathbf{R})$ and $u = d^2(\mathrm{vp}(1/x))/dx^2$.

Exercise 3.1.10. Define $\log(x + i0) = \log|x| + \pi i H(-x)$, $x \in \mathbf{R}$, calculate the derivative, and compare it with $\mathrm{vp}(1/x)$.

Exercise 3.1.11. Show that the function

$$x \mapsto \frac{e^x}{(x + i\varepsilon)^2} + \frac{e^{-x}}{(x - i\varepsilon)^2} - \frac{2(x^2 - \varepsilon^2)}{(x^2 + \varepsilon^2)^2}$$

has a limit in $\mathscr{D}'(\mathbf{R})$ when $\varepsilon \to 0$ and give it in a simple form.

Exercise 3.1.12. Compute the nth derivative of $x \mapsto f(|x|)$ when $f \in C^n(\overline{\mathbf{R}}_+)$.

Exercise 3.1.13. Compute the nth derivative of $x \mapsto |x|f(x)$ when $f \in C^n(\mathbf{R})$.

Exercise 3.1.14. Let I be an open interval $\subset \mathbf{R}$, and let $a \in I$. a) Show that for every $f \in \mathscr{D}'(I)$ there is a solution $u \in \mathscr{D}'(I)$ to the equation $(x - a)u = f$, and that two solutions of this division problem differ by a multiple of δ_a. b) Give a solution when $f = \delta_a^{(j)}$.

Exercise 3.1.15. Show that if I is an open interval on $\mathbf{R}$ and $F \in C^\infty(I)$ has no zero of infinite order, then the equation $Fu = g$ has a solution $u \in \mathscr{D}'(I)$ for every $g \in \mathscr{D}'(I)$. Describe the solutions when $g = 0$.

Exercise 3.1.16. Show that if $F \in C^\infty(\mathbf{R})$ and the equation $Fu = 1$ has a solution $u \in \mathscr{D}'(\mathbf{R})$, then F cannot have a zero of infinite order.

Exercise 3.1.17. Compute $x^j \delta_0^{(k)}$ for all integers $j \geq 0$, $k \geq 0$.

Exercise 3.1.18. Compute $f\delta_0^{(k)}$ when $f \in C^k(\mathbf{R})$.

Exercise 3.1.19. Determine all primitive distributions of a) $H(x)$; b) $xH(x)$; c) $e^x H(x)$; d) δ_0; e) $\mathrm{vp}(1/x)$.

Exercise 3.1.20. Determine all $u \in \mathscr{D}'(\mathbf{R})$ satisfying the equations

a) $xu' = \delta_0$,	b) $xu' + u = 0$,	c) $x^2u' + u = 0$,
d) $x^2u' + xu = \delta_0$,	e) $(x-1)u = \delta_0'$,	f) $(x^2-1)u = \delta_0$,
g) $(\exp(2\pi i x) - 1)u = 0$,	h) $u'' = \delta_0' - 2\delta_1$,	i) $xu' = \delta_1 - \delta_{-1}$,
j) $(x+1)^3u' + u = \delta$,	k) $x^4u' + u = 0$,	l) $\cos^2 x\, u' = 0$.

Exercise 3.1.21. Determine all distributions in $\mathbf{R}^n$ with $x_n^N u = 0$ where N is a fixed positive integer.

Exercise 3.1.22. Determine all distributions in $\mathbf{R}^2$ with $(x_1^2 - x_2^2)u = 0$ and $x_1x_2u = 0$.

Exercise 3.1.23. Let $\varphi, \varphi \in C^\infty(\mathbf{R}^n)$ be real valued, $\partial\varphi/\partial x_1 \neq 0$ when $\varphi = 0$. When does $u \in \mathscr{D}'(\mathbf{R}^n)$, $\varphi u = 0$, $\varphi \partial u/\partial x_1 = 0$ imply $u = 0$?

Exercise 3.1.24. Show that $u_N = \lim_{\varepsilon\to+0}(x_1^2 + \varepsilon + ix_2)^{-N}$ exists in $\mathscr{D}'(\mathbf{R}^2)$ for every integer $N > 0$. Calculate $f_N = \partial u_N/\partial x_1 + 2ix_1\partial u_N/\partial x_2$, determine the order of u_N and $\operatorname{sing\,supp} u_N$.

Exercise 3.1.25. Assume that $0 \leq g \in C^\infty(\mathbf{R}^2)$ and set $u_\varepsilon(x) = \varepsilon(g(x) - i\varepsilon)^{-2}$, $x \in \mathbf{R}^2$, $\varepsilon > 0$. Prove that $\lim_{\varepsilon\to 0} u_\varepsilon$ exists in $\mathscr{D}'(\mathbf{R}^2)$ if (and only if)

$$g(x) = 0 \implies \det g''(x) > 0;\ \text{where } g'' = (\partial^2 g/\partial x_j \partial x_k)_{j,k=1}^2.$$

What is the limit?

Exercise 3.1.26. Let f and g be real valued functions in $C^\infty(\mathbf{R}^n)$, $n \geq 2$, and assume that $f'(x) \neq 0$, $g'(x) \neq 0$, $x \in \mathbf{R}^n$. Show that $\lim_{\varepsilon\to+0} H(f(x))/(g(x) + i\varepsilon)$ exists in $\mathscr{D}'(\mathbf{R}^n)$ if $\log|g(x)|$ is locally integrable on the surface $f^{-1}(0)$.

Exercise 3.1.27. Set $s_n(z) = 1 + z + \cdots + z^n$, $z \in \mathbf{C} = \mathbf{R}^2$. Show that $u = \lim_{n\to\infty} n^{-1} \log|s_n(z)|$ exists in $\mathscr{D}'(\mathbf{R}^2)$, and calculate u and Δu.

Exercise 3.1.28. Let u be the characteristic function of the unit disc in $\mathbf{R}^2$. Calculate $x\partial u/\partial x + y\partial u/\partial y$.

Exercise 3.1.29. u is the characteristic function of the unit disc in $\mathbf{R}^2$. Determine the order of the distributions

a) $\partial u/\partial x$, b) $\partial^2 u/\partial x \partial y$, c) $x^2 \partial^2 u/\partial y^2 + y^2 \partial^2 u/\partial x^2 - 2xy \partial^2 u/\partial x \partial y$.

Exercise 3.1.30. For which positive numbers a and b are the first derivatives of the characteristic function u of

$$\{(x, y) \in \mathbf{R}^2; y < x^a \sin(x^{-b}), x > 0\}$$

of order 0?

Exercise 3.1.31. Put $F(x, y) = x^2 + f(y)$, where $0 \leq f \in C^\infty(\mathbf{R})$. What is the condition for the existence of a distribution $u \in \mathscr{D}'(\mathbf{R}^2)$ with $Fu = 1$.

Exercise 3.1.32. Determine all real valued functions $\varphi \in C^1(\mathbf{R}_+)$ such that

$$u(t, x) = \begin{cases} x/t, & \text{when } x \leq \varphi(t) \\ -1, & \text{when } x > \varphi(t) \end{cases}$$

satisfies the equation $\partial u/\partial t + \partial(u^2/2)/\partial x = 0$ in the distribution sense when $t > 0$.

Exercise 3.1.33. Set $f(z) = (z-1)^{-3} \log z$, where $-\pi \leq \operatorname{Im} \log z < \pi$. Determine the limit of $x \mapsto f(x + i\varepsilon) - f(x - i\varepsilon)$ in $\mathscr{D}'(\mathbf{R})$ as $\varepsilon \to +0$.

Section 3.2

Exercise 3.2.1. For $a \in \mathbf{C}$ define $Z(a) = \{u \in \mathscr{D}'(\mathbf{R}); xu' = au\}$ and prove

(i) If $u \in Z(a)$ then $|x|^{-a}u$ is a constant in $\mathbf{R}_+ \setminus 0$ and in $\mathbf{R}_- \setminus 0$.
(ii) If $u \in Z(a)$ and $\operatorname{supp} u = \{0\}$, then $a = -j - 1$ and $u = C\delta_0^{(j)}$ for some integer $j \geq 0$.
(iii) If $u \in Z(a)$ then $xu \in Z(a+1)$ and $u' \in Z(a-1)$.
(iv) The maps $\frac{d}{dx} : Z(a) \to Z(a-1)$ and $\frac{x}{a} : Z(a-1) \to Z(a)$ are bijective and each other's inverses if $a \neq 0$, while $\frac{d}{dx}$ maps $Z(0)$ to the multiples of δ_0 and x maps $Z(-1)$ to the constants.
(v) The dimension of $Z(a)$ is equal to 2.
(vi) If $0 \neq u \in Z(a)$ then the order of u is the smallest integer $k \geq 0$ such that $k + \operatorname{Re} a + 1 > 0$, unless $\operatorname{supp} u = \{0\}$ and the order is $-a - 1$.

Exercise 3.2.2. For $a \in \mathbf{C}$ and a positive integer k define

$$Z(a, k) = \{u \in \mathscr{D}'(\mathbf{R}); (xd/dx - a)^k u = 0\}.$$

Prove that

(i) If $u \in Z(a,k)$ then $|x|^{-a}u$ is a polynomial in $\log|x|$ of degree $< k$ on $\mathbf{R}_+$ and on $\mathbf{R}_-$.

(ii) If $u \in Z(a,k)$ and $\operatorname{supp} u = \{0\}$, then $a = -j-1$ and $u = C\delta_0^{(j)}$ where j is a non-negative integer.

(iii) If $u \in Z(a,k)$ then $xu \in Z(a+1,k)$ and $u' \in Z(a-1,k)$.

(iv) The composition of the maps $\frac{d}{dx} : Z(a,k) \to Z(a-1,k)$ and $x : Z(a-1,k) \to Z(a,k)$ is the sum of a nilpotent map and a times the identity, so the maps are bijective if $a \neq 0$. $x\frac{d}{dx}$ maps $Z(0,k)$ onto $Z(0,k-1)$, $\frac{d}{dx}Z(0,k) = \{v \in \mathscr{D}'(\mathbf{R}); xv \in Z(0,k-1)\}$; $\frac{d}{dx}x$ maps $Z(-1,k)$ onto $Z(-1,k-1)$, and $xZ(-1,k) = \{u \in \mathscr{D}'(\mathbf{R}); u' \in Z(-1,k-1)\}$.

(v) $\dim Z(a,k) = 2k$, and $\frac{d}{dx}Z(0,k) \subset Z(-1,k) \subset \frac{d}{dx}Z(0,k+1)$ with codimension 1 in each inclusion.

(vi) If $0 \neq u \in Z(a,k)$ then the order is the smallest integer $m \geq 0$ such that $m + \operatorname{Re} a + 1 > 0$ unless $\operatorname{supp} u = \{0\}$ and the order is $-a-1$.

Exercise 3.2.3. Determine the dimension of the space of solutions $u \in \mathscr{D}'(\mathbf{R})$ of the differential equation

$$\sum_0^m a_j(xd/dx)^j u = 0,$$

where a_j are constants and $a_m \neq 0$.

Exercise 3.2.4. Show that if $u \in \mathscr{D}'^k(\mathbf{R}^n)$ and the restriction of u to $\mathbf{R}^n \setminus 0$ is homogeneous of degree a and not $\equiv 0$, then $k + \operatorname{Re} a + n > 0$.

Exercise 3.2.5. Determine the order and the degree of homogeneity of the distributions

$$u(\varphi) = \int_0^\infty \int_0^\infty (\varphi(x,y) - \varphi(-x,y) - \varphi(x,-y) + \varphi(-x,-y))\, dx\, dy/xy, \quad \varphi \in C_0^\infty(\mathbf{R}^2),$$

$$v(\varphi) = \int_0^\infty (\varphi_y'(x,0) - \varphi_y'(-x,0))\, dx/x, \quad \varphi \in C_0^\infty(\mathbf{R}^2).$$

Section 3.3

Exercise 3.3.1. Calculate $\mu = \Delta \log|f|$ when f is a meromorphic function in a connected open set $Z \subset \mathbf{C} = \mathbf{R}^2$ and $f \not\equiv 0$.

Exercise 3.3.2. Assuming that $A(x) = a_{11}x_1^2 + 2a_{12}x_1x_2 + a_{22}x_2^2 \neq 0$ and that $\operatorname{Re} A(x) \geq 0$ when $0 \neq x \in \mathbf{R}^2$, determine a constant C such that $C \log A(x)$ is a fundamental solution of $a_{22}\partial_1^2 - 2a_{12}\partial_1\partial_2 + a_{11}\partial_2^2$.

Exercise 3.3.3. Determine a fundamental solution E of Δ^2 in $\mathbf{R}^n$, $n > 2$.

Exercise 3.3.4. Compute Δu where $u(x) = e^{a|x|}/|x|$, $x \in \mathbf{R}^3$; $a \in \mathbf{C}$.

Exercise 3.3.5. Compute Δu when $u(x) = (\sin|x|)/|x|$, $x \in \mathbf{R}^3$.

Exercise 3.3.6. Determine a fundamental solution of $\Delta + a^2$ in $\mathbf{R}^3$.

Exercise 3.3.7. Let $A \in C^1(\mathbf{R})$ and $A(x) \neq 0$ if $x \neq 0$. Prove that

$$E(\varphi) = \lim_{\varepsilon \to 0} \iint_{|x|>\varepsilon} \varphi(x,y)/(A(x) + iy)\, dxdy$$
$$= \int dx(\int \varphi(x,y)/(A(x) + iy)\, dy)$$

exists when $\varphi \in C_0^1(\mathbf{R}^2)$ and that $E \in \mathscr{D}'^1(\mathbf{R}^2)$. Calculate $f = \partial E/\partial x + iA'(x)\partial E/\partial y$.

Exercise 3.3.8. Prove for every positive integer N the existence of the limit

$$u_N(\varphi) = \lim_{\varepsilon \to 0} \iint_{|x+iy|>\varepsilon} (x+iy)^{-N} \varphi(x,y)\, dx\, dy, \quad \varphi \in \mathbf{C}_0^\infty(\mathbf{R}^2),$$

and that $u_N \in \mathscr{D}'(\mathbf{R}^2)$ is homogeneous of degree $-N$. Calculate $f = \partial u_N/\partial x + i\partial u_N/\partial y$.

Exercise 3.3.9. Show that if f is an analytic function in an open connected set $Z \subset \mathbf{C} = \mathbf{R}^2$ and $f \neq 0$, then the limit

$$u(\varphi) = \lim_{\varepsilon \to 0} \iint_{|f(x+iy)|>\varepsilon} \frac{\varphi(x,y)}{f(x+iy)}\, dx\, dy, \quad \varphi \in C_0^\infty(Z),$$

exists and defines a distribution with $fu = 1$.

Exercise 3.3.10. Find a fundamental solution E of ∂^α in $\mathbf{R}^n$ with support in the first quadrant, when all α_j are positive.

Exercise 3.3.11. Determine a constant C such that

$$E(x,t) = C \text{ when } c^2t^2 - x^2 \geq 0,\ t \geq 0; \qquad E(x,t) = 0 \text{ otherwise};$$

is a fundamental solution of the wave operator $c^{-2}\partial^2/\partial t^2 - \partial^2/\partial x^2$.

Exercise 3.3.12. Determine an entire analytic function F such that

$$E(x, y) = F(cxy) \text{ when } x \geq 0, y \geq 0; \qquad E(x, y) = 0 \text{ otherwise};$$

is a fundamental solution of the differential operator $\partial^2/\partial x \partial y - c$ where $c \in \mathbf{C}$.

Exercise 3.3.13. In $\mathbf{R}^4$ with coordinates denoted (t, x), $t \in \mathbf{R}$ and $x \in \mathbf{R}^3$, let u be the characteristic function of the light cone $\{(t, x); t > |x|\}$, and calculate $v = \Box u$, $w = \Box v$, where $\Box = \partial_t^2 - \Delta$ is the wave operator.

Section 4.1

Exercise 4.1.1. Show that if $u_\varepsilon = \operatorname{sgn} t\, \chi_+^a(t^2 - \varepsilon^2)$, $0 \neq \varepsilon \in \mathbf{R}$, then $\lim_{\varepsilon \to 0} u_\varepsilon$ exists in $\mathscr{D}'(\mathbf{R})$ for any $a \in \mathbf{C}$. (The distributions χ_+^a were defined in Section 3.2. The definition of the composition is obvious when $\operatorname{Re} a > -1$ and is given by analytic continuation otherwise.) Calculate the limit when a is a negative integer.

Exercise 4.1.2. Let u be subharmonic in $\{z \in \mathbf{C}; |z| < R\}$ and set $\mu = \Delta u$. Prove that

$$\int_0^{2\pi} u(re^{i\theta})\, d\theta - 2\pi u(0) = \int_{|z|<r} \log \frac{r}{|z|}\, d\mu(z), \quad 0 < r < R.$$

Exercise 4.1.3. Calculate $\Delta |\operatorname{Im} \sqrt{f(z)}\,|$ where $f(z) = z^2 + a$, $a \in \mathbf{R}$.

Exercise 4.1.4. Let $f_n(z) = \max \operatorname{Re} w$ taken over all $w \in \mathbf{C}$ with $w^n = z$, where n is a positive integer. Calculate $\Delta f_n(z)$.

Section 4.2

Exercise 4.2.1. Calculate $f * f * \cdots * f$ (with n factors) if a) $f(t) = H(t)$ b) $f(t) = e^{-t} H(t)$.

Exercise 4.2.2. Calculate $\delta_0^{(k)} * H$ where k is a positive integer.

Exercise 4.2.3. Let f_a be the characteristic function of $(0, a) \subset \mathbf{R}$, where $a > 0$. Determine a distribution u_a with support on $\overline{\mathbf{R}}_+$ such that $f_a * u_a = \delta_0$.

Exercise 4.2.4. Calculate $(1 * \delta_0') * H$ and $1 * (\delta_0' * H)$.

Exercise 4.2.5. Prove that $(\delta_h * u - u)/h \to -u'$ in $\mathscr{D}'(\mathbf{R})$ as $h \to 0$, if $u \in \mathscr{D}'(\mathbf{R})$.

Exercise 4.2.6. Recall that the distributions $\chi_+^\lambda \in \mathscr{D}'(\mathbf{R})$ depend analytically on $\lambda \in \mathbf{C}$ and are defined by $\chi_+^\lambda(x) = x^\lambda/\Gamma(\lambda+1)$, $x > 0$, $\chi_+^\lambda(x) = 0$, $x \le 0$, when $\operatorname{Re}\lambda > -1$. Determine $\chi_+^\lambda * \chi_+^\mu$ for arbitrary $\lambda, \mu \in \mathbf{C}$.

Exercise 4.2.7. Solve Abel's integral equation $\chi_+^\lambda * u = f$ where f is a given distribution with support on $\overline{\mathbf{R}}_+$ and the solution u is also required to have its support there.

Exercise 4.2.8. u and v are the surface measures on the spheres $\{x; |x| = a\}$ and $\{x; |x| = b\}$ in $\mathbf{R}^3$. Compute the convolution $u * v$ and determine its singular support.

Section 4.3

Exercise 4.3.1. Show that $\operatorname{supp}(u * v) = \operatorname{supp} u + \operatorname{supp} v$ if u and v are positive measures in $\mathbf{R}^n$, one of which has compact support.

Exercise 4.3.2. If $u, v \in \mathscr{E}'(\mathbf{R}^n)$ and $u * v = 0$, it follows from the theorem of supports that $u = 0$ or $v = 0$. Is this true if only one of the factors u and v has compact support?

Exercise 4.3.3. Show that if $u, v \in \mathscr{E}'(\mathbf{R}^n)$ and $\operatorname{supp} u * v$ is contained in an affine subspace V of $\mathbf{R}^n$, then the supports of u and of v are contained in affine subspaces parallel to V.

Exercise 4.3.4. Let u be the characteristic function of the square in $\mathbf{R}^2$ defined by $|x_1| < 1$, $|x_2| < 1$, and let $f = P(\partial)u$ where $P(\partial)$ is a differential operator with constant coefficients. Describe the possible sets $\operatorname{supp} f$ which can occur and the corresponding polynomials.

Section 4.4

Exercise 4.4.1. Calculate $\mu = \Delta \log|f(z)|$ where f is analytic outside $[-1, 1]$ and $f(z)^2 - 2zf(z) + 1 = 0$, $f(2) > 1$. Calculate $\mu * E$ where $E(z) = (2\pi)^{-1} \log|z|$.

Exercise 4.4.2. Set $f(x) = |x|^{-5} \sum_{j,k=1}^3 a_{jk}\, x_j x_k$, $x \in \mathbf{R}^3 \setminus 0$, where (a_{jk}) is a constant symmetric matrix. a) What is the condition for the existence of the limit

$$F(\varphi) = \lim_{\varepsilon \to 0} \int_{|x|>\varepsilon} f(x)\varphi(x)\, dx$$

for arbitrary $\varphi \in C_0^\infty(\mathbf{R}^3)$? b) Calculate ΔF when this condition is fulfilled and prove that $F = E * \Delta F$ where $E(x) = -1/(4\pi|x|)$.

Exercise 4.4.3. V is a function on $\mathbf{R}^3$ such that $V(x) \to 0$ as $x \to \infty$, and V is harmonic outside a compact set. Show that $-4\pi|x|V(x) \to \langle \Delta V, 1\rangle$, $x \to \infty$. Is the hypothesis $V(x) \to 0$ essential?

Exercise 4.4.4. Let $u \in \mathscr{D}'(\mathbf{R})$ and assume that for some integer $k \geq 0$ we have $u * f * f \in C^\infty(\mathbf{R})$ for every $f \in C_0^k(\mathbf{R})$. Show that $u \in C^\infty(\mathbf{R})$.

Exercise 4.4.5. Let u be a distribution in $\mathbf{R}^n$ with compact support, and assume that $f_k = (\partial_1 \dots \partial_n)^k u$ is a continuous function for $k = 1, 2, 3, \dots$. Show that $u \in C_0^\infty(\mathbf{R}^n)$.

Exercise 4.4.6. Show that a differential equation $P(d/dx)u = f$, where $f \in \mathscr{E}'(\mathbf{R})$ and P is a polynomial, has a solution $u \in \mathscr{E}'(\mathbf{R})$ if and only if $\langle f, \varphi\rangle = 0$ for every solution φ of the adjoint differential equation $P(-d/dx)\varphi = 0$.

Exercise 4.4.7. Let χ be the characteristic function of $(-1, 1)$. Determine a number $a \in (0, 1)$ and a function $u \in C_0^2(\mathbf{R})$ such that

$$\chi - \delta_a - \delta_{-a} = d^4u/dx^4.$$

Show that $u \geq 0$, compute $I = \int u\, dx$ and show that

$$|\int_{-1}^{1} f\, dx - f(a) - f(-a)| \leq I \max|f^{(4)}|, \quad f \in C^4([-1, 1]).$$

Exercise 4.4.8. Let $f \in \mathscr{E}'(\mathbf{R}^n)$ and let $\alpha = (\alpha_1, \dots, \alpha_n)$ be a multi-index. Show that there exists some $u \in \mathscr{E}'(\mathbf{R}^n)$ with $\partial^\alpha u = f$ if and only if $\langle f, x^\beta\rangle = 0$ for all multi-indices not satisfying the condition $\beta \geq \alpha$.

Exercise 4.4.9. Show that if $f \in \mathscr{E}'(\mathbf{R}^2)$ then the equation $\Delta u = f$ has a solution $u \in \mathscr{E}'(\mathbf{R}^2)$ if and only if $\langle f, \varphi\rangle = 0$ when $\varphi(x, y) = (x \pm iy)^n$, $n = 0, 1, 2, \dots$

Section 5.1

Exercise 5.1.1. Show that if $u \in \mathscr{D}'^k$, $v \in \mathscr{D}'^l$, then $u \otimes v \in \mathscr{D}'^{k+l}$, and that $u \otimes v \in \mathscr{D}'^N$ implies $u \in \mathscr{D}'^N$, $v \in \mathscr{D}'^N$ unless u or v equals 0.

Exercise 5.1.2. Construct for given positive integers k and l two distributions $u \in \mathscr{D}'^k(\mathbf{R})$ and $v \in \mathscr{D}'^l(\mathbf{R})$ such that $u \otimes v$ is not of order $k+l-1$.

Exercise 5.1.3. Construct for a given positive integer N two distributions u_0, u_1 on $\mathbf{R}$ which are not of order $N-1$ such that $u_0 \otimes u_1$ is of order N.

Section 5.2

Exercise 5.2.1. Let f be a continuous function from $\mathbf{R}$ to $\mathbf{R}$. Which operator has the distribution kernel $\partial H(y - f(x))/\partial y$?

Exercise 5.2.2. What is the kernel of the operator

$$\mathscr{K}\varphi(x) = \varphi(x) + \int_{\mathbf{R}} a(x,y)\varphi'(y)\,dy, \quad \varphi \in C_0^\infty(\mathbf{R}),$$

where $a \in C(\mathbf{R}^2)$?

Exercise 5.2.3. K is a measurable function in $X_1 \times X_2$ where X_j is an open subset of $\mathbf{R}^{n_j}$, such that

$$\int_{X_1} |K(x,y)|\,dx \le A, \text{ for almost all } y \in X_2;$$

$$\int_{X_2} |K(x,y)|\,dy \le B, \text{ for almost all } x \in X_1.$$

Prove that for the corresponding operator $\mathscr{K}$

$$\|\mathscr{K}\varphi\|_{L^p} \le A^{1/p}B^{1-1/p}\|\varphi\|_{L^p}, \text{ if } \varphi \in C_0^\infty(X_2), \ 1 \le p \le \infty.$$

Section 6.1

Exercise 6.1.1. Calculate $\delta_a(\cos x)$ when $-1 < a < 1$.

Exercise 6.1.2. Calculate $u = \delta_0'(f)$ in $\mathbf{R}^2 \setminus 0$ when $f(x) = x_1x_2$.

Exercise 6.1.3. Determine the limit of $\varphi_\varepsilon(x^2 - y^2)\varphi_\varepsilon(y-1)$ in $\mathscr{D}'(\mathbf{R}^2)$ as $\varepsilon \to +0$, where $\varphi \in C_0^\infty$, $\int \varphi(x)\,dx = 1$, and $\varphi_\varepsilon(t) = \varphi(t/\varepsilon)/\varepsilon$.

Exercise 6.1.4. Let f, g be real valued functions in $C^\infty(X)$, X open in $\mathbf{R}^n$, such that df and dg are linearly independent when $f = g = 0$. Determine $u = \delta(f, g)$.

Exercise 6.1.5. Let $f, g \in C^\infty(\mathbf{R}^n)$ be real valued, $df \neq 0$ when $f = 0$ and $dg \neq 0$ when $g = 0$. Show that if $\varphi \in C_0^\infty(\mathbf{R}^n)$, $\varphi \in C_0^\infty(\mathbf{R}^n)$ and $u = (\varphi\delta(f)) * (\varphi\delta(g))$, then sing supp u is contained in

$$\{x + y; x \in f^{-1}(0) \cap \operatorname{supp}\varphi, y \in g^{-1}(0) \cap \operatorname{supp}\varphi \text{ and } df(x), dg(y) \text{ are linearly dependent}\}.$$

Give an integral formula for u valid in the complement of this set.

Exercise 6.1.6. Set $u_\varepsilon(x) = (f(x) + i\varepsilon)^{-1}$ where $f \in C^\infty(\mathbf{R})$ is real valued. Determine the condition on f required for the existence of the limits $u_\pm = \lim_{\varepsilon \to \pm 0} u_\varepsilon$, and calculate $u_+ - u_-$ then.

Exercise 6.1.7. Show that if $s > 0$ and k is a positive integer, then the function $x \mapsto (x^{2k} - s^{2k} + i\varepsilon)^{-1}$ has a limit $f_s \in \mathscr{D}'(\mathbf{R})$ as $\varepsilon \to +0$. Show that one can find $u_0, \ldots, u_k \in \mathscr{D}'(\mathbf{R})$ such that

$$f_s - \sum_0^{k-1} s^{2j+1-2k} u_j \to u_k, \quad s \to 0,$$

and determine support and order for these distributions.

Section 6.2

Exercise 6.2.1. In $\mathbf{R} \times \mathbf{R}^n$, with variables denoted (t, x), let $\Box = \partial^2/\partial t^2 - \Delta_x$ be the wave operator. Calculate the fundamental solution E_k of $\Box^{k+1}$ with support in the forward light cone $\{(t, x); t \geq |x|\}$ for every integer $k \geq 0$.

Exercise 6.2.2. Find the forward fundamental solution F_a of $\Box - a$ for every $a \in \mathbf{C}$, with notation as in the preceding exercise.

Exercise 6.2.3. Find the forward fundamental solution F of the operator $\Box + 2b_0\partial_t + 2\sum_1^n b_j\partial_j + c$ for arbitrary complex $b_0, \ldots, b_n, c$, with notation as in the preceding exercises.

Section 7.1

Exercise 7.1.1. For which even positive integers m and n is $f(x) = \exp(x^n + i\exp(x^m))$ in $\mathscr{S}'(\mathbf{R})$?

Exercise 7.1.2. Let M be an unbounded subset of $\mathbf{R}^n$. Show that for every integer m there is a distribution $u \in \mathscr{S}'(\mathbf{R}^n)$ with $\operatorname{supp} u \subset M$ such that the order of $\hat{u}$ in the unit ball is $> m$.

Exercise 7.1.3. Show that if u is a measurable function on $\mathbf{R}^n$ and m is a positive integer, then $u \in \mathscr{S}'$ and $\hat{u} \in \mathscr{D}'^m$ if $\int |u(x)|^2(1+|x|^2)^{-m}\,dx < \infty$.

Exercise 7.1.4. Prove that if K is a compact subset of $\mathbf{R}^n$ and $\xi_j \in \mathbf{R}^n$, $|\xi_j - \xi_k| \geq 1$, $j \neq k$, then

$$\sum_1^\infty |\hat{\varphi}(\xi_j)|^2 \leq C_K \int |\varphi(x)|^2\,dx, \quad \varphi \in C_0^\infty(K),$$

with C_K independent of the sequence ξ_j.

Exercise 7.1.5. Show that if $\xi_j \in \mathbf{R}^n$, $|\xi_j - \xi_k| \geq 1$, $j \neq k$, and m is a non-negative integer, then $\sum a_j e^{i\langle x,\xi_j\rangle}$ converges in $\mathscr{S}'$ to a sum of order $\leq m$ if $a_j \in \mathbf{C}$ and $\sum |a_j|^2(1+|\xi_j|^2)^{-m} < \infty$.

Exercise 7.1.6. Let $u \in \mathscr{S}'(\mathbf{R}^n)$. When does there exist a function $f \in \mathscr{S}$ such that $u = u * f$?

Exercise 7.1.7. Show that if $u \in L^p(\mathbf{R}^n)$ and $|\xi| \leq \lambda$ when $\xi \in \operatorname{supp} \hat{u}$, then $\|u'\|_{L^p} \leq C\lambda \|u\|_{L^p}$ where C only depends on n.

Exercise 7.1.8. Show that if $u \in L^\infty(\mathbf{R}^n)$ and $\xi \in \operatorname{supp} \hat{u}$, then one can find a sequence $\varphi_j \in \mathscr{S}$ such that $|u * \varphi_j| \leq 1$ and $u * \varphi_j(x) \to e^{i\langle x,\xi\rangle}$ uniformly on every compact set.

Exercise 7.1.9. When does a differential equation $P(D)u = 0$ with constant coefficients have a solution $\neq 0$ in a) $\mathscr{D}'$ b) $\mathscr{S}'$ c) $\mathscr{E}'$ d) C^∞ e) $\mathscr{S}$.

Exercise 7.1.10. Let $f \in L^1(\mathbf{R}^n)$ and $f * f = f$. Find f.

Exercise 7.1.11. Show that the equation $u - u * f = f$ for a given $f \in \mathscr{S}(\mathbf{R}^n)$ has a solution $u \in \mathscr{S}$ if and only if $\hat{f} \neq 1$.

Exercise 7.1.12. Show that if $u, v \in \mathscr{S}'(\mathbf{R})$ have supports on the positive half axis, then $u * v \in \mathscr{S}'(\mathbf{R})$.

Exercise 7.1.13. Let $u_a(x) = 1/|\log x|^a$ when $0 < x < \frac{1}{2}$, $u(x) = 0$ when $x < 0$ or $x > 1$, and $u \in C^\infty$ when $x > 0$; here $a > 0$. Determine the limit of $\hat{u}(\xi)\xi(\log|\xi|)^a$ as $\xi \to \infty$.

Exercise 7.1.14. What is the Fourier transform of the space $Z(a,k)$ in Exercise 3.2.2?

Exercise 7.1.15. Set $\mathscr{F}u = (2\pi)^{-n/2}\hat{u}$ when $u \in \mathscr{S}(\mathbf{R}^n)$. Prove that a) $\mathscr{F}^4 = I$, the identity, and that every $u \in \mathscr{S}(\mathbf{R}^n)$ has a unique decomposition

$$u = \sum_0^3 u_k; \qquad u_k \in \mathscr{S}(\mathbf{R}^n),\ \mathscr{F} u_k = i^k u_k.$$

b) Show that the differential operators $L_\nu u = x_\nu u + \partial_\nu u,\ \nu = 1, \dots, n$, are surjective on $\mathscr{S}(\mathbf{R}^n)$, determine the kernels and show that $\mathscr{F} L_\nu u_k = i^{k+1} L_\nu u_k$ for the terms in the decomposition.

Exercise 7.1.16. Show that if K is a continuous function in $\mathbf{R}^n$ then the following conditions are equivalent:

(i) The convolution operator $\varphi \mapsto K * \varphi$ is positive on C_0^∞, that is, $(K * \varphi, \varphi) \ge 0$, $\varphi \in C_0^\infty(\mathbf{R}^n)$, where $(\cdot, \cdot)$ is the L^2 scalar product.
(ii) $\sum_{j,k=1}^N K(x_j - x_k) t_j \bar{t}_k \ge 0$ for all $x_1, \dots, x_N \in \mathbf{R}^n$ and $t_1, \dots, t_N \in \mathbf{C}$, $N = 1, 2, \dots$.
(iii) $K = \hat{\mu}$ where μ is a positive measure with finite total mass, $\langle \mu, 1 \rangle = K(0)$.

Exercise 7.1.17. Show that if $K \in \mathscr{D}'(\mathbf{R}^n)$ then the following conditions are equivalent:

(i) $(K * \varphi, \varphi) \ge 0$ for all $\varphi \in C_0^\infty(\mathbf{R}^n)$.
(ii) $K = \hat{\mu}$ where μ is a positive measure and $\int (1 + |x|)^{-N} d\mu < \infty$ for some N.

Exercise 7.1.18. Let f be a bounded continuous function on $\mathbf{R}$ with $\hat{f} = 0$ in a neighborhood of 0. Show that the primitive functions u of f are bounded. What is the support of $\hat{u}$?

Exercise 7.1.19. What is the Fourier transform of $\mathbf{R}^n \ni x \mapsto e^{i\langle x, \theta \rangle}$?

Exercise 7.1.20. Find the Fourier transform of the following functions on $\mathbf{R}$: a) $x \mapsto x \sin x$ b) $x \mapsto x \sin^2 x$ c) $x \mapsto x^{-1} \sin x$ d) $x \mapsto (\sin x)^k$ (k positive integer) e) $x \mapsto xH(x)$ f) $x \mapsto H(1+x) + H(1-x)$ g) $x \mapsto \operatorname{sgn} x = H(x) - H(-x)$ h) $x \mapsto \sin|x|$ i) $x \mapsto 1/(1+x^2)$ j) $x \mapsto x/(1+x^2)$ k) $x \mapsto x^3/(1+x^2)$ l) $x \mapsto \arctan x$ m) $x \mapsto x^3/(1+x^4)$.

Exercise 7.1.21. Use Parseval's formula to calculate the integrals

$$\int_{-\infty}^{\infty} dx/(x^2+1)^2 \quad \text{and} \quad \int_{-\infty}^{\infty} x^2\, dx/(x^2+1)^2.$$

Exercise 7.1.22. Find the Fourier transform of the function $x \mapsto |x^2 - 1|$ on $\mathbf{R}$.

Exercise 7.1.23. Find the Fourier transform of the distribution $x \mapsto (x^2 - s^2 + i0)^{-1}$ on $\mathbf{R}$, where $s > 0$.

Exercise 7.1.24. Calculate the Fourier transform of the function $f_t(x) = (\cos x - e^{-itx})/x$, $x \in \mathbf{R}$, and then $\int_{-\infty}^{\infty} |f_t(x)|^2\, dx$.

Exercise 7.1.25. Determine the limit as $\varepsilon \to +0$ of the fundamental solution $E_\varepsilon \in \mathscr{S}'$ of the differential operator $i\varepsilon(d/dx)^4 + (d/dx)^2 + 1$ on $\mathbf{R}$.

Exercise 7.1.26. Find all solutions $u \in \mathscr{S}(\mathbf{R}^n)$ of the differential equation

$$\Delta u + \sum_1^n x_j \partial u/\partial x_j + nu = 0.$$

Describe the solutions in $\mathscr{S}'(\mathbf{R}^n)$ also.

Exercise 7.1.27. Calculate the Fourier transforms of the following functions in $\mathbf{R}^2$: a) $x \mapsto H(x_1)H(x_2)$ b) $x \mapsto x_2/((1+x_1^2)(1+x_2^2))$ c) $x \mapsto x_1 e^{-\pi x_2^2}$ d) $\delta_1'(x_1) \otimes e^{-x_2^2/2}$.

Exercise 7.1.28. Let f be a continuous function on $\mathbf{R}$ with $f(x) = 1/x + O(|x|^{-2})$, as $x \to \infty$. Show that $\hat{f}$ is a function which is continuous except at the origin where left and right limits exist. Determine the jump $\hat{f}(+0) - \hat{f}(-0)$.

Exercise 7.1.29. Extend the preceding exercise to functions f with

$$f(x) = \sum_0^k a_j x^{-j} + O(x^{-k-1}), \quad x \to \infty,$$

where k is a positive integer.

Exercise 7.1.30. Let f be a continuous function on $\mathbf{R}$ with $f(x) = a_\pm/x + O(|x|^{-2})$ as $x \to \pm\infty$. Prove that $g(\xi) = \hat{f}(\xi) + (a_+ - a_-)\log|\xi|$ has a limit as $\xi \to \pm 0$, and determine $g(+0) - g(-0)$.

Exercise 7.1.31. For which $a \in \mathbf{C}$ does the differential equation

$$xu'' + 2u' + (a - x)u = 0$$

have a solution $\neq 0$ in $\mathscr{S}'(\mathbf{R})$ such that the limits $u(\pm 0)$ exist?

Exercise 7.1.32. Determine the Fourier transform of $\mathbf{R} \ni x \mapsto f(e^{ix})$ where f is an analytic function in a neighborhood of the unit circle. Work out the special case $f(z) = z/((2z-1)(z-2))$ explicitly.

Exercise 7.1.33. What is the Fourier transform of $\mathbf{R} \ni x \mapsto |x|^{-a}$, where $a \in \mathbf{C}$ and $0 < \operatorname{Re} a < 1$.

Exercise 7.1.34. What is the Fourier transform of $\mathbf{R} \ni x \mapsto |x+1|^{-\frac{1}{2}}$?

Exercise 7.1.35. Calculate the Fourier transform of $f_a(x) = |x|^{-a}$, $x \in \mathbf{R}^n$, where $0 < \operatorname{Re} a < n$.

Exercise 7.1.36. Show that if $u_\alpha(x) = x^\alpha |x|^{-n}$, $x \in \mathbf{R}^n$, where α is a multi-index $\neq 0$, then $\hat{u}_\alpha(\xi) = c_n(i\partial_\xi)^\alpha \log|\xi|$, and calculate the constant c_n.

Exercise 7.1.37. What is the Fourier transform of the function $\mathbf{R}^2 \ni x \mapsto A(x)^{-\frac{1}{2}}$ where A is a positive definite quadratic form.

Exercise 7.1.38. Calculate the Fourier transform of a function $u \in C(\mathbf{R}^2 \setminus 0)$ which is even and homogeneous of degree -1.

Exercise 7.1.39. Find the Fourier transform of

$$f(x) = (x_1^2 + x_2^2 + x_3^2 + 2ix_1x_2)^{-1}, \quad x \in \mathbf{R}^3.$$

Exercise 7.1.40. Find the Fourier transform of the distribution $(x, y) \mapsto 1/(x+iy)$ in $\mathbf{R}^2$.

Exercise 7.1.41. Find the Fourier transform of $\mathbf{R}^3 \ni x \mapsto e^{-|x|}$.

Exercise 7.1.42. Let $u \in \mathscr{E}'(\mathbf{R}^3)$ be the surface measure on the unit sphere. Find the Fourier transform.

Exercise 7.1.43. Find the Fourier transform of the distribution $u = \delta_0'(|x|^2 - 1)$ in $\mathbf{R}^3$.

Exercise 7.1.44. Find the Fourier transform of the distribution u_N in Exercise 3.1.24.

Exercise 7.1.45. Find the Fourier transform of the distribution u_N in Exercise 3.3.8.

Exercise 7.1.46. Find the Fourier transform of the distribution F in Exercise 4.4.2.

Section 7.2

Exercise 7.2.1. Develop in Fourier series the function B_1 on $\mathbf{R}$ with period 1 and $B_1(x) = x - \frac{1}{2}$, $0 < x < 1$, and deduce the Poisson summation formula.

Exercise 7.2.2. f is periodic on $\mathbf{R}$ with period 2π and $f(x) = \cos ax$, $|x| \le \pi$. Calculate $f'' + a^2 f$ and develop f in Fourier series.

Exercise 7.2.3. Evaluate the sum $S(x) = \sum_1^\infty (\cos nx)/(1+n^2)$.

Exercise 7.2.4. Find the Fourier transform of the distributions on $\mathbf{R}$ defined by

$$\text{a)} \quad \lim_{\varepsilon \to +0} \tan(x + i\varepsilon) \qquad \text{b)} \quad \lim_{\varepsilon \to +0} (\tan(x + i\varepsilon))^2.$$

Exercise 7.2.5. Show that if $f \in L^1(\mathbf{R})$ and $\varphi \in \mathscr{S}(\mathbf{R})$, then

$$2\pi \sum_{-\infty}^{\infty} f * \varphi(2\pi n) = \sum_{-\infty}^{\infty} \hat{f}(n)\hat{\varphi}(n),$$

with absolute convergence on both sides. Deduce that if $f, f', f'' \in L^1(\mathbf{R})$ then this remains true with φ replaced by δ_0.

Exercise 7.2.6. Find the Fourier transform of $f_z(x) = (x^2 - z^2)^{-1}$ where $\operatorname{Im} z > 0$, and calculate $\sum_{-\infty}^{\infty} f_z(n)$.

Exercise 7.2.7. Find the Fourier transform of $\mathbf{R} \ni x \mapsto 1/(1+x^2)$ and determine the limit

$$\lim_{\varepsilon \to +0} e^{2\pi/\varepsilon} \Big(\sum_{-\infty}^{\infty} \varepsilon/(1 + \varepsilon^2 n^2) - \pi\Big).$$

Exercise 7.2.8. Find the Fourier transform of $\varphi(x) = (\sin x)^2/x^2$, $x \in \mathbf{R}$, and calculate $\sum_{-\infty}^{\infty} \varphi(x + \pi n)$.

Exercise 7.2.9. Show that if $u \in L^\infty(\mathbf{R})$ and φ is the function in the preceding exercise, then the series

$$u_\varepsilon(x) = \sum_{-\infty}^{\infty} u(x + \pi n/\varepsilon)\varphi(\varepsilon x + \pi n)$$

converges, the range of u_ε is contained in the closed convex hull of that of u, u_ε is periodic with period π/ε, and

$$\operatorname{supp} \hat{u}_\varepsilon \subset \{\xi + \eta; \xi \in \operatorname{supp} \hat{u}, |\eta| \le 2\varepsilon\},$$
$$|u(x) - u_\varepsilon(x)| \le 2 \sup |u| (1 - \varphi(\varepsilon x)).$$

Exercise 7.2.10. Show that if $u \in L^\infty(\mathbf{R})$ is real valued and $\sup |u| < 1$, $\operatorname{supp} \hat{u} \subset [-\lambda, \lambda]$, then $u(x) - \cos(\lambda x)$ has precisely one zero in each interval $(n\pi/\lambda, (n+1)\pi/\lambda)$, $n \in \mathbf{Z}$; prove that it is simple and that there are no other zeros in $\mathbf{C}$.

Exercise 7.2.11. Show that if u is a real valued function with $u' \in L^\infty$, $\sup|u'| < 1$, and $\operatorname{supp} \hat{u} \cap (-\lambda, \lambda) = \emptyset$, and if $h(x) = \min_{n\in\mathbf{Z}} |x - 2\pi n/\lambda| - \pi/2\lambda$, then $h - u$ has the same sign as h at every maximum or minimum point of h. Deduce that $\sup|u| < \sup|h| = \pi/2\lambda$. (Bohr's inequality.)

Section 7.3

Exercise 7.3.1. Determine all $u \in \mathscr{E}'(\mathbf{R}) \setminus 0$ such that every factorization $u = v * w$ with $v, w \in \mathscr{E}'(\mathbf{R})$ is trivial in the sense that one factor is a Dirac measure.

Exercise 7.3.2. Prove that if $u \in L^p(\mathbf{R}^n)$ and $\hat{u}$ has compact support, then u can be extended to an entire analytic function in $\mathbf{C}^n$ such that

$$\|u(\cdot + iy)\|_{L^p} \le e^{H(-y)}\|u\|_{L^p}, \quad y \in \mathbf{R}^n,$$

where H is the supporting function of $\operatorname{supp} \hat{u}$.

Exercise 7.3.3. Let $u \in L^\infty(\mathbf{R})$, $\operatorname{supp} \hat{u} \subset (-1, 1)$. Prove that $u' = K * u$ where $\widehat{K}(\xi) = i\xi$, $-1 \le \xi \le 1$, $\widehat{K}(\xi) = i(2-\xi)$, $1 \le \xi \le 3$, and $\widehat{K}$ has period 4. Calculate K and deduce an estimate for $\sup|u'|$ in terms of $\sup|u|$.

Exercise 7.3.4. Prove that if $u \in L^p(\mathbf{R})$ for some $p \in [1, \infty]$ and $\operatorname{supp} \hat{u} \subset [-\lambda, \lambda]$ then

$$\|\sin\alpha u'/\lambda + \cos\alpha u\|_{L^p} \le \|u\|_{L^p}, \quad \alpha \in \mathbf{R}.$$

Exercise 7.3.5. Prove that if $P(D)$, $D = -i\partial/\partial x$ is a *homogeneous* polynomial then the polynomials $h(x)$ in $\mathbf{R}^n$ satisfying the differential equation $P(D)h(x) = 0$ are dense in the set of all solutions of the equation in $C^\infty(\mathbf{R}^n)$. Prove that the equation $P(D)u = f$, $f \in \mathscr{E}'$, has a solution $u \in \mathscr{E}'$ if and only if $\langle f, h\rangle = 0$ for all such polynomials h.

Section 7.4

Exercise 7.4.1. Let $f \in L^2((0, \infty))$. Show that the Fourier-Laplace transform

$$\hat{f}(\zeta) = \int_0^\infty e^{-ix\zeta} f(x)\, dx$$

is an analytic function in the half plane where $\operatorname{Im}\zeta < 0$ and that

$$\int_0^\infty |f(x)|^2\,dx = \sup_{\eta<0}(2\pi)^{-1}\int |\hat f(\xi+i\eta)|^2\,d\xi$$
$$= \lim_{\eta\to 0}(2\pi)^{-1}\int |\hat f(\xi+i\eta)|^2\,d\xi.$$

Characterize the functions $\hat f$ obtained in this way.

Exercise 7.4.2. Show that if $-\infty < a_1 < a_2 < \infty$ and $f(x)e^{a_j x} \in L^2(\mathbf{R})$, $j = 1, 2$, then the Fourier-Laplace transform $\hat f$ is analytic in the strip $\{\zeta;\ a_1 < \operatorname{Im}\zeta < a_2\}$ and

$$\max_j \int |f(x)|^2 e^{2a_j x}\,dx = \sup_{a_1<\eta<a_2}(2\pi)^{-1}\int |\hat f(\xi+i\eta)|^2\,d\xi.$$

Show that every analytic function $\hat f$ in the strip, such that the right-hand side is finite, is the Fourier-Laplace transform of a function f with $f(x)e^{a_j x} \in L^2$, $j = 1, 2$.

Exercise 7.4.3. Find an analytic function F in the strip $\{z \in \mathbf{C};\ |\operatorname{Im} z| < 1\}$ such that $(1+z^2)F(z)$ is bounded in the strip and

$$F(x+i-i0) + F(x-i+i0) = \delta(x).$$

Determine the Fourier transform G of $x \mapsto F(x)$ and prove that F is unique.

Exercise 7.4.4. Let A be a positive definite real $n\times n$ matrix. Show that if $f(x)e^{\langle Ax,x\rangle/2}$ is in $L^2(\mathbf{R}^n)$, then the Fourier-Laplace transform $\hat f(\zeta)$ is an entire analytic function with

$$\iint |\hat f(\xi+i\eta)|^2 e^{-\langle A^{-1}\eta,\eta\rangle}\,d\xi\,d\eta = \pi^{n/2}(2\pi)^n\sqrt{\det A}\int |f(x)|^2 e^{\langle Ax,x\rangle}\,dx.$$

Show also that every entire analytic function such that the left-hand side is finite is such a Laplace transform.

Exercise 7.4.5. Let $\varphi(z,y) = \frac12\langle Az,z\rangle + \langle Bz,y\rangle + \frac12\langle Cy,y\rangle$, $z \in \mathbf{C}^n$, $y \in \mathbf{R}^n$, where A and C are symmetric matrices, $\operatorname{Im} C$ positive definite and B non-singular. Show that if $u \in L^2$ then

$$U(z) = 2^{-\frac{n}{2}}\pi^{-\frac{3n}{4}}(\det\operatorname{Im} C)^{-\frac14}|\det B|\int e^{i\varphi(z,y)}u(y)\,dy$$

is an entire analytic function and that

$$\int |u(y)|^2\,dy = \int |U(z)|^2 e^{-2\Phi(z,\bar z)}\,d\lambda(z),$$

where $\Phi(z,\bar z) = \max_{y\in\mathbf{R}^n} - \operatorname{Im}\varphi(z,y)$ and $d\lambda$ is the Lebesgue measure in $\mathbf{C}^n$. Show that all entire functions for which the right-hand side is finite can be obtained in this way. Characterize the functions U obtained if $u \in \mathscr{S}$ or $u \in \mathscr{S}'$ instead.

Exercise 7.4.6. Let u be an entire analytic function in $\mathbf{C}^n$ such that

$$|u(z)| \le C e^{a|\operatorname{Im} z|^2/2 - b|\operatorname{Re} z|^2/2}, \quad z \in \mathbf{C}^n,$$

where $a > 0$ and $b > 0$. Prove that the Fourier-Laplace transform

$$U(\zeta) = \int e^{-i\langle x+iy,\zeta\rangle} u(x+iy)\,dx$$

is independent of y and that

$$|U(\zeta)| \le C(2\pi/b)^{\frac{n}{2}} e^{|\operatorname{Im}\zeta|^2/2b - |\operatorname{Re}\zeta|^2/2a}.$$

Show that $u = 0$ if $b > a$.

Section 7.6

Exercise 7.6.1. Find the Fourier transform of $\mathbf{R}^3 \ni x \mapsto \exp\, i(x_1^2 + x_2^2 - x_3^2)$

Exercise 7.6.2. Let $f \in C_0^\infty(\mathbf{R}^n)$. Prove that for every $t > 0$ there is a function $f_t \in \mathscr{S}$ such that $\hat f_t(\xi) = \hat f(\xi)\exp(it|\xi|^2)$, and prove that $|f_t(x)| \le Ct^{-n/2}$ for $x \in \mathbf{R}^n$ and $t > 1$. Use this to decide for which $p \in [1,\infty]$ that the Fourier transform of L^p consists of measures.

Exercise 7.6.3. Find the Fourier transform of the distribution $u = \delta(x_2 - x_1^2)$ in $\mathbf{R}^2$.

Exercise 7.6.4. Find a fundamental solution $E \in \mathscr{S}'(\mathbf{R}^3)$ of the differential operator

$$\partial/\partial x_1 + i\partial/\partial x_2 + (\partial/\partial x_3)^2.$$

Exercise 7.6.5. Find a real number a and $u \in \mathscr{D}'(\mathbf{R})$ such that the sequence $f_n(x) = n^a \sin(nx^2)$, $n = 1, 2, \dots$ has the limit $u \ne 0$ in $\mathscr{D}'(\mathbf{R})$ as $n \to \infty$.

Exercise 7.6.6. Find a real number a and $u \in \mathscr{D}'(\mathbf{R}^2)$ such that the sequence $u_n(x) = n^a \sin(nx_1x_2)$, $n = 1, 2, \ldots$, has the limit $u \neq 0$ in $\mathscr{D}'(\mathbf{R}^2)$ as $n \to \infty$.

Exercise 7.6.7. For which positive real numbers a and which $p \in [1, \infty]$ is the Fourier transform of $f_a(x) = (1 + x^2)^{-a/2} e^{ix^2}$ in L^p?

Exercise 7.6.8. Let p be a polynomial in $x \in \mathbf{R}$ of degree $m > 1$ and real coefficients. Prove that the Fourier transform F of $e^{ip(x)}$ is an entire analytic function, and determine a homogeneous differential equation of order $m - 1$ with linear coefficients which it satisfies.

Exercise 7.6.9. Let A be a symmetric $n \times n$ matrix with $\operatorname{Re} A$ positive semi-definite and $\| \operatorname{Im} A \| \leq 1$. Prove that if $n < \mu < n + 1$ and u is Hölder continuous of order μ in $\mathbf{R}^n$, then $e^{-\langle AD,D\rangle} u$ is continuous and with C independent of A and u

$$\sup |e^{-\langle AD,D\rangle} u| \leq C(\mu - n)^{-1} |u|_\mu;$$

$$|u|_\mu = \sum_{|\alpha| \leq n} \sup |\partial^\alpha u| + \sum_{|\alpha| = n} \sup_{x \neq y} |\partial^\alpha u(x) - \partial^\alpha u(y)| |x - y|^{n-\mu}.$$

Exercise 7.6.10. Prove that if in addition to the assumptions in the preceding lemma we know that $A(D)^j u$ is Hölder continuous of order μ for $0 \leq j \leq N$, then

$$|e^{-\langle AD,D\rangle} u(x) - \sum_{j<N} (-\langle AD, D\rangle)^j u(x)/j!| \leq C(\mu - n)^{-1} |\langle AD, D\rangle^N u|_\mu / N!.$$

Answers and Hints to all the Exercises

Chapter I

1.1. Since $f^{(k)}(x)$ is odd when k is odd we have $f^{(k)}(0) = 0$ then. By Theorem 1.2.6 we can choose $g_0 \in C^\infty(\mathbf{R})$ with the Taylor expansion $\sum f^{(2k)}(0)t^k/(2k)!$. All derivatives of $f_1(x) = f(x) - g_0(x^2)$ vanish at 0 then. Show that

$$g(x) = \begin{cases} g_0(x) + f_1(\sqrt{x}), & \text{if } x > 0, \\ g_0(x), & \text{if } x \le 0, \end{cases}$$

has the required properties. Alternatively, prove that if $g(x) = f(\sqrt{x})$, $x \ge 0$, then

$$g^{(n)}(x) = 2^{1-2n} \int_0^1 (1-t^2)^{n-1} f^{(2n)}(t\sqrt{x})dt/(n-1)!, \quad x > 0.$$

Conclude that $g \in C^\infty$ when $x \ge 0$ and extend g to $\mathbf{R}$.

1.2. Use the preceding exercise.

1.3. Review the solution of the preceding exercises.

1.4. Iterate the result in the preceding exercise.

1.5. Introduce $x_1 \pm x_2$ as new coordinates and use Exercise 1.3.

1.6. Choose for example $b_k = -2^k$ and solve the equations (iii) with $n \le N$, $k \le N$ first, which gives

$$a_k^{(N)} = \prod_{k<j\le N} \frac{b_j - 1}{b_j - b_k} \prod_{0\le j<k} \frac{1-b_j}{b_k - b_j}, \quad 0 \le k \le N.$$

Show that $|a_k^{(N)}| \le C2^{-k^2/2}$ and that $a_k = \lim_{N\to\infty} a_k^{(N)}$ exists. (The procedure is called the Seeley extension; see Seeley [2].)

1.7. Replacing $f(x)$ by $\sup_{t\le x} f(t)$ we may assume that f is increasing. Then take $F(x) = \int_1^2 f(tx)\varphi(t)\,dt$, $x > 0$, with $0 \le \varphi \in C_0^\infty((1,2))$, $\int \varphi(t)\,dt = 1$.

Chapter II

2.1. The condition is $\operatorname{Re} b > -1 - k$. Necessity: We must have

$$|\int \varphi(x)x^b\,dx| \le \sum_0^k \sup|\varphi^{(j)}|, \quad \varphi \in C_0^\infty((0,1)).$$

Testing with $\varphi(x) = \varphi(x/\varepsilon)$, $\varphi \in C_0^\infty((1,2))$ shows when $\varepsilon \to 0$ that we must have $\operatorname{Re} b + 1 + k \ge 0$. In case of equality testing with $\varphi(x) = \sum_1^N \varphi(2^\nu x)2^{\nu(i \operatorname{Im} b - k)}$ shows that

$$N|\int \varphi(x)x^b\,dx| \le C,$$

although the left-hand side $\to \infty$ as $N \to \infty$. – The sufficiency follows by taking

$$u(\varphi) = (-1)^k \int_0^\infty \varphi^{(k)}(x)x^{b+k}\,dx/(b+k)\dots(b+1), \quad \varphi \in C_0^\infty(\mathbf{R}),$$

if $b \ne -k,\dots,-1$; replace $x^{b+k}/(b+k)$ by $\log x$ if $b = -k < 0$.

2.2. No. Test as in the preceding exercise!

2.3. The condition is now $k(a+1) + \operatorname{Re} b + 1 > 0$. Modify the solution of the preceding exercises! Note that the oscillating factor may keep the order down.

2.4. Choose g with $g'(x)$ increasing so rapidly when $x \to 0$ that $f(x)/g'(x) \to 0$ as $x \to 0$ and $\int_0^1 |f'(x)/g'(x)|\,dx < \infty$, and define

$$u(\varphi) = i\int_0^\infty e^{ig}(f\varphi/g')'\,dx, \quad \varphi \in C_0^\infty(\mathbf{R}).$$

(Verify that $\int_0^1 |f|d(1/g') < \infty$.) Note that the amplitude may be very large if the oscillation is fast!

2.5. The limits are a) $\pi\delta_0$, b) $2\sqrt{\pi}\delta_0$.

2.6. a) 0 b) $-2\delta_0$ c) $\pi\delta_0$. (Use that residue calculus gives $\int_\infty^\infty \sin x\,dx/x = \pi$.)

2.7. $-\pi\delta_0$. (Use the preceding exercise.)

2.8. a) For $a + k < 0$. b) For all a.

2.9. Hint: Prove first that there is a uniform bound for u_j on every closed ball $\subset K$; use Arzela-Ascoli's theorem to extract a convergent subsequence and identify the limit with u. Conclude that it was not necessary to take a subsequence.

2.10. $a = -\frac{1}{2}$ and $u = \sqrt{2}\sum \delta_{2k\pi}$, $v = 2\sqrt{\pi}\sum \delta_{(2k+1)\pi}$. (Hint: By the periodicity it suffices to study u_α when $|t| < 2\pi$. Taylor expansion of $1-\cos t$ at 0 reduces the first question to a study of $(\alpha - a)t^{2\alpha}$ at 0. The dominating contributions to v occur when $\cos t = -1$, so look at the Taylor expansion there, which leads to Exercise 2.5 b).)

2.11. Show that every $\varphi \in C_0^\infty(\mathbf{R}^n)$ can be written in the form

$$\varphi(x) = \varphi_1(x) + \cdots + \varphi_n(x) + x_1 \cdots x_n \phi(x)$$

with $\varphi_j \in C_0^\infty(\mathbf{R}^n)$ even in x_j and $\phi \in C_0^\infty(\mathbf{R}^n)$.

2.12. $a_0(w) = \pi w^{-3/4}/\sqrt{2}$, $a_1(w) = 0$, $a_2(w) = \pi w^{-1/4}/\sqrt{8}$, where $|\arg w| < \pi/2$. Hint: It suffices to study $\int f_w(x)\varphi(x)\,dx$ when φ is an even test function. Write $\varphi(x) = \varphi(0) + x^2\varphi''(0)/2 + \phi(x)$ and note that $|\phi(x)/(x^4+w)| \le |\phi(x)/x^4|$ which is integrable. Use residue calculus to handle the other two terms.

2.13. For $\varphi \in C_0^\infty(\mathbf{R}^2)$ choose a rectangle $Q = [-a, a] \times [-b, b]$ containing the support, and write

$$\iint \chi_\varepsilon(xy)\varphi(x, y)\,dx\,dy = I_1 + I_2 + I_3 + I_4, \qquad \text{where}$$

$$I_1 = \iint_Q \chi_\varepsilon(xy)(\varphi(x, y) - \varphi(x, 0) - \varphi(0, y) + \varphi(0, 0))\,dx\,dy \to 0,$$

$$I_2 = \iint_Q \chi_\varepsilon(xy)(\varphi(x, 0) - \varphi(0, 0))\,dx\,dy \to \int_{|x|<a} (\varphi(x, 0) - \varphi(0, 0))\,dx/|x|,$$

$$I_3 = \iint_Q \chi_\varepsilon(xy)(\varphi(0, y) - \varphi(0, 0))\,dx\,dy \to \int_{|y|<b} (\varphi(0, y) - \varphi(0, 0))\,dy/|y|,$$

$$I_4 = \varphi(0, 0) \iint_Q \chi(xy/\varepsilon)\,dx\,dy/\varepsilon = 2\varphi(0, 0) \int_{|t|<ab/\varepsilon} \chi(t)\log(ab/\varepsilon|t|)\,dt.$$

(Verify these claims!) Hence $C = 2$ and

$$u = 2\int(-\log|t|)\chi(t)\,dt\,\delta_0(x, y) + v,$$

where

$$v(\varphi) = \int_{-a}^{a} (\varphi(x,0) - \varphi(0,0))\, dx/|x|$$

$$+ \int_{-b}^{b} (\varphi(0,y) - \varphi(0,0))\, dy/|y| + 2\log(ab)\varphi(0,0)$$

$$= - \int_{0}^{\infty} (\varphi'_x(t,0) - \varphi'_x(-t,0) + \varphi'_y(0,t) - \varphi'_y(0,-t)) \log t\, dt.$$

2.14. $u_{a,\varepsilon} \to \pi\delta_0$ and $u_{a,\varepsilon}u_{b,\varepsilon} \to 2\pi(b-a)^{-2}\delta_0$ if $a \neq b$ but the limit does not exist when $a = b$. (Hint: Look separately at the contributions to $\langle u_{a,\varepsilon}u_{b,\varepsilon}, \varphi\rangle$ when $|x - a\sqrt{\varepsilon}| \ll \sqrt{\varepsilon}$, when $|x - b\sqrt{\varepsilon}| \ll \sqrt{\varepsilon}$, and from the rest of $\mathbf{R}$.)

2.15. If $\varphi \in C_0^\infty(\mathbf{R}^2 \setminus 0)$ and $\varphi(x,y) = \Phi(r,\theta)$ with polar coordinates then

$$\langle f_t, \varphi\rangle = t \int_0^{2\pi} \int_0^\infty \sin(t|r^2 - 1|)\Phi(r,\theta) r\, dr d\theta$$

$$= t \int_0^{2\pi} \int_{-1}^\infty \sin(t|s|)\Phi(\sqrt{s+1},\theta) ds d\theta/2.$$

Integration by parts for $s \lessgtr 0$ gives the limit $\int \Phi(1,\theta)d\theta$, that is, $f_t \to ds$, the arc length measure on the unit circle. If we take instead $\varphi = \varphi(x^2+y^2)$ where $\varphi \in C_0^\infty((-1,1))$, then

$$\langle f_t, \varphi\rangle = 2\pi t \int_0^1 \sin(t(1-r^2))\varphi(r^2) r\, dr = \pi t \int_0^1 \sin(ts)\varphi(1-s) ds$$

$$= \pi[-\cos(ts)\varphi(1-s)]_0^1 - \pi \int_0^1 \cos(ts)\varphi'(1-s) ds = -\pi\varphi(0)\cos t + o(1).$$

If $\varphi(0) \neq 0$ the oscillation as $t \to \infty$ shows that there is no limit in $\mathscr{D}'(\mathbf{R}^2)$.

2.16. Already the inner limit in a) is equal to 0. If $\varphi \in C_0^\infty$ and $\varphi(0) = 0$, we can write $\varphi(x) = x\varphi(x)$, $\varphi \in C_0^\infty$, and see that the second limit of $\langle f_{t,\varepsilon}, \varphi\rangle$ is also 0. If $\varphi = 1$ in $(-r,r)$ then Cauchy's integral formula gives

$$\lim_{\varepsilon \to +0} \langle f_{t,\varepsilon}, \varphi\rangle = -2\pi i + \Big(\int_{|x|>r} + \int_{|x|=r, \operatorname{Im} x<0}\Big) e^{-itx}\varphi(x)\, dx/x \to -2\pi i, \quad t \to +\infty.$$

The limit b) is therefore $-2\pi i\delta_0$; the order of the limits is essential!

Section 3.1

3.1.1. $u' = f$ means that $u(\varphi') = -f(\varphi)$, $\varphi \in C_0^\infty(I)$. If $\chi \in C_0^\infty(I)$ is fixed with $\int \chi\, dx = 1$ then every $\varphi \in C_0^\infty(I)$ has a unique decomposition $\varphi = a\chi - \varphi'$ with $a \in \mathbf{C}$ and $\varphi \in C_0^\infty(I)$; we have $a = \int \varphi\, dx$, and $u(\varphi) = f(\varphi) + Ca$ where $C = u(\chi)$. Now defining u in this way we see at once for every C that $u \in \mathscr{D}'(I)$ and that $u' = f$.

3.1.2. It is clear that the order is at most $k+1$; the solution to the preceding exercise proves the opposite inequality: if u' is of order k, then u is of order $k-1$.

3.1.3. Use the answer to Exercise 3.1.1 if $k > 0$.

3.1.4. Immediate consequence of the preceding exercise.

3.1.5. Yes; if $f(x) = 0$, $x \le 0$, and $f(x) = \exp(ie^{1/x})$, $x > 0$, then we can take $u = ix^2 f'$.

3.1.6. Elaboration of the preceding answers gives the condition $ak \ge 1$.

3.1.7. The limit is $\int_0^\infty (\varphi(x) - \varphi(-x))\, dx/x$; the order is 1.

3.1.8. Show first that $f_t(x) = tf(tx)$ converges to $\mathrm{vp}(1/x)$; then it follows that $g_t \to d(\mathrm{vp}(1/x))/dx$.

3.1.9. Convergence requires $a_1 + 2a_2 = 0$, which allows two partial integrations giving the desired result when $a_1 + 4a_2 = 2$, that is, $a_1 = -2$ and $a_2 = 1$. – Generalize this example to higher derivatives of $\mathrm{vp}(1/x)$!

3.1.10. On one hand, $\log(x + i0)$ is the limit in L^1_{loc} of $\log(x + i\varepsilon)$, so the derivative is the limit $1/(x + i0)$ of $1/(x + i\varepsilon)$, as $\varepsilon \to +0$. On the other hand, differentiation of the two terms in the given definition shows that the derivative is also equal to $\mathrm{vp}(1/x) - \pi i \delta_0$.

3.1.11. The limit is $2(\cosh x - 1)/x^2 - 2\pi i \delta_0$. Hint: Note that the expression can be simplified to $(e^x - 1)(x + i\varepsilon)^{-2} + (e^{-x} - 1)(x - i\varepsilon)^{-2}$.

3.1.12. By induction: $f^{(n)}(|x|)(\operatorname{sgn} x)^n + 2\sum_{2j+2 \le n} f^{(2j+1)}(0)\delta_0^{(n-2j-2)}$.

3.1.13. By induction:
$|x| f^{(n)}(x) + n \operatorname{sgn} x f^{(n-1)}(x) + 2\sum_0^{n-2} (k+1) f^{(k)}(0)\delta_0^{(n-k-2)}$.

3.1.14. a) Modify the answer to Exercise 3.1.1. b) $u = -\delta_a^{(j+1)}/(j+1)$.

3.1.15. Use the preceding exercise. When $g = 0$ we get $u = \sum c_{a,j}\delta_a^{(j)}$ where $c_{a,j}$ are constants and a is a zero of F of order $> j$.

3.1.16. Hint: Use test functions $\varphi((x - a)/\varepsilon)$ if a is a zero of infinite order.

3.1.17. $(-1)^j \delta_0^{(k-j)} k!/(k-j)!$ if $j \le k$ and 0 otherwise.

3.1.18. $\sum_{j=0}^k (-1)^j \binom{k}{j} f^{(j)}(0) \delta_0^{(k-j)}$.

3.1.19. a) $x_+ + C$; b) $\frac{1}{2}x_+^2 + C$; c) $(e^x - 1)H(x) + C$;
d) $H(x) + C$; e) $\log|x| + C$.

3.1.20. a) $u = -\delta_0 + C_1 + C_2 H$
b) $u = C_1 \operatorname{vp}(1/x) + C_2 \delta$
c) $u = Cf(x)$ where $f(x) = \exp(1/x)$ when $x < 0$ and $f(x) = 0$ when $x \ge 0$.
d) $u = \delta_0' + C_1 x_+^{-1} + C_2 x_-^{-1} + C_3 \delta_0$
e) $u = \delta_0 - \delta_0' + C\delta_1$
f) $u = -\delta_0 + C_1\delta_1 + C_2\delta_{-1}$
g) $u = \sum c_j \delta_j$
h) $u = H(x) - 2(x-1)H(x-1) + C_1 x + C_2$
i) $u = H(x-1) + H(x+1) + C_1 H(x) + C_2$
j) $u = H(x)\exp(((x+1)^{-2} - 1)/2)$. (As in c) we have $u = 0$ in $(-\infty, -1)$ and $(-1, 0)$, and Exercise 3.1.17 shows that there is no contribution with support at -1.)
k) $u = Cf(x)$ where $f(x) = \exp(1/3x^3)$, $x < 0$ and $f(x) = 0$ when $x \ge 0$ (compare with c)).
l) $u = \sum c_j \delta_{\pi/2 + j\pi} + \sum d_j H(\pi/2 - |x - (j+1)\pi|)$.

3.1.21. $u(\varphi) = \sum_{j<N} c_j(\partial_n^j \varphi(\cdot, 0))$, $\varphi \in C_0^\infty(\mathbf{R}^n)$, where $c_j \in \mathscr{D}'(\mathbf{R}^{n-1})$.

3.1.22. $u = (C_0 + C_1\partial_1 + C_2\partial_2 + C_3(\partial_1^2 + \partial_2^2))\delta_0$. Hint: Prove first that $x_1^3 u = 0$, $x_2^3 u = 0$, deduce that $u = \sum C_\alpha \partial^\alpha \delta_0$ with $\alpha_1 \le 2$, $\alpha_2 \le 2$, and use Exercise 3.1.17.

3.1.23. The condition is that $\varphi = \varphi = 0$ implies $\partial\varphi/\partial x_1 \ne 0$. Hint: Note that $0 = \varphi\partial(\varphi u)/\partial x_1 = \varphi\partial\varphi/\partial x_1 u = 0$, hence $\varphi u = 0$ and $\partial\varphi/\partial x_1 u = 0$ to prove sufficiency. Choose a Dirac measure to prove necessity.

3.1.24. $f_N = 0$, the order of u_N is $N-1$, and $\operatorname{sing\,supp} u_N = \{0\}$. Hint: Use that $(t + ix_2)^{-N}$ is a continuous function of $t \ge 0$ with values in $\mathscr{D}'$. Since u_1 is locally integrable and $\partial u_N/\partial x_2 = -iN u_{N+1}$, the order of u_N is at most $N-1$. Use suitable test functions to show that it cannot be $N-2$.

3.1.25. The limit is $\sum_{g(x)=0} 2\pi i \delta_x / \sqrt{\det g''(x)}$. Hint: Only zeros of g are important. If $g(0) = 0$ then $g(x) = Q(x) + O(|x|^3)$ where Q is a positive semidefinite quadratic form. If it is positive definite we have $g(x) \ge Q(x)/2$ for small $|x|$, and taking $x/\sqrt{\varepsilon}$ as new variable in the integral $\langle u_\varepsilon, \varphi\rangle$ then gives the answer. To prove necessity consider $\operatorname{Im}\langle u_\varepsilon, \varphi\rangle$ with $\varphi \ge 0$, conclude that $m\{x \in K; \frac{1}{2}\varepsilon \le g(x) \le \varepsilon\} \le C\varepsilon$ for every compact set K and that if K is a ball with $g \neq 0$ in the interior then $\det g''(x) \neq 0$ at zeros of g on ∂K. Hence the necessity unless $g \equiv 0$, which is obviously excluded.

3.1.26. Hint: This is obvious near any point where f' and g' are linearly independent or $f \neq 0$ or $g \neq 0$. Changing coordinates shows that it suffices to prove that the limit exists in a neighborhood of 0 when $f(x) = x_1$, $g(0) = 0$, and dg is proportional to dx_1 at 0, hence $\partial_1 g(0) \neq 0$. Integrate by parts in $\int_{x_1>0} \varphi(x)/(g(x) + i\varepsilon)^{-1}\, dx$ and note that $|\log(g + i\varepsilon)| \leq C + |\log|g||$ which is locally integrable both in $\mathbf{R}^n$ and when $x_1 = 0$, which allows one to use the dominated convergence theorem.

3.1.27. $u(z) = 0$ if $|z| \leq 1$, $u(z) = \log|z|$ if $|z| > 1$. (Motivate the distribution convergence near the unit circle carefully!) Δu is the arc length measure on the unit circle.

3.1.28. This is minus the arc length measure on the unit circle.

3.1.29. The order is 0 in cases a) and c); it is 1 in case b). (Use polar coordinates in case c) to show that the distribution is -1 times the arc length measure on the unit circle.)

3.1.30. We must have $a > b$. If $f(x) = x^a \sin(x^{-b})$ then $\langle \partial u/\partial y, \varphi\rangle = -\int_0^\infty \varphi(x, f(x))\, dx$, $\varphi \in C_0^\infty$, so $\partial u/\partial y$ is always a measure. When $x > 0$ then $\partial u/\partial x = -f'(x)\partial u/\partial y$ has infinite measure near the origin unless $a > b$. When this condition is fulfilled verify that $\partial u/\partial x = -f'(x)\partial u/\partial y + \delta(x)H(-y)$.

3.1.31. Either f vanishes identically or else all zeros of f have finite order. The sufficiency is close to the one dimensional case (Exercise 3.1.15). To prove necessity test with functions of the form $\varphi(x)\varphi((y-a)/\varepsilon)$ where a is an endpoint of an interval where $f > 0$, and estimate the order of a as a zero of f by means of the order of u as a distribution!

3.1.32. $\varphi(t) = C\sqrt{t} - t$ where C is an arbitrary constant. – Direct calculation shows that we have a solution outside the curve $x = \varphi(t)$. Taking the jumps into account we obtain $\varphi(t) = -t$ or the differential equation $\varphi'(t) + (1 - \varphi(t)/t)/2 = 0$.

3.1.33. The limit is $2\pi i((x-1)^{-3}H(-x) + \delta_1'(x) + \frac{1}{2}\delta_1(x))$. Hint: Take the Taylor expansion at $z = 1$.

Section 3.2

3.2.1. Verify (i),(ii),(iii) by direct computation. Then (v),(vi) follow if $\operatorname{Re} a > -1$. The first part of (iv) is clear. $Z(0)$ consists of functions constant on each half axis. Now $u \in Z(-1)$ means that $xu = C_0$ and $u = C_0 \operatorname{vp}(1/x) + C_1\delta_0$ which proves that $\dim Z(-1) = 2$. The other statements follow now from the first part of (iv). (See also Exercises 2.1 and 3.1.2.)

3.2.2. Argue as in the preceding exercise. To prove (iv) note that since $x\frac{d}{dx}Z(0,k) \subset Z(0,k-1)$ and $x\frac{d}{dx}Z(0,1) = \{0\}$, the dimensions of the spaces show that the inclusion is an equality, which implies the statement on $\frac{d}{dx}Z(0,k)$. Since $\frac{d}{dx}xZ(-1,k) \subset Z(-1,k-1)$ we conclude from (v) for lower k that the dimension of $Z(-1,k)$ is at most $2k$. We have

$$xZ(-1,k) \supset x\tfrac{d}{dx}Z(0,k) = Z(0,k-1).$$

The inclusion is strict, for if $w_k(x) = (\log|x|)^k \in Z(0,k+1)$ then $xw_k' = kw_{k-1}$ is in $Z(0,k) \setminus Z(0,k-1)$ although $w_k' \in Z(-1,k)$ since we have $(x\frac{d}{dx}+1)^k\frac{d}{dx}w_k = \frac{d}{dx}(x\frac{d}{dx})^k w_k = 0$. Hence $\dim Z(-1,k) = 2k$ and the other statements follow.

3.2.3. The dimension is $2m$. Hint: Write $\sum a_j\tau^j = a_m \prod(\tau-\lambda_\nu)^{k_\nu}$ and use the preceding exercise.

3.2.4. Hint: Elaborate the solution of Exercise 2.1.

3.2.5. The order of u is 1, the order of v is 2, and the degree of homogeneity is -2 for both u and v. That the order is at most 2 is obvious. That the order of v is not 1 follows using test functions of the form $\varphi_1(x/\delta)\varphi_2(y/\varepsilon)$ where $\varepsilon \to 0$ first and then $\delta \to 0$. To prove that the order of u is 1, note that the integrand can be estimated by $\sup|\varphi'| \min(x,y)/xy$.

Section 3.3

3.3.1. $\mu = 2\pi \sum m_j \delta_{z_j}$ where z_j are the zeros and poles, with multiplicity m_j and $-m_j$ respectively. Hint: $\log|f|$ is harmonic except at the zeros and poles, and in a neighborhood of such a point z_j we can write $f(z) = (z-z_j)^{m_j}g(z)$ where g is analytic and $g(z_j) \neq 0$.

3.3.2. $1/C = 4\pi(a_{11}a_{22} - a_{12}^2)^{\frac{1}{2}}$ with the square root analytic outside $\mathbf{R}_-$. First change the coordinates to make the coefficient a_{12} vanish. If a_{11} and a_{22} are then positive, taking $x_j a_{jj}^{\frac{1}{2}}$ as new coordinates reduces to the

Euclidean case. The general formula follows by analytic continuation. Write down the statement explicitly and do the argument in detail!

3.3.3. $E(x) = |x|^{4-n}/((4-n)(4-2n)c_n)$ if $n \neq 4$; $E(x) = -(\log r)/(4c_4)$ if $n = 4$; here c_n is the area of S^{n-1}. Note that the Laplacian in $\mathbf{R}^n$ acts on functions of $r = |x|$ as $\partial_r^2 + (n-1)r^{-1}\partial_r$. Consider singularities at 0 carefully!

3.3.4. $\Delta u = a^2 u - 4\pi\delta_0$. Use the expression for Δ in the preceding answer when $x \neq 0$, and use the Taylor expansion and the known fundamental solution of Δ to examine the singularity at 0.

3.3.5. $\Delta u = -u$. (Special case of preceding exercise!)

3.3.6. $(C_+ e^{ia|x|} + C_- e^{-ia|x|})/|x|$ where $C_+ + C_- = -1/(4\pi)$. When a is real we can take $-\cos(a|x|)/(4\pi|x|)$; when $\operatorname{Im} a > 0$ the fundamental solution $-e^{ia|x|}/(4\pi|x|)$ is often preferred because of its decrease at infinity.

3.3.7. $f = \pi(\operatorname{sgn}(A(+0)) - \operatorname{sgn}(A(-0))\delta_0$. Hint: Note that $f_t(y) = 1/(t+iy)$ is bounded in $\mathscr{D}'^1(\mathbf{R})$ when $t \neq 0$, depends continuously on t and has limits as $t \to \pm 0$ with $f_+ - f_- = 2\pi\delta_0$. The inner integral is $\langle f_{A(x)}, \varphi(x, \cdot)\rangle$. – Note that if A is smooth and does not change sign, we get a solution of a *homogeneous* differential equation with singular support reduced to a point.

3.3.8. $f = 2\pi(-\partial/\partial z)^{N-1}\delta_0/(N-1)!$ where $\partial/\partial z = \frac{1}{2}(\partial/\partial x - i\partial/\partial y)$. Hint: By Taylor's formula we can write

$$\varphi(z) = \sum_{j+k<N} a_{jk} z^j \bar{z}^k + \varphi(z)$$

where $a_{jk} = (\partial/\partial z)^j(\partial/\partial\bar{z})^k\varphi(0)/j!k!$ and $|\varphi(z)| \leq C|z|^N \sum_{|\alpha|=N} \sup|\partial^\alpha\varphi|$. Show that φ may be replaced by φ when integrating over an annulus $\{z; \varepsilon < |z| < R\}$, and that $\langle \partial u_N/\partial x + i\partial u_N/\partial y, \varphi\rangle = 2\pi a_{N-1,0}$.

3.3.9. Near a zero a of f of order m we can write $f(z) = g(z)^m$ where $g(a) = 0$, $g'(a) \neq 0$. With the new variables $g(z)$ we can use the preceding exercise.

3.3.10. $E(x) = \prod_1^n x_j^{\alpha_j - 1} H(x_j)/(\alpha_j - 1)!$.

3.3.11. $C = c/2$; a change of variables reduces to the preceding exercise.

3.3.12. $F(z) = \sum_0^\infty z^j/(j!)^2$, a Bessel function. We have a fundamental solution if $F(0) = 1$ and $zF''(z) + F'(z) = F(z)$, and this determines the coefficients of the power series.

3.3.13. v is the measure $v(\varphi) = 2\int \varphi(|x|, x)\, dx/|x|$ supported by the light cone boundary, and $w = 8\pi\delta_0$, which means that $v/8\pi$ is a fundamental solution. To prove this first note that if $\varrho = t^2 - |x|^2$, then $\Box U(\varrho) = (4\varrho U'(\varrho))' + 4U'(\varrho)$. Letting $U_\varepsilon \to H$ we can calculate $v(\varphi)$ as the limit of $\langle \Box U_\varepsilon, \varphi \rangle$ after taking ϱ as a new variable instead of t in a neighborhood of a point with $t = |x|$ containing $\operatorname{supp} \varphi$. This gives the assertion on v in $\mathbf{R}^4 \setminus 0$, which is enough since the degree of homogeneity is $-2 > -4$. In the same way it follows that $w = 0$ in $R^4 \setminus 0$, so $w = C\delta$ by the homogeneity. One obtains $C = 8\pi$ by testing with a function of t only. – The tools introduced in Section 6.1 simplify such arguments since the change of variables is built into the theory.

Section 4.1

4.1.1. If $a = -N$ then the limit is $(-1)^N((N-1)!/(2N-1)!)\delta_0^{(2N-1)}$. Hint: Note that u_ε is orthogonal to even test functions. For an odd test function $\varphi(t) = t\varphi(t^2)$, $\varphi \in C_0^\infty$, (cf. Exercise 1.2) we have $u_\varepsilon(\varphi) = \langle \varphi, \chi_+^a(t-\delta)\rangle$, $\delta = \varepsilon^2$. (Motivate by analytic continuation from the easy case where $\operatorname{Re} a > 0$.) The right-hand side is a convolution, hence a continuous function of δ. Use that $\chi_+^a = \delta_0^{(N-1)}$ if $a = -N$.

4.1.2. Hint: Prove this first when $u \in C^\infty$ by using Green's formula and the fact that $\Delta \log |z| = 2\pi\delta_0$. Note the special case where $u = \log |f|$, f analytic.

4.1.3. The measure is equal to

$$2\sqrt{x^2/(x^2+a)}\delta_0(y) + 2\sqrt{y^2/(a-y^2)}\delta_0(x)$$

where $x^2 + a > 0$ in the first term and $y^2 < a$ in the second one. Hint: Since the preceding exercise shows that no point carries a positive mass it suffices to prove this at points where $f(z) \neq 0$, $f'(z) \neq 0$, taking $\sqrt{f(z)}$ as new coordinates. (This will become easier to do in Section 6.1.)

4.1.4. The Laplacian is $2n^{-1} \sin(\pi/n) x_-^{(1-n)/n} \delta_0(y)$. Hint: f_n is continuous and $\partial f_n/\partial z = z^{(1-n)/n}/2n$ in $\mathbf{C} \setminus \overline{\mathbf{R}}_-$; use Theorem 3.1.12.

Section 4.2

4.2.1. The convolutions are a) $t^{n-1}H(t)/(n-1)!$ b) $e^{-t}t^{n-1}H(t)/(n-1)!$

4.2.2. The convolution is $H^{(k)} = \delta_0^{(k-1)}$.

4.2.3. $u_a = \sum_0^\infty \delta'_{ka}$.

4.2.4. The convolutions are 0 and 1. Note that the convolution need not be associative unless all factors except one have compact support.

4.2.5. $\langle(\delta_h * u - u)/h, \varphi\rangle = \langle u, (\delta_{-h} * \varphi - \varphi)/h\rangle \to \langle u, \varphi'\rangle$, if $\varphi \in C_0^\infty(\mathbf{R})$.

4.2.6. The convolution is $\chi_+^{\lambda+\mu+1}$, which follows by analytic continuation from the special case in Section 3.4.

4.2.7. $u = \chi_+^{-2-\lambda} * f$ by the preceding exercise since $\chi_+^{-1} = \delta_0$.

4.2.8. The convolution is the function $x \mapsto 2\pi ab/|x|$ when $|a-b| \le |x| \le a+b$ and 0 otherwise. The singular support consists of the spheres with radius $a+b$ and $|a-b|$ and center at 0. To verify this consider first two continuous functions $u(|x|)$ and $v(|x|)$. The convolution is a continuous function of $|x|$ which can be determined by using a radial test function $\varphi(x) = \varphi(|x|^2)$. Note that $|r\omega + r'\omega'|^2 = r^2 + r'^2 + 2rr'\langle\omega, \omega'\rangle$ if ω and ω' are unit vectors, and use the fact, known to Archimedes, that the surface measure of $\{\omega; |\omega| = 1, \langle\omega, \omega'\rangle < t\}$ is $2\pi(1+t)$ if $-1 < t < 1$.

Section 4.3

4.3.1. If $0 \le \varphi \in C_0^\infty$, then $\langle u * v, \varphi\rangle = \langle u, \check{v} * \varphi\rangle$ where $\check{v} * \varphi(x) > 0$ if $x = y - z$ for some y with $\varphi(y) > 0$ and some $z \in \operatorname{supp} v$; if $x \in \operatorname{supp} u$, that is, $y \in \operatorname{supp} u + \operatorname{supp} v$, it follows that $\langle u * v, \varphi\rangle > 0$.

4.3.2. No, we have for example $\delta_0' * 1 = 0$.

4.3.3. Hint: A convex set K is contained in an affine hyperplane with normal ξ if and only if $\mathbf{R} \ni t \mapsto H(t\xi)$ is linear, where H is the supporting function of K. Now apply the theorem of supports.

4.3.4. $\operatorname{supp} f$ is the square when $P(0) \ne 0$; when $P(0) = 0$ it is the boundary with the interior of the sides parallel to the x_j axis removed if $P(\partial)$ is divisible by ∂_j, and it is empty when $P = 0$. Generalize to an arbitrary polygon!

Section 4.4

4.4.1. $f(z) = z + \sqrt{z^2 - 1} \ne 0$ outside $[-1, 1]$ so $\mu = 0$ there; $\log|f(z)|$ is continuous and

$$(\partial/\partial x - i\partial/\partial y)\log|f| = f'/f = 1/\sqrt{z^2-1}$$

also in the sense of distribution theory. The distribution boundary values at $x \pm i0$ are $\mp i/\sqrt{1-x^2}$, hence $\mu = iu(x)\delta(y)$ where $u(x) = -2i/\sqrt{1-x^2}$ if $|x| < 1$, $u(x) = 0$ if $|x| > 1$, that is, $\mu(\varphi) = 2\int_{-1}^{1} \varphi(x,0)\,dx/\sqrt{1-x^2}$. Since $v = \log|f| - \mu * E$ is harmonic,

$$f(z) = 2z(1 + O(z^{-2})), \quad \mu * E(z) = \log|z| \int d\mu/2\pi + O(1/|z|), \quad z \to \infty,$$

where $\int d\mu = 2\pi$, the maximum principle gives $v = \log 2$ identically, so $\mu * E(z) = \log|f(z)/2|$.

4.4.2. Taking $\varphi = 1$ near 0 shows that we must have $\sum_1^3 a_{jj} = 0$, and this implies that the limit exists and that

$$F(\varphi) = \int_{|x|<R} f(x)(\varphi(x) - \varphi(0))\,dx$$

if $|x| < R$ when $x \in \operatorname{supp}\varphi$. Direct computation gives $\Delta f(x) = 0$; hence ΔF has support at 0 and homogeneity -5, so there is a symmetric matrix (b_{jk}) such that

$$\Delta F = \sum b_{jk}\partial_j\partial_k\delta_0.$$

We have $F = E * \Delta F$ since the difference is harmonic and $\to 0$ at ∞. If φ is a function of $|x|$ equal to $|x|^2$ near 0 then $0 = \langle F, \Delta\varphi\rangle = \langle \Delta F, \varphi\rangle = 2\sum b_{jj}$ and we obtain $-4\pi F = 3\sum b_{jk}x_jx_k|x|^{-5}$, so $b_{jk} = -4\pi a_{jk}/3$ and $\Delta F = -4\pi\sum a_{jk}\partial_j\partial_k\delta_0/3$.

4.4.3. $h = V - E * \Delta V$ is harmonic, if $E(x) = -1/(4\pi|x|)$, and if $V \to 0$ at ∞ then $h \to 0$ there so $h = 0$. This gives the claim, and $V \equiv 1$ is a counterexample showing that the hypothesis is essential.

4.4.4. Set $F(x) = \varphi(x)x^{k+1}H(x)/(k+1)!$ where $\varphi \in C_0^\infty(\mathbf{R})$ is equal to 1 near 0. Then $F \in C_0^k$ and $F^{(k+2)} = \delta_0 + g$ with $g \in C_0^\infty$. Hence

$$d^{2k+4}(u * F * F)/dx^{2k+4} = u * (\delta_0 + g) * (\delta_0 + g) = u + 2u * g + u * g * g$$

where all terms except u are already known to be in C^∞.

4.4.5. Express u in terms of f_k by convolution with a fundamental solution.

4.4.6. Use a fundamental solution of $P(d/dx)$ to prove the sufficiency.

4.4.7. $a = 1/\sqrt{3}$, $I = 1/135$. Hint: By the preceding exercise a is determined by $\langle x^2, \chi - \delta_a - \delta_{-a}\rangle = 0$. Looking successively at sign changes of u''', u'', u' gives $u \geq 0$; we have $I = \langle x^4/4!, \chi - \delta_a - \delta_{-a}\rangle$. – The result is the lowest order case of Gauss integration.

4.4.8. Elaborate the similar exercise above on an ordinary differential equation. – Theorem 7.3.2 gives a general result.

4.4.9. To prove sufficiency examine $u = E * f$ where $E(z) = (2\pi)^{-1} \log|z|$ is the usual fundamental solution of Δ. Expand $E(z-\zeta)$ in a power series in ζ when $|z| > R$ and $|\zeta| < R$ where $R > \sup_{\zeta \in \operatorname{supp} f} |\zeta|$ and conclude that $u(z) = 0$ when $|z| > R$.

Section 5.1

5.1.1. Immediate consequence of the definitions.

5.1.2. Try $u = \sum_1^\infty v^{-2} 2^{kv} e^{i2^v x}$ and a similar definition of v; test with $\varphi(x, y) e^{-i(x+y)2^v}$.

5.1.3. Take $u_j = \sum_1^\infty \lambda_{2v+j}^{N-\frac{1}{2}} e^{i\lambda_{2v+j}x}$ where $\lambda_v = 2^{v!}$. The idea is that $\lambda_{v-1}^N \ll \lambda_v^{\frac{1}{2}}$ for large v which makes one amplitude factor dominate in any term in the product. (To prove that u_j is not of order $N-1$ note that if $\sum \lambda_{2v+j}^{\frac{1}{2}} e^{i\lambda_{2v+j}x}$ were a measure $d\mu$ then $\int_0^{2\pi} e^{-i\lambda_{2v+j}x}\, d\mu = 2\pi \lambda_{2v+j}^{\frac{1}{2}}$ which is absurd.)

Section 5.2

5.2.1. The composition $\varphi \mapsto \varphi \circ f$.

5.2.2. $\partial(H(y-x) - a(x,y))/\partial y$.

5.2.3. This is obvious by Fubini's theorem if $p = 1$ or $p = \infty$. For $1 < p < \infty$ write $|K(x,y)\varphi(y)| = |K(x,y)|^{1-1/p}(|K(x,y)|^{1/p}|\varphi(y)|)$ and use Hölder's inequality.

Section 6.1

6.1.1. $\cos x = a$ when $x = 2k\pi \pm \arccos a$ with integer k, and then we have $\sin x = \pm\sqrt{1-a^2}$. The answer is therefore

$$(1-a^2)^{-\frac{1}{2}} \sum_{-\infty}^{\infty} (\delta_{2k\pi + \arccos a} + \delta_{2k\pi - \arccos a}).$$

6.1.2. $u = \delta'(x_1) \otimes (1/x_2|x_2|) + (1/x_1|x_1|) \otimes \delta'(x_2)$. Hint: Calculate $v = \delta_0(x_1x_2)$ first and note that $\partial_1 v = x_2 u$, $\partial_2 v = x_1 u$.

6.1.3. The limit is $\frac{1}{2}(\delta_{(1,1)} + \delta_{(-1,1)})$. (Take $x^2 - y^2$ and $y - 1$ as local coordinates.)

6.1.4. $u = d\sigma/\sqrt{|f'|^2|g'|^2 - (f',g')^2}$ where $d\sigma$ is the Euclidean surface measure on $\Sigma = \{x; f(x) = g(x) = 0\}$. Hint: Assume coordinates labelled so that $y_1 = f(x)$, $y_2 = g(x)$, $y_j = x_j$, $j > 2$, is a local coordinate system at a chosen point $x^0 \in \Sigma$. Then

$$\langle u, \varphi \rangle = \int \varphi(x(0,0,y'))|Dx/Dy|\, dy', \quad y' = (y_3, \dots, y_n); \quad \varphi \in C_0^\infty;$$

if $\operatorname{supp} \varphi$ is close to x^0. Here $|Dy/Dx| = |\partial f/\partial x_1 \partial g/\partial x_2 - \partial f/\partial x_2 \partial g/\partial x_1|$. At a point where the tangent plane is $dx_1 = dx_2 = 0$, this is equal to $\sqrt{|f'|^2|g'|^2 - (f',g')^2}$ since $\partial f/\partial x_j = \partial g/\partial x_j = 0$, $j > 2$, so we have the asserted density at such a point. Orthogonal invariance proves that it is true everywhere.

6.1.5. If $\Sigma_x = \{y; f(y) = g(x-y) = 0\}$, then

$$u(x) = \int_{\Sigma_x} \varphi(y)\varphi(x-y)\, d\sigma/\sqrt{|f'(y)|^2|g'(x-y)|^2 - (f'(y), g'(x-y))^2},$$

where $d\sigma$ is the surface measure of Σ_x. (Use the preceding example and compare with Exercise 4.2.8.)

6.1.6. The condition is that $f(x) = 0$ implies $f'(x) \neq 0$. (Consider $\operatorname{Im} \langle u_\varepsilon, \varphi \rangle$ with $\varphi \geq 0$.) The limits are then $f^* v_\pm$ where $v_\pm = (t \pm i0)^{-1}$, so $u_+ - u_- = -2\pi i \sum_{f(x)=0} \delta_x/|f'(x)|$.

6.1.7. The first assertion follows since $t \mapsto (t + i\varepsilon)^{-1} \to (t + i0)^{-1}$ in $\mathscr{D}'$ as $\varepsilon \to +0$ and since $x \mapsto x^{2k} - s^{2k}$ does not have 0 as a critical value when $s > 0$. Examination of $\langle f_s, \varphi \rangle$ for even test functions φ shows after Taylor expansion that the stated formula holds with u_j equal to a constant times $\delta_0^{(2j)}$, of order $2j$ and support $\{0\}$, when $j < k$, whereas u_k is a constant times $(d/dx)^{2k-1} \operatorname{vp}(1/x)$, thus of order $2k$ and support equal to $\mathbf{R}$.

Section 6.2

6.2.1. $E_k = \frac{1}{2}\pi^{(1-n)/2} 4^{-k} \chi_+^{k+(1-n)/2}(A)/k!$ for $t > 0$, where $A = t^2 - |x|^2$; E_k is extended as a homogeneous distribution of degree $2k + 1 - n$. – Note that E_k is a constant times the characteristic function of the forward light cone if n is odd and $k = (n-1)/2$. (Cf. Exercise 3.3.13.)

6.2.2. $F_a = \sum_0^\infty a^k E_k$ with the notation in the preceding exercise. The sum converges, for the terms with $k + (1-n)/2 > 0$ are continuous and

$$\Phi(A) = \sum_0^\infty 4^{-k} A^k / (k!\Gamma(k+1+(1-n)/2))$$

converges to an entire analytic function. We have $F_a = 0$ when $t < |x|$ and $F_a = \frac{1}{2}\pi^{(1-n)/2}\Phi(aA)A^{(1-n)/2}$ when $t > |x|$, and the singularities at the light cone are described by $\sum_{k<(n-1)/2} a^k E_k$. (Cf. Exercise 3.3.12.)

6.2.3. $F = e^{-b_0 t + \sum_1^n b_j x_j} F_{-a}$, $a = \sum_1^n b_j^2 - b_0^2 + c$, with the notation in the preceding exercise.

Section 7.1

7.1.1. The condition is $m \geq n$. Hint: Compare with exercise 3.1.5. Note that a function may be in $\mathscr{S}'$ although the absolute value is very large.

7.1.2. Hint: Choose a sequence $x_j \in M$ with $|x_j| > j$ and set $u = \sum |x_j|^{m+1} \delta_{x_j}$. Prove that $u \in \mathscr{S}'$ and derive a contradiction if $\hat{u}$ is of order m in the unit ball by looking at $|\varphi|^2 * u$ where $\varphi \in \mathscr{S}$ and $|\xi| < \frac{1}{2}$ if $\xi \in \operatorname{supp} \hat{\varphi}$.

7.1.3. Hint: Write $u(x) = \sum_{|\alpha| \leq m} x^\alpha u_\alpha(x)$ with $u_\alpha \in L^2$.

7.1.4. Hint: Choose $\chi \in C_0^\infty(\mathbf{R}^n)$ equal to 1 in K and use that $\varphi = \chi\varphi$ implies $(2\pi)^n \hat{\varphi} = \hat{\chi} * \hat{\varphi}$,

$$\sum |\hat{\chi} * \hat{\varphi}(\xi_j)|^2 \leq C \int |\hat{\varphi}(\xi)|^2 \sum |\hat{\chi}(\xi_j - \xi)| \, d\xi.$$

7.1.5. Hint: Reduce to $m = 0$ and apply the preceding exercise.

7.1.6. $\operatorname{supp} \hat{u}$ must be compact, for the equation is equivalent to $(1-\hat{f})\hat{u} = 0$, and $1 - \hat{f} \neq 0$ outside a compact set. Conversely, we can always take for f the inverse Fourier transform of a function in C_0^∞ equal to 1 on $\operatorname{supp} \hat{u}$ if this is a compact set.

7.1.7. Use the answer to the preceding exercise.

7.1.8. Hint: Reduce to the case $\xi = 0$ by passing to $ue^{-i\langle x, \xi\rangle}$. Choose $\chi \in \mathscr{S}$ so that $\hat{\chi}$ has support in the unit ball and $\hat{\chi}(0) \neq 0$. Then

$u_\delta = \chi_\delta * u \neq 0$, if $\chi_\delta(x) = \chi(\delta x)$. Now take $\varphi_j(x) = c_j \chi_{1/j}(x + x_j)$ where $|c_j| = 1/\sup|u_{1/j}|$ and x_j is chosen so that $|u * \varphi_j(0) - 1| < 1/j$. Conclude using the preceding exercise that $|u * \varphi_j(x) - 1| < (1 + C|x|)/j$. (The result is due to Beurling.)

7.1.9. a) and d) P not a constant $\neq 0$ b) When P has a real zero c) and e) Only when $P = 0$.

7.1.10. $f = 0$. Since $\hat{f}$ is continuous and $\to 0$ at ∞, the equation $\hat{f}(1 - \hat{f}) = 0$ implies $\hat{f} \equiv 0$.

7.1.11. $\hat{u} = \hat{f}/(1 - \hat{f})$ is in $\mathscr{S}$ precisely when the denominator never vanishes. – The statement is also valid with $\mathscr{S}$ replaced by L^1, but the sufficiency of the condition is a much harder theorem of Wiener then.

7.1.12. Hint: Use that $\langle u * v, \varphi\rangle = \langle u \otimes v, \Phi\rangle$ where $\Phi(\xi, \eta) = \varphi(\xi + \eta)$, $\varphi \in C_0^\infty$. Replace Φ by $\chi\Phi$ for a suitable $\chi \in C^\infty$ which is 1 in the first quadrant and vanishes when $|\xi| + |\eta| > 2(1 + |\xi + \eta|)$. Generalize to higher dimensions!

7.1.13. The limit is $-i$. Hint: We may assume that u is real valued; then $\hat{u}(-\xi) = \overline{\hat{u}(\xi)}$ so we may take $\xi > 0$. Using (3.1.13) we obtain

$$\begin{aligned}\hat{u}(\xi) &= \int_0^1 \frac{e^{-\xi y}(-i)dy}{(-\log y + \pi i/2)^a} + O(1/\xi^2)\\ &= -\frac{i}{\xi}\int_0^\xi \frac{e^{-t}\,dt}{(\log\xi - \log t + \pi i/2)^a} + O(1/\xi^2);\end{aligned}$$

use dominated convergence to get the result. Note that u is continuous but $\hat{u} \notin L^1$ if $a \le 1$.

7.1.14. $Z(-1 - a, k)$. Use this to simplify the answer to Exercise 3.2.2!

7.1.15. That $\mathscr{F}^4 = I$ is a consequence of the Fourier inversion formula. Let p_k, $k = 0, 1, 2, 3$, be the interpolation polynomials defined by $p_k(\tau) = (\tau^4 - 1)/(4i^{3k}(\tau - i^k))$ and show that the decomposition holds precisely when $u_k = p_k(\mathscr{F})u$. Show that $\mathscr{F}L_\nu = iL_\nu\mathscr{F}$ and solve the differential equation $L_\nu u = f \in \mathscr{S}$ explicitly. The kernel of L_ν is the set of functions $u \in \mathscr{S}$ such that $u(x)e^{x_\nu^2/2}$ is independent of x_ν.

7.1.16. Hint: Approximate $\sum t_j \delta_{x_j}$ by functions $\varphi \in C_0^\infty$ to prove that (i) $\Longrightarrow$ (ii), and approximate integrals by Riemann sums to prove the converse. Note that (ii) with $N = 2$ implies that $K(x) = \overline{K(-x)}$ and that $|K(x)| \le K(0)$, so $K \in \mathscr{S}'$. (iii) $\Longrightarrow$ (i) since

$$(K * \varphi, \varphi) = (K, \varphi * \tilde{\varphi}) = (\mu, |\hat{\varphi}|^2), \quad \tilde{\varphi}(x) = \overline{\varphi(-x)}.$$

(i),(ii) $\Longrightarrow$ (iii), for $(\widehat{K}, |\varphi|^2) \geq 0$, $\varphi \in \mathscr{S}$ implies $(\widehat{K}, \chi) \geq 0$ for all $\chi \in C_0^\infty$ with $\chi \geq 0$ (approximate χ by the square of $(\chi + \varepsilon e^{-|x|^2})^{\frac{1}{2}} \in \mathscr{S}$). Hence $\widehat{K}$ is a positive measure. If $\varphi_\delta(x) = \varphi(x/\delta)\delta^{-n}$, $\varphi \in C_0^\infty$, $\int \varphi\, dx = 1$, then

$$K(0) = \lim_{\delta \to 0}(K * \varphi_\delta, \varphi_\delta) = \lim_{\delta \to 0}(2\pi)^{-n}\langle \widehat{K}, |\hat{\varphi}(\delta \cdot)|^2\rangle \geq (2\pi)^{-n}\langle \widehat{K}, 1\rangle$$

so $\widehat{K}$ has finite mass and equality holds. – K is called positive definite and the result is Bochner's theorem.

7.1.17. Hint: Choose φ_δ as in the preceding answer. Show that (i) implies that $K_\delta = K * \varphi_\delta * \tilde{\varphi}_\delta = \hat{\mu}_\delta$ where μ_δ is a positive measure with total mass $K_\delta(0) = O(\delta^{-N})$ if K is of order N in a neighborhood of 0. Show that $\mu_\delta|\hat{\varphi}(\varepsilon\cdot)|^2 = \mu_\varepsilon|\hat{\varphi}(\delta\cdot)|^2$ and conclude that $\mu_\delta = |\hat{\varphi}(\delta\cdot)|^2\mu$ where μ is a measure with $\int_{|\xi|<1/\delta} d\mu(\xi) = O(\delta^{-N})$, which gives (ii) with N replaced by $N+1$. The converse is straightforward. – The result is due to L. Schwartz.

7.1.18. The primitive functions are of course $O(x)$, hence in $\mathscr{S}'$, and $i\xi\hat{u} = \hat{f}$, which means that $\hat{u} = (i\xi)^{-1}\hat{f}$ outside the origin, where there may be a multiple of the Dirac measure. Thus $\operatorname{supp}\hat{u} = \operatorname{supp}\hat{f}$ or $\{0\} \cup \operatorname{supp}\hat{f}$. Choose $\widehat{K} \in C^\infty$ so that $i\xi\widehat{K}(\xi) = 1 + \hat{\varphi}(\xi)$ where $\hat{\varphi}\hat{f} = 0$ and $\hat{\varphi} \in C_0^\infty$. Since $\widehat{K}^{(j)}$ is integrable for $j > 0$, it follows that $x^j K(x)$ is continuous and bounded for every $j > 0$, and since $K' - \delta = \varphi \in \mathscr{S}$ it follows that K is bounded at the origin, hence that $K \in L^1$. Since $v = K * f$ is bounded and $v' = f + \varphi * f = f$ this proves the assertion.

7.1.19. $(2\pi)^n\delta_\theta$ (by Fourier's inversion formula).

7.1.20. a) $\pi(\delta_1' - \delta_{-1}')$ b) $\pi i(2\delta_0' - \delta_2' - \delta_{-2}')/2$ c) π times the characteristic function of $(-1,1)$ (use Fourier's inversion formula!) d) $2\pi(2i)^{-k}\sum_0^k \binom{k}{j}\delta_{k-2j}(-1)^j$. e) $-(\xi - i0)^{-2}$ f) $2\pi\delta_0 + 2\sin\xi/\xi$ g) $-2i\,\mathrm{vp}\,1/\xi$ h) $\mathrm{vp}\,1/(\xi+1) - \mathrm{vp}\,1/(\xi-1)$ i) $\pi e^{-|\xi|}$ (residue calculus) j) $\pi\operatorname{sgn}\xi e^{-|\xi|}/i$ k) $2\pi i\delta_0' + \pi i \operatorname{sgn}\xi e^{-|\xi|}$ l) $\mathrm{vp}\,\pi e^{-|\xi|}/i\xi$ (differentiate the function!) m) $-\pi i \operatorname{sgn}\xi e^{-|\xi|/\sqrt{2}}\cos(\xi/\sqrt{2})$ (residue calculus).

7.1.21. Both integrals are $\pi/2$. (Use i) and j) in the preceding exercise.)

7.1.22. $-2\pi(\delta_0'' + \delta_0) - 8(\xi\cos\xi - \sin\xi)/\xi^3$. (Hint: Consider $x \mapsto x^2 - 1$ first and then the difference.)

7.1.23. $-\pi i s^{-1}e^{-is|\xi|}$. (Hint: Determine the Fouriertransform of $x \mapsto (x^2 - s^2 + i\varepsilon)^{-1}$ by residue calculus and let $\varepsilon \to +0$.)

7.1.24. $\hat{f}_t(\xi) = -\pi i(H(\xi-1) + H(\xi+1) - 2H(\xi+t))$, $\int_{\mathbf{R}} |f_t(x)|^2\, dx = \pi\max(1, 2|t| - 1)$.

7.1.25. The limit is $-ie^{i|x|}/2$. Hint: Prove that

$$E_\varepsilon(x) = (2\pi)^{-1} \int_{-\infty}^{\infty} e^{ix\xi}/(i\varepsilon\xi^4 - \xi^2 + 1)\, d\xi \to (2\pi)^{-1} \int_\Gamma e^{ix\xi}/(1-\xi^2)\, d\xi,$$

where Γ is the real axis with a neighborhood of $+1$ (of -1) replaced by a half circle in the lower (upper) half plane.

7.1.26. The solutions are $u(x) = Ce^{-|x|^2/2}$ with a constant C. The solutions in $\mathscr{S}'$ are characterized by the fact that $e^{|\xi|^2/2}\hat{u}$ is a homogeneous distribution of degree 0, so u is the convolution of $e^{-|x|^2/2}$ and an arbitrary distribution homogeneous of degree $-n$.

7.1.27. a) $-1/(\xi_1 - i0) \otimes 1/(\xi_2 - i0)$ b) $-i\pi^2 \operatorname{sgn} \xi_2 e^{-|\xi_1|-|\xi_2|}$ c) $2\pi i\delta_0'(\xi_1) \otimes \exp(-\xi_2^2/4\pi)$ d) $i\sqrt{2\pi}\xi_1 e^{-i\xi_1 - \xi_2^2/2}$.

7.1.28. The jump is $-2\pi i$. Note that $f(x) = x/(x^2+1) + g(x)$ where g is integrable, so $\hat{f}(\xi) - \pi \operatorname{sgn} \xi e^{-|\xi|}/i = \hat{g}(\xi)$ is continuous.

7.1.29. $\hat{f} = 2\pi a_0 \delta_0 + g$ where $g \in C^{k-1}$ on each closed half axis and $g^{(j)}$ has the jump $2\pi i^{-j-1} a_{j+1}$ at 0 when $0 \le j \le k-1$.

7.1.30. $g(+0) - g(-0) = -\pi i(a_+ + a_-)$. Hint: By Exercise 7.1.28 it suffices to study the Fourier transform of $f(x) = H(x-1)/x$.

7.1.31. a must be an even integer. Hint: Solve the differential equation obtained by Fourier transformation, examine the solution at infinity, and use the preceding exercise.

7.1.32. If $f(z) = \sum a_n z^n$ then the Fourier transform is $2\pi \sum a_n \delta_n$. In the special case we have $a_n = -2^{-|n|}/3$.

7.1.33. $\xi \mapsto |\xi|^{a-1}\sqrt{\pi}2^{1-a}\Gamma((1-a)/2)/\Gamma(a/2)$. Note that the Fourier transform has to be even and homogeneous of degree $a-1$, hence $\xi \mapsto C|\xi|^{a-1}$, and C can be determined using a Gaussian as test function. Note the special case $a = \frac{1}{2}$ where the value of the constant follows from Fourier's inversion formula.

7.1.34. $e^{i\xi}\sqrt{2\pi/|\xi|}$.

7.1.35. $\hat{f}_a = c_a f_{n-a}$ where $c_a = \pi^{n/2}2^{n-a}\Gamma((n-a)/2)/\Gamma(a/2)$. (Argue as in the case $n = 1$ in Exercise 7.1.33.)

7.1.36. $c_n = -2\pi^{n/2}/\Gamma(n/2)$. Hint: Use the preceding exercise with $a \to n$ or the known fundamental solution of the Laplacian.

7.1.37. $\xi \mapsto 2\pi B(\xi)^{-\frac{1}{2}}$ where $B(\xi_1, \xi_2) = A(\xi_2, -\xi_1)$. Hint: Reduce to the Euclidean form by a linear transformation.

7.1.38. $\hat{u}(x, y) = 2\pi u(y, -x)$, which generalizes the preceding exercise. Hint: Express $\hat{u}(\varphi) = u(\hat{\varphi})$ in terms of polar coordinates and do the radial integration using Fourier's inversion formula.

7.1.39. $\hat{f}(\xi) = 2\pi^2(\xi_1^2 + \xi_2^2 - 2i\xi_1\xi_2 + 2\xi_3^2)^{-\frac{1}{2}}$. Hint: Discuss the Fourier transform of $x \mapsto 1/A(x)$ first when A is a positive definite quadratic form and use analytic continuation to pass to forms with positive definite real part.

7.1.40. $2\pi/(i\xi - \eta)$. (Use that we have essentially a fundamental solution of the Cauchy-Riemann operator.)

7.1.41. $\xi \mapsto 8\pi(1 + |\xi|^2)^{-2}$. Hint: If $f \in C_0(\mathbf{R})$ then the Fourier transform of $\mathbf{R}^3 \ni x \mapsto f(|x|)$ is a function of $|\xi|$ given by

$$\int_0^\infty r^2 f(r)\,dr \int_{|\omega|=1} e^{-ir\langle\omega,\xi\rangle}\,d\omega = \int_0^\infty r^2 f(r)\,dr \int_{-1}^{1} e^{-ir|\xi|s} 2\pi\,ds$$
$$= \frac{4\pi}{|\xi|}\int_0^\infty r f(r) \sin(r|\xi|)\,dr.$$

(See the answer to Exercise 4.2.8. There is an analogue of this formula in every dimension. It involves a Bessel function which is less elementary for even dimensions.)

7.1.42. $\hat{u}(\xi) = 4\pi|\xi|^{-1} \sin|\xi|$. (Use the answer to the preceding exercise.)

7.1.43. $\hat{u}(\xi) = -\pi \cos|\xi|$. Hint: By the second exercise above the Fourier transform is

$$\langle r\delta'(r^2 - 1), 4\pi \sin(r|\xi|)\rangle/|\xi| = 2\pi\langle\delta'(t-1), \sin(\sqrt{t}|\xi|)\rangle/|\xi|.$$

7.1.44. $\hat{u}_N(\xi) = 2\pi^{\frac{3}{2}}((N-1)!)^{-1}|\xi_2|^{N-\frac{3}{2}} \exp(\xi_1^2/4\xi_2)$ when $\xi_2 < 0$ and 0 when $\xi_2 \geq 0$. (Take the Fourier transform in x_2 first.)

7.1.45. $\hat{u}_N(\xi, \eta) = \pi i^{-N} 2^{2-N}(\xi - i\eta)^{N-1}(\xi + i\eta)^{-1}/(N-1)!$. (Use the differential equation established.)

7.1.46. $-4\pi \sum a_{jk}\xi_j\xi_k/(3|\xi|^2)$. (Use the calculation of ΔF in the earlier exercise.)

Section 7.2

7.2.1. $B_1(x) = -\sum_{n\neq 0} e^{2\pi inx}/2\pi in$, hence by differentiation $1 - \sum\delta_n = -\sum_{n\neq 0} e^{2\pi inx}$, that is, $\sum\delta_n = \sum e^{2\pi inx}$, which means that the Fourier transform of $\sum\delta_{2\pi n}$ is $\sum\delta_n$.

7.2.2. $f'' + a^2 f = 2a \sin(a\pi) \sum_{-\infty}^{\infty} \delta_{(2k+1)\pi}$. Calculation of the Fourier coefficients gives

$$f(x) = \sum_{-\infty}^{\infty} a \sin(a\pi)/\pi(-1)^n e^{inx}/(a^2 - n^2);$$

a term with $a = n$ should be read as $e^{inx}/2$. Thus $f'' + a^2 f = a \sin(a\pi)/\pi \sum (-1)^n e^{inx}$, so $\sum (-1)^n \delta_n$ has $2\pi \sum \delta_{(2k+1)\pi}$ as Fourier transform, which again proves Poisson's summation formula. Show on the other hand how Poisson's summation formula allows one to obtain the Fourier coefficients without any calculation!

7.2.3. $S(x) = \frac{1}{2}(\pi \cosh(x - \pi)/\sinh \pi - 1)$ when $0 \le x \le 2\pi$. Hint: $S(x) - S''(x) = \sum_1^{\infty} \cos nx = \pi \sum_{-\infty}^{\infty} \delta_{2k\pi} - \frac{1}{2}$ by Poisson's summation formula. Determine S from this differential equation and the fact that S is even and periodic.

7.2.4. a) $2\pi i(\delta_0 + 2\sum_1^{\infty}(-1)^n \delta_{2n})$; b) $-8\pi \sum_1^{\infty}(-1)^n n\delta_{2n} - 2\pi\delta_0$. Hint: First expand $\tan(x + i\varepsilon)$ in a power series in $e^{2ix-2\varepsilon}$; for b) use that $(\tan z)^2 = d(\tan z)/dz - 1$.

7.2.5. Note that $g(x) = \sum (f * \varphi)(x + 2\pi n)$ converges absolutely and locally uniformly, and is in C^∞ with period 2π. The Fourier coefficients are $\hat{f}(k)\hat{\varphi}(k)/2\pi$ which gives the first statement. If $\chi \in C_0^\infty(-\pi, \pi)$ is equal to 1 near 0, then $(H\chi)' = \delta_0 + H\chi'$, hence $f = f' * (H\chi) - f * (H\chi')$ which proves the absolute and locally uniform convergence of $\sum f(2\pi n - x)$ when $f, f' \in L^1$. When also $f'' \in L^1$ then $\hat{f}(\xi)(1 + \xi^2)$ is bounded, so the absolute convergence of $\sum \hat{f}(n)$ is trivial. Letting $\varphi \to \delta_0$ in the usual way we obtain the second statement.

7.2.6. $\hat{f}_z(\xi) = \pi i z^{-1} e^{iz|\xi|}$; the sum is $-\pi z^{-1} \cot(\pi z)$ by Poisson's summation formula.

7.2.7. The Fourier transform is $\xi \mapsto \pi e^{-|\xi|}$ (cf. Exercise 7.1.20 i)). By Poisson's summation formula it follows that

$$\sum \varepsilon/(1 + \varepsilon^2 n^2) = \pi \sum e^{-2\pi|n/\varepsilon|} = \pi(1 + 2e^{-2\pi/\varepsilon} + O(e^{-4\pi/\varepsilon})).$$

The limit is therefore 2π. – Note the extremely good approximation to the integral given by the Riemann sums.

7.2.8. $\hat{\varphi}(\xi) = \max(0, \pi(1 - |\xi|/2))$; by Poisson's summation formula $\sum \varphi(x + \pi n) = \hat{\varphi}(0)/\pi = 1$.

7.2.9. Note that the partial sums of the series are bounded by $\sup |u|$ and that they converge uniformly on compact sets, hence in $\mathscr{S}'$. The approximation procedure will be used in the following two exercises.

7.2.10. Hint: Use the preceding exercise to reduce the proof to the case where u is a trigonometrical polynomial. (A trigonometrical polynomial $\sum_{-v}^{v} a_j e^{i\lambda jx}$ of degree v has at most $2v$ zeros $e^{i\lambda x}$, and the values can be prescribed at $2v$ such points.) Note that the result implies that $-u'(x) + \lambda \sin(\lambda x)$ has the same sign as $\sin(\lambda x)$ if $u(x) = \cos(\lambda x)$ and deduce that $|u'|^2/\lambda^2 + |u|^2 \le \sup|u|^2$ for every $u \in L^\infty(\mathbf{R})$ with $\operatorname{supp}\hat{u} \subset [-\lambda, \lambda]$. (Bernstein's inequality.)

7.2.11. u is bounded by Example 7.1.18. Consider first the case where u is periodic with period $2\pi N/\lambda$, and extend to the non-periodic case using Exercise 7.2.9.

Section 7.3

7.3.1. u must be a linear combination of δ_a and δ_a' for some $a \in \mathbf{R}$. Hint: The sufficiency of this condition follows from Exercise 4.3.3, for example. To prove the necessity note that if $\hat{u}(c) = 0$ for some $c \in \mathbf{C}$ then we can take $\hat{v}(\zeta) = \zeta - c$ and $\hat{w} = \hat{u}(\zeta)/(\zeta - c)$, so w must be a Dirac measure and u has the stated form. If $\hat{u}$ has no zero at all then $\log\hat{u}(\zeta)$ is an entire function with real part $\le C(1+|\operatorname{Im}\zeta|+\log(1+|\zeta|))$ which implies that $\hat{u}(\zeta) = Ce^{-ia\zeta}$ for some $a \in \mathbf{R}$, so $u = C\delta_a$.

7.3.2. Hint: Assume first that $p = \infty$. By the Paley-Wiener-Schwartz theorem we have an analytic extension $u(z)$ with

$$|u(z)| \le C(1+|z|)^N e^{H(-\operatorname{Im} z)}, \quad z \in \mathbf{C}^n.$$

The Phragmén-Lindelöf theorem applied to $w \mapsto u(x+wy)e^{-H(-y)\operatorname{Im} w}$ when $\operatorname{Im} w > 0$ proves that the absolute value is $\le \|u\|_{L^\infty}$, proving the claim when $p = \infty$. If $p < \infty$, put $q = p/(p-1)$ and apply the result already proved to $u * v$ where $v \in L^q$ has compact support, $\|v\|_{L^q} \le 1$.

7.3.3. $K = 4\pi^{-2}\sum_{-\infty}^{\infty}(-1)^k(2k-1)^{-2}\delta_{(2k-1)\pi/2}$. (Compute the Fourier coefficients of $\widehat{K}$ noting that $\widehat{K}''$ consists of Dirac measures.) The formula $u' = K * u$ is clear if $\hat{u} \in C_0^\infty((-1,1))$ and follows in general by regularization since K has finite mass. (It is true when $\operatorname{supp}\hat{u} \subset [-1,1]$ since we can apply it to $u(tx)$ and let $t \to 1-0$.) The total mass of K is $\widehat{K}(1)/i = 1$, so $\sup|u'| \le \sup|u|$, which is Bernstein's inequality. Equality occurs when $u(x) = ae^{ix} + be^{-ix}$.

7.3.4. Reduce to $\lambda = 1$ by a dilation. Then choose $\widehat{K}(\xi) = i\xi\sin\alpha + \cos\alpha$, $-1 \le \xi < 1$, so that $\Phi(\xi) = \widehat{K}(\xi)e^{-i\alpha\xi}$ is periodic with period 2. Develop Φ in Fourier series and show that K is a measure with support

$\{k\pi - \alpha; k \in \mathbf{Z}\}$ with total mass $e^{-i\alpha}\widehat{K}(1) = 1$, and argue as in the preceding exercise.

7.3.5. Hint: Apply Theorems 7.3.6, 7.3.2 and Lemma 7.3.7. If an entire function h satisfies the equation $P(D)h = 0$ then $P(D)h_k = 0$ for every k if h_k is the sum of the terms of order k in the Taylor expansion of h, for all the polynomials $P(D)h_k$ have different degrees of homogeneity. – The result is true if (and only if) every irreducible factor of P vanishes at the origin, but the proof is harder then (see Malgrange [1]).

Section 7.4

7.4.1. Parseval's formula gives $\int |\hat{f}(\xi + i\eta)|^2\, d\xi = 2\pi \int_0^\infty |f(x)|^2 e^{2x\eta}\, dx$ which proves the claim. Every analytic function F in the lower half plane such that $\|F(\cdot + i\eta)\|_{L^2}$ is bounded when $\eta < 0$ is the Laplace transform of a function in $L^2((0,\infty))$. Use Theorem 7.4.2 or prove directly that if f_η is the inverse Fourier transform of $F(\cdot + i\eta)$ then $e^{-x\eta} f_\eta(x)$ is independent of η.

7.4.2. This exercise and the preceding one together are the Paley-Wiener theorem in the strict sense.

7.4.3. $2F(z) = 1/(e^{\pi z/2} + e^{-\pi z/2})$ and $G(\xi) = (e^{\xi} + e^{-\xi})^{-1}$. The Fourier transform of $x \mapsto F(x + iy)$ is $\xi \mapsto e^{-\xi y} G(\xi)$ is, so $G(\xi)(e^{\xi} + e^{-\xi}) = 1$. Use residue calculus to get F.

7.4.4. Hint: Express $\int |\hat{f}(\xi + i\eta)|^2\, d\xi$ by Parseval's formula.

7.4.5. The first part is a reformulation of the preceding exercise. When $u \in \mathscr{S}'$ one obtains the functions with $|U(z)|(1 + |z|)^N e^{-\Phi(z,\bar{z})}$ bounded for some $N < 0$, and when $u \in \mathscr{S}$ one obtains those for which this is true for all N; then we have

$$u(y) = 2^{-\frac{n}{2}} \pi^{-\frac{3n}{4}} (\det \operatorname{Im} C)^{-\frac{1}{4}} |\det B| \int e^{-i\overline{\varphi(z,y)} - 2\Phi(z,\bar{z})} U(z)\, d\lambda(z),$$

which implies $u \in \mathscr{S}$. In the case of $\mathscr{S}'$ we obtain u by duality.

7.4.6. Make an optimal choice of y when estimating $U(\zeta)$. The last statement follows by using Fourier's inversion formula or Liouville's theorem. – A distribution theory with the usual properties apart from the notion of support can be based on the test functions in this exercise. (See Gelfand and Šilov [1,2].)

Section 7.6

7.6.1. $\xi \mapsto \pi^{3/2}e^{\pi i/4}\exp(-i(\xi_1^2+\xi_2^2-\xi_3^2)/4)$.

7.6.2. The condition is $1 \le p \le 2$. (Cf. Exercise 7.1.13 for a dual special case.) – The estimate of f_t follows from the fact that f_t is the convolution of f and $ce^{-i|\xi|^2/4t}t^{-n/2}$ where c is a constant. By Parseval's formula we conclude that $\int |f_t|^p\,dx \le Ct^{n(1-p/2)}$ if $p > 2$ whereas the L^1 norm of $\hat f_t$ is independent of t over any compact set. If $p > 2$ it follows by the closed graph theorem and Baire's theorem that one can find $g \in L^p$ such that $\hat g$ is not a measure on any open subset of $\mathbf{R}^n$. On the other hand, if $1 \le p \le 2$ then the Fourier transform of L^p is contained in L^q, where $1/p + 1/q = 1$.

7.6.3. $\hat u(\xi) = e^{i\xi_1^2/4\xi_2}\sqrt{\pi/i\xi_2}$ with the square root in the right half plane. Hint: In the integral

$$\langle u, \hat\varphi\rangle = \int \hat\varphi(t,t^2)\,dt = \int dt \int e^{-i(t\xi_1+t^2\xi_2)}\varphi(\xi)\,d\xi, \quad \varphi \in \mathscr{S},$$

we can change the order of integration if we just integrate for $-T \le t \le T$. The integral of the exponential with respect to t can be estimated by $C|\xi_2|^{-\frac12}$ independently of T. When $\xi_2 \ne 0$ it converges to the result given, and we can use the dominated convergence theorem to justify the result.

7.6.4. $E = (4\pi|x_2|)^{-\frac12}e^{ix_3^2/4x_2-\pi i(\operatorname{sgn} x_2)/4}/(2\pi(x_1+ix_2))$ is one solution. Hint: Take the Fourier transform with respect to x_3 and choose a fundamental solution for the resulting operator, which is essentially the Cauchy-Riemann operator, so that it is Gaussian in ξ_3.

7.6.5. $a = \frac12$ and $u = \delta_0\sqrt{\pi/2}$. Hint: Express $\langle e^{inx^2}, \varphi\rangle$ in terms of $\hat\varphi$ when $\varphi \in \mathscr{S}$ is real valued.

7.6.6. $a = 2$ and $u = 2\pi\partial^2\delta_0/\partial x_1\partial x_2$. Hint: This time we obtain for real valued $\varphi \in \mathscr{S}$

$$\iint e^{inx_1x_2}\varphi(x)\,dx = (2\pi n)^{-1}\iint e^{-in^{-1}\xi_1\xi_2}\hat\varphi(\xi)\,d\xi$$
$$= (2\pi/n)\varphi(0) + (2\pi i/n^2)\partial_1\partial_2\varphi(0) + O(n^{-3}).$$

The first term drops out when we take the imaginary part.

7.6.7. $ap > 1$. Hint: Prove first that

$$\hat f_a(\xi) = e^{-i\xi^2/4}\int_\Gamma (1+(z+\xi/2)^2)^{-a/2}e^{iz^2}\,dz$$

where Γ consists of the lines $\operatorname{Im} z = \pm\frac{1}{2}$, $\pm\operatorname{Re} z > 1$ and the interval between $(1, \frac{i}{2})$ and $(-1, -\frac{i}{2})$. Conclude that $\hat{f}_a$ is continuous and that

$$|\hat{f}_a(\xi)|(|\xi|/2)^a \to |\int_\Gamma e^{iz^2}\, dz| = \int e^{-t^2}\, dt = \sqrt{\pi}.$$

7.6.8. $p'(-D)F(\xi) = \xi F(\xi)$, which implies that F is an entire analytic function. Alternatively, prove first that

$$F(\xi) = \int_\Gamma e^{i(p(x)-x\xi)}\, dx,$$

where Γ consists of an interval from $c_+ i$ to $c_- i$ on the imaginary axis and the half axes $\{x + c_\mp i; x \lessgtr 0\}$ where $a_m(\pm 1)^{m-1} c_\pm > 0$ if a_m is the leading coefficient in p. (Compare with the discussion of the Airy function.)

7.6.9. Hint: Reduce first to the case where $A = iC$ with C real and of diagonal form with diagonal elements ± 1. Decompose u by the partition of unity in Theorem 1.4.6 and integrate by parts n times in each term as in the proof of Lemma 7.6.4. Regularize the terms obtained where u is differentiated n times and integrate by parts once more in the smooth part.

7.6.10. Hint: Apply Taylor's formula to $e^{-t\langle AD,D\rangle}u$ as a function of t.

Bibliography

Agmon, S.: [1] The coerciveness problem for integro-differential forms. J. Analyse Math. 6, 183–223 (1958).
– [2] Spectral properties of Schrödinger operators. Actes Congr. Int. Math. Nice 2, 679–683 (1970).
– [3] Spectral properties of Schrödinger operators and scattering theory. Ann. Scuola Norm. Sup. Pisa (4) 2, 151–218 (1975).
– [4] Unicité et convexité dans les problèmes différentiels. Sém. Math. Sup. No 13, Les Presses de l'Univ. de Montreal, 1966.
– [5] Lectures on elliptic boundary value problems. van Nostrand Mathematical Studies 2, Princeton, N.J. 1965.
– [6] Problèmes mixtes pour les équations hyperboliques d'ordre supérieur. Coll. Int. CNRS 117, 13–18, Paris 1962.
– [7] Some new results in spectral and scattering theory of differential operators on $\mathbb{R}^n$. Sém. Goulaouic-Schwartz 1978–1979, Exp. II, 1–11.
Agmon, S., A. Douglis and L. Nirenberg: [1] Estimates near the boundary for solutions of elliptic partial differential equations satisfying general boundary conditions. I. Comm. Pure Appl. Math. 12, 623–727 (1959); II. Comm. Pure Appl. Math. 17, 35–92 (1964).
Agmon, S. and L. Hörmander: [1] Asymptotic properties of solutions of differential equations with simple characteristics. J. Analyse Math. 30, 1–38 (1976).
Agranovich, M.S.: [1] Partial differential equations with constant coefficients. Uspehi Mat. Nauk 16:2, 27–94 (1961). (Russian; English translation in Russian Math. Surveys 16:2, 23–90 (1961).)
Ahlfors, L. and M. Heins: [1] Questions of regularity connected with the Phragmén-Lindelöf principle. Ann. of Math. 50, 341–346 (1949).
Airy, G.B.: [1] On the intensity of light in a neighborhood of a caustic. Trans. Cambr. Phil. Soc. 6, 379–402 (1838).
Alinhac, S.: [1] Non-unicité du problème de Cauchy. Ann. of Math. 117, 77–108 (1983).
– [2] Non-unicité pour des opérateurs différentiels à caractéristiques complexes simples. Ann. Sci. École Norm. Sup. 13, 385–393 (1980).
– [3] Uniqueness and non-uniqueness in the Cauchy problem. Contemporary Math. 27, 1–22 (1984).
Alinhac, S. and M.S. Baouendi: [1] Uniqueness for the characteristic Cauchy problem and strong unique continuation for higher order partial differential inequalities. Amer. J. Math. 102, 179–217 (1980).
Alinhac, S. and C. Zuily: [1] Unicité et non-unicité du problème de Cauchy pour des opérateurs hyperboliques à caractéristiques doubles. Comm. Partial Differential Equations 6, 799–828 (1981).
Alsholm, P.K.: [1] Wave operators for long range scattering. Mimeographed report, Danmarks Tekniske Højskole 1975.
Alsholm, P.K. and T. Kato: [1] Scattering with long range potentials. In Partial Diff. Eq., Proc. of Symp. in Pure Math. 23, 393–399. Amer. Math. Soc. Providence, R.I. 1973.

Amrein, W.O., Ph.A. Martin and P. Misra: [1] On the asymptotic condition of scattering theory. Helv. Phys. Acta 43, 313–344 (1970).

Andersson, K.G.: [1] Propagation of analyticity of solutions of partial differential equations with constant coefficients. Ark. Mat. 8, 277–302 (1971).

Andersson, K.G. and R.B. Melrose: [1] The propagation of singularities along glidings rays. Invent. Math. 41, 197–232 (1977).

Arnold, V.I.: [1] On a characteristic class entering into conditions of quantization. Funkcional. Anal. i Priložen. 1, 1–14 (1967) (Russian); also in Functional Anal. Appl. 1, 1–13 (1967).

Aronszajn, N.: [1] Boundary values of functions with a finite Dirichlet integral. Conference on Partial Differential Equations 1954, University of Kansas, 77–94.

– [2] A unique continuation theorem for solutions of elliptic partial differential equations or inequalities of second order. J. Math. Pures Appl. 36, 235–249 (1957).

Aronszajn, N., A. Krzywcki and J. Szarski: [1] A unique continuation theorem for exterior differential forms on Riemannian manifolds. Ark. Mat. 4, 417–453 (1962).

Asgeirsson, L.: [1] Über eine Mittelwerteigenschaft von Lösungen homogener linearer partieller Differentialgleichungen 2. Ordnung mit konstanten Koeffizienten. Math. Ann. 113, 321–346 (1937).

Atiyah, M.F.: [1] Resolution of singularities and division of distributions. Comm. Pure Appl. Math. 23, 145–150 (1970).

Atiyah, M.F. and R. Bott: [1] The index theorem for manifolds with boundary. Proc. Symp. on Differential Analysis, 175–186. Oxford 1964.

– [2] A Lefschetz fixed point formula for elliptic complexes. I. Ann. of Math. 86, 374–407 (1967).

Atiyah, M.F., R. Bott and L. Gårding: [1] Lacunas for hyperbolic differential operators with constant coefficients. I. Acta Math. 124, 109–189 (1970).

– [2] Lacunas for hyperbolic differential operators with constant coefficients. II. Acta Math. 131, 145–206 (1973).

Atiyah, M.F., R. Bott and V.K. Patodi: [1] On the heat equation and the index theorem. Invent. Math. 19, 279–330 (1973).

Atiyah, M.F. and I.M. Singer: [1] The index of elliptic operators on compact manifolds Bull. Amer. Math. Soc. 69, 422–433 (1963).

– [2] The index of elliptic operators. I, III. Ann. of Math. 87, 484–530 and 546–604 (1968).

Atkinson, F.V.: [1] The normal solubility of linear equations in normed spaces. Mat. Sb. 28 (70), 3–14 (1951) (Russian).

Avakumovič, V.G.: [1] Über die Eigenfunktionen auf geschlossenen Riemannschen Mannigfaltigkeiten. Math. Z. 65, 327–344 (1956).

Bang, T.: [1] Om quasi-analytiske funktioner. Thesis, Copenhagen 1946, 101 pp.

Baouendi, M.S. and Ch. Goulaouic: [1] Nonanalytic-hypoellipticity for some degenerate elliptic operators. Bull. Amer. Math. Soc. 78, 483–486 (1972).

Beals, R.: [1] A general calculus of pseudo-differential operators. Duke Math. J. 42, 1–42 (1975).

Beals, R. and C. Fefferman: [1] On local solvability of linear partial differential equations. Ann. of Math. 97, 482–498 (1973).

– [2] Spatially inhomogeneous pseudo-differential operators I. Comm. Pure Appl. Math. 27, 1–24 (1974).

Beckner, W.: [1] Inequalities in Fourier analysis. Ann. of Math. 102, 159–182 (1975).

Berenstein, C.A. and M.A. Dostal: [1] On convolution equations I. In L'anal. harm. dans le domaine complexe. Springer Lecture Notes in Math. 336, 79–94 (1973).

Bernstein, I.N.: [1] Modules over a ring of differential operators. An investigation of the fundamental solutions of equations with constant coefficients. Funkcional. Anal. i Priložen. 5:2, 1–16 (1971) (Russian); also in Functional Anal. Appl. 5, 89–101 (1971).

Bernstein, I.N. and S.I. Gelfand: [1] Meromorphy of the function P^λ. Funkcional. Anal. i Priložen. 3:1, 84–85 (1969) (Russian); also in Functional Anal. Appl. 3, 68–69 (1969).

Bernstein, S.: [1] Sur la nature analytique des solutions des équations aux dérivées partielles du second ordre. Math. Ann. 59, 20–76 (1904).

Beurling, A.: [1] Quasi-analyticity and general distributions. Lectures 4 and 5, Amer. Math. Soc. Summer Inst. Stanford 1961 (Mimeographed).

– [2] Sur les spectres des fonctions. Anal. Harm. Nancy 1947, Coll. Int. XV, 9–29.

– [3] Analytic continuation across a linear boundary. Acta Math. 128, 153–182 (1972).

Björck, G.: [1] Linear partial differential operators and generalized distributions. Ark. Mat. 6, 351–407 (1966).

Björk, J.E.: [1] Rings of differential operators. North-Holland Publ. Co. Math. Library series 21 (1979).

Bochner, S.: [1] Vorlesungen über Fouriersche Integrale. Leipzig 1932.

Boman, J.: [1] On the intersection of classes of infinitely differentiable functions. Ark. Mat. 5, 301–309 (1963).

Bonnesen, T. and W. Fenchel: [1] Theorie der konvexen Körper. Erg. d. Math. u. ihrer Grenzgeb. 3, Springer Verlag 1934.

Bony, J.M.: [1] Une extension du théorème de Holmgren sur l'unicité du problème de Cauchy. C.R. Acad. Sci. Paris 268, 1103–1106 (1969).

– [2] Extensions du théorème de Holmgren. Sém. Goulaouic-Schwartz 1975–1976, Exposé no. XVII.

– [3] Equivalence des diverses notions de spectre singulier analytique. Sém. Goulaouic-Schwartz 1976–1977, Exposé no. III.

Bony, J.M. and P. Schapira: [1] Existence et prolongement des solutions holomorphes des équations aux dérivées partielles. Invent. Math. 17, 95–105 (1972).

Borel, E.: [1] Sur quelques points de la théorie des fonctions. Ann. Sci. École Norm. Sup. 12 (3), 9–55 (1895).

Boutet de Monvel, L.: [1] Comportement d'un opérateur pseudo-différentiel sur une variété à bord. J. Analyse Math. 17, 241–304 (1966).

– [2] Boundary problems for pseudo-differential operators. Acta Math. 126, 11–51 (1971).

– [3] On the index of Toeplitz operators of several complex variables. Invent. Math. 50, 249–272 (1979).

– [4] Hypoelliptic operators with double characteristics and related pseudo-differential operators. Comm. Pure Appl. Math. 27, 585–639 (1974).

Boutet de Monvel, L., A. Grigis and B. Helffer: [1] Parametrixes d'opérateurs pseudo-différentiels à caractéristiques multiples. Astérisque 34–35, 93–121 (1976).

Boutet de Monvel, L. and V. Guillemin: [1] The spectral theory of Toeplitz operators. Ann. of Math. Studies 99 (1981).

Brézis, H.: [1] On a characterization of flow-invariant sets. Comm. Pure Appl. Math. 23, 261–263 (1970).

Brodda, B.: [1] On uniqueness theorems for differential equations with constant coefficients. Math. Scand. 9, 55–68 (1961).

Browder, F.: [1] Estimates and existence theorems for elliptic boundary value problems. Proc. Nat. Acad. Sci. 45, 365–372 (1959).

Buslaev, V.S. and V.B. Matveev: [1] Wave operators for the Schrödinger equation with a slowly decreasing potential. Theor. and Math. Phys. 2, 266–274 (1970). (English translation.)

Calderón, A.P.: [1] Uniqueness in the Cauchy problem for partial differential equations. Amer. J. Math. 80, 16–36 (1958).

– [2] Existence and uniqueness theorems for systems of partial differential equations. Fluid Dynamics and Applied Mathematics (Proc. Symp. Univ. of Maryland 1961), 147–195. New York 1962.

– [3] Boundary value problems for elliptic equations. Outlines of the joint Soviet-American symposium on partial differential equations, 303–304, Novosibirsk 1963.

Calderón, A.P. and R. Vaillancourt: [1] On the boundedness of pseudo-differential operators. J. Math. Soc. Japan 23, 374–378 (1972).

– [2] A class of bounded pseudo-differential operators. Proc. Nat. Acad. Sci. U.S.A. 69, 1185–1187 (1972).

Calderón, A.P. and A. Zygmund: [1] On the existence of certain singular integrals. Acta Math. 88, 85–139 (1952).

Carathéodory, C.: [1] Variationsrechnung und partielle Differentialgleichungen Erster Ordnung. Teubner, Berlin, 1935.

Carleman, T.: [1] Sur un problème d'unicité pour les systèmes d'équations aux dérivées partielles à deux variables indépendentes. Ark. Mat. Astr. Fys. 26B No 17, 1–9 (1939).

– [2] L'intégrale de Fourier et les questions qui s'y rattachent. Publ. Sci. Inst. Mittag-Leffler, Uppsala 1944.

– [3] Propriétés asymptotiques des fonctions fondamentales des membranes vibrantes. C.R. Congr. des Math. Scand. Stockholm 1934, 34–44 (Lund 1935).

Catlin, D.: [1] Necessary conditions for subellipticity and hypoellipticity for the $\bar{\partial}$Neumann problem on pseudoconvex domains. In Recent developments in several complex variables. Ann. of Math. Studies 100, 93–100 (1981).

Cauchy, A.: [1] Mémoire sur l'intégration des équations linéaires. C.R. Acad. Sci. Paris 8 (1839). In Œuvres IV, 369–426, Gauthier-Villars, Paris 1884.

Cerezo, A., J. Chazarain and A. Piriou: [1] Introduction aux hyperfonctions. Springer Lecture Notes in Math. 449, 1–53 (1975).

Chaillou, J.: [1] Hyperbolic differential polynomials and their singular perturbations. D. Reidel Publ. Co. Dordrecht, Boston, London 1979.

Charazain, J.: [1] Construction de la paramétrix du problème mixte hyperbolique pour l'équation des ondes. C.R. Acad. Sci. Paris 276, 1213–1215 (1973).

– [2] Formules de Poisson pour les variétés riemanniennes. Invent. Math. 24, 65–82 (1974).

Chazarain, J. and A. Piriou: [1] Introduction à la théorie des équations aux dérivées partielles linéaires. Gauthier-Villars 1981.

Chester, C., B. Friedman and F. Ursell: [1] An extension of the method of steepest descent. Proc. Cambr. Phil. Soc. 53, 599–611 (1957).

Cohen, P.: [1] The non-uniqueness of the Cauchy problem. O.N.R. Techn. Report 93, Stanford 1960.

– [2] A simple proof of the Denjoy-Carleman theorem. Amer. Math. Monthly 75, 26–31 (1968).

– [3] A simple proof of Tarski's theorem on elementary algebra. Mimeographed manuscript, Stanford University 1967, 6 pp.

Colin de Verdière, Y.: [1] Sur le spectra des opérateurs elliptiques à bicharactéristiques toutes périodiques. Comment. Math. Helv. 54, 508–522 (1979).

Cook, J.: [1] Convergence to the Møller wave matrix. J. Mathematical Physics 36, 82–87 (1957).

Cordes, H.O.: [1] Über die eindeutige Bestimmtheit der Lösungen elliptischer Differentialgleichungen durch Anfangsvorgaben. Nachr. Akad. Wiss. Göttingen Math.-Phys. Kl. IIa, No. 11, 239–258 (1956).

Cotlar, M.: [1] A combinatorial inequality and its application to L^2 spaces. Rev. Math. Cuyana 1, 41–55 (1955).

Courant, R. and D. Hilbert: [1] Methoden der Mathematischen Physik II. Berlin 1937.

Courant, R. and P.D. Lax: [1] The propagation of discontinuities in wave motion. Proc. Nat. Acad. Sci. 42, 872–876 (1956).

De Giorgi, E.: [1] Un esempio di non-unicitá della soluzione del problema di Cauchy relativo ad una equazione differenziale lineare a derivate parziali ti tipo parabolico. Rend. Mat. 14, 382–387 (1955).

– [2] Solutions analytiques des équations aux dérivées partielles à coefficients constants. Sém. Goulaouic-Schwartz 1971–1972, Exposé 29.

Deič, V.G., E.L. Korotjaev and D.R. Jafaev: [1] The theory of potential scattering with account taken of spatial anisotropy. Zap. Naučn. Sem. Leningrad Otdel. Mat. Inst. Steklov 73, 35–51 (1977).

Dencker, N.: [1] On the propagation of singularities for pseudo-differential operators of principal type. Ark. Mat. 20, 23–60 (1982).

– [2] The Weyl calculus with locally temperate metrics and weights. Ark. Mat. 24, 59–79 (1986).

Dieudonné, J.: [1] Sur les fonctions continus numériques définies dans un produit de deux espaces compacts. C.R. Acad. Sci. Paris 205, 593–595 (1937).

Dieudonné, J. and L. Schwartz: [1] La dualité dans les espaces $(\mathscr{F})$ et $(\mathscr{L}\mathscr{F})$. Ann. Inst. Fourier (Grenoble) 1, 61–101 (1949).

Dollard, J.D.: [1] Asymptotic convergence and the Coulomb interaction. J. Math. Phys. 5, 729–738 (1964).

– [2] Quantum mechanical scattering theory for short-range and Coulomb interactions. Rocky Mountain J. Math. 1, 5–88 (1971).

Douglis, A. and L. Nirenberg: [1] Interior estimates for elliptic systems of partial differential equations. Comm. Pure Appl. Math. 8, 503–538 (1955).

Duistermaat, J.J.: [1] Oscillatory integrals, Lagrange immersions and unfolding of singularities. Comm. Pure Appl. Math. 27, 207–281 (1974).

Duistermaat, J.J. and V.W. Guillemin: [1] The spectrum of positive elliptic operators and periodic bicharacteristics. Invent. Math. 29, 39–79 (1975).

Duistermaat, J.J. and L. Hörmander: [1] Fourier integral operators II. Acta Math. 128, 183–269 (1972).

Duistermaat, J.J. and J. Sjöstrand: [1] A global construction for pseudo-differential operators with non-involutive characteristics. Invent. Math. 20, 209–225 (1973).

DuPlessis, N.: [1] Some theorems about the Riesz fractional integral. Trans. Amer. Math. Soc. 80, 124–134 (1955).

Egorov, Ju.V.: [1] The canonical transformations of pseudo-differential operators. Uspehi Mat. Nauk 24:5, 235–236 (1969).

– [2] Subelliptic pseudo-differential operators. Dokl. Akad. Nauk SSSR 188, 20–22 (1969); also in Soviet Math. Doklady 10, 1056–1059 (1969).

– [3] Subelliptic operators. Uspehi Mat. Nauk 30:2, 57–114 and 30:3, 57–104 (1975); also in Russian Math. Surveys 30:2, 59–118 and 30:3, 55–105 (1975).

Ehrenpreis, L.: [1] Solutions of some problems of division I. Amer. J. Math. 76, 883–903 (1954).

– [2] Solutions of some problems of division III. Amer. J. Math. 78, 685–715 (1956).

– [3] Solutions of some problems of division IV. Amer. J. Math. 82, 522–588 (1960).

– [4] On the theory of kernels of Schwartz. Proc. Amer. Math. Soc. 7, 713–718 (1956).

– [5] A fundamental principle for systems of linear differential equations with constant coefficients, and some of its applications. Proc. Intern. Symp. on Linear Spaces, Jerusalem 1961, 161–174.

– [6] Fourier analysis in several complex variables. Wiley-Interscience Publ., New York, London, Sydney, Toronto 1970.

– [7] Analytically uniform spaces and some applications. Trans. Amer. Math. Soc. 101, 52–74 (1961).

– [8] Solutions of some problems of division V. Hyperbolic operators. Amer. J. Math. 84, 324–348 (1962).

Enqvist, A.: [1] On fundamental solutions supported by a convex cone. Ark. Mat. 12, 1–40 (1974).

Enss, V.: [1] Asymptotic completeness for quantum-mechanical potential scattering. I. Short range potentials. Comm. Math. Phys. 61, 285–291 (1978).

– [2] Geometric methods in spectral and scattering theory of Schrödinger operators.

In Rigorous Atomic and Molecular Physics, G. Velo and A. Wightman ed., Plenum, New York, 1980–1981 (Proc. Erice School of Mathematical Physics 1980).

Eškin, G.I.: [1] Boundary value problems for elliptic pseudo-differential equations. Moscow 1973; Amer. Math. Soc. Transl. of Math. Monographs 52, Providence, R.I. 1981.

– [2] Parametrix and propagation of singularities for the interior mixed hyperbolic problem. J. Analyse Math. 32, 17–62 (1977).

– [3] General initial-boundary problems for second order hyperbolic equations. In Sing. in Boundary Value Problems. D. Reidel Publ. Co., Dordrecht, Boston, London 1981, 19–54.

– [4] Initial boundary value problem for second order hyperbolic equations with general boundary conditions I. J. Analyse Math. 40, 43–89 (1981).

Fedosov, B.V.: [1] A direct proof of the formula for the index of an elliptic system in Euclidean space. Funkcional. Anal. i Priložen. 4:4, 83–84 (1970) (Russian); also in Functional Anal. Appl. 4, 339–341 (1970).

Fefferman, C.L.: [1] The uncertainty principle. Bull. Amer. Math. Soc. 9, 129–206 (1983).

Fefferman, C. and D.H. Phong: [1] On positivity of pseudo-differential operators. Proc. Nat. Acad. Sci. 75, 4673–4674 (1978).

– [2] The uncertainty principle and sharp Gårding inequalities. Comm. Pure Appl. Math. 34, 285–331 (1981).

Fredholm, I.: [1] Sur l'intégrale fondamentale d'une équation différentielle elliptique à coefficients constants. Rend. Circ. Mat. Palermo 25, 346–351 (1908).

Friedlander, F.G.: [1] The wave front set of the solution of a simple initial-boundary value problem with glancing rays. Math. Proc. Cambridge Philos. Soc. 79, 145–159 (1976).

Friedlander, F.G. and R.B. Melrose: [1] The wave front set of the solution of a simple initial-boundary value problem with glancing rays. II. Math. Proc. Cambridge Philos. Soc. 81, 97–120 (1977).

Friedrichs, K.: [1] On differential operators in Hilbert spaces. Amer. J. Math. 61, 523–544 (1939).

– [2] The identity of weak and strong extensions of differential operators. Trans. Amer. Math. Soc. 55, 132–151 (1944).

– [3] On the differentiability of the solutions of linear elliptic differential equations. Comm. Pure Appl. Math. 6, 299–326 (1953).

– [4] On the perturbation of continuous spectra. Comm. Pure Appl. Math. 1, 361–406 (1948).

Friedrichs, K. and H. Lewy: [1] Über die Eindeutigkeit und das Abhängigkeitsgebiet der Lösungen beim Anfangswertproblem linearer hyperbolischer Differentialgleichungen. Math. Ann. 98, 192–204 (1928).

Fröman, N. and P.O. Fröman: [1] JWKB approximation. Contributions to the theory. North-Holland Publ. Co. Amsterdam 1965.

Fuglede, B.: [1] A priori inequalities connected with systems of partial differential equations. Acta Math. 105, 177–195 (1961).

Gabrielov, A.M.: [1] A certain theorem of Hörmander. Funkcional. Anal. i Priložen. 4:2, 18–22 (1970) (Russian); also in Functional Anal. Appl. 4, 106–109 (1970).

Gårding, L.: [1] Linear hyperbolic partial differential equations with constant coefficients. Acta Math. 85, 1–62 (1951).

– [2] Dirichlet's problem for linear elliptic partial differential equations. Math. Scand. 1, 55–72 (1953).

– [3] Solution directe du problème de Cauchy pour les équations hyperboliques. Coll. Int. CNRS, Nancy 1956, 71–90.

– [4] Transformation de Fourier des distributions homogènes. Bull. Soc. Math. France 89, 381–428 (1961).

– [5] Local hyperbolicity. Israel J. Math. 13, 65–81 (1972).

– [6] Le problème de la dérivée oblique pour l'équation des ondes. C.R. Acad. Sci. Paris 285, 773–775 (1977). Rectification C.R. Acad. Sci. Paris 285, 1199 (1978).
– [7] On the asymptotic distribution of the eigenvalues and eigenfunctions of elliptic differential operators. Math. Scand. 1, 237–255 (1953).
Gårding, L. and J.L. Lions: [1] Functional analysis. Nuovo Cimento N. 1 del Suppl. al Vol. (10)14, 9–66 (1959).
Gårding, L. and B. Malgrange: [1] Opérateurs différentiels partiellement hypoelliptiques et partiellement elliptiques. Math. Scand. 9, 5–21 (1961).
Gask, H.: [1] A proof of Schwartz' kernel theorem. Math. Scand. 8, 327–332 (1960).
Gelfand, I.M. and G.E. Šilov: [1] Fourier transforms of rapidly increasing functions and questions of uniqueness of the solution of Cauchy's problem. Uspehi Mat. Nauk 8:6, 3–54 (1953) (Russian); also in Amer. Math. Soc. Transl. (2) 5, 221–274 (1957).
– [2] Generalized functions. Volume 1: Properties and operations. Volume 2: Spaces of fundamental and generalized functions. Academic Press, New York and London 1964, 1968.
Gevrey, M.: [1] Démonstration du théorème de Picard-Bernstein par la méthode des contours successifs; prolongement analytique. Bull. Sci. Math. 50, 113–128 (1926).
Glaeser, G.: [1] Etude de quelques algèbres Tayloriennes. J. Analyse Math. 6, 1–124 (1958).
Godin, P.: [1] Propagation des singularités pour les opérateurs pseudo-différentiels de type principal à partie principal analytique vérifiant la condition (*P*), en dimension 2. C.R. Acad. Sci. Paris 284, 1137–1138 (1977).
Gorin, E.A.: [1] Asymptotic properties of polynomials and algebraic functions of several variables. Uspehi Mat. Nauk 16:1, 91–118 (1961) (Russian); also in Russian Math. Surveys 16:1, 93–119 (1961).
Grubb, G.: [1] Boundary problems for systems of partial differential operators of mixed order. J. Functional Analysis 26, 131–165 (1977).
– [2] Problèmes aux limites pseudo-différentiels dépendant d'un paramètre. C.R. Acad. Sci. Paris 292, 581–583 (1981).
Grušin, V.V.: [1] The extension of smoothness of solutions of differential equations of principal type. Dokl. Akad. Nauk SSSR 148, 1241–1244 (1963) (Russian); also in Soviet Math. Doklady 4, 248–252 (1963).
– [2] A certain class of hypoelliptic operators. Mat. Sb. 83, 456–473 (1970) (Russian); also in Math. USSR Sb. 12, 458–476 (1970).
Gudmundsdottir, G.: [1] Global properties of differential operators of constant strength. Ark. Mat. 15, 169–198 (1977).
Guillemin, V.: [1] The Radon transform on Zoll surfaces. Advances in Math. 22, 85–119 (1976).
– [2] Some classical theorems in spectral theory revisited. Seminar on sing. of sol. of diff. eq., Princeton University Press, Princeton, N.J., 219–259 (1979).
– [3] Some spectral results for the Laplace operator with potential on the n-sphere. Advances in Math. 27, 273–286 (1978).
Guillemin, V. and D. Schaeffer: [1] Remarks on a paper of D. Ludwig. Bull. Amer. Math. Soc. 79, 382–385 (1973).
Guillemin, V. and S. Sternberg: [1] Geometrical asymptotics. Amer. Math. Soc. Surveys 14, Providence, R.I. 1977.
Gurevič, D.I.: [1] Counterexamples to a problem of L. Schwartz. Funkcional. Anal. i Priložen. 9:2, 29–35 (1975) (Russian); also in Functional Anal. Appl. 9, 116–120 (1975).
Hack, M.N.: [1] On convergence to the Møller wave operators. Nuovo Cimento (10) 13, 231–236 (1959).
Hadamard, J.: [1] Le problème de Cauchy et les équations aux dérivées partielles linéaires hyperboliques. Paris 1932.
Haefliger, A.: [1] Variétés feuilletées. Ann. Scuola Norm. Sup. Pisa 16, 367–397 (1962).

Hanges, N.: [1] Propagation of singularities for a class of operators with double characteristics. Seminar on singularities of sol. of linear partial diff. eq., Princeton University Press, Princeton, N.J. 1979, 113–126.

Hardy, G.H. and J.E. Littlewood: [1] Some properties of fractional integrals. (I) Math. Z. 27, 565–606 (1928); (II) Math. Z. 34, 403–439 (1931–32).

Hausdorff, F.: [1] Eine Ausdehnung des Parsevalschen Satzes über Fourierreihen. Math. Z. 16, 163–169 (1923).

Hayman, W.K. and P.B. Kennedy: [1] Subharmonic functions I. Academic Press, London, New York, San Francisco 1976.

Hedberg, L.I.: [1] On certain convolution inequalities. Proc. Amer. Math. Soc. 36, 505–510 (1972).

Heinz, E.: [1] Über die Eindeutigkeit beim Cauchyschen Anfangswertproblem einer elliptischen Differentialgleichung zweiter Ordnung. Nachr. Akad. Wiss. Göttingen Math.-Phys. Kl. IIa No. 1, 1–12 (1955).

Helffer, B.: [1] Addition de variables et applications à la régularité. Ann. Inst. Fourier (Grenoble) 28:2, 221–231 (1978).

Helffer, B. and J. Nourrigat: [1] Caractérisation des opérateurs hypoelliptiques homogènes invariants à gauche sur un groupe de Lie nilpotent gradué. Comm. Partial Differential Equations 4:8, 899–958 (1979).

Herglotz, G.: [1] Über die Integration linearer partieller Differentialgleichungen mit konstanten Koeffizienten I–III. Berichte Sächs. Akad. d. Wiss. 78, 93–126, 287–318 (1926); 80, 69–114 (1928).

Hersh, R.: [1] Boundary conditions for equations of evolution. Arch. Rational Mech. Anal. 16, 243–264 (1964).

– [2] On surface waves with finite and infinite speed of propagation. Arch. Rational Mech. Anal. 19, 308–316 (1965).

Hirzebruch, F.: [1] Neue topologische Methoden in der algebraischen Geometrie. Springer Verlag, Berlin-Göttingen-Heidelberg 1956.

Hlawka, E.: [1] Über Integrale auf konvexen Körpern. I. Monatsh. Math. 54, 1–36 (1950).

Holmgren, E.: [1] Über Systeme von linearen partiellen Differentialgleichungen. Öfversigt af Kongl. Vetenskaps-Akad. Förh. 58, 91–103 (1901).

– [2] Sur l'extension de la méthode d'intégration de Riemann. Ark. Mat. Astr. Fys. 1, No 22, 317–326 (1904).

Hörmander, L.: [1] On the theory of general partial differential operators. Acta Math. 94, 161–248 (1955).

– [2] Local and global properties of fundamental solutions. Math. Scand. 5, 27–39 (1957).

– [3] On the regularity of the solutions of boundary problems. Acta Math. 99, 225–264 (1958).

– [4] On interior regularity of the solutions of partial differential equations. Comm. Pure Appl. Math. 11, 197–218 (1958).

– [5] On the division of distributions by polynomials. Ark. Mat. 3, 555–568 (1958).

– [6] Differentiability properties of solutions of systems of differential equations. Ark. Mat. 3, 527–535 (1958).

– [7] Definitions of maximal differential operators. Ark. Mat. 3, 501–504 (1958).

– [8] On the uniqueness of the Cauchy problem I, II. Math. Scand. 6, 213–225 (1958); 7, 177–190 (1959).

– [9] Null solutions of partial differential equations. Arch. Rational Mech. Anal. 4, 255–261 (1960).

– [10] Differential operators of principal type. Math. Ann. 140, 124–146 (1960).

– [11] Differential equations without solutions. Math. Ann. 140, 169–173 (1960).

– [12] Hypoelliptic differential operators. Ann. Inst. Fourier (Grenoble) 11, 477–492 (1961).

– [13] Estimates for translation invariant operators in L^p spaces. Acta Math. 104, 93–140 (1960).
– [14] On the range of convolution operators. Ann. of Math. 76, 148–170 (1962).
– [15] Supports and singular supports of convolutions. Acta Math. 110, 279–302 (1963).
– [16] Pseudo-differential operators. Comm. Pure Appl. Math. 18, 501–517 (1965).
– [17] Pseudo-differential operators and non-elliptic boundary problems. Ann. of Math. 83, 129–209 (1966).
– [18] Pseudo-differential operators and hypoelliptic equations. Amer. Math. Soc. Symp. on Singular Integrals, 138–183 (1966).
– [19] An introduction to complex analysis in several variables. D. van Nostrand Publ. Co., Princeton, N.J. 1966.
– [20] Hypoelliptic second order differential equations. Acta Math. 119, 147–171 (1967).
– [21] On the characteristic Cauchy problem. Ann. of Math. 88, 341–370 (1968).
– [22] The spectral function of an elliptic operator. Acta Math. 121, 193–218 (1968).
– [23] Convolution equations in convex domains. Invent. Math. 4, 306–317 (1968).
– [24] On the singularities of solutions of partial differential equations. Comm. Pure Appl. Math. 23, 329–358 (1970).
– [25] Linear differential operators. Actes Congr. Int. Math. Nice 1970, 1, 121–133.
– [26a] The calculus of Fourier integral operators. Prospects in math. Ann. of Math. Studies 70, 33–57 (1971).
– [26] Fourier integral operators I. Acta Math. 127, 79–183 (1971).
– [27] Uniqueness theorems and wave front sets for solutions of linear differential equations with analytic coefficients. Comm. Pure Appl. Math. 24, 671–704 (1971).
– [28] A remark on Holmgren's uniqueness theorem. J. Diff. Geom. 6, 129–134 (1971).
– [29] On the existence and the regularity of solutions of linear pseudo-differential equations. Ens. Math. 17, 99–163 (1971).
– [30] On the singularities of solutions of partial differential equations with constant coefficients. Israel J. Math. 13, 82–105 (1972).
– [31] On the existence of real analytic solutions of partial differential equations with constant coefficients. Invent. Math. 21, 151–182 (1973).
– [32] Lower bounds at infinity for solutions of differential equations with constant coefficients. Israel J. Math. 16, 103–116 (1973).
– [33] Non-uniqueness for the Cauchy problem. Springer Lecture Notes in Math. 459, 36–72 (1975).
– [34] The existence of wave operators in scattering theory. Math. Z. 146, 69–91 (1976).
– [35] A class of hypoelliptic pseudo-differential operators with double characteristics. Math. Ann. 217, 165–188 (1975).
– [36] The Cauchy problem for differential equations with double characteristics. J. Analyse Math. 32, 118–196 (1977).
– [37] Propagation of singularities and semiglobal existence theorems for (pseudo-) differential operators of principal type. Ann. of Math. 108, 569–609 (1978).
– [38] Subelliptic operators. Seminar on sing. of sol. of diff. eq. Princeton University Press, Princeton, N.J., 127–208 (1979).
– [39] The Weyl calculus of pseudo-differential operators. Comm. Pure Appl. Math. 32, 359–443 (1979).
– [40] Pseudo-differential operators of principal type. Nato Adv. Study Inst. on Sing. in Bound. Value Problems. Reidel Publ. Co., Dordrecht, 69–96 (1981).
– [41] Uniqueness theorems for second order elliptic differential equations. Comm. Partial Differential Equations 8, 21–64 (1983).
– [42] On the index of pseudo-differential operators. In Elliptische Differentialgleichungen Band II, Akademie-Verlag, Berlin 1971, 127–146.
– [43] L^2 estimates for Fourier integral operators with complex phase. Ark. Mat. 21, 297–313 (1983).
– [44] On the subelliptic test estimates. Comm. Pure Appl. Math. 33, 339–363 (1980).

Hurwitz, A.: [1] Über die Nullstellen der Bessel'schen Funktion. Math. Ann. 33, 246–266 (1889).

Iagolnitzer, D.: [1] Microlocal essential support of a distribution and decomposition theorems - an introduction. In Hyperfunctions and theoretical physics. Springer Lecture Notes in Math. 449, 121–132 (1975).

Ikebe, T.: [1] Eigenfunction expansions associated with the Schrödinger operator and their applications to scattering theory. Arch. Rational Mech. Anal. 5, 1–34 (1960).

Ikebe, T. and Y. Saito: [1] Limiting absorption method and absolute continuity for the Schrödinger operator. J. Math. Kyoto Univ. 12, 513–542 (1972).

Ivrii, V.Ja: [1] Sufficient conditions for regular and completely regular hyperbolicity. Trudy Moskov. Mat. Obšč. 33, 3–65 (1975) (Russian); also in Trans. Moscow Math. Soc. 33, 1–65 (1978).

– [2] Wave fronts for solutions of boundary value problems for a class of symmetric hyperbolic systems. Sibirsk. Mat. Ž. 21:4, 62–71 (1980) (Russian); also in Sibirian Math. J. 21, 527–534 (1980).

– [3] On the second term in the spectral asymptotics for the Laplace-Beltrami operator on a manifold with boundary. Funkcional. Anal. i Priložen. 14:2, 25–34 (1980) (Russian); also in Functional Anal. Appl. 14, 98–106 (1980).

Ivrii, V.Ja and V.M. Petkov: [1] Necessary conditions for the correctness of the Cauchy problem for non-strictly hyperbolic equations. Uspehi Mat. Nauk 29:5, 3–70 (1974) (Russian); also in Russian Math. Surveys 29:5, 1–70 (1974).

Iwasaki, N.: [1] The Cauchy problems for effectively hyperbolic equations (general case). J. Math. Kyoto Univ. 25, 727–743 (1985).

Jauch, J.M. and I.I. Zinnes: [1] The asymptotic condition for simple scattering systems. Nuovo Cimento (10) 11, 553–567 (1959).

Jerison D. and C.E. Kenig: [1] Unique continuation and absence of positive eigenvalues for Schrödinger operators. Ann. of Math. 121, 463–488 (1985).

John, F.: [1] On linear differential equations with analytic coefficients. Unique continuation of data. Comm. Pure Appl. Math. 2, 209–253 (1949).

– [2] Plane waves and spherical means applied to partial differential equations. New York 1955.

– [3] Non-admissible data for differential equations with constant coefficients. Comm. Pure Appl. Math. 10, 391–398 (1957).

– [4] Continuous dependence on data for solutions of partial differential equations with a prescribed bound. Comm. Pure Appl. Math. 13, 551–585 (1960).

– [5] Linear partial differential equations with analytic coefficients. Proc. Nat. Acad. Sci. 29, 98–104 (1943).

Jörgens, K. and J. Weidmann: [1] Zur Existenz der Wellenoperatoren. Math. Z. 131, 141–151 (1973).

Kashiwara, M.: [1] Introduction to the theory of hyperfunctions. In Sem. on microlocal analysis, Princeton Univ. Press, Princeton, N.J., 1979, 3–38.

Kashiwara, M. and T. Kawai: [1] Microhyperbolic pseudo-differential operators. I. J. Math. Soc. Japan 27, 359–404 (1975).

Kato, T.: [1] Growth properties of solutions of the reduced wave equation with a variable coefficient. Comm. Pure Appl. Math. 12, 403–425 (1959).

Keller, J.B.: [1] Corrected Bohr-Sommerfeld quantum conditions for nonseparable systems. Ann. Physics 4, 180–188 (1958).

Kitada, H.: [1] Scattering theory for Schrödinger operators with long-range potentials. I: Abstract theory. J. Math. Soc. Japan 29, 665–691 (1977). II: Spectral and scattering theory. J. Math. Soc. Japan 30, 603–632 (1978).

Knapp, A.W. and E.M. Stein: [1] Singular integrals and the principal series. Proc. Nat. Acad. Sci. U.S.A. 63, 281–284 (1969).

Kohn, J.J.: [1] Harmonic integrals on strongly pseudo-convex manifolds I, II. Ann. of Math. 78, 112–148 (1963); 79, 450–472 (1964).

– [2] Pseudo-differential operators and non-elliptic problems. In Pseudo-differential operators, CIME conference, Stresa 1968, 157–165. Edizione Cremonese, Roma 1969.
Kohn, J.J. and L. Nirenberg: [1] On the algebra of pseudo-differential operators. Comm. Pure Appl. Math. 18, 269-305 (1965).
– [2] Non-coercive boundary value problems. Comm. Pure Appl. Math. 18, 443-492 (1965).
Kolmogorov, A.N.: [1] Zufällige Bewegungen. Ann. of Math. 35, 116–117 (1934).
Komatsu, H.: [1] A local version of Bochner's tube theorem. J. Fac. Sci. Tokyo Sect. I-A Math. 19, 201–214 (1972).
– [2] Boundary values for solutions of elliptic equations. Proc. Int. Conf. Funct. Anal. Rel. Topics, 107–121. University of Tokyo Press, Tokyo 1970.
Kreiss, H.O.: [1] Initial boundary value problems for hyperbolic systems. Comm. Pure Appl. Math. 23, 277-298 (1970).
Krzyżański, M. and J. Schauder: [1] Quasilineare Differentialgleichungen zweiter Ordnung vom hyperbolischen Typus. Gemischte Randwertaufgaben. Studia Math. 6, 162–189 (1936).
Kumano-go, H.: [1] Factorizations and fundamental solutions for differential operators of elliptic-hyperbolic type. Proc. Japan Acad. 52, 480–483 (1976).
Kuroda, S.T.: [1] On the existence and the unitary property of the scattering operator. Nuovo Cimento (10) 12, 431–454 (1959).
Lascar, B. and R. Lascar: [1] Propagation des singularités pour des équations hyperboliques à caractéristiques de multiplicité au plus double et singularités Masloviennes II. J. Analyse Math. 41, 1–38 (1982).
Lax, A.: [1] On Cauchy's problem for partial differential equations with multiple characteristics. Comm. Pure Appl. Math. 9, 135–169 (1956).
Lax, P.D.: [2] On Cauchy's problem for hyperbolic equations and the differentiability of solutions of elliptic equations. Comm. Pure Appl. Math. 8, 615–633 (1955).
– [3] Asymptotic solutions of oscillatory initial value problems. Duke Math. J. 24, 627–646 (1957).
Lax, P.D. and L. Nirenberg: [1] On stability for difference schemes: a sharp form of Gårding's inequality. Comm. Pure Appl. Math. 19, 473–492 (1966).
Lebeau, G.: [1] Fonctions harmoniques et spectre singulier. Ann. Sci. École Norm. Sup. (4) 13, 269–291 (1980).
Lelong, P.: [1] Plurisubharmonic functions and positive differential forms. Gordon and Breach, New York, London, Paris 1969.
– [2] Propriétés métriques des variétés définies par une équation. Ann. Sci. École Norm. Sup. 67, 22–40 (1950).
Leray, J.: [1] Hyperbolic differential equations. The Institute for Advanced Study, Princeton, N.J. 1953.
– [2] Uniformisation de la solution du problème linéaire analytique de Cauchy près de la variété qui porte les données de Cauchy. Bull. Soc. Math. France 85, 389–429 (1957).
Lerner, N.: [1] Unicité de Cauchy pour des opérateurs différentiels faiblement principalement normaux. J. Math. Pures Appl. 64, 1–11 (1985).
Lerner, N. and L. Robbiano: [1] Unicité de Cauchy pour des opérateurs de type principal. J. Analyse Math. 44, 32–66 (1984/85).
Levi, E.E.: [1] Caratterische multiple e problema di Cauchy. Ann. Mat. Pura Appl. (3) 16, 161–201 (1909).
Levinson, N.: [1] Transformation of an analytic function of several variables to a canonical form. Duke Math. J. 28, 345–353 (1961).
Levitan, B.M.: [1] On the asymptotic behavior of the spectral function of a self-adjoint differential equation of the second order. Izv. Akad. Nauk SSSR Ser. Mat. 16, 325–352 (1952).
– [2] On the asymptotic behavior of the spectral function and on expansion in eigenfunctions of a self-adjoint differential equation of second order II. Izv. Akad. Nauk SSSR Ser. Mat. 19, 33–58 (1955).

Lewy, H.: [1] An example of a smooth linear partial differential equation without solution. Ann. of Math. 66, 155–158 (1957).
– [2] Extension of Huyghen's principle to the ultrahyperbolic equation. Ann. Mat. Pura Appl. (4) 39, 63–64 (1955).
Lions, J.L.: [1] Supports dans la transformation de Laplace. J. Analyse Math. 2, 369–380 (1952–53).
Lions, J.L. and E. Magenes: [1] Problèmes aux limites non homogènes et applications I–III. Dunod, Paris, 1968–1970.
Łojasiewicz, S.: [1] Sur le problème de division. Studia Math. 18, 87–136 (1959).
Lopatinski, Ya.B.: [1] On a method of reducing boundary problems for a system of differential equations of elliptic type to regular integral equations. Ukrain. Mat. Ž. 5, 123–151 (1953). Amer. Math. Soc. Transl. (2) 89, 149–183 (1970).
Ludwig, D.: [1] Exact and asymptotic solutions of the Cauchy problem. Comm. Pure Appl. Math. 13, 473–508 (1960).
– [2] Uniform asymptotic expansions at a caustic. Comm. Pure Appl. Math. 19, 215–250 (1966).
Luke, G.: [1] Pseudodifferential operators on Hilbert bundles. J. Differential Equations 12, 566–589 (1972).
Malgrange, B.: [1] Existence et approximation des solutions des équations aux dérivées partielles et des équations de convolution. Ann. Inst. Fourier (Grenoble) 6, 271–355 (1955–56).
– [2] Sur une classe d'opérateurs différentiels hypoelliptiques. Bull. Soc. Math. France 85, 283–306 (1957).
– [3] Sur la propagation de la régularité des solutions des équations à coefficients constants. Bull. Math. Soc. Sci. Math. Phys. R.P. Roumanie 3 (53), 433–440 (1959).
– [4] Sur les ouverts convexes par rapport à un opérateur différentiel. C. R. Acad. Sci. Paris 254, 614–615 (1962).
– [5] Sur les systèmes différentiels à coefficients constants. Coll. CNRS 113–122, Paris 1963.
– [6] Ideals of differentiable functions. Tata Institute, Bombay, and Oxford University Press 1966.
Mandelbrojt, S.: [1] Analytic functions and classes of infinitely differentiable functions. Rice Inst. Pamphlet 29, 1–142 (1942).
– [2] Séries adhérentes, régularisations des suites, applications. Coll. Borel, Gauthier-Villars, Paris 1952.
Martineau, A.: [1] Les hyperfonctions de M. Sato. Sém. Bourbaki 1960–1961, Exposé No 214.
– [2] Le "edge of the wedge theorem" en théorie des hyperfonctions de Sato. Proc. Int. Conf. Funct. Anal. Rel. Topics, 95–106. University of Tokyo Press, Tokyo 1970.
Maslov, V.P.: [1] Theory of perturbations and asymptotic methods. Moskov. Gos. Univ., Moscow 1965 (Russian).
Mather, J.: [1] Stability of C^∞ mappings: I. The division theorem. Ann. of Math. 87, 89–104 (1968).
Melin, A.: [1] Lower bounds for pseudo-differential operators. Ark. Mat. 9, 117–140 (1971).
– [2] Parametrix constructions for right invariant differential operators on nilpotent groups. Ann. Global Analysis and Geometry 1, 79–130 (1983).
Melin, A. and J. Sjöstrand: [1] Fourier integral operators with complex-valued phase functions. Springer Lecture Notes in Math. 459, 120–223 (1974).
– [2] Fourier integral operators with complex phase functions and parametrix for an interior boundary value problem. Comm. Partial Differential Equations 1:4, 313–400 (1976).
Melrose, R.B.: [1] Transformation of boundary problems. Acta Math. 147, 149–236 (1981).

– [2] Equivalence of glancing hypersurfaces. Invent. Math. 37, 165–191 (1976).
– [3] Microlocal parametrices for diffractive boundary value problems. Duke Math. J. 42, 605–635 (1975).
– [4] Local Fourier-Airy integral operators. Duke Math. J. 42, 583–604 (1975).
– [5] Airy operators. Comm. Partial Differential Equations 3:1, 1–76 (1978).
– [6] The Cauchy problem for effectively hyperbolic operators. Hokkaido Math. J. To appear.
– [7] The trace of the wave group. Contemporary Math. 27, 127–167 (1984).
Melrose, R.B. and J. Sjöstrand: [1] Singularities of boundary value problems I, II. Comm. Pure Appl. Math. 31, 593–617 (1978); 35, 129–168 (1982).
Mihlin, S.G.: [1] On the multipliers of Fourier integrals. Dokl. Akad. Nauk SSSR 109, 701–703 (1956) (Russian).
Mikusiński, J.: [1] Une simple démonstration du théorème de Titchmarsh sur la convolution. Bull. Acad. Pol. Sci. 7, 715–717 (1959).
– [2] The Bochner integral. Birkhäuser Verlag, Basel and Stuttgart 1978.
Minakshisundaram, S. and Å. Pleijel: [1] Some properties of the eigenfunctions of the Laplace operator on Riemannian manifolds. Canad. J. Math. 1, 242–256 (1949).
Mizohata, S.: [1] Unicité du prolongement des solutions des équations elliptiques du quatrième ordre. Proc. Jap. Acad. 34, 687–692 (1958).
– [2] Systèmes hyperboliques. J. Math. Soc. Japan 11, 205–233 (1959).
– [3] Note sur le traitement par les opérateurs d'intégrale singulière du problème de Cauchy. J. Math. Soc. Japan 11, 234–240 (1959).
– [4] Solutions nulles et solutions non analytiques. J. Math. Kyoto Univ. 1, 271–302 (1962).
– [5] Some remarks on the Cauchy problem. J. Math. Kyoto Univ. 1, 109–127 (1961).
Møller, C.: [1] General properties of the characteristic matrix in the theory of elementary particles. I. Kongl. Dansk. Vidensk. Selsk. Mat.-Fys. Medd. 23, 2–48 (1945).
Morrey, C.B.: [1] The analytic embedding of abstract real-analytic manifolds. Ann. of Math. 68, 159–201 (1958).
Morrey, C.B. and L. Nirenberg: [1] On the analyticity of the solutions of linear elliptic systems of partial differential equations. Comm. Pure Appl. Math. 10, 271–290 (1957).
Moyer, R.D.: [1] Local solvability in two dimensions: Necessary conditions for the principle-type case. Mimeographed manuscript, University of Kansas 1978.
Müller, C.: [1] On the behaviour of the solutions of the differential equation $\Delta U = F(x, U)$ in the neighborhood of a point. Comm. Pure Appl. Math. 7, 505–515 (1954).
Münster, M.: [1] On A. Lax's condition of hyperbolicity. Rocky Mountain J. Math. 8, 443–446 (1978).
– [2] On hyperbolic polynomials with constant coefficients. Rocky Mountain J. Math. 8, 653–673 (1978).
von Neumann, J. and E. Wigner: [1] Über merkwürdige diskrete Eigenwerte. Phys. Z. 30, 465–467 (1929).
Nirenberg, L.: [1] Remarks on strongly elliptic partial differential equations. Comm. Pure Appl. Math. 8, 648–675 (1955).
– [2] Uniqueness in Cauchy problems for differential equations with constant leading coefficients. Comm. Pure Appl. Math. 10, 89–105 (1957).
– [3] A proof of the Malgrange preparation theorem. Liverpool singularities I. Springer Lecture Notes in Math. 192, 97–105 (1971).
– [4] On elliptic partial differential equations. Ann. Scuola Norm. Sup. Pisa (3) 13, 115–162 (1959).
– [5] Lectures on linear partial differential equations. Amer. Math. Soc. Regional Conf. in Math., 17, 1–58 (1972).

Nirenberg, L. and F. Treves: [1] Solvability of a first order linear partial differential equation. Comm. Pure Appl. Math. 16, 331–351 (1963).
– [2] On local solvability of linear partial differential equations. I. Necessary conditions. II. Sufficient conditions. Correction. Comm. Pure Appl. Math. 23, 1–38 and 459–509 (1970); 24, 279–288 (1971).
Nishitani, T.: [1] Local energy integrals for effectively hyperbolic operators I, II. J. Math. Kyoto Univ. 24, 623–658 and 659–666 (1984).
Noether, F.: [1] Über eine Klasse singulärer Integralgleichungen. Math. Ann. 82, 42–63 (1921).
Olejnik, O.A.: [1] On the Cauchy problem for weakly hyperbolic equations. Comm. Pure Appl. Math. 23, 569–586 (1970).
Olejnik, O.A. and E.V. Radkevič: [1] Second order equations with non-negative characteristic form. In Matem. Anal. 1969, ed. R.V. Gamkrelidze, Moscow 1971 (Russian). English translation Plenum Press, New York-London 1973.
Oshima, T.: [1] On analytic equivalence of glancing hypersurfaces. Sci. Papers College Gen. Ed. Univ. Tokyo 28, 51–57 (1978).
Palamodov, V.P.: [1] Linear differential operators with constant coefficients. Moscow 1967 (Russian). English transl. Grundl. d. Math. Wiss. 168, Springer Verlag, New York, Heidelberg, Berlin 1970.
Paley, R.E.A.C. and N. Wiener: [1] Fourier transforms in the complex domain. Amer. Math. Soc. Coll. Publ. XIX, New York 1934.
Pederson, R.: [1] On the unique continuation theorem for certain second and fourth order elliptic equations. Comm. Pure Appl. Math. 11, 67–80 (1958).
– [2] Uniqueness in the Cauchy problem for elliptic equations with double characteristics. Ark. Mat. 6, 535–549 (1966).
Peetre, J.: [1] Théorèmes de régularité pour quelques classes d'opérateurs différentiels. Thesis, Lund 1959.
– [2] Uniqueness in the Cauchy problem for elliptic equations with double characteristics. tels". Math. Scand. 8, 116–120 (1960).
– [3] Another approach to elliptic boundary problems. Comm. Pure Appl. Math. 14, 711–731 (1961).
– [4] New thoughts on Besov spaces. Duke Univ. Math. Series I. Durham, N.C. 1976.
Persson, J.: [1] The wave operator and P-convexity. Boll. Un. Mat. Ital. (5) 18-B, 591–604 (1981).
Petrowsky, I.G.: [1] Über das Cauchysche Problem für Systeme von partiellen Differentialgleichungen. Mat. Sb. 2 (44), 815–870 (1937).
– [2] Über das Cauchysche Problem für ein System linearer partieller Differentialgleichungen im Gebiete der nichtanalytischen Funktionen. Bull. Univ. Moscow Sér. Int. 1, No. 7, 1–74 (1938).
– [3] Sur l'analyticité des solutions des systèmes d'équations différentielles. Mat. Sb. 5 (47), 3–70 (1939).
– [4] On the diffusion of waves and the lacunas for hyperbolic equations. Mat. Sb. 17 (59), 289–370 (1945).
– [5] Some remarks on my papers on the problem of Cauchy. Mat. Sb. 39 (81), 267–272 (1956). (Russian.)
Pham The Lai: [1] Meilleures estimations asymptotiques des restes de la fonction spectrale et des valeurs propres relatifs au laplacien. Math. Scand. 48, 5–31 (1981).
Piccinini, L.C.: [1] Non surjectivity of the Cauchy-Riemann operator on the space of the analytic functions on $\mathbb{R}^n$. Generalization to the parabolic operators. Bull. Un. Mat. Ital. (4) 7, 12–28 (1973).
Pliš, A.: [1] A smooth linear elliptic differential equation without any solution in a sphere. Comm. Pure Appl. Math. 14, 599–617 (1961).
– [2] The problem of uniqueness for the solution of a system of partial differential equations. Bull. Acad. Pol. Sci. 2, 55–57 (1954).

– [3] On non-uniqueness in Cauchy problem for an elliptic second order differential equation. Bull. Acad. Pol. Sci. 11, 95–100 (1963).

Poincaré, H.: [1] Sur les propriétés du potentiel et les fonctions abeliennes. Acta Math. 22, 89–178 (1899).

Povzner, A.Ya.: [1] On the expansion of arbitrary functions in characteristic functions of the operator $-\Delta u + cu$. Mat. Sb. 32 (74), 109–156 (1953).

Radkevič, E.: [1] A priori estimates and hypoelliptic operators with multiple characteristics. Dokl. Akad. Nauk SSSR 187, 274–277 (1969) (Russian); also in Soviet Math. Doklady 10, 849–853 (1969).

Ralston, J.: [1] Solutions of the wave equation with localized energy. Comm. Pure Appl. Math. 22, 807–823 (1969).

– [2] Gaussian beams and the propagation of singularities. MAA Studies in Mathematics 23, 206–248 (1983).

Reed, M. and B. Simon: [1] Methods of modern mathematical physics. III. Scattering theory. Academic Press 1979.

Rempel, S. and B.-W. Schulze: [1] Index theory of elliptic boundary problems. Akademie-Verlag, Berlin (1982).

de Rham, G.: [1] Variétés différentiables. Hermann, Paris, 1955.

Riesz, F.: [1] Sur l'existence de la dérivée des fonctions d'une variable réelle et des fonctions d'intervalle. Verh. Int. Math. Kongr. Zürich 1932, I, 258–269.

Riesz, M.: [1] L'intégrale de Riemann-Liouville et le problème de Cauchy. Acta Math. 81, 1–223 (1949).

– [2] Sur les maxima des formes bilinéaires et sur les fonctionnelles linéaires. Acta Math. 49, 465–497 (1926).

– [3] Sur les fonctions conjuguées. Math. Z. 27, 218–244 (1928).

– [4] Problems related to characteristic surfaces. Proc. Conf. Diff. Eq. Univ. Maryland 1955, 57–71.

Rothschild, L.P.: [1] A criterion for hypoellipticity of operators constructed from vector fields. Comm. Partial Differential Equations 4:6, 645–699 (1979).

Saito, Y.: [1] On the asymptotic behavior of the solutions of the Schrödinger equation $(-\Delta + Q(y) - k^2)V = F$. Osaka J. Math. 14, 11–35 (1977).

– [2] Eigenfunction expansions for the Schrödinger operators with long-range potentials $Q(y) = O(|y|^{-\varepsilon})$, $\varepsilon > 0$. Osaka J. Math. 14, 37–53 (1977).

Sakamoto, R.: [1] E-well posedness for hyperbolic mixed problems with constant coefficients. J. Math. Kyoto Univ. 14, 93–118 (1974).

– [2] Mixed problems for hyperbolic equations I. J. Math. Kyoto Univ. 10, 375–401 (1970), II. J. Math. Kyoto Univ. 10, 403–417 (1970).

Sato, M.: [1] Theory of hyperfunctions I. J. Fac. Sci. Univ. Tokyo I, 8, 139–193 (1959).

– [2] Theory of hyperfunctions II. J. Fac. Sci. Univ. Tokyo I, 8, 387–437 (1960).

– [3] Hyperfunctions and partial differential equations. Proc. Int. Conf. on Funct. Anal. and Rel. Topics, 91–94, Tokyo University Press, Tokyo 1969.

– [4] Regularity of hyperfunction solutions of partial differential equations. Actes Congr. Int. Math. Nice 1970, 2, 785–794.

Sato, M., T. Kawai and M. Kashiwara: [1] Hyperfunctions and pseudodifferential equations. Springer Lecture Notes in Math. 287, 265–529 (1973).

Schaefer, H.H.: [1] Topological vector spaces. Springer Verlag, New York, Heidelberg, Berlin 1970.

Schapira, P.: [1] Hyperfonctions et problèmes aux limites elliptiques. Bull. Soc. Math. France 99, 113–141 (1971).

– [2] Propagation at the boundary of analytic singularities. Nato Adv. Study Inst. on Sing. in Bound. Value Problems. Reidel Publ. Co., Dordrecht, 185–212 (1981).

– [3] Propagation at the boundary and reflection of analytic singularities of solutions of linear partial differential equations. Publ. RIMS, Kyoto Univ., 12 Suppl., 441–453 (1977).

Schechter, M.: [1] Various types of boundary conditions for elliptic equations. Comm. Pure Appl. Math. 13, 407–425 (1960).
– [2] A generalization of the problem of transmission. Ann. Scuola Norm. Sup. Pisa 14, 207–236 (1960).
Schwartz, L.: [1] Théorie des distributions I, II. Hermann, Paris, 1950–51.
– [2] Théorie des noyaux. Proc. Int. Congr. Math. Cambridge 1950, I, 220–230.
– [3] Sur l'impossibilité de la multiplication des distributions. C. R. Acad. Sci. Paris 239, 847–848 (1954).
– [4] Théorie des distributions à valeurs vectorielles I. Ann. Inst. Fourier (Grenoble) 7, 1-141 (1957).
– [5] Transformation de Laplace des distributions. Comm. Sém. Math. Univ. Lund, Tome suppl. dédié à Marcel Riesz, 196–206 (1952).
– [6] Théorie générale des fonctions moyenne-périodiques. Ann. of Math. 48, 857–929 (1947).
Seeley, R.T.: [1] Singular integrals and boundary problems. Amer. J. Math. 88, 781–809 (1966).
– [2] Extensions of C^∞ functions defined in a half space. Proc. Amer. Math. Soc. 15, 625–626 (1964).
– [3] A sharp asymptotic remainder estimate for the eigenvalues of the Laplacian in a domain of $\mathbb{R}^3$. Advances in Math. 29, 244–269 (1978).
– [4] An estimate near the boundary for the spectral function of the Laplace operator. Amer. J. Math. 102, 869-902 (1980).
– [5] Elliptic singular integral equations. Amer Math. Soc. Symp. on Singular Integrals, 308–315 (1966).
Seidenberg, A.: [1] A new decision method for elementary algebra. Ann. of Math. 60, 365–374 (1954).
Shibata, Y.: [1] E-well posedness of mixed initial-boundary value problems with constant coefficients in a quarter space. J. Analyse Math. 37, 32–45 (1980).
Siegel, C.L.: [1] Zu den Beweisen des Vorbereitungssatzes von Weierstrass. In Abhandl. aus Zahlenth. u. Anal., 299–306. Plenum Press, New York 1968.
Sjöstrand, J.: [1] Singularités analytiques microlocales. Prépublications Université de Paris-Sud 82-03.
– [2] Analytic singularities of solutions of boundary value problems. Nato Adv. Study Inst. on Sing. in Bound. Value Prob., Reidel Publ. Co., Dordrecht, 235–269 (1981).
– [3] Parametrices for pseudodifferential operators with multiple characteristics. Ark. Mat. 12, 85–130 (1974).
– [4] Propagation of analytic singularities for second order Dirichlet problems I, II, III. Comm. Partial Differential Equations 5:1, 41–94 (1980), 5:2, 187–207 (1980), and 6:5, 499–567 (1981).
– [5] Operators of principal type with interior boundary conditions. Acta Math. 130, 1-51 (1973).
Sobolev, S.L.: [1] Méthode nouvelle à résoudre le problème de Cauchy pour les équations linéaires hyperboliques normales. Mat. Sb. 1 (43), 39–72 (1936).
– [2] Sur un théorème d'analyse fonctionnelle. Mat. Sb. 4 (46), 471–497 (1938). (Russian; French summary.) Amer. Math. Soc. Transl. (2) 34, 39–68 (1963).
Sommerfeld, A.: [1] Optics. Lectures on theoretical physics IV. Academic Press, New York, 1969.
Stein, E.M.: [1] Singular integrals and differentiability properties of functions. Princeton Univ. Press 1970.
Sternberg, S.: [1] Lectures on differential geometry. Prentice-Hall Inc. Englewood Cliffs, N.J., 1964.
Stokes, G.B.: [1] On the numerical calculation of a class of definite integrals and infinite series. Trans. Cambridge Philos. Soc. 9, 166–187 (1850).

Svensson, L.: [1] Necessary and sufficient conditions for the hyperbolicity of polynomials with hyperbolic principal part. Ark. Mat. 8, 145–162 (1968).

Sweeney, W.J.: [1] The *D*-Neumann problem. Acta Math. 120, 223–277 (1968).

Szegö, G.: [1] Beiträge zur Theorie der Toeplitzschen Formen. Math. Z. 6, 167–202 (1920).

Täcklind, S.: [1] Sur les classes quasianalytiques des solutions des équations aux dérivées partielles du type parabolique. Nova Acta Soc. Sci. Upsaliensis (4) 10, 1–57 (1936).

Tarski, A.: [1] A decision method for elementary algebra and geometry. Manuscript, Berkeley, 63 pp. (1951).

Taylor, M.: [1] Gelfand theory of pseudodifferential operators and hypoelliptic operators. Trans. Amer. Math. Soc. 153, 495–510 (1971).

– [2] Grazing rays and reflection of singularities of solutions to wave equations. Comm. Pure Appl. Math. 29, 1–38 (1976).

– [3] Diffraction effects in the scattering of waves. In Sing. in Bound. Value Problems, 271–316. Reidel Publ. Co., Dordrecht 1981.

– [4] Pseudodifferential operators. Princeton Univ. Press, Princeton, N.J., 1981.

Thorin, O.: [1] An extension of a convexity theorem due to M. Riesz. Kungl Fys. Sällsk. Lund. Förh. 8, No 14 (1939).

Titchmarsh, E.C.: [1] The zeros of certain integral functions. Proc. London Math. Soc. 25, 283–302 (1926).

Treves, F.: [1] Solution élémentaire d'équations aux dérivées partielles dépendant d'un paramètre. C. R. Acad. Sci. Paris 242, 1250–1252 (1956).

– [2] Thèse d'Hörmander II. Sém. Bourbaki 135, 2ᵉ éd. (Mai 1956).

– [3] Relations de domination entre opérateurs différentiels. Acta Math. 101, 1–139 (1959).

– [4] Opérateurs différentiels hypoelliptiques. Ann. Inst. Fourier (Grenoble) 9, 1–73 (1959).

– [5] Local solvability in L^2 of first order linear PDEs. Amer. J. Math. 92, 369–380 (1970).

– [6] Fundamental solutions of linear partial differential equations with constant coefficients depending on parameters. Amer. J. Math. 84, 561–577 (1962).

– [7] Un théorème sur les équations aux dérivées partielles à coefficients constants dépendant de paramètres. Bull. Soc. Math. France 90, 473–486 (1962).

– [8] A new method of proof of the subelliptic estimates. Comm. Pure Appl. Math. 24, 71–115 (1971).

– [9] Introduction to pseudodifferential and Fourier integral operators. Volume 1: Pseudodifferential operators. Volume 2: Fourier integral operators. Plenum Press, New York and London 1980.

Vauthier, J.: [1] Comportement asymptotique des fonctions entières de type exponentiel dans $\mathbb{C}^n$ et bornées dans le domaine réel. J. Functional Analysis 12, 290–306 (1973).

Vekua, I.N.: [1] Systeme von Differentialgleichungen erster Ordnung vom elliptischen Typus und Randwertaufgaben. Berlin 1956.

Veselič, K. and J. Weidmann: [1] Existenz der Wellenoperatoren für eine allgemeine Klasse von Operatoren. Math. Z. 134, 255–274 (1973).

– [2] Asymptotic estimates of wave functions and the existence of wave operators. J. Functional Analysis. 17, 61–77 (1974).

Višik, M.I.: [1] On general boundary problems for elliptic differential equations. Trudy Moskov. Mat. Obšč. 1, 187–246 (1952) (Russian). Also in Amer. Math. Soc. Transl. (2) 24, 107–172 (1963).

Višik, M.I. and G.I. Eškin: [1] Convolution equations in a bounded region. Uspehi Mat. Nauk 20:3 (123), 89–152 (1965). (Russian.) Also in Russian Math. Surveys 20:3, 86–151 (1965).

– [2] Convolution equations in a bounded region in spaces with weighted norms. Mat. Sb. 69 (111), 65–110 (1966) (Russian). Also in Amer. Math. Soc. Transl. (2) 67, 33–82 (1968).
– [3] Elliptic convolution equations in a bounded region and their applications. Uspehi Mat. Nauk 22:1 (133), 15–76 (1967). (Russian.) Also in Russian Math. Surveys 22:1, 13–75 (1967).
– [4] Convolution equations of variable order. Trudy Moskov. Mat. Obšč. 16, 25–50 (1967). (Russian.) Also in Trans. Moskov. Mat. Soc. 16, 27–52 (1967).
– [5] Normally solvable problems for elliptic systems of convolution equations. Mat. Sb. 74 (116), 326–356 (1967). (Russian.) Also in Math. USSR-Sb. 3, 303–332 (1967).
van der Waerden, B.L.: [1] Einführung in die algebraische Geometrie. Berlin 1939.
– [2] Algebra I–II. 4. Aufl. Springer Verlag, Berlin-Göttingen-Heidelberg 1959.
Wang Rou-hwai and Tsui Chih-yung: [1] Generalized Leray formula on positive complex Lagrange-Grassmann manifolds. Res. Report, Inst. of Math., Jilin Univ. 8209, 1982.
Warner, F.W.: [1] Foundations of differentiable manifolds and Lie Groups. Scott, Foresman and Co., Glenview, Ill., London, 1971.
Weinstein, A.: [1] The order and symbol of a distribution. Trans. Amer. Math. Soc. 241, 1–54 (1978).
– [2] Asymptotics of eigenvalue clusters for the Laplacian plus a potential. Duke Math. J. 44, 883–892 (1977).
– [3] On Maslov's quantization condition. In Fourier integral operators and partial differential equations. Springer Lecture Notes in Math. 459, 341–372 (1974).
Weyl, H.: [1] The method of orthogonal projection in potential theory. Duke Math. J. 7, 411–444 (1940).
– [2] Die Idee der Riemannschen Fläche. 3. Aufl., Teubner, Stuttgart, 1955.
– [3] Über gewöhnliche Differentialgleichungen mit Singularitäten und die zugehörigen Entwicklungen willkürlicher Funktionen. Math. Ann. 68, 220–269 (1910).
– [4] Das asymptotische Verteilungsgesetz der Eigenwerte linearer partieller Differentialgleichungen (mit einer Anwendung auf die Theorie der Hohlraumstrahlung). Math. Ann. 71, 441–479 (1912).
Whitney, H.: [1] Analytic extensions of differentiable functions defined in closed sets. Trans. Amer. Math. Soc. 36, 63–89 (1934).
Widom, H.: [1] Eigenvalue distribution in certain homogeneous spaces. J. Functional Analysis 32, 139–147 (1979).
Yamamoto, K.: [1] On the reduction of certain pseudo-differential operators with non-involution characteristics. J. Differential Equations 26, 435–442 (1977).
Zeilon, N.: [1] Das Fundamentalintegral der allgemeinen partiellen linearen Differentialgleichung mit konstanten Koeffizienten. Ark. Mat. Astr. Fys. 6, No 38, 1–32 (1911).
Zerner, M.: [1] Solutions de l'équation des ondes présentant des singularités sur une droite. C. R. Acad. Sci. Paris 250, 2980–2982 (1960).
– [2] Solutions singulières d'équations aux dérivées partielles. Bull. Soc. Math. France 91, 203–226 (1963).
– [3] Domaine d'holomorphie des fonctions vérifiant une équation aux dérivées partielles. C. R. Acad. Sci. Paris 272, 1646–1648 (1971).
Zuily, C.: [1] Uniqueness and non-uniqueness in the Cauchy problem. Progress in Math. 33, Birkhäuser, Boston, Basel, Stuttgart 1983.
Zygmund, A.: [1] On a theorem of Marcinkiewicz concerning interpolation of operators. J. Math. Pures Appl. 35, 223–248 (1956).

Index

Index of Notation

Spaces of Functions and Distributions

Special Symbols

Special Distributions

Grundlehren der mathematischen Wissenschaften

A Series of Comprehensive Studies in Mathematics

A Selection

190. Faith: Algebra: Rings, Modules, and Categories I
191. Faith: Algebra II, Ring Theory
192. Mal'cev: Algebraic Systems
193. Pólya/Szegö: Problems and Theorems in Analysis I
194. Igusa: Theta Functions
195. Berberian: Baer*-Rings
196. Arthreya/Ney: Branching Processes
197. Benz: Vorlesungen über Geometrie der Algebren
198. Gaal: Linear Analysis and Representation Theory
199. Nitsche: Vorlesungen über Minimalflächen
200. Dold: Lectures on Algebraic Topology
201. Beck: Continuous Flows in the Plane
202. Schmetterer: Introduction to Mathematical Statistics
203. Schoeneberg: Elliptic Modular Functions
204. Popov: Hyperstability of Control Systems
205. Nikol'skiĭ: Approximation of Functions of Several Variables and Imbedding Theorems
206. André: Homologie des Algébres Commutatives
207. Donoghue: Monotone Matrix Functions and Analytic Continuation
208. Lacey: The Isometric Theory of Classical Banach Spaces
209. Ringel: Map Color Theorem
210. Gihman/Skorohod: The Theory of Stochastic Processes I
211. Comfort/Negrepontis: The Theory of Ultrafilters
212. Switzer: Algebraic Topology - Homotopy and Homology
215. Schaefer: Banach Lattices and Positive Operators
217. Stenström: Rings of Quotients
218. Gihman/Skorohod: The Theory of Stochastic Processes II
219. Duvant/Lions: Inequalities in Mechanics and Physics
220. Kirillov: Elements of the Theory of Representations
221. Mumford: Algebraic Geometry I: Complex Projective Varieties
222. Lang: Introduction to Modular Forms
223. Bergh/Löfström: Interpolation Spaces. An Introduction
224. Gilbarg/Trudinger: Elliptic Partial Differential Equations of Second order
225. Schütte: Proof Theory
226. Karoubi: K-Theory. An Introduction
227. Grauert/Remmert: Theorie der Steinschen Räume
228. Segal/Kunze: Integrals and Operators
229. Hasse: Number Theory
230. Klingenberg: Lectures on Closed Geodesics
231. Lang: Elliptic Curves: Diophantine Analysis
232. Gihman/Skorohod: The Theory of Stochastic Processes III
233. Stroock/Varadhan: Multidimensional Diffusion Processes
234. Aigner: Combinatorial Theory
235. Dynkin/Yushkevich: Controlled Markov Processes
236. Grauert/Remmert: Theory of Stein Spaces
237. Köthe: Topological Vector Spaces II
238. Graham/McGehee: Essays in Commutative Harmonic Analysis
239. Elliott: Probabilistic Number Theory I
240. Eliott: Probabilistic Number Theory II

241. Rudin: Function Theory in the Unit Ball of C^n
242. Huppert/Blackburn: Finite Groups II
243. Huppert/Blackburn: Finite Groups III
244. Kubert/Lang: Modular Units
245. Cornfeld/Fomin/Sinai: Ergodic Theory
246. Naimark/Štern: Theory of Group Representations
247. Suzuki: Group Theory I
248. Suzuki: Group Theory II
249. Chung: Lectures from Markov Processes to Brownian Motion
250. Arnold: Geometrical Methods in the Theory of Ordinary Differential Equations
251. Chow/Hale: Methods of Bifurcation Theory
252. Aubin: Nonlinear Analysis on Manifolds. Monge-Ampère Equations
253. Dwork: Lectures on p-adic Differential Equations
254. Freitag: Siegelsche Modulfunktionen
255. Lang: Complex Multiplication
256. Hörmander: The Analysis of Linear Partial Differential Operators I
257. Hörmander: The Analysis of Linear Partial Differential Operators II
258. Smoller: Shock Waves and Reaction-Diffusion Equations
259. Duren: Univalent Functions
260. Freidlin/Wentzell: Random Perturbations of Dynamical Systems
261. Bosch/Güntzer/Remmert: Non Archimedian Analysis – A Systematic Approach to Rigid Analytic Geometry
262. Doob: Classical Potential Theory and Its Probabilistic Counterpart
263. Krasnosel'skiĭ/Zabreĭko: Geometrical Methods of Nonlinear Analysis
264. Aubin/Cellina: Differential Inclusions
265. Grauert/Remmert: Coherent Analytic Sheaves
266. de Rham: Differentiable Manifolds
267. Arbarello/Cornalba/Griffiths/Harris: Geometry of Algebraic Curves, Vol. I
268. Arbarello/Cornalba/Griffiths/Harris: Geometry of Algebraic Curves, Vol. II
269. Schapira: Microdifferential Systems in the Complex Domain
270. Scharlau: Quadratic and Hermitian Forms
271. Ellis: Entropy, Large Deviations, and Statistical Mechanics
272. Elliott: Arithmetic Functions and Integer Products
273. Nikol'skiĭ: Treatise on the Shift Operator
274. Hörmander: The Analysis of Linear Partial Differential Operators III
275. Hörmander: The Analysis of Linear Partial Differential Operators IV
276. Liggett: Interacting Particle Systems
277. Fulton/Lang: Riemann-Roch Algebra
278. Barr/Wells: Toposes, Triples and Theories
279. Bishop/Bridges: Constructive Analysis
280. Neukirch: Class Field Theory
281. Chandrasekharan: Elliptic Functions
282. Lelong/Gruman: Entire Functions of Several Complex Variables
283. Kodaira: Complex Manifolds and Deformation of Complex Structures
284. Finn: Equilibrium Capillary Surfaces
285. Burago/Zalgaller: Geometric Inequalities
286. Andrianov: Quadratic Forms and Hecke Operators
287. Maskit: Kleinian Groups
288. Jacod/Shiryaev: Limit Theorems for Stochastic Processes
289. Manin: Gauge Field Theory and Complex Geometry
290. Conway/Sloane: Sphere Packings, Lattices and Groups
291. Hahn/O'Meara: The Classical Groups and K-Theory

Springer-Verlag
Berlin Heidelberg New York London Paris Tokyo Hong Kong

Druck: STRAUSS OFFSETDRUCK, MÖRLENBACH
Verarbeitung: SCHÄFFER, GRÜNSTADT